Characterization of Nanophase Materials

Edited by
Zhong Lin Wang

WILEY-VCH

Other titles of interest:

Janos H. Fendler
Nanoparticles and Nanostructured Films

S. Amelinckx, D. van Dyck, J. van Landuyt, G. van Tendeloo
Handbook of Microscopy

N. John DiNardo
Nanoscale Characterization of Surfaces and Interfaces

Characterization of Nanophase Materials

Edited by
Zhong Lin Wang

WILEY-VCH

Weinheim · New York · Chichester · Brisbane · Singapore · Toronto

Prof. Z. L. Wang
School of Materials Science and Engineering
Georgia Institute of Technology
Atlanta, GA 30332-0245
USA

First Edition 2000

Library of Congress Card No. applied for

A catalogue record for this book is available from the Britsh Library

Deutsche Bibliothek Cataloguing-in-Publication Data:
Ein Titeldatensatz für diese Publikation ist bei Der Deutschen Bibliothek verfügbar.

Printed on acid-free and chlorine-free paper.

Composition: Kühn & Weyh, D-79111 Freiburg
Printing: Strauss Offsetdruck, D-69509 Mörlenbach
Bookbinding: Wilhelm Osswald & Co., D-67433 Neustadt

Printed in the Federal Republic of Germany.

List of Contributers

S. Amelinckx
EMAT
University of Antwerp (RUCA)
Groenenborgerlaan 171
Antwerp B-2020
Belgium

Moungi G. Bawendi
Department of Chemistry, 6-223
Massachusetts Institute of Technology
Cambridge, MA 02139
USA

C. Burda
School of Chemistry and Biochemistry
Georgia Institute of Technology
Atlanta GA 30332-0400
USA

A. Chemseddine
Physikal Chemistry Department (CK)
Hahn-Meitner-Institut
Glienicker Straße 100
14109 Berlin
Germany

Lifeng Chi
Physikalisches Institut
Westfälische Wilhelms-Universität
Münster
Wilhelm-Klemm-Straße 10
48149 Münster
Germany

Walt de Heer
School of Physics
Georgia Institute of Technology
Atlanta GA 30332-0430
USA

Mostafa A. El-Sayed
Laser Dynamics Laboratory
School of Chemistry and Biochemistry
Georgia Institute of Technology
Atlanta GA 30332-0400
USA

Stephen Empedocles
Department of Chemistry, 6-223
Massachusetts Institute of Technology
Cambridge, MA 02139
USA

Gregory J. Exarhos
Pacific Northwest National Laboratory
Battelle Blvd.
Richland, Washington 99352
USA

Travis Green
Laser Dynamics Laboratory
School of Chemistry and Biochemistry
Georgia Institute of Technology
Atlanta GA 30332-0400
USA

Blair D. Hall
Measurement Standards Laboratory
Caixa Postal 6192 – CEP 13083-970
Campinas, São Paulo
Brasil (Brazil)

C. Landes
Laser Dynamics Laboratory
School of Chemistry and Biochemistry
Georgia Institute of Technology
Atlanta GA 30332-0400
USA

S. Link
Laser Dynamics Laboratory
School of Chemistry and Biochemistry
Georgia Institute of Technology
Atlanta GA 30332-0400
USA

R. Little
Laser Dynamics Laboratory
School of Chemistry and Biochemistry
Georgia Institute of Technology
Atlanta GA 30332-0400
USA

Jingyue Liu
Monsanto Company
Analytical Sciences Center
800 N. Lindbergh Blvd., U1E
St. Louis, Missouri 63167
USA

Jun Liu
Pacific Northwest National Laboratory
Battelle Blvd.
Richland, Washington 99352
USA

Meilin Liu
School of Materials Science
and Engineering
Georgia Institute of Technology
Atlanta GA 30332-0245
USA

Robert Neuhauser
Department of Chemistry, 6-223
Massachusetts Institute of Technology
Cambridge, MA 02139
USA

Janet M. Petroski
Laser Dynamics Laboratory
School of Chemistry and Biochemistry
Georgia Institute of Technology
Atlanta GA 30332-0400
USA

Christian Röthig
Physikalisches Institut
Westfälische Wilhelms-Universität
Münster
Wilhelm-Klemm-Straße 10
48149 Münster
Germany

Zhong Shi
School of Materials Science
and Engineering
Georgia Institute of Technology
Atlanta GA 30332-0245
USA

Kentaro Shimizu
Department of Chemistry, 6-223
Massachusetts Institute of Technology
Cambridge, MA 02139
USA

Daniel Ugarte
Laboratorio Nacional de Luz Sincrontron
Caixa Postal 6192 – CEP 13083-970
Campinas, São Paulo
Brasil (Brazil)

G. Van Tendeloo
EMAT
University of Antwerp (RUCA)
Groenenborgerlaan 171
Antwerp B-2020
Belgium

Li-Qiong Wang
Pacific Northwest National Laboratory
Battelle Blvd.
Richland, Washington 99352
USA

Zhong Lin Wang
School of Materials Science
and Engineering
Georgia Institute of Technology
Atlanta GA 30332-0245
USA

Daniela Zanchet
Laboratorio Nacional de Luz Sincrontron
Caixa Postal 6192 – CEP 13083-970
Campinas, São Paulo
Brasil (Brazil)

Contents

8 Nuclear Magnetic Resonance – Characterization of Self-Assembled Nanostructural Materials

Li-Qiong Wang, Gregory J. Exarhos, and Jun Liu

9 Photoluminescence from Single Semiconductor Nanostructures

Stephen Empedocles, Robert Neuhauser, Kentaro Shimizu and Moungi Bawendi

10 Nanomagnetism

Wal A. de Heer

11 Metal-oxide and -sulfide Nanocrystals and Nanostructures

A. Chemseddine

12 Electron Microscopy of Fullerenes and Related Materials

G. Van Tendeloo and S. Amelinckx

List of Symbols and Abbreviations

a	lattice parameter
A	area of an electrode
A_d	area on the CRT display
$A(\mathbf{K})$	aperture function
$A(k)$	backscattering amplitude
A_s	area scanned on the sample
B	magnetic field
c	lattice parameter
c	spring constant of the cantilever
c	elastic constant
C_0	capacitance of the empty cell used for transfer function measurement, $C_0 = \varepsilon_0 A/d$.
$C_{1/2}$	capacities
C_{dl}	double-layer capacitance
C_f	sensitivity constant derived from the Sauerbrey relationship
c_j	concentration of species j
c_j^*	bulk concentration of species j
Cs	spherical abberation coefficient
d	distance
d	resonator thickness
d	separation between two parallel electrode in an impedance measurement
d	thickness
D	Debye-Waller factor
D_j	diffusion coefficient of species j
$D(\mathbf{K})$	transmission function of the detector
D_Q	dissipation coefficient corresponding to the energy losses during oscillation
E	photoelectron energy
E	polarization of the emitted light
E	voltage or electric potential
ΔE	Stark shift
E_0	accelerating voltage
E_o	threshold energy
$E_{1/2}$	half-wave potential (in voltammetry)
E_b	biexciton binding energy
E_i	initial potential
E_p	peak potential
ΔE_P	$\lvert E_P{}^A - E_P{}^C \rvert$ in CV
$E_{p/2}$	potential where $I = I_p/2$ in LSV or CV
F	electric field
F	faraday constant, $F = 96{,}485$ C/equiv
F	net force
Δf	lens defocus
Δf	measured frequency change

f_0	frequency of a quartz resonator prior to a mass change
$F'(d)$	force gradient
$F(\bar{k})$.	structure amplitude
$f(s)$	scattering factor
$FT[V_p(\mathbf{b})]$	Fourier transform of the crystal potential
Fz	attractive force
$\mathbf{G}$	reciprocal lattice vector
h	piezoelectric stress constant
H	total Hamiltonian
I	transmitted intensity
I	tunneling current
I_o	incident beam intensity
$I_0(\mathbf{x})$	intensity distribution of the incident probe
$I_0(\Delta)$	integrated intensity of the low-loss region including the zero-loss peak for an energy window Δ
i_f	faraday current
$I_N(s)$	power scattered per unit solid angle in the direction defined by s
I_p	peak current
i_r	current during reversal step
I_{SE}	total integrated SE intensity
$I(\mathbf{X})$	image intensity
j	coordination shell index
j	imaginary unit, $j = (-1)^{1/2}$
$\mathbf{J}$	net electronic angular momentum
J_0	exchange current density
J_{ij}	exchange energy constants
Jn	Bessel functions of order n.
k	electron wave-vector
k	spring constant
K	anisotropy energy
$\bar{K}$	wavevector of the scattered wave
K_0	cut-off wave-vector
$\bar{K}_o$	wavevector of the incident wave
k^o	standard heterogeneous rate constant
k_F	Fermi wave vector
ℓ	length
L	average escape-depth
$\mathbf{L}$	total orbital angular momentum
L_d	thickness of a Nernst diffusion layer
m	electron mass
M	magnetization
M	magnification
Δm	mass change
$M_r(\omega)$	modulus function, $M_r(\omega) = [\varepsilon_r(\omega)]^{-1}$
MW	apparent molar mass ($g\ mol^{-1}$)
$M_{\mu\nu}$	tunneling matrix element
n	density
n	number of electrons involved in an electrochemical process
N	number of identical atoms in the same coordination shell
p	momentum

$P(\mathbf{b},\Delta z)$	propagation function
P_{Lm}	associated Legendre function
P_j	depolarization factors for the three axes A, B, C of the nanorod with A > B = C
Q	charge
$Q(\mathbf{b},z+\Delta z)$	phase grating function of the slice
Q_{dl}	charge due to double layer charging
$Q(\mathbf{K})$	Fourier transform of the object transmission function
$q(\mathbf{x})$	transmission function of the object
r	distance between absorbing and neighbor atoms
R	gas constant
R	radius
R	resistance
R_b	bulk resistance of a electrolyte
R_{ct}	resistance to charge transfer at electrolyte-electrode interfaces
r_{m}	radius
r_{mn}	distance between atom m and atom n
R_{mt}	steady state mass-transfer resistance
$\mathbf{S}$	total spin angular momentum
$S_o^2(k)$	amplitude reduction factor due to many-body effects
$\mathbf{S}_{\mathrm{i}}$	spin operator of ith electron
t	time
T	absolute temperature
T	material thickness
T_1	energy relaxation
T_2	dephasing time
T_c	Curie temperature
$T(\mathbf{K})$	transfer function of the microscope
$t_{\mathrm{obj}}(x,y)$	inverse Fourier transform of $T(\mathbf{K})$
$t(\mathbf{x})$	amplitude distribution of the incident probe
$\mathbf{u}$	reciprocal space vector
U	tunneling voltage
U_0	acceleration voltage
v	linear potential scan rate
v	electron velocity
V	volume
V_m	molar volume
$V\mathrm{p}(\mathbf{b})$	hickness-projected potential of the crystal
W	distance between tip and sample
$\mathbf{X}$	beam position
Δx	rms atomic displacement
x_0	impact parameter
$Y(\omega)$	$Y(\omega)=[Z(\omega)]^{-1}$, admittance function
Δz	displacement of the cantilever and piezo
z_i	charge carried by species i signed units of electronic charge
Z_w	Warburg impedance
$Z(\omega)$	impedance function
α, β	angle
α, β	parameters
$\alpha_{\mathrm{a}}, \alpha_{\mathrm{c}}$	anodic and cathodic charge transfer coefficient

β	asymmetry parameter for a one-electron process
$\chi(k)$	EXAFS oscillations
$\chi(\mathbf{K})$	aberration function of the objective lens
$\chi(T)$	magnetic susceptibility
$\chi(\sigma t)$	tabulated number
Δ	defocus value
δ	temporal phase angle between the charging current and the total current
ε_0	absolute permettivity (or the permittivity of free space)
ε_{m}	dielectric constant
ε_Q	dielectric constant of quartz
ε_r	relative permittivity of a material
ε_r'	dielectric constant
$\varepsilon(\omega)$	dielectric function
ϕ	tilt angle between μ and sample plane
ϕ	total photoelectron phase shift
Φ	workfunction
$\phi(k)$	total phase shift
$\phi(r)$	electronic ground state wave function
$\phi(\mathbf{x})$	projected specimen potential along the incident beam direction
λ	wavelength
$\lambda(k)$	photoelectron mean free path
μ	absorption coefficient
μ	paramagnetic atom
μ	transition dipole vector
μ_o	atomic absorption coefficient
μ_{B}	Bohr magneton
$\mu(E)$	absorption coefficient associated with a particular edge
$\Delta\mu(E)$	change in the atomic absorption across the edge
$\mu_o(E)$	absorption coefficient of an isolated gold atom
$\mu_{\mathrm{Exc.}}$	exciton dipole moment
μ_Q	shear modulus of AT-cut quartz
μ_{S}	net surface dipole moment
v_{tr}	transverse velocity of sound in AT-cut quartz (3.34×10^4 m s^{-1})
θ	angle between emission polarization and projection of μ onto the sample plane
θ	scattering angle
θ	temporal phase angle
θ_{B}	Bragg diffraction angle
ρ_Q	density of quartz
ρ_{S}	density of states of sample
$\rho_{\mathrm{S}}(z,E)$	local density of states of the sample
ρ_T	density of states of tip
σ	atomic scattering cross-section
σ	interaction constant
σ	total Debye-Waller factor (including static and dynamic contributions)
$\sigma_{\mathrm{i,el}}$	ionic conductivity (Ω^{-1}cm^{-1}) of an electrolyte
$\sigma\mathrm{A}(\Delta,\beta)$	energy and angular integrated ionization cross-section
σ_{ext}	total extinction coefficient
τ	forward step duration time in a double-step experiment

τ	relaxation time
ω	angular frequency
Ω	atomic volume
Ψ	total electronic wave function
$\Psi(\mathbf{K})$	exit wave function
$\Psi(\mathbf{K}, \mathbf{X})$	amplitude function
$\psi(\mathbf{u})$	Fourier transform of the wave
$\Psi(x, y)$	transmitted wave function
ADF	annular dark-field
AE	Auger electron
AFM	atomic force microscopy
bcc	body-centered cubic
BF	bright-field
CA	chronoamperometry
CB	conduction band
CBED	convergent beam electron-diffraction
CCD	charge coupled device
CCM	constant current mode
CE	counter electrode
CEND	coherent electron nanodiffraction
CHA	concentric hemispherical analyzer
CHM	constant height mode
CID	chemical interface damping
C.L	cathodoluminescence
CMA	cylindrical mirror analyzer
CP	cross-polarization
CPR	current pulse relaxation
CRT	cathode-ray-tube
CTAB	cetyltrimethylammonium bromide
CTAC	cetyltrimethylammonium chloride
CTF	contrast transfer function
CV	cyclic voltammetry
DAS	dynamic-angle spinning
dec	decahedron
DiI	1,1′-dioctadecyl-3,3,3′,3′-tetramethylindocarbocyanine
DFA	Debye function analysis
DOR	double-rotation
DSTEM	dedicated scanning transmission electron microscopy
EDS	energy dispersive x-ray spectroscopy
EELS	energy-loss spectroscopy
EFM	electric force microscopy
EF-TEM	energy-filtered transmission electron microscopy
ELD	electroless deposition
ELNES	energy-loss near edge structure
EQCM	electrochemical quartz crystal microbalance
EXAFS	extended x-ray absorption fine structure
fcc	face-centered-cubic
FEG	field-emission gun
FE-SAM	field emission scanning Auger microscope

FE-TEM	field-emission transmission electron microscopy
FFM	frictional force microscopy
FLDOS	local density of states near the Fermi energy
FM	frequency modulation
FMM	force modulation microscopy
FWHM	full-width-at-half-maximum
GITT	galvanostatic intermittent titration technique
GMR	giant magnetoresistance
HAADF	high-angle annular dark-field
HOMO	highest occupied molecular orbital
HOPG	highly oriented pyrolytic graphite
HRTEM	high resolution transmission electron microscopy
ico	icosahedron
IR	infrared spectroscopy
IS	impedance spectroscopy
LABF	large angle bright-field
LB	Langmuir-Blodgett
LNLS	Brazilian National Synchrotron Laboratory
LO	longitudinal-optical
LSV	linear sweep voltammetry
LTS	local tunneling spectroscopy
LT-STM	low-temperature scanning tunneling microscopy
LUMO	lowest unoccupied molecular orbital
MAS	magic angle spinning
MECS	multiple expansion cluster source
MFM	magnetic force microscopy
MIDAS	microscope for imaging, diffraction, and analysis of surfaces
MIEC	mixed ionic-electronic conductor
MTP	multiply-twinned particles
NCA	nanocrystal arrays
NCS	nanocrystal superlattices
NMR	nuclear magnetic resonance
NQ	napthoquinone
NSOM	near-field scanning optical microscopy
OCV	open-circuit voltage
OD	optical density
ODPA	octadecylphosphonate
PCTF	phase-contrast transfer function
PEELS	parallel electron energy-loss spectroscopy
PL	photoluminescence
POA	phase object approximation
PS	polystyrene
PSD	position-sensitive detector
PSP	poly(styrenephosphonate diethyl ester)
PVK	polyvinylcarbazole
P″VP	poly(2-vinylpyridine)
QCM	quartz crystal microbalance
QCNB	quartz crystal nanobalance
QDQW	quantum-dot quantum-well
QDs	quantum dots

RE	reference electrode
REDOR	rotational-echo double resonance
ROMP	ring-opening metathesis polymerization
SA	self-assembly
SAM	scanning Auger microscopy
SAMs	self-assembled monolayers
SAXS	small-angle elastic x-ray scattering
SCAM	scanning capacitance microscopy
SE	secondary electron
SEMPA	scanning electron microscopy with polarization analysis
SEDOR	spin-echo double resonance
SES	lower case Secondary electron spectroscopy
SET	single-electron-tunneling
SFM	scanning force microscopy
SNOM	scanning near-field optical microscopy
SP	single-pulse
SPM	scanning probe microscopes
SPs	surface plasmons
STEM	scanning transmission electron microscopy
STM	scanning tunneling microscopy
STS	scanning tunneling spectroscopy
T3	2,5‴-bis(acetylthio)-5,2′,5′,2″-terthienyl
TAD	thin annular detector
TADBF	thin annular detector for bright-field
TADDF	thin annular detector for dark-field
TEM	transmission electron microscopy
TDS	thermal diffuse scattering
TO	truncated octahedral
TP	thiophenol
UHV	ultrahigh vacuum
VB	valence band
VOA	virtual objective aperture
WE	working electrode
WPOA	weak scattering object approximation
XANES	x-ray absorption near edge structure
XAS	x-ray absorption spectroscopy
XEDS	x-ray energy-dispersive spectroscopy
XPS	x-ray photoelectron spectroscopy
XRD	x-ray diffraction

Part I Technical Approaches

1 Nanomaterials for Nanoscience and Nanotechnology

Zhong Lin Wang

Technology in the twenty first century requires the miniaturization of devices into nanometer sizes while their ultimate performance is dramatically enhanced. This raises many issues regarding to new materials for achieving specific functionality and selectivity. Nanophase and nanostructured materials, a new branch of materials research, are attracting a great deal of attention because of their potential applications in areas such as electronics [1], optics [2], catalysis [3], ceramics [4], magnetic data storage [5, 6], and nanocomposites. The unique properties and the improved performances of nanomaterials are determined by their sizes, surface structures and interparticle interactions. The role played by particle size is comparable, in some cases, to the particle chemical composition, adding another flexible parameter for designing and controlling their behavior. To fully understand the impacts of nanomaterials in nanoscience and nanotechnology and answer the question of why nanomaterials is so special, this chapter reviews some of the unique properties of nanomaterials, aiming at elucidating their distinct characteristics.

1.1 Why nanomaterials?

Nanomaterials are classified into nanostructured materials and nanophase/nanoparticle materials. The former refer to condensed bulk materials that are made of grains with grain sizes in the nanometer size range, while the latter are usually the dispersive nanoparticles. The nanometer size here covers a wide range which can be as large as 100–200 nm. To distinguish nanomaterials from bulk, it is vitally important to demonstrate the unique properties of nanomaterials and their prospective impacts in science and technology.

1.1.1 Transition from fundamental elements to solid states

Nanomaterials are a bridge that links single elements with single crystalline bulk structures. Quantum mechanics has successfully described the electronic structures of single elements and single crystalline bulks. The well established bonding, such as ionic, covalent, metallic and secondary, are the basis of solid state structure. The theory for transition in energy levels from discrete for fundamental elements to continuous bands for bulk is the basis of many electronic properties. This is an outstanding question in basic science. Thus, a thorough understanding on the structure of nanocrystals can provide deep insight in the structural evolution from single atoms to crystalline solids.

Nucleation and growth are two important processes in synthesizing thin solid films. Nucleation is a process in which an aggregation of atoms is formed, and is the first step of phase transformation. The growth of nuclei results in the formation of large crystalline particles. Therefore, study of nanocrystals and its size-dependent structures and properties is a key in understanding the nucleation and growth of crystals.

1.1.2 Quantum confinement

A specific parameter introduced by nanomaterials is the surface/interface-to-volume ratio. A high percentage of surface atoms introduces many size-dependent phenomena. The finite size of the particle confines the spatial distribution of the electrons, leading to the quantized energy levels due to size effect. This quantum confinement has applications in semiconductors, optoelectronics and non-linear optics. Nanocrystals provide an ideal system for understanding quantum effects in a nanostructured system, which could lead to major discoveries in solid state physics.

The spherical-like shape of nanocrystals produces surface stress (positive or negative), resulting in lattice relaxation (expansion or contraction) and change in lattice constant [7]. It is known that the electron energy band structure and bandgap are sensitive to lattice constant. The lattice relaxation introduced by nanocrystal size could affect its electronic properties.

1.1.3 Size and shape dependent catalytic properties

The most important application of nanocrystals has been in catalysis. A larger percentage of surface atoms greatly increases surface activities. The unique surface structure, electronic states and largely exposed surface area are required for stimulating and promoting chemical reactions. The size-dependent catalytic properties of nanocrystals have been widely studied, while investigations on the shape (facet)-dependent catalytic behavior are cumbersome. The recent success in synthesizing shape-controlled nanocrystals, such as the ones dominated by {100}, {111} [8] and even {110} facets [9], is a step forward in this field.

1.1.4 Novel mechanical properties

It is known that mechanical properties of a solid depend strongly on the density of dislocations, interface-to-volume ratio and grain size. An enhancement in damping capacity of a nanostructured solid may be associated with grain-boundary sliding [10] or with energy dissipation mechanism localized at interfaces [11] A decrease in grain size significantly affects the yield strength and hardness [12]. The grain boundary structure, boundary angle, boundary sliding and movement of dislocations are important factors that determine the mechanical properties of the nanostructured materials. One of the most important applications of nanostructured materials is in superplasticity, the capability of a polycrystalline material to undergo extensive tensible deformation without necking or fracture. Grain boundary diffusion and sliding are the two key requirements for superplasticity.

1.1.5 Unique magnetic properties

The magnetic properties of nano-size particles differ from those of bulk mainly in two points. The large surface-to-volume ratio results in a different local environment for the surface atoms in their magnetic coupling/interaction with neighboring atoms, leading to the mixed volume and surface magnetic characteristics. Unlike bulk ferromagnetic materials, which usually form multiple magnetic domains, several small ferromagnetic particles could consist of only a single magnetic domain. In the case of a single particle being a single domain, the superparamagnetism occurs, in which the magnetizations of the particles are randomly distributed and they are aligned only under an applied magnetic field, and the alignment disappears once the external field is withdrawn. In ultra-compact information storage [13, 14], for example, the size of the domain determines the limit of storage density. Magnetic nanocrystals have other important applications such as in color imaging [15], bioprocessing [16], magnetic refrigeration [17], and ferrofluids [18].

Metallic heterostructured multilayers comprised of alternating ferromagnetic and nonmagnetic layers such as Fe-Cr and Co-Cu have been found to exhibit giant magnetoresistance (GMR), a significant change in the electrical resistance experienced by current flowing parallel to the layers when an external magnetic field is applied [19]. GMR has important applications in data storage and sensors.

1.1.6 Crystal-shape-dependent thermodynamic properties

The large surface-to-volume ratio of nanocrystals greatly changes the role played by surface atoms in determining their thermodynamic properties. The reduced coordination number of the surface atoms greatly increases the surface energy so that atom diffusion occurs at relatively lower temperatures. The melting temperature of Au particles drops to as low as ~ 300 °C for particles with diameters of smaller than 5 nm, much lower than the bulk melting temperature 1063 °C for Au [20]. Nanocrystals usually have faceted shape and mainly enclosed by low index crystallographic planes. It is possible to control the particle shape, for example, cubic Pt nanocrystals bounded by {100} facets and tetrahedral Pt nanocrystals enclosed by {111} facets [8]. The rod-like Au nanocrystals have also been synthesized, which are enclosed by {100} and {110} facets [9]. The density of surface atoms changes significantly for different crystallographic planes, possibly leading to different thermodynamic properties.

1.1.7 Semiconductor quantum dots for optoelectronics

Semiconductor nanocrystals are zero-dimensional quantum dots, in which the spatial distribution of the excited electron-hole pairs are confined within a small volume, resulting in the enhanced non-linear optical properties [21–24]. The density of states concentrates carriers in a certain energy range, which is likely to increase the gain for electro-optical signals. The quantum confinement of carriers converts the density of states to a set of discrete quantum levels. This is fundamental for semiconductor lasers.

With consideration the small size of a semiconductor nanocrystal, its electronic properties are significantly affected by the transport of a single electron, giving the possibility of producing single electron devices [25]. This is likely to be important in quantum devices and nanoelectronics, in which the size of the devices are required to be in the nanometer range.

Nanostructured porous silicon has been found to give visible photoluminescence [27, 28]. The luminescence properties of silicon can be easily integrated with its electronic properties, possibly leading to a new approach for opto-electronic devices. The mechanism has been proposed to be associated with either quantum confinement or surface properties linked with silica. This is vitally important to integrate optical circuits with silicon based electronics. The current research has been concentrated on understanding the mechanism for luminescence and improving its efficiency.

1.1.8 Quantum devices for nanoelectronics

As the density of logic circuits per chip approaching 10^8, the average distance between circuits is 1.7 μm, between which all of the circuit units and interconnects must be accommodated. The size of devices is required to be less than 100 nm and the width of the interconnects is less than 10 nm. The miniaturization of devices breaks the fundamentals set by classical physics based on the motion of particles. Quantum mechanical phenomena are dominant, such as the quantization of electron energy levels (e.g., the 'particle in a box' quantum confinement problem), electron wave function interference, quantum tunneling between the energy levels belonging to two adjacent nanostructures, and discreteness of charge carriers (e.g., single electron conductance). The quantum devices rely on tunneling through the classically forbidden energy barriers. With an appropriate voltage bias across two nanostructures, the electron energy levels belonging to the two nanostructures line up and resonance tunneling occurs, resulting in an abrupt increase in tunneling current. The single-electron electronics uses the energy required to transport a single electron to operate a switche, transistor or memory element.

These new effects not only raise fundamental questions in physics, but also call on challenges in new materials. There are two outstanding material's issues. One is the semiconductor nanocrystals suitable for nanoelectronics. Secondly, for the operation of high density electronics system, new initiatives must be made to synthesize interconnects, with minimum heat dissipation, high strength and high resistance to electromigration. The most challenging problem is how to manipulate the nanostructures in assembling devices. This is not only an engineering question but rather a science question because of the small size of the nanostructures.

Semiconductor heterostructures are usually referred to as one-dimensional artificially structured materials composed of layers of different phases/compositions. This multilayered material is particularly interesting if the layer thickness is less than the mean-free-pathlength of the electron, providing an ideal system for quantum well structure. The semiconductor heterostructured material is the optimum candidate for fabricating electronic and photonic nanodevices [28].

1.1.9 Carbon fullerences and nanotubes

Research in nanomaterials opens many new challenges both in fundamental science and technology. The discovery of C_{60} fullerence [29], for example, has sparked a great research effort in carbon related nanomaterials. Besides diamond and graphite structures, fullerence is a new state of carbon. The current most stimulating research focuses on carbon nanotubes, a long-rod-like structure comprised of cylindrical concentric graphite sheets [30]. The finite dimension of the carbon nanotube and the chirality angle following which the graphite sheet is rolled result in unique electronic properties, such as the ballistic quantum conductance [31], the semiconductor junctions [32], electron field emission [33] etc. The unique tube structure is also likely to produce extraordinarily strong mechanical strength and extremely high elastic limit. The reversible buckling of the tube results in high mechanical flexibility.

Fullerence and carbon nanotubes can be chemically functionalized and they can serve as the sites/cells for nano-chemical reaction [34]. The long, smooth and uniform cylindrical structure of the nanotube is ideal for probe tips in scanning tunneling microscopy and atomic force microscopy [35]. The covalent bonding of the carbon atoms and the functionalized nanotube tip gives the birth of the chemical microscopy [36], which can be used to probe the local bonding, bond-to-bond interactions and chemical forces.

1.1.10 Ordered self-assembly of nanocrystals

Size and even shape selected nanocrystals behave like a molecular matter, and are ideal building blocks for two- and three-dimensional cluster self-assembled superlattice structures [37–40]. The electric, optical, transport and magnetic properties of the structures depend not only on the characteristics of individual nanocrystals, but also on the coupling and interaction among the nanocrystals arranged with long-range translational and even orientational order [41, 42]. Self-assembled arrays involve self-organization into monolayers, thin films, and superlattices of size-selected nanocrystals encapsulated in a protective compact organic coating. Nanocrystals are the hard cores that preserve ordering at the atomic scale; the organic molecules adsorbed on their surfaces serve as the interparticle molecular bonds and as protection for the particles in order to avoid direct core contact with a consequence of coalescing. The interparticle interaction can be changed via control over the length of the molecular chains. Quantum transitions and insulator to conductor transition could be introduced, possibly resulting in tunable electronic, optical and transport properties [43].

1.1.11 Photonic crystals for optically-active devices and circuits

Photonic crystals are synthetic materials that have a patterned periodic dielectric constant that creates an optical bandgap in the material [44]. To understand the mechanism of photonic crystals, one starts from the energy band structure of electrons in a crystalline solid. Using the Fermi velocity of the electrons in a solid, it can be found that the electron wavelength is compatible to the spacing between the atoms. Electron motion in a periodic potential results in the quantized energy levels. In the energy regions filled with energy levels, bands are formed. An energy gap between the conduction band and the valence band would be formed, which is a key factor in determining the conductivity of the solid. If the bandgap is zero, the material is conductive; for a small bandgap, the material is semiconductor; and the material is insulator if the bandgap is large.

The wavelength of photons is in the order of a few hundreds of nanometers. It is necessary to artificially create a dielectric system which has periodically modulated dielectric function at a periodicity compatible with the wavelength of the photon. This can be done by processing materials that are comprised of patterned structures. As a result, photons with energies lying within the bandgap cannot be propagated unless a defect causes an allowed propagation state within the bandgap (similar to a defect state), leading to the possibility of fabricating photon conductors, semiconductors and insulators. Thus point, line, or planar defects can be shown to act as optical cavities, waveguides, or mirrors and offer a completely different mechanism for the control of light and advancement of all-optical integrated circuits [45–47]. By using particles sizes in the nanometer regime with different refractive indices than the host material, these effects should be observable in the near infrared and visible spectral regions.

1.1.12 Mesoporous materials for low-loss dielectrics and catalysis

Mesoporous materials can be synthesized by a wide range of techniques such as chemical etching, sol gel processing and template-assisted techniques. Ordered self-assembly of hollow structures of silica [48], carbon [49] and titania [50, 51] has drawn much attention recently because of their applications in low-loss dielectrics, catalysis, filtering and photonics. The ordered hollow structure is made through a template-assisted technique. The monodispersive polystyrene (PS) particles are used as the template to form an ordered, self-assembled periodic structure. Infiltrating the template by metal-organic liquid and a subsequent heat treatment form the ordered pores, whose walls are metal oxides. The structure is ordered on the length-scale of the template spheres and the pore sizes are in submicron to micron range. Alternatively, ordered porous silica with much smaller pore sizes in nanosize range (< 30 nm), produced deliberately by introducing surfactant, has also been processed [52, 53], in which the porosity is created by surfactants. A combination of the template assisted pore structure and the surfactant introduced in the infiltration liquid results in a new structure that have porousity at double-length scales [54]. The low density (~ 10% of the bulk density) of the material results in very low dielectric constant, a candidate for low-loss electronic devices. The large surface area of the porous materials is ideal for catalysis. The synthesis of mesoporous materials can be useful for environmental cleaning [55] and energy storage [56].

1.2 Characterization of nanophase materials

There are three key steps in the development of nanoscience and nanotechnology: materials preparation, property characterization, and device fabrication. Preparation of nanomaterials is being advanced by numerous physical and chemical techniques. The purification and size selection techniques developed can produce nanocrystals with well defined structure and morphology. The current most challenging tasks are property characterization and device fabrication. Characterization contains two main categories: structure analysis and property measurements. Structure analysis is carried out using a variety of microscopy and spectroscopy techniques, while the property characterization is rather challenging.

Due to highly size and structure selectivity of the nanomaterials, the physical properties of nanomaterials could be quite diverse. To maintain and utilize the basic and technological advantages offered by the size specificity and selectivity of nanomaterials, it is imperative to understand the principles and methodologies for characterization the physical properties of individual nanoparticles/nanotubes. It is known that the properties of nanostructures depend strongly on their size and shape. The properties measured from a large quantity of nanomaterials could be an average of the over all properties, so that the unique characteristics of individual nanostructure could be embedded. The ballistic quantum conductance of a carbon nanotube [31] (Fig. 1-1), for example, was observed only from defect-free carbon nanotubes grown by an arc-discharge technique, while such an effect vanishes in the catalytically grown carbon nanotubes [57, 58] because of high density of defects. This effect may have great impact on molecular electronics, in which carbon nanotubes could be used as interconnects for molecular devices with no heat dissipation, high mechanical strength and flexibility. The covalent bonding of the carbon atoms is also a plus for molecular device because of the chemical and bonding compatibility. Therefore, an essential task in nanoscience is property characterization of an individual nanostructure with well defined atomic structure.

Characterizing the properties of individual nanoparticle/nanotube/nanofiber (e.g., nanostructure) is a challenge to many existing testing and measuring techniques because of the following constrains. First, the size (diameter and length) is rather small, prohibiting the applications of the well-established testing techniques. Tensile and creep testing of a fiber-like material, for example, require that the size of the sample be sufficiently large to be clamped rigidly by the sample holder without sliding. This is impossible for nanostructured fibers using conventional means. Secondly, the small size of the nanostructures makes their manipulation rather difficult, and specialized techniques are needed for picking up and installing individual nanostructure. Finally, new methods and methodologies must be developed to quantify the properties of individual nanostructures.

Mechanical characterization of individual carbon nanotubes is a typical example. By deflecting on one-end of the nanofiber with an AFM tip and holding the other end fixed, the mechanical strength has been calculated by correlating the lateral displacement of the fiber as a function of the applied force [59, 60]. Another technique that

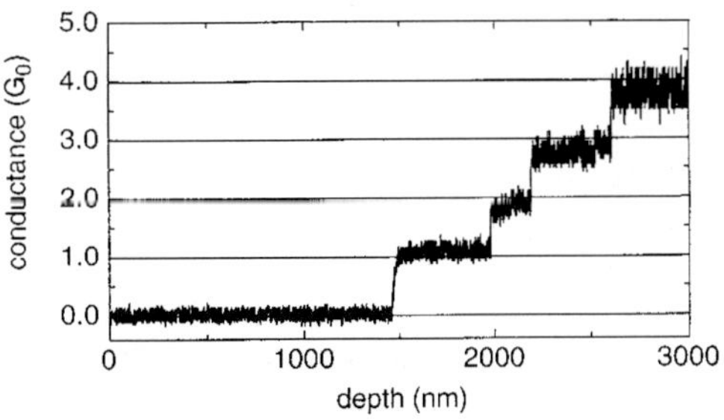

Figure 1-1. Quantized conductance of a multiwalled carbon nanotube observed as a function of the depth into the liquid mercury the nanotube was inserted in an atomic force microscope, where $G_0 = 2e^2/h = (12.9\ k\Omega)^{-1}$ is the quantum mechanically predicted conductance for a single channel (Courtesy of Walt de Heer, Georgia Institute of Technology).

has been previously used involves measurement of the bending modulus of carbon nanotubes by measuring the vibration amplitude resulting from thermal vibrations [61], but the experimental error is quite large. A new technique has been demonstrated recently for measurement the mechanical strength of single carbon nanotubes using in-situ TEM [62]. The measurements is done on a specific nanotube whose microstructure is determined by transmission electron imaging and diffraction. If an oscillating voltage is applied on the nanotube with ability to tune the frequency of the applied voltage, the periodic force acting on the nanotube induces electro-mechanical resonance (Fig. 1-2). The resonance frequency is an accurate measure of the mechanical modulus.

In analogous to a spring oscillator, the mass of the particle attached at the end of the spring can be determined if the vibration frequency is measured, provided the spring constant is calibrated. This principle can be adopted to determine the mass of a very tiny particle attached at the tip of the free end of the nanotube (Fig. 1-3). This newly discovered "nanobalance" has been shown to be able to measure the mass of a particle as small as 22 ± 6 fg (1f = 10^{-15}) [62]. *The most sensitive and smallest balance in the world.* The nanobalance is anticipated to be useful for measuring the mass of a large biomolecule or biomedical particle, such as virus.

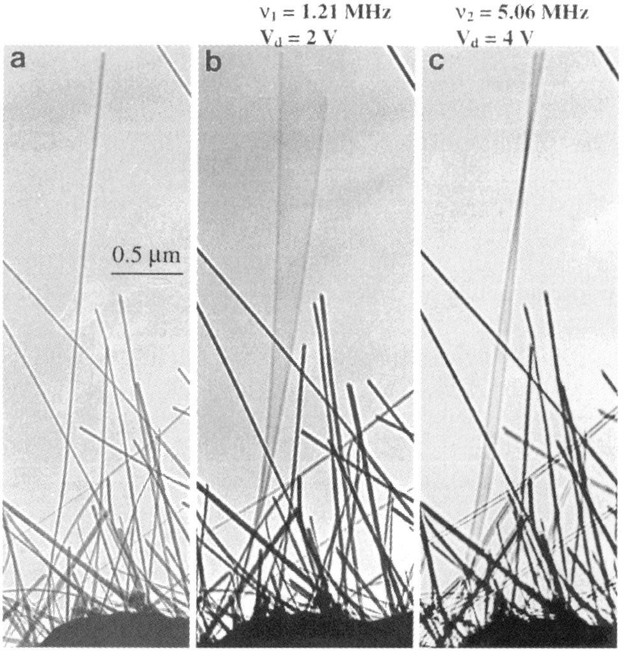

Figure 1-2. Electro-mechanical resonance of carbon nanotubes. A selected carbon nanotube at (a) stationary, (b) the first harmonic resonance (ν_1 = 1.21 MHz) and (c) the second harmonic resonance (ν_2 = 5.06 MHz).

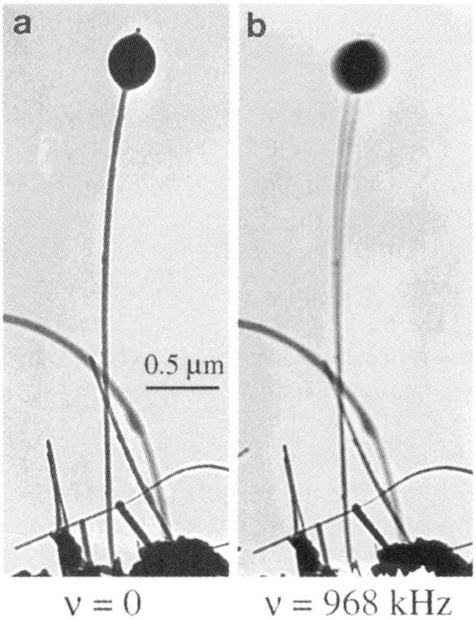

Figure 1-3. A small particle attached at the end of a carbon nanotube at (a) stationary and (b) first harmonic resonance (ν = 0.968 MHz). The effective mass of the particle is measured to be ~ 22 fg (1 f = 10^{-15}).

1.3 Scope of the book

Development of nanotechnology involves several key steps. First, synthesis of size and even shape controlled nanoparticles is the key for developing nanodevices. Second, characterization of nanoparticles is indispensable to understand the behavior and properties of nanoparticles, aiming at implementing nanotechnology, controlling their behavior and designing new nanomaterial systems with super performance. Third, theoretical modeling is vitally important to understand and predict material's performance. Finally, the ultimate goal is to develop devices using nanomaterials. With consideration the large diversity of research in nanomaterials, this book concentrates on the structure and property characterization of nanomaterials.

The book emphasizes the techniques used for characterizing nanophase materials, including x-ray diffraction, transmission electron microscopy, scanning transmission electron microscopy, scanning probe microscopy, optical, electrical and electrochemical characterizations. The book aims at describing the physical mechanisms and detailed applications of these techniques for characterizations of nanophase materials to fully understand the morphology, surface and the atomic level microstructures of nanomaterials and their relevant properties. It is also intended as a guidance with introduction of the fundamental techniques for structure analysis. The book focuses also on property characterization of nanophase materials in different systems, such as the family of metal, semiconductor, magnetic, oxide and carbon systems. It is the key to illustrate the unique properties of the nanocrystals and emphasize how the structures and the properties are characterized using the techniques presented in the book.

References

[1] M.A. Kastner, *Phys. Today*, **1993**, *46*, 24.
[2] L. Brus, *Appl. Phys. A*, **1991**, *53*, 465.
[3] L.N. Lewis, *Chem. Rev.*, **1993**, *93*, 2693.
[4] R. Freer, *Nanoceramics* London, Institute of Materials, **1993**.
[5] D.D. Awschalom and D.P. DiVincenzo, *Phys. Today*, **1995**, *48*, 43.
[6] J.F. Smyth, *Science*, **1992**, *258*, 414.
[7] J. Woltersdorf,, A.S. Nepijko, and E. Pippel, *Surf. Sci.*, **1981**, *106*, 64.
[8] T.S. Ahmadi, Z.L. Wang, T.C. Green, A. Henglein and M.A. El-Sayed, *Science*, **1996**, *272*, 1924.
[9] M. Mahamed, Z.L. Wang and M.A. El-Sayed, *J. Phys. Chem. B*, **1999**, submitted.
[10] B.S. Berry and W.C. Pritchett, *Thin Solid Films*, **1976**, *33*, 19.
[11] C.M. Su, R.R. Oberle, M. Wuttig, and R.C. Cammarata, *Mater. Res. Soc. Symp. Proc.* **1993**, *280*, 527.
[12] for a review see J.R. Weertman and R.S. Averback, in *Nanomaterials: Synthesis, Properties and Applications*, eds. A.S. Edelstein and R.C. Cammarata, London, Institue of Phys. Publ., **1996,** Chapter 13, 323, and references therein.
[13] L. Gunther, *Phys. World*, **1990**, *3*, 28.
[14] R.G. Audran and A.P. Huguenard, *U.S. Patent*, **1981**, 4,302,523.
[15] R.F. Ziolo, *U.S. Patent*, **1984**, 4,474,866.
[16] R.H. Marchessault, S. Ricard and P. Rioux, *Carbohydrate Res.*, **1992**, *224*, 133.
[17] R.D. McMichael, R.D. Shull, L.J. Swartzendruber, L.H. Bennett, R.E. Watson, *J. Magn. Magnsm Mater.*, **1992**, *111*, 29.
[18] I. Anton et al., *J. Magn. Magnsm Mater.*, **1990**, *85*, 219.
[19] M.N. Baibich, J.M. Broto, A. Fert, F. Nguyen Van dau, F. Petroff, P. Etienne, G. Greuzet, A. Friederich, and J. Chazelas, *Phys. Rev. Letts.*, **1988**, *61*, 2472.
[20] Ph. Buffat, J.P. Borel, *Phys. Rev. A*, **1976**, *13*, 2287.
[21] S.M. Prokes, *Appl. Phys. Lett.*, **1993**, *62*, 3244.
[22] L. Brus, *Appl. Phys. – Mater. Sci. & Processing*, **1991**, *53*, 465.
[23] A.P. Alivisatos, *Science*, **1996**, *271*, 933.
[24] C.B., Murray, D.J. Norris, M.G. Bawendi, *J. of American Chemical Society*, **1993**, *115*, 8706.
[25] D.L. Klein, R. Roth, A.K.L. Lim, A.P. Alivisatos, P.L. McEuen, *Nature*, **1997**, *389*, 699.
[26] C. Pickering, M.I.J. Beale, D.J. Robbins, P.J. Pearson and R. Greef, *J. Phys. C: Solid State Phys.*, **1984**, *17*, 6536.
[27] L.T. Canham, *Appl. Phys. Lett.*, **1990**, *57*, 1046.
[28] for a review see R.C. Cammarata, in *Nanomaterials: Synthesis, Properties and Applications*, eds. A.S. Edelstein and R.C. Cammarata, London, Institue of Phys. Publ., **1996**, Chapter 13, 323, and references therein.
[29] H.W. Kroto, J.R. Heath, S.C. O'Brien, R.F. Curl and R.F. Smalley, *Nature*, **1985**, *318*, 162.
[30] S. Iijima, *Nature*, **1991**, *354*, 56.
[31] S. Frank, P. Poncharal, Z.L. Wang and W.A. de Heer, *Science*, **1998**, *280*, 1744.
[32] S.J. Tans, A.R.M. Verschueren, C. Dekker, *Nature*, **1998**, *393*, 49.
[33] W. A. de Heer, A. Chatelain, D. Ugarte, *Science*, **1995**, *270*, 1179.
[34] W.Q. Han, S.S. Fan, Q.Q. Li and Y.D. Hu, *Science*, **1997**, *277*, 1287.
[35] H. Dai, J.H. Hafner, A.G. Rinzler, D.T. Colbert, R.E. Smalley, *Nature*, **1996**, *384*, 147.
[36] S.S. Wong, E. Joselevich, A.T. Woolley, C.L. Cheung, C.M. Lieber, *Nature*, **1998**, *394*, 52.
[37] R.L. Whetten, J.T. Khoury, M.M. Alvarez, S. Murthy, I. Vezmar, Z.L. Wang, C.C. Cleveland, W.D. Luedtke, U. Landman, *Adv. Materials*, **1996**, *8*, 428.
[38] J.S. Yin and Z.L. Wang, *Phys. Rev. Lett.*, **1997**, *79*, 2570.
[39] C.B. Murray, C.R. Kagan, M.G. Bawendi, *Science*, **1995**, *270*, 1335.
[40] S. Sun and C.B. Murray, *J. Appl. Phys.* **1999**, *85*, 4325.
[41] Z.L. Wang, *Adv. Mater.*, **1998**, *10*, 13.
[42] S.A. Harfenist, Z.L.Wang, M.M.Alvarez, I.Vezmar and R.L.Whetten, *J. Phys. Chem.*, **1996**, *100*, 13904.
[43] C.P. Collier, R.J. Saykally, J.J. Shiang, S.E. Henrichs and J.R. Heath, *Science*, **1997**, *277*, 1978.
[44] J.D. Joannopoulos, P.R. Villeneuve and S. Fan, *Nature*, **1997**, *386*, 143.
[45] F. Gadot F, A. Chelnokov, A. DeLustrac, P. Crozat, J.M. Lourtioz, D. Cassagne, C. Jouanin, *Appl. Phys. Letts.*, **1997**, *71*, 1980.
[46] T.F. Krauss, R.M. DeLaRue, S. Brand, *Nature*, **1996**, *383*, 699.
[47] D.F. Sievenpiper, M.E. Sickmiller, E. Yablonovitch E, *Phys. Rev. Letts.*, **1996**, *76*, 2480.
[48] O.D. Velev, T.A. Jede, R.F. Lobo and A.M. Lenhoff, *Nature*, **1997**, *389*, 448.

[49] A.Z. Zakhidov, R.H. Baughman, Z. Iqbal, C. Cui, I. Khayrullin, S.O. Dantas, J. Marti, and V.G. Ralchenko, *Science*, **1998**, *282*, 897.
[50] J.E.G.J. Wijnhoven and W.L. Vos, *Science*, **1998**, *281*, 802.
[51] B.T. Holland, C.F. Blanford and A. Stein, *Science*, **1998**, *281*, 538.
[52] C.T. Kresge, M.E. Leonowioz, W.J. Roth, J.C. Vartuli and J.S. Beck, *Nature*, **1992**, *359*, 710.
[53] D. Zhao, J. Feng, Q. Huo, N. Melosh, G.H. Fredrickson, B.F. Chmelka and G.D. Stucky, *Science*, **1998**, *279*, 548.
[54] J.S. Yin and Z.L. Wang, *Appl. Phys. Letts.*, **1999**, *74*, 2629.
[55] X. Feng, G.E. Fryxell, L.Q. Wang, A.Y. Kim, J. Liu, K.M. Kemner, *Science*, **1997**, *276*, 923.
[56] U. Bach, D. Lupo, P. Comte, J.E. Moser, F. Weissortel, J. Salbeck, H. Spreitzer, M. Gratzel, *Nature*, **1998**, *395*, 583.
[57] H.J. Dai, E.W. Wong, C.M. Lieber, *Science*, **1996**, *272*, 523.
[58] T.W. Ebbesen, H.J. Lezec, H. Hiura, J.W. Bennett, H.F. Ghaemi, T. Thio, *Nature*, **1996**, *382*, 54.
[59] E. Wong, P. Sheehan and C. Lieber, *Science*, **1997**, *277*, 1971.
[60] J.P. Salvetat, A.J. Kulik, J.M. Bonard, G.A.D. Briggs, T. Stockli, K. Metenier, S. Bonnamy, F. Beguin, N.A. Burnham, L. Forro, *Adv. Materials*, **1999**, *11*, 161.
[61] M.M. Treacy, T.W. Ebbesen, J.M. Gibson, *Nature*, **1996**, *38*, 678.
[62] P. Poncharal, Z.L. Wang, D. Ugarte and W.A. de Heer, *Science*, **1999**, *283*, 1513.

2 X-ray Characterization of Nanoparticles

Daniela Zanchet, Blair D. Hall, and Daniel Ugarte

2.1 Introduction

The hunt for new applications of nanostructured systems is now a major area of research in materials science and technology. To exploit the full potential that nanosystems offer, it is important that novel methods of manipulation and fabrication be developed, in addition to extending current techniques of sample characterization to smaller sizes. Success in devising and assembling systems on the scale of nanometers will require a deeper understanding of the basic processes and phenomena involved. Hence, one of the current key objectives is to adapt and develop a range of techniques that can characterize the structural, electronic, magnetic and optical properties of nanostructured systems. High-resolution techniques, that provide local information on the nanometer scale (such as electron or scanning probe microscopies), as well as those that provide only ensemble-average measurements, are all important in obtaining a complete picture of material properties.

One of the most fundamental characteristics of nanometer-sized particles is their very high surface-to-volume ratio. This can lead to novel and unexpected atomic arrangements, and may also have dramatic effects on other physical or chemical attributes. Because of this, the precise determination of nanoparticle structure, both medium-range order and/or the existence of local distortion, is a fundamental issue. Methods of structure determination can be broadly classified in two categories, depending on the use of real or reciprocal space data. Direct space methods allow the visualization of the atomic arrangement in nanometer-sized regions; the most vivid examples are: High Resolution Transmission Electron Microscopy (HRTEM) and Scanning Probe Microscopies (Scanning Tunneling Microscopy; Atomic Force Microscopy; etc). Reciprocal space-based methods exploit interference and diffraction effects of photons or electrons, to provide sample-averaged information about structure. For most bulk material-related studies, reciprocal space methods are much easier to apply than direct methods, disposing of numerous, flexible, mathematical tools to fully exploit the experimental data.

In fact, it must be recognized that X-ray diffraction (XRD), based on wide-angle elastic scattering of X-rays, has been the single most important technique for determining the structure of materials characterized by long-range order [1]. However, for other systems, such as disordered materials, XRD has been of limited use, and other experimental techniques have had to be developed. A particularly powerful example is the technique of EXAFS (Extended X-ray Absorption Fine Structure) [2], which probes the local environment of a particular element. Although this method is, as XRD, reciprocal space-based, it is essentially a spectroscopic technique, exploiting the energy-dependence of X-ray absorption due to interference effects in the individual photoelectron scattering process. This fact allows precise measurement of a local environment without the necessity of long-range order in the material.

Small-angle elastic X-ray scattering (SAXS) can provide direct information about the external form of nanoparticles or macromolecules, by measuring the typical size of the electron density variations [3]. For example, SAXS measurements can be used to estimate the radius of gyration of particles, giving information related to the average particle diameter. The applicability of this technique depends on both detection range and X-ray beam divergence: in general, it can be applied to determine the size, or even size distribution, of nanoparticles in the 1–200 nm range.

This chapter will discuss structural characterization of nanostructured materials using X-rays. Many techniques, of varying degrees of complexity, could be presented here, however, it is not our intention to review all available experimental methods. Instead, we wish to highlight problems that can arise when well-established methods of measurement, or treatment of data, for the bulk are applied to nanosystems. To do this, we will analyze the structure of a sample of 2 nm gold particles using both XRD and EXAFS. SAXS studies are not included in our discussion as their size-range of applicability is already well suited to nanoparticle studies; readers are referred to existing texts describing this technique [3, 4].

XRD and EXAFS are both reciprocal space-based methods usually applied to comparatively large amounts of sample (~ mm^3). They are both capable of providing information on the average behavior of nanoparticle samples, however, they differ in the nature of X-ray interaction with matter (elastic or inelastic), and give two relatively complementary types of information (long-to-medium range, versus local order, respectively). We will concentrate on the application of these techniques to nanosystems, and the special considerations that must be taken into account when doing so. The discussion should enable the reader to get an idea of the general aspects involved in characterizing nanosized-volumes of matter, and thereby understand how to optimize experiments and data processing to fully exploit the capacities of these powerful techniques.

2.2 X-ray sources

Since their discovery, at the end of last century, X-ray tubes have not changed in their basic principle of operation. A beam of energetic electrons is directed onto a solid target (Copper, Molybdenum, etc), generating X-ray photons. The resulting X-ray energy spectrum is made up of intense narrow fluorescent lines (white lines), characteristic of the target material, and a less intense continuous spectrum (Bremsstrahlung) [1].

This simple type of device has allowed the development of extremely powerful crystallographic methods that have been used extensively to determine material structure by diffraction experiments. As already mentioned, XRD involves the elastic scattering of photons; it requires a collimated, and rather monochromatic, incident X-ray beam. These conditions can be reasonably well met by using the characteristic lines of X-ray tubes. For many years, experimental methods have been limited by the discrete nature of the energy distribution of the conventional X-ray tubes. However, in the last few decades, synchrotron facilities characterized by high intensity, enhanced brightness, and a continuous energy spectrum, have been developed. Such sources, combined with efficient and flexible X-ray optics (mirrors, monochromators, slits, etc.), can supply a monochromatic beam of X-rays for which continuous variation of the energy is possible. This has stimulated the emergence of new techniques of analysis, among them spectroscopic techniques, such as EXAFS [5].

Usually, in nanophase materials or nanoparticles, the actual amount of matter constituting the nanometric sized volumes is extremely small. As a consequence, most experimental techniques used for characterization are limited by the poor quality of signal that can be obtained. For X-ray methods, this becomes critical and the use of the modern synchrotron sources is almost a prerequisite to obtain good quality data. In the sections that follow, we describe the results of experiments performed at the Brazilian National Synchrotron Laboratory (LNLS), where conventional X-ray optics and experimental set-ups for powder diffraction, and X-ray absorption spectroscopy were used; no special apparatus was employed for these measurements on nanosystems.

2.3 Wide-angle X-ray diffraction

2.3.1 Diffraction from small particles

The distribution of X-ray intensity scattered by a finite-sized atomic aggregate takes on a simple form, provided that aggregates within the sample are uniformly and randomly oriented with respect to the incident beam. Under these conditions, a radially symmetric powder diffraction pattern can be observed.

Powder diffraction patterns for individual particle structures can be calculated using the Debye equation of kinematic diffraction [6]. For aggregates containing only one type of atom, the Debye equation is expressed as:

$$I_N(s) = I_0 N f^2(s)\left(1 + \frac{D}{N}\sum_{n\neq m}\frac{(\sin 2\pi s r_{mn})}{2\pi s r_{mn}}\right) \tag{2-1}$$

where I_0 is the intensity of the incident beam and $I_N(s)$ is the power scattered per unit solid angle in the direction defined by $s = 2\sin(\theta)/\lambda_p$, with θ equal to half the scattering angle and λ_p the wavelength of the radiation. The scattering factor, $f(s)$, determines the single-atom contribution to scattering, and is available in tabulated form [7]. N is the number of atoms in the cluster and r_{mn} is the distance between atom m and atom n. The Debye-Waller factor, D, damps the interference terms and so expresses a degree of disorder in the sample. This disorder may be dynamic, due to thermal vibrations, or static, as defects in the structure. A simple model assumes that the displacement of atoms is random and isotropic about their equilibrium positions. In this case,

$$D = \exp\left(-(2\pi s \Delta x)^2\right) \tag{2-2}$$

where Δx is the rms atomic displacement from equilibrium along one Cartesian coordinate [6].

The Debye equation is actually in the form of a three dimensional Fourier transform, in the case where spherical symmetry can be assumed. As such, some feeling for the details of powder diffraction patterns can be borrowed from simple one-dimensional Fourier theory. Firstly, at very low values of s, the reciprocal space variable, one can expect to find the low-frequency components of the scatterer structure. This is the small-angle scattering region, in which information about the particle size and external form is concentrated. No details of the internal atomic arrangements are contained in the small-angle intensity distribution (see [3,4,6] for a description of small-angle diffraction). Secondly, beyond the small-angle region, intensity fluctuations arise from interference due to the internal structure of the particle, or domains of structure within it. The size of domains and the actual particle size need not to be the same, so small-angle and wide-angle diffraction data can be considered as complementary sources of information. Thirdly, the finite size of domains will result in the convolution of domain-size information with the more intense features of the internal domain structure. This domain size information will tend to have oscillatory lobes that decay slowly on either side of intense features (examples will be shown below).

2.3.2 Distinctive aspects of nanoparticle diffraction

The very small grain size of clusters in nanophase materials gives their diffraction pattern the appearance of an amorphous material. Of course they are not amorphous: the problem of accurately describing nanoparticle structure is one of the central themes of this text. The difficulty in determining their structure by X-ray diffraction, however, is imposed at a fundamental level by two features of these systems: the small size of structural domains that characterize the diffraction pattern; and the occurrence of highly symmetric, but, non-crystalline structures. In short, the common assumption that there exists some kind of underlying long-range order in the system under study does not apply to nanophase materials. This is most unfortunate because the wealth of techniques available to the X-ray crystallographer must largely be put aside.

Size-dependent and structure-specific features in diffraction patterns can be quite striking in nanometer-sized particles. Small particles have fairly distinct diffraction patterns, both as a function of size and as a function of structure type. In general, regardless of structure, there is a steady evolution in the aspect of diffraction profiles: as particles become larger, abrupt changes do not occur, features grow continuously from the diffraction profile and more detail is resolved. These observations form the basis for a direct technique of diffraction pattern analysis that can be used to obtain structural information from experimental diffraction data. This will be outlined in Section 2.3.3 below.

2.3.2.1 Crystalline particles

Single crystal nanoparticles exhibit features in diffraction that are size-dependent, including slight shifts in the position of Bragg peaks, anomalous peak heights and widths [8]. Figure 2-1 shows the diffraction patterns for three sizes of face-centered-cubic (fcc) particles, spanning a diameter range of 1.6–2.8 nm, and containing from 147 to 561 atoms. The intensities have been normalized, so that the first maximum in each profile has the same height, and shifted vertically, so that the features of each can be clearly seen. Also shown are the positions of the bulk (Bragg) diffraction lines for gold, indexed at the top.

It should be immediately apparent from Fig. 2-1 that there is considerable overlap in the peaks of the particle profiles. In fact, the familiar concept of a diffraction peak begins to loose meaning when considering diffraction from such small particles. On the contrary, Eq. 2-1 shows the diffraction from a small body to be made up of a combination of continuous oscillating functions. This actually has several important consequences, which have been known for some time [8]:

- not all peaks associated with a particular structure are resolved in small crystalline particles;

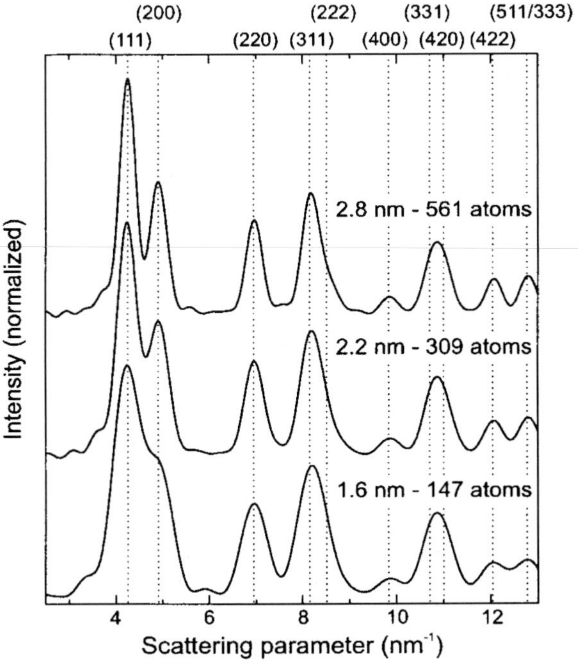

Figure 2-1. Calculated diffraction pattern of three successive sizes of cuboctahedral (fcc) particles. The intensities of the main (111) peak have been normalized to the same value for display, in reality their intensity increases rapidly with size. The baselines of the profiles have also been shifted vertically. At the top of the figure the indices for the Bragg diffraction peaks are shown. The number of atoms per model and the approximate diameters are inset.

- those peaks that are resolved may have maxima that do not align with expected bulk peak positions;
- peak shapes, peak intensities and peak widths may differ from extrapolated bulk estimates;
- few minima in intensity between peaks actually reach zero;
- small, size-related, features appear in the diffraction pattern.

Clearly, to extract quantitative information based on size-limited bulk structure formulae is fraught with difficulty. It means, for example, that an apparent lattice contraction, or expansion, due to a single peak shift may be size related. Also, the familiar Scherrer formula [6], relating particle-size to peak-width, will be difficult to apply accurately [9].

2.3.2.2 Non-crystalline structures

In many metals, distinct structures can occur that are not characteristic of the bulk crystal structure. For most fcc metals, the preferred structures of sufficiently small particles exhibit axes of five-fold symmetry (see schematic representation in Fig. 2-2), which is forbidden in crystals. These Multiply-Twinned Particles (MTPs) were first identified in clusters of gold [10], and have since been well documented in a range of metals [11]. Although MTPs exhibit distinct diffraction patterns, the interpretation of diffraction data can not be made by applying conventional methods of analysis because they lack a uniform crystal structure (e.g.: the location of the maximum in the diffraction peak does not give precise information about the nearest-neighbor distances within the particles [12]. Furthermore, MTPs often co-exist with small fcc parti-

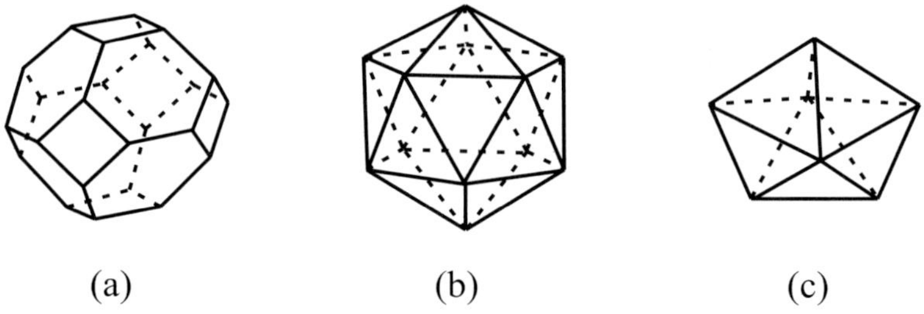

Figure 2-2. Schematic representation of the three possible structures of metal nanoparticles: a) cuboctahedron, formed by a fcc crystal truncated at (100) and (111) atomic planes; b) icosahedron and c) decahedron. b) and c) are known as Multiply-Twinned-Particles (MTPs), characterized by five-fold axes of symmetry.

cles in a single sample. This greatly complicates the problem of determining nanoparticle structure because, in general, it will not be possible to characterize an experimental diffraction pattern by a single particle structure.

The extent to which size and structure can impact on the MTP diffraction patterns can be seen in Fig. 2-3 and Fig. 2-4. These figures show the two distinct MTP structures: the icosahedron [13, 14] and the truncated decahedron [14, 15]. Both of these can be constructed by adding complete shells of atoms to a basic geometry. Individual shells repeat the geometrical form, but with an increasing number of atoms used in the construction. The models below used the inter-atomic distance in bulk gold, with the MTP structures given a uniform relaxation as prescribed by Ino [14]. No Debye-Waller factor was included.

Figure 2-3 shows three sizes of icosahedral particles, in a similar presentation to Fig. 2-1. The icosahedral structure cannot be considered as a small piece of a crystal lattice. However, an icosahedron can be assembled from twenty identical tetrahedral

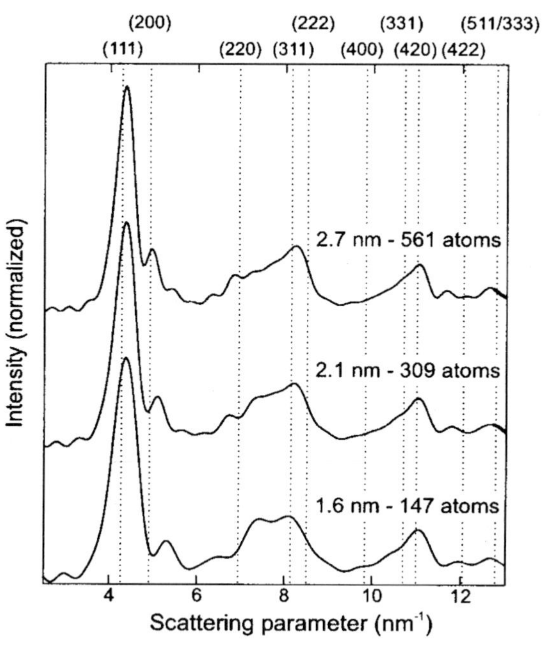

Figure 2-3. The diffraction patterns of three successively larger icosahedra. The intensities of the main peak have been normalized to the same value for display, and the profile baseline shifted. The Bragg indices, that apply to the bulk fcc structure, are shown here only for reference.

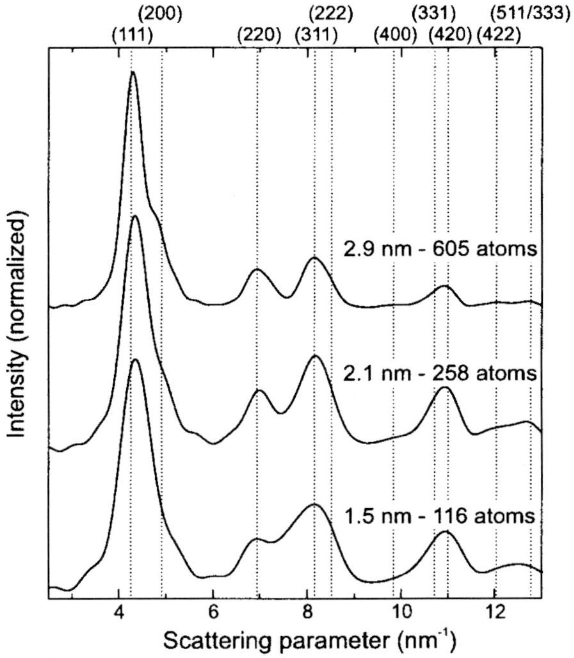

Figure 2-4. Three diffraction patterns from successive sizes of decahedral particles. The fine edges of the decahedra are truncated with (100) planes as suggested by Ino [14].

units, brought together at a common apex in the center of the particle. These tetrahedra are arranged as twins with their three immediate neighbors, so that the complete structure contains thirty twin planes. The individual tetrahedra are exact sub-units of a rhombohedral lattice [16], although it is common to regard them as slightly distorted fcc tetrahedra.

The oscillatory features in Fig. 2-3 (clearly visible at $s \approx 3$ and $s \approx 6$ nm^{-1}) are size-related. In particular, the illusive, shoulder peak, to the right of the diffraction maximum, carries no internal-structure information [17]. It is clear in Fig. 2-3 that these oscillations increase in frequency as the particle size increases. As a result, the prominent shoulder peak moves in, towards the diffraction maximum, as the particle size increases. These size-effects are most obvious in the icosahedral diffraction profiles but occur for each structure type. In general, it is important to consider the consequences of a sample size distribution (something that in practice is almost impossible to avoid) when interpreting diffraction data.

In Fig. 2-3 the bulk fcc Bragg peak locations have been reported for reference again. It can be seen that, in a mixture of structures, an icosahedral component may not clearly distinguish itself, because its contributions to the diffracted intensity profile will be most important in the same regions in which fcc peaks occur. This will be exacerbated by a distribution of sizes: the feature on the right of the main peak will broaden and may form a shoulder on the principal peak, making the latter appear as a single asymmetrical diffraction peak.

Given that the tetrahedral structure of the icosahedral sub-units is crystalline, one might expect the diffraction pattern of icosahedra to be characteristic of that structure. This does happen, but only at much larger sizes: when much larger models are constructed, the rhombohedral lattice peaks start to be resolved [18].

The last in this series of figures shows the diffraction patterns of three truncated decahedral particles (Fig. 2-4). The decahedral structure is the one originally proposed by Ino [14]. The decahedron can be considered as five tetrahedral sub-units, arranged to share a common edge, which forms the five-fold axis of the particle. The five component tetrahedra are in fact perfect sub-units of an orthorhombic crystal, and can be derived from a slightly distorted fcc structure [16]. The narrow external edges formed by the tetrahedra of a perfect geometrical structure are unfavorable energetically. Here, these edges have been truncated by (100) planes according to Ino [14], the more complex surface features proposed by Marks have not been modeled, as they cannot be observed by diffraction [11].

While the general remarks made previously in relation to Figs. 2-1 and 2-3 apply here too, it is apparent that the degree to which Fig. 2-4 differs from Fig. 2-1 is not as great as in the icosahedral case. This is because on the one hand, the distortion of the tetrahedra from fcc to orthorhombic involves less change than does the distortion in the icosahedral structure. On the other hand, there are only five twin planes in the decahedral structure, meaning that the sub-units are relatively larger, compared to the particle diameter, and their structure is more apparent in the diffraction profile.

2.3.3 Direct analysis of nanoparticle diffraction patterns

The Debye equation shows that the diffracted intensity depends on the distribution of inter-atomic distances within the scattering volume. While the internal structure can be used to calculate a set of inter-atomic distances, the converse is not necessarily true. This is a ubiquitous problem in diffraction, but it is worth drawing attention to the present context: if a model diffraction pattern agrees well with observed data, then there is strong evidence that the structure represented by the model characterizes the actual structure of the sample. However, if more than one model structure actually have very similar distributions of inter-atomic distances then diffraction measurements will not be able to distinguish between them.

With this in mind, a direct approach can be used to analyze experimental diffraction data from nanoparticles. We have seen above that quite distinct features are associated with both particle size and structure. Here, there is an opportunity to extract both the structure type and the domain size distribution from an experimental diffraction profile. This is achieved by comparing combinations of model structure diffraction patterns with the experimental data, until a satisfactory match is obtained. Such a procedure neatly handles the problem of sample size dispersion and makes only such assumptions about particle structure as are necessary to construct a series of trial model structures. This method was originally proposed as a way to interpret electron diffraction data of unsupported silver particles [19, 20]. It has come to be known as Debye Function Analysis (DFA) and has been applied since by two groups independently [21–24]. A more complete description of DFA can be found in [25].

It is important to realize that the DFA does not alter the structure of the models in any way: it does not attempt to refine nanoparticle structure. The DFA uses a finite set of fixed-structure diffraction patterns to assemble the best possible approximation to an experimental diffraction pattern. If the physical sample differs significantly in structure from the models, for example because of lattice contraction or other relaxation, then the results of the DFA will show this up in the quality of the fit.

DFA analysis is sensitive to the domain structure within the (often) imperfect particles. It remains an open question as to how particle imperfections will contribute to a diffraction pattern: it has been suggested that certain defects can form domains with local atomic arrangements similar to small icosahedra or decahedra, while the particle as a whole may not appear to have this structure [26].

To illustrate DFA, we apply it to a diffraction pattern obtained from a sample of gold particles. These nanoparticles have been synthesized by chemical methods [27] and consist of gold clusters covered by thiol molecules ($C_{12}H_{25}SH$), which are attached to the surface by the sulfur atom. From transmission electron microscopy, the size distribution was estimated to have a mean diameter of 2 nm and a half width of 1 nm [28]. Powder X-ray diffraction studies were performed using 8.040 keV photons. As nanoparticles diffraction peaks are rather large (FWHM $\approx$ 5 degrees for the (111) peak of 2 nm particles), we used 2 mm detection slits before the scintillation detector. To set up the DFA, complete-shelled models for the three structure types (fcc, ico, dec) were used to calculate diffraction patterns. A total of twelve diffraction patterns were calculated, covering the diameter range between approximately 1 and 3 nm for each structure type (see Table 2-1). In addition, two parameters were assigned for background scattering: one to the substrate contribution, the other a constant offset. The Debye-Waller parameter was kept fixed during optimization, however the rms atomic displacement (see Eq. 2-2) was estimated to be $19 \pm 3 \times 10^{-12}$ m, by repeating the fitting procedure with a range of values for D.

Table 2-1. Numerical values associated with the fit of Fig. 2-5. Note that the proportions of each structure have been rounded to integers, and when estimated values round to zero the associated uncertainties are not reported. The quoted diameter values are obtained from the distribution of inter-atomic distances. The abrevations fcc, ico and dec, refer to cuboctahedral, icosahedral and decahedral model structures, respectively.

Struct.	No. of atoms	Diam. (nm)	Percentage by number	Percentage by weight
fcc	55	1.2	0	0
fcc	147	1.6	0	0
fcc	309	2.2	0	0
fcc	561	2.8	0	0
ico	55	1.1	40 ± 14	29 ± 10
ico	147	1.6	7 ± 7	14 ± 13
ico	309	2.1	0	1 ± 1
ico	561	2.7	0	1 ± 1
dec	39	1.0	30 ± 3	15 ± 2
dec	116	1.5	20 ± 14	31 ± 21
dec	258	2.1	0	0
dec	605	2.9	1 ± 2	8 ± 14

The results of the DFA are presented in Table 2-1 summarizing both the models used and the values estimated for each parameter. In particular, relative proportions are reported both as a number fraction and as a fraction of the total sample weight. This is done because the intensity of diffraction from particles increases in proportion to the number of atoms, and hence larger particles will dominate an observation. The upper window in Fig. 2-5 shows the experimental data (dotted line) superimposed on the DFA best-fit (solid line), and once again, fcc peak positions are indicated for reference. The lower window shows the difference between the two curves. The almost complete absence of structure in the difference curve indicates that a good fit has been obtained.

The uncertainties reported in Table 2-1 are obtained by collecting statistics from repeated runs. These runs simulate the variability of the measurement process by adding a random component (noise) to the experimental data before each fit starts. The random noise is added here assuming a Poisson process with a rate equal to that of the actual measured intensity. The uncertainty estimates here are obtained from the standard deviations of individual parameters by analyzing values from ten runs. A more complete discussion of this approach to estimating uncertainty in parameters is given by Press et al. [29].

Table 2-1 shows that the sample is composed only of MTP structures with a roughly equal split between icosahedra and decahedra, both types of structure have sizes mainly less than 2 nm. A greater proportion of the icosahedra are the smallest size in the fit, while the decahedral contribution is dominated by the second smallest (1.5 nm) size domains. This is not immediately apparent from a visual inspection of the raw data. Certainly, comparing the profiles of Fig. 2-3 with the experimental diffraction pattern in Fig. 2-5 one would not necessarily expect small icosahedra to be present. Nevertheless, the quality of the fit is definitely made worse if icosahedral structures are excluded. Another remarkable fact is the total absence of fcc nanoparticles. This probably indicates a high proportion of imperfect structures in the sample, which may be showing up as MTPs in the fit.

While the uncertainty associated with the relative proportions of each model is rather high, it must be remembered that these parameters are not independent: at each run the balance of individual structure types is being varied – a little less of one

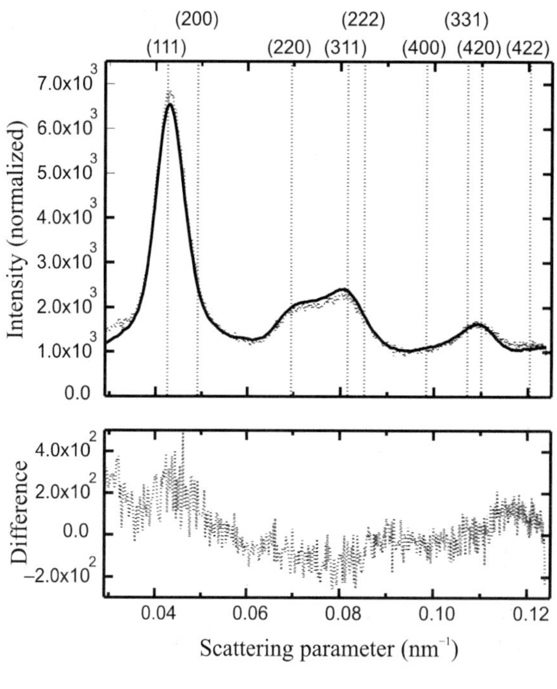

Figure 2-5. Fitting results of the DFA on a sample of 2 nm gold particles. The upper window shows the experimental data (dotted line) superimposed on the fit (solid line). The lower window shows the simple difference between the two curves. Intensity values are arbitrary, but close to the actual count rates that occurred during the experiment.

structure means a little more of another – and this correlation is not captured by our simple calculation of the statistics from several runs. Also, it is important to understand that the parameters estimated by DFA are describing the *shape* of domain size distributions (the uncertainty in the average domain size, for example, is much less). It should not be surprising, therefore, that in the tails of the distributions (where parameter values are small) the relative uncertainty is higher than near the distribution centers. What these results do show is that it is becoming difficult to distinguish clearly between the smallest decahedra and icosahedra. Improving the signal-to-noise ratio, accounting better for systematic contributions (background, and gas scattering), and extending measurements to higher values of scattering parameter, s, would all help to aleviate this problem.

We conclude that DFA analysis has clearly identified the presence of MTP structures, and strongly suggests that both icosahedral and decahedral domains are contained in the particles of the sample. Furthermore, given the closeness in size of the decahedral domains to the size observed by TEM, we are confident that single-domain decahedral particles are present in this sample.

2.3.3.1 Technical considerations

The measurement of diffraction patterns from nanoparticle samples differs somewhat from standard powder diffraction work. The intensity of scattering per atom is weak, and the amount of sample material is often limited, so the detection of a low level of diffraction signal can be critical. It is important to reduce unwanted sources of diffraction, such as scattering from gas in the diffraction chamber, and diffraction from the substrate material. Also, because diffraction features are very broad, only low angular resolution is required at the detector and very broad collection angle should be used, increasing the intensity and improving the statistical uncertainty of readings. It must also be remembered that the entire diffraction profile contains structural information about the nanoparticles and therefore that good quality data should be collected for the whole profile, not just in the more intense regions (peaks) of the diffraction pattern.

If good results are to be obtained by DFA, it is important to account for all systematic contributions to the diffraction pattern. DFA tries to recreate an exact match to the experimental data, and is therefore quite sensitive to effects that change the profile shape or components that introduce some structure of their own. It is hard to predict how such systematic errors will show up in the results of the analysis. We therefore recommend that careful measurement of background scattering terms and careful determination of the origin for the diffraction pattern ($s = 0$) be carried out.

An implementation of DFA requires a certain effort in the development of software. A more interactive approach may be valuable in preliminary work, and has sometimes been used in the past [18]. This can usually be done with any general purpose computer mathematical package with graphics. The process is, surprisingly, reasonably quick; indeed, it is an excellent manner in which to convince oneself of the necessity to both incorporate size distributions and include MTP structures in the analysis of metal nanoparticle structure.

An interactive method cannot, however, provide convincing, unbiased results, nor estimate uncertainties in a large number of parameters: some form of automatic optimization is desirable. Unfortunately, the problem to be solved is not easy. Experience has shown that a technique suitable for global optimization is necessary, as one can easily be trapped in local minima. In the work presented here, a simple form of the simulated annealing algorithm is used and found to be quick and reliable [22, 29], the code used for this work is written in C and it runs on a desktop PC.

2.4 Extended X-ray absorption spectroscopy

2.4.1 X-ray absorption spectroscopy

The technique of X-ray Absorption Spectroscopy (XAS) explores variations in the absorption coefficient of matter with photon energy. When a monochromatic X-ray beam passes through a material, its intensity is reduced by various interaction processes (scattering, absorption, etc.). For hard X-rays (more than 1000 eV), the photoelectric effect dominates, in which a core atomic electron is ejected by photon absorption. The absorption coefficient, μ, can be defined as [2, 30]:

$$I = I_0 e^{-\mu t} \tag{2-3}$$

where I is the transmitted intensity, t is the material thickness traversed and I_0 is the incident beam intensity. The coefficient μ depends both on material properties and photon energy (E).

In general, as photon energy increases, the absorption coefficient decreases gradually until critical energies are attained, whereupon it changes abruptly. These discontinuities, known as absorption edges, occur when the photon energy corresponds to a threshold (E_0) for core electron excitation. The energy of the absorption edge is specific to each chemical element, since it corresponds to the binding energy of the photoelectron.

When this absorption process occurs in condensed matter, the ejected photoelectron interacts with atoms in the immediate neighborhood, resulting in a modulation of the absorption coefficient beyond the edge. These modulations can easily be identified in an experimental spectrum, for example that shown in Fig. 2-6a. Based on the energy of the ejected electron ($E - E_0$), it is possible to roughly divide the absorption spectrum in two regions, according to different interaction regimes with the surrounding atoms:

- XANES (X-ray Absorption Near Edge Structure): ≈ 0–40 eV above E_0, where multiple scattering events take place, yielding information about symmetries and chemical state.
- EXAFS (Extended X-ray Absorption Fine Structure): ≈ 40–1000 eV, where single scattering events dominate, providing structural information, such as coordination numbers and inter-atomic distances.

To understand and model the XANES spectral region usually requires heavy, and complicated, multiple scattering calculations. On the other hand, EXAFS oscillations, dominated by single electron scattering process, can be handled in a simpler mathematical treatment. The availability of reliable and simplified data processing has transformed EXAFS into a widely used structural characterization technique, and also explains why most XAS studies have been done in this spectral region. In particular, EXAFS has been used as an alternative (and sometimes a complement) to XRD, since it probes the local environment of the excited atom. Indeed, EXAFS can be

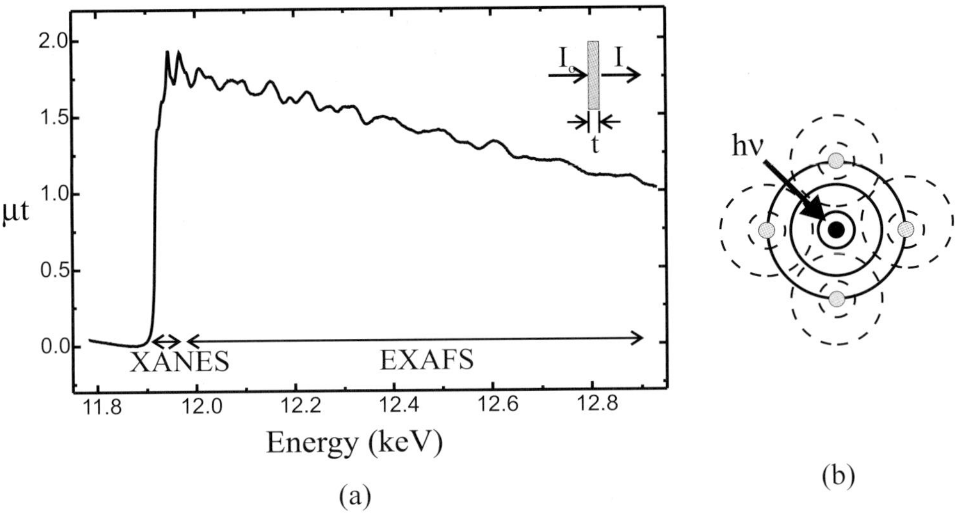

Figure 2-6. a) X-ray absorption spectrum of Au-L_3 edge of a gold film. Note the modulation of the absorption coefficient above the edge. b) Pictorial representation of the interference process between ejected (solid line) and backscattered (dashed line) photoelectron waves that gives rise to the EXAFS oscillations. See text for explanations.

seen as a low-energy electron diffraction process, where the photelectron comes from an energy-selected atomic element. Concerning nanostructured materials, many authors have tackled the problem of determining the inter-atomic distances by EXAFS [31–33]. This measurement is rather difficult to obtain in these systems by other means, due to the intrinsic lack of long-range order in small particles. In the following sections we describe the basic process and the special features of EXAFS experiments on nanoparticles, and illustrate these with a particular example.

2.4.2 EXAFS

In order to understand the physical origin of EXAFS oscillations, we must first remember that the probability for a core electron to absorb an X-ray photon depends on both the initial and final states. Above the edge, the final state can be described by an outgoing spherical wave, originating at the absorbing atom. This wave may be scattered by neighboring atoms, resulting in an interference pattern (see Fig. 2-6b). The final state will depend on both (outgoing and scattered) wave phases, which in turn will depend on the electron wavevector (k), or equivalently on the ejection energy. Hence, the exact position of neighboring atoms can affect the probability of exciting a core electron and gives rise to the oscillatory behavior of the absorption coefficient as a function of photoelectron energy.

Mathematically, the interference term arising from scattering by a single neighbor can be expressed as $A(k)\sin[2kr+\phi(k)]$, where k is the modulus of the wavevector, r is the distance between absorbing and neighbor atoms, and $\phi(k)$ represents the total photoelectron phase-shift and depends on both photoabsorber and scattering atoms. $A(k)$ is the backscattering amplitude and is mainly a characteristic of the scatterer atom. The total EXAFS signal consists of the superposition of individual pair-wise contributions from all neighboring atoms; these can be grouped into coordination shells composed of atoms found at similar distances from the absorbing atom.

The EXAFS oscillations, $\chi(k)$, are given by [30]:

$$\chi(k)=\sum_{j} S_0^2(k)\frac{N_j}{kr_j^2}\,e^{-2\sigma_j^2k^2}\,e^{-2\frac{r_j}{\lambda_j(k)}}A_j(k)\sin\left[2kr_j+\phi_j(k)\right] \tag{2-4}$$

where

$$k=\sqrt{\frac{2m}{\hbar^2}\left(E-E_0\right)} \tag{2-5}$$

k	=	photoelectron wavevector modulus;
j	=	coordination shell index;
r	=	distance between the absorbing atom and a neighbor;
N	=	number of identical atoms in the same coordination shell;
$A(k)$	=	backscattering amplitude;
σ	=	total Debye-Waller factor (including static and dynamic contributions);
$\phi(k)$	=	total phase shift;
$\lambda(k)$	=	photoelectron mean free path;
$S_0^2(k)$	=	amplitude reduction factor due to many-body effects;
E	=	photon energy;
m	=	electron mass;
E_0	=	threshold energy.

The structural parameters involved in the EXAFS equation are the coordination number (N), the inter-atomic distance (r), and the Debye-Waller factor (σ). The latter includes two contributions: dynamic, arising from atomic vibration, and static, which is caused by structural disorder in a given coordination shell. This expression also includes atomic parameters such as $\lambda(k)$, $A(k)$, $\phi(k)$, $S_0^2(k)$. The EXAFS equation supposes that a harmonic approximation applies to atomic vibrations and that the pair-distribution function for inter-atomic distances, $P(r)$, is assumed to be Gaussian. The term $\exp(-2r/\lambda(k))$ accounts for the finite photoelectron lifetime and represents the probability for the photoelectron to travel to, and from, the backscatterer without additional scattering and before the core hole is filled.

Since it has been assumed that $\chi(k)$ can be represented by a linear combination of sine waves from each coordination shell, it is possible, in principle, to separate each contribution by applying a Fourier transform. By extracting and analyzing the EXAFS signal we can obtain estimates for the structural parameters: N, r and σ for each shell, however, this requires, prior knowledge of the atomic parameters: $\lambda(k)$, $A(k)$, $\phi(k)$, $S_0^2(k)$. To obtain the required set of atomic parameters, two approaches are possible: they can be either calculated theoretically, or determined experimentally by using a standard reference compound; in principle both methods seem to have the same limits of accuracy [2].

2.4.2.1 Special features of EXAFS in nanoparticle systems

The wide use of EXAFS in materials research has led to the development of robust procedures for the use of this technique for bulk systems. As mentioned in Section 2.4.2, the origin of EXAFS oscillations lies in the interaction of the ejected photoelectron with the neighboring atoms. Since only the local environment of the excited atom is probed, both experimental methods and data analysis procedures can be applied to nanostructured systems. However, as might be expected, the structural parameters (N, r, σ) one obtains depend on the characteristic size of the system.

Intuitively, the very small size of particles will lead to a decrease in the mean coordination number, and an absence of higher order coordination shells. Deviation in the measured value of N, from the bulk, may be used as a rough estimate of the mean particle diameter [34, 35], however other factors may also influence this measurement, such as size-induced modification of the particle structure.

One of the main results that can be obtained by EXAFS is a precise determination of inter-atomic distance (in practice, to better than 0.002 nm). EXAFS is well suited to this measurement because it does not require long-range order and can be used to directly determine changes to the nearest-neighbor distance in disordered or finite systems.

EXAFS can also provide valuable information on nanoparticle structure. Although structure cannot be fully determined, it has been proposed that through the analysis of the first and second coordination shell distances, and in particular the ratio r_I/r_{II} (I – first and II – second shells), it should be possible to distinguish between MTPs and bulk-like structures in metals [36]. In other systems, such as semiconductors, a careful comparison among the parameters has been used to differentiate hexagonal from cubic structures [35].

Experimental measurements of the Debye-Waller factor can be used to obtain further structural information. Firstly, the stiffening or softening of the chemical bonds due to the modification of the surface atomic potential can be observed, because of modifications in the atomic vibrations [32, 37]. Secondly, we expect, based on surface studies, that the large surface-to-volume ratio will result in a higher structural disorder in nanosystems. It is well known that the lack of translational symmetry in surfaces leads to atomic rearrangement, changing the bond distribution. In nanostructured materials, this effect will tend to grow as the particle diameter decreases and will enhance the average static contribution to the Debye-Waller factor. Careful analysis of EXAFS data at different temperatures can be used to separate the dynamic and static contributions to the Debye-Waller factor and hence measure the structural disorder and deduce vibrational properties (such as the Debye temperature) [35, 36, 38].

In summary, structural information can be obtained by the EXAFS through the parameters N, r and σ. When dealing with nanosystems, these parameters are dependent on the size of the system. One of the key measurements that EXAFS can perform is an accurate estimate of nearest-neighbor distance. The technique can also provide indirect information on: particle dimensions; vibrational properties; and structural defects.

2.4.3 Data analysis applied to nanoparticles

We now present an example of an EXAFS study that shows the main aspects of interest in nanomaterials research. We have selected a system where it is possible to observe most of the advantages and some specific difficulties of the technique. The experimental sample consists of the same thiol-capped 2 nm gold nanoparticles described in Section 2.3.3. The experiment set out to investigate the possibility of changes in inter-atomic distances, since a contraction is expected for metal clusters due to surface stress [39]. Measurements were performed on the Au-L_3 edge (11.919 keV), at low temperature (8 K), in transmission mode and with an energy resolution of 1.8 eV. Detection used two gas ionization chambers.

From a qualitative point of view, we can expect to observe certain differences between an absorption spectrum from a bulk sample and one from a sample of thiol-capped gold nanoparticles (Fig. 2-7). Firstly, a stronger attenuation in the EXAFS oscillations is expected. The reduction of the mean coordination number, and the absence of higher order coordination shells, results in a homogeneous attenuation of the EXAFS oscillation amplitude. In addition, higher structural disorder and the surface metal-ligand bonds contribute to the damping of the EXAFS oscillations at high k-values. It must be remembered that due to the passivation by thiol molecules, surface gold atoms are coordinated with both sulfur and gold. The metal-ligand damping effect can be understood because the EXAFS interference term is multiplied by the backscattering amplitude factor, $A(k)$, which strongly depends on the atomic number of the scatterer atom. For heavy atoms, $A(k)$ has a significant contribution over the whole k-range of interest, whereas for light elements, such as sulfur, it decreases monotonically at high k-values.

Secondly, the characteristic frequency of the EXAFS oscillations may be expected to change if the nearest-neighbor inter-atomic distance does so. Both, the existence of Au-Au inter-atomic distance contraction and the presence of the Au-S bond (shorter

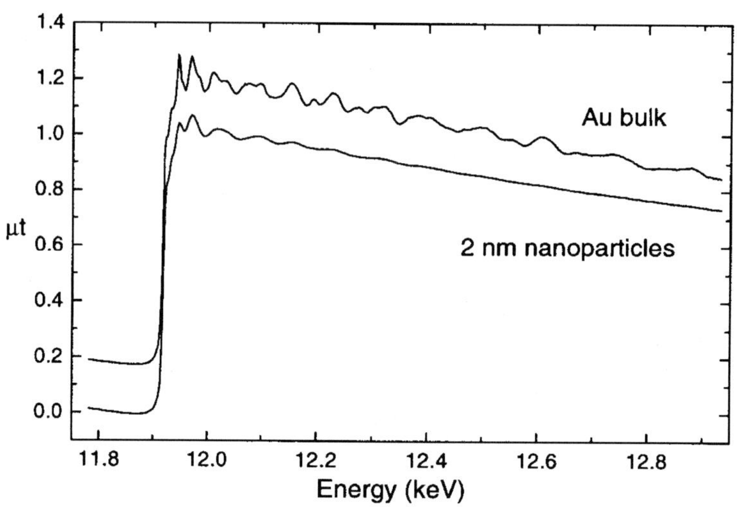

Figure 2-7. Comparison of the absorption spectrum of 2 nm thiol-capped gold nanoparticles and bulk gold. The attenuation and damping of the nanoparticle EXAFS oscillations can be clearly observed. Measurements were performed at low temperature (8 K).

than the Au-Au one) should decrease the oscillation frequency in the k-space. Although it is not easy to see this effect directly in the EXAFS spectrum, it becomes visible when the signal from the first coordination shell is isolated.

In the remainder of this section, we present an example of practical EXAFS data analysis in some detail, the individual steps have been assigned to subsections [2, 30, 40].

2.4.3.1 Extraction of EXAFS signal

The absorption spectrum is composed of EXAFS oscillations superposed on a smooth background, which comes from the other absorption edges and from the other elements in the sample. In terms of the measured signal, $\mu(E)$, the EXAFS oscillations $\chi(E)$ are defined as:

$$\chi(E) = \frac{\mu(E) - \mu_0(E)}{\Delta\mu(E)} \tag{2-6}$$

where $\mu(E)$ is the absorption coefficient associated with a particular edge (Au-L_3, in this case) and $\mu_0(E)$ is the absorption coefficient of an isolated gold atom;$\Delta\mu(E)$ is the change in the atomic absorption across the edge, and provides normalization. Consequently, the first step in the analysis involves (see Fig. 2-8):

a) subtraction of the pre-edge background: obtained by a least squares fit with a straight line or Victoree-like function ($a\lambda_p^3 - b\lambda_p^4$, λ_p= X-ray wavelength);
b) subtraction of the atomic absorption coefficient (μ_0): estimated by a polynomial fit in the EXAFS region, by cubic-splines or other interpolation functions;
c) normalization: difference between μ_0 and the extrapolation of the pre-edge region. This extrapolation is usually performed by fitting a standard analytical function, it requires an estimate of the threshold value, E_0. The choice of this parameter is arbitrary, as it is refined during the fitting procedure. Usually it is chosen as a characteristic point in the absorption curve, for example, the first inflection point.

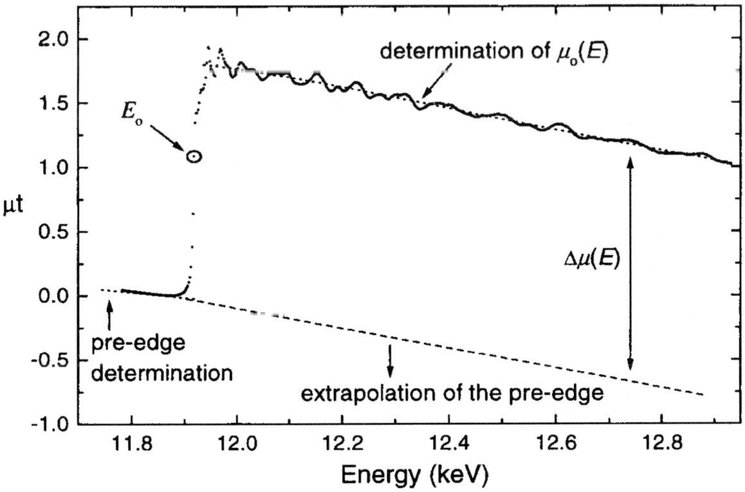

Figure 2-8. Representation of the EXAFS signal extraction from the X-ray absorption spectrum ($\mu(k)$ = absorption coefficient; $\mu_0(k)$ = atomic absorption coefficient, $\Delta\mu(k)$ = normalization). See text for explanations.

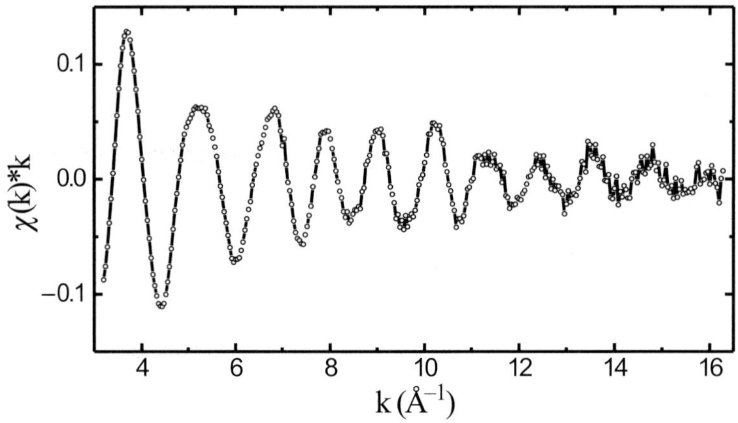

Figure 2-9. EXAFS oscillations of 2 nm gold nanoparticles.

Because the absorption spectrum is measured as a function of energy, but the photoelectron interference is better described in terms of k, a change of variable is made to transform from E- to k-space using Eq. 2-5.

A check on the quality with which the EXAFS signal has been extracted can be made by verifying that the oscillations are symmetrically distributed around zero. Another point to check is that the function representing the atomic absorption coefficient does not match the EXAFS oscillations, this can be seen by taking its derivative. Finally, we obtain the EXAFS oscillations shown in Fig. 2-9.

2.4.3.2 Fourier Transform

To separate the EXAFS contributions from individual coordination shells, we can Fourier Transform (FT) $\chi(k)$ into r-space. The FT of the EXAFS oscillations corresponds to a pseudo-radial function, where peak positions, R, are related to inter-atomic distances (although they also include phase-shift effects), and the area under peaks

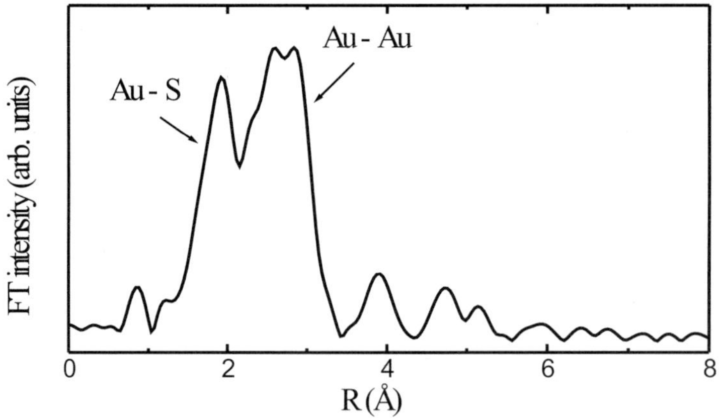

Figure 2-10. Fourier transform of the EXAFS signal using $\Delta k = 12.6$ Å^{-1}; $k_{min} = 3.4$ Å^{-1}. Note the contribution of Au-S at $R \approx 2$ Å, where R is the raw inter-atomic distance, not corrected by phase shift.

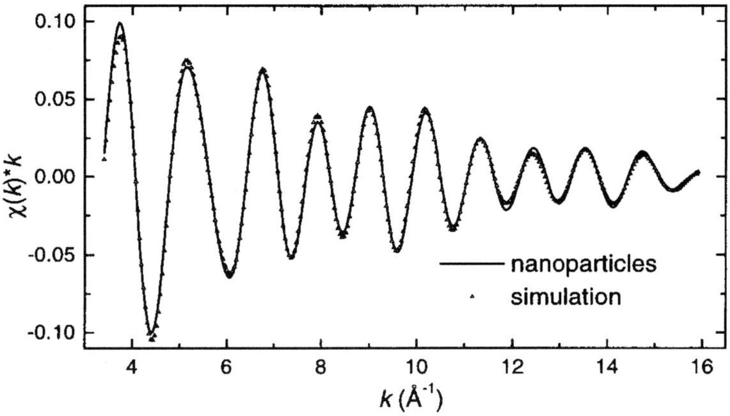

Figure 2-11. Fourier filtering of the nanoparticles first coordination shell and corresponding simulation, which includes both Au-Au and Au-S contributions (ΔR = 1.85 Å, R_{min} = 1.55 Å).

can be associated with the number and type of backscatterers. It is usual to truncate the FT at a lower limit ~ k_{min} = 3 Å^{-1}, in order to avoid the contamination by multiple scattering effects (XANES region). Figure 2-10 shows the resulting FT curve; in this particular case, only the first Au-Au coordination shell can be clearly identified due to the small size of particles (2 nm).

2.4.3.3 Isolation of a specific coordination shell

A selected r-space range can be transformed again, from r-space to k-space. This has been done in Fig. 2-11, where the k-space signal now refers to what is the first coordination shell. In this particular case, the difference between Au-Au and Au-S inter-atomic distances is too close to separate out the two contributions by FT, and as a result, it will be necessary to treat the two shells together.

2.4.3.4 Fitting procedure

The last step consists of a procedure to estimate the structural parameters (N, r, σ) by least-squares. Values for the atomic parameters: $\lambda(k)$, $A(k)$, $\phi(k)$ and $S_0^2(k)$ must be provided. Because there are two contributions (Au-Au, Au-S) to be modeled, two sets of atomic parameters are necessary. We have used a thin gold film to derive the Au-Au shell atomic parameters (N = 12, r = 2.865 Å, σ = 0), whereas Au-S parameters were obtained theoretically (McKale tables [41], $r_{Au\text{-}S}$ = 2.32 Å). In the first case, an identical EXAFS procedure is applied to experimental data of the standard compound however now, in the fit, the known parameters are the structural ones. The resulting parameters are presented in Table 2-2; uncertainty estimates were obtained by doubling the residual at the minimum [42].

Table 2-2. Fit results of EXAFS spectrum for 2.0 nm nanoparticles. Bulk gold structural parameters are: $r_{Au\text{-}Au}$ = 2.865 Å and $N_{Au\text{-}Au}$ = 12; $\sigma_{Au\text{-}Au}$ is assumed to be equal zero. ΔE_0 takes into account the threshold energy refinement.

	Au-Au	Au-S
r [Å]	2.843 ± 0.007	2.34 ± 0.02
N	7.2 ± 0.8	0.8 ± 0.2
σ [Å]	0.078 ± 0.004	0.06 ±0.02
ΔE_0 [eV]	0.0 ± 0.6	10 ± 2

In this example, the EXAFS analysis reveals two important effects: the existence of a slight Au-Au inter-atomic distance contraction of about 0.8% ($r_{Au\text{-}Au}$ = 2.843 Å) and a short Au-S distance ($r_{Au\text{-}S}$ = 2.34 Å), comparable to other gold-sulfur systems [43]. These results indicate that the expected contraction, verified in gold particles immersed in a weakly interacting matrix (~ 1.4% [37]), may be partially compensated by the bonding with thiol molecules. It has already been established that the presence of sulfur on (100) surfaces of fcc metals induces an expansion between surface planes [44]. As for the coordination number (N), the estimated value (7.2) is smaller than the expected one (≈ 9.3) for an ideal 2 nm fcc particle formed by ≈ 200 atoms. Several experimental factors, such as thickness variations in the sample, may affect the determination of N, and account for this discrepancy. However, it is important to realize that the error in N is usually quite large (10–20%) and its relationship to other sample parameters (such as nanoparticle size) needs to be handled with care.

2.4.4 Troublesome points in the data treatment

In the example above, we have chosen to use the conventional method of analysis, by Fourier transforms, because of its mathematical simplicity, which makes it easier to understand the processes involved. Although we have only analyzed the first coordination shell, it is also possible to treat higher order coordination shells in systems where these can be easily identified, such as bigger particles or when the Debye-Waller factor is small enough. However, the analysis of higher-order shells is usually more complex, because of possible multiple scattering processes (ex. focusing); also, additional care must be taken to account for possible mean free path effects [2].

The formulation of the EXAFS spectrum presented in Section 2.4.2 is valid in the small structural disorder limit, or the harmonic approximation, where the Debye-Waller factor is given by $\exp(-2\sigma^2k^2)$. However, this expression is not valid for systems with a high degree of disorder or high-temperature experiments, where anharmonic contributions are no longer negligible. In fact, for nanosystems, due to the intrinsic asymmetry of the surface atomic potential, the temperature effect can be much more pronounced than in bulk materials. In many cases, it is necessary to use a more general equation, where the probability of finding the jth species in the range r_j to r_j+dr_j is represented by the distribution function $P(r)$, which may be asymmetric. However, in such cases the data treatment becomes more complex due to the inclusion of additional free parameters in the fitting procedure [2, 45, 46]. It is worth pointing out that the existence of an asymmetric distribution function, if not properly taken into account, leads to serious errors in calculated structural parameters. In general, the main consequences will be a reduction in the coordination number and a fictitious inter-atomic distance contraction [46].

One alternative to solve the more general EXAFS problem is to use the method of cumulants, which involves the expansion of a function associated with $P(r)$ in a moment-series; deviations from a Gaussian distribution are represented by the higher-order terms of the expansion [2]. Unfortunately, the inclusion of these extra terms may lead to trouble in the data analysis, such as strong correlations among the parameters [37]. In addition, for the system where two types of atom constitute the first shell, as studied here, this method has to be applied with care.

Another approach is to analyze the absorption spectrum directly, instead of just the EXAFS oscillations. In this method, the background and all the coordination shells are fitted simultaneously. This provides a more complete description of the absorption phenomenon but requires more complex theoretical calculations. In particular, it is necessary to use a cluster model that is close to the actual structure. This method has been implemented in a software package called GNXAS, which was initially developed for disordered structures and it has also been applied to nanosystems [45, 47]. The EXAFS technique, as presented above, has been used widely in materials research because of its relatively straightforward data treatment.

The use of more sophisticated packages, although essential in particular cases, make it more difficult to ascribe physical meaning to the additional parameters included in an analysis. In general, in a first attempt, one should try to analyze and evaluate possible sources of asymmetry in the distribution function. In particular, performing measurements at low temperature avoids problems with anharmonic terms [48]. In our example, this was clearly demonstrated: room temperature experiments suggested an inter-atomic distance (2.82 ± 0.01 Å) shorter than the one measured at 8 K (2.843 ± 0.007 Å). Another source of problems is high structural disorder, it is prudent to use complementary techniques, such as XRD and HRTEM to evaluate the importance of this in a particular study.

As in all experimental data analysis, results from the EXAFS fits should be qualified with uncertainty estimates for each parameter, as well as the correlations between them. There are some well-established and specific guidelines for EXAFS analysis that must be considered in the data treatment [42]. In particular, it is necessary to take care about the correlation between N and σ (related to the oscillation amplitude) and r and E_0 (related to the oscillation frequency).

In this section, we have attempted to give a basic introduction to EXAFS as well as the treatment of data when applied to nanosystems. There are, however, new analytical procedures that have been implemented and it was not intended to cover all possible methodologies here. Readers will find ample details in the specialized literature, where there are many other examples related to nanostructured materials [31–38, 47, 48].

2.5 Conclusions

X-ray methods of characterization represent a powerful approach to the study of nanophase materials. The advantage of these techniques is to provide meaningful ensemble-averaged information about both medium range, and local, atomic structure in nanosystems. By characterizing the sample as a whole, they are an essential complement to other high-resolution methods, which provide rather detailed information on only a few particles.

In particular, we have discussed the application of two of the most popular X-ray based structural probes: diffraction and EXAFS. We have shown that it is possible to acquire detailed information about the structure of very small gold particles (2 nm in diameter), identifying and characterizing non-crystallographic atomic arrangements in the sample, as well as making a precise measurement of the nearest-neighbor distance in the clusters.

We do not wish the reader to conclude that the application of EXAFS and XRD are limited by the parameters of our two examples. Concerning EXAFS, we have only discussed the details of inter-atomic distance determination in small metal clusters. However, in favorable cases, a more extensive study, could yield reliable estimates of other parameters, such as structure [32, 35, 36], mean particle diameter [34] and structural disorder [32, 35, 36]. It is important to note that our examples dealt with gold, where the very interesting occurrence of MTP structures greatly complicates the analysis. MTPs are multi-domain structures in which the constituent units represent slight distortions of the fcc lattice. Both EXAFS and XRD methods could be expected to perform much better in systems of nanoparticles where the structural variety of clusters is better differentiated, such as semiconductors (e.g.: the transition between wurzite and zinblende structures [35]).

In dealing with nanoparticle samples, it must be stressed that results of high-quality can only be acquired if very careful attention is paid to both the measurement of raw data and its subsequent processing and interpretation. Synchrotron radiation sources are virtually a necessity to obtain measurements of sufficient quality; such data can then be subjected to detailed analysis, and valuable information derived. Modern optics and the existence of special devices, such as wigglers allow measurements of small quantities of sample material at good signal levels.

For EXAFS in particular, the association of more intense sources with an optimized setup for low temperature measurements would allow inclusion of the additional structural information in the high k-range (> 16 Å^{-1}); thereby improving the precision in parameter estimates through the use of the Fourier transform in EXAFS analysis.

For diffraction, it would be desirable to include a greater range of scattering parameter in future studies. The range should be extended to collect data both at larger values of s, and in the low-angle region. The latter provides direct information about particle size and shape, complementing the results of DFA. This would allow a more definitive statement to be made about the numbers of defective structures in the sample, by allowing comparison of particle size information (from small-angles) with domain size information (from wide-angles). Experiments performed with higher energy photons would permit measurements to be extended out to higher values of s. Such measurements would capture more structural information and thus improve the selectivity of the DFA method. If the range of s can be made sufficiently great, then it is also possible to use Fourier techniques to invert the diffraction pattern, thereby obtaining the distribution of inter-atomic distances in the clusters. This information could be compared directly with estimates of nearest-neighbor distance determined by EXAFS.

We hope that the examples presented will have shown the reader the possibilities offered by X-ray techniques in probing nanosystems, and that the information provided may assist in the correct application of the methods described.

Acknowledgement

The authors acknowledge the invaluable help of the LNLS staff, and thank H. Tolentino for stimulating discussions. DZ and DU are indebted to FAPESP (Contracts 96/12550–8, 97/04236-4) and CNPq for funding. BDH gratefully acknowledges financial support for travel under the NZ/BRAP STC Agreement Programme (99-BRAP-11-HALL).

References

[1] B. D. Cullity, *Elements of X-Ray Diffraction*, Reading, Addison-Wesley Publishing Co. Inc., **1978**.
[2] D. C. Koningsberger and R. Prins, *X-Ray Absorption: Principles, Applications, Techniques of EXAFS, SEXAFS and XANES*, New York, John Wiley & Sons, **1988**, Chapters 1, 6 and 9.
[3] A. Guinier and G. Fournet, *Small-Angle Scattering of X-Rays*, New York, Wiley, **1955**.
[4] O. Glatter and O. Kratky, *Small-Angle X-Ray Scattering*, London, Academic Press, **1982**.
[5] G. Margaritondo, *Introduction to Synchrotron Radiation*, Oxford, Oxford University Press, **1988**, Chapter 1.
[6] A. Guinier, *X-ray diffraction in crystals, imperfect crystals, and amorphous bodies*, New York, Dover, **1994**.
[7] A. J. C. Wilson (Ed.), *International Tables for Crystallography*, Dordrecht, Netherlands, Kluwer Academic, **1992**.
[8] C. W. B. Grigson and E. Barton, *Brit. J. Appl. Phys.* **1967**, *18*, 175.
[9] J. B. Cohen, *Ultramicroscopy* **1990**, *34*, 41.
[10] S. Ino, *J. Phys. Soc. Japan* **1966**, *21*, 346; S. Ino and D. Ogawa, *J. Phys. Soc. Japan* **1967**, *22*, 1365; J. G. Allpress and J. V. Sanders, *Surf. Sci.* **1967**, *7*, 1.
[11] L. D. Marks, *Rep. Prog. Phys.* **1994**, *57*, 603.
[12] C. L. Briant and J. J. Burton, *Surf. Sci.* **1975**, *51*, 345.
[13] A. L. Mackay, *Acta Cryst.* **1962**, *15*, 916.
[14] S. Ino, *J. Phys. Soc. Jpn.* **1969**, *26*, 1559; S. Ino, *J. Phys. Soc. Jpn.* **1969**, *27*, 941.
[15] B. G. Bagley, *Nature* **1965**, *208*, 674.
[16] C. Y. Yang, *J. Cryst. Growth* **1979**, *47*, 274.
[17] J. W. Lee and G. D. Stein, *J. Phys. Chem.* **1987**, *91*, 2450.
[18] D. Reinhard, B. D. Hall, D. Ugarte, and R. Monot, *Phys. Rev. B* **1997**, *55*, 7868.
[19] B. D. Hall, M. Flüeli, R. Monot and J.-P. Borel, *Z. Phys. D* **1989**, *12*, 97.
[20] B. D. Hall, M. Flüeli, R. Monot, and J. -P. Borel, *Phys. Rev. B* **1991**, *43*, 3906.
[21] D. Reinhard, B. D. Hall, P. Berthoud, S. Valkealahti and R. Monot, *Phys. Rev. Lett.* **1997**, *79*, 1459.
[22] D. Reinhard, B. D. Hall, P. Berthoud, S. Valkealahti and R. Monot, *Phys. Rev. B* **1998**, *58*, 4917.
[23] V. Gnutzmann and W. Vogel, *J. Phys. Chem.* **1990**, *94*, 4991.
[24] W. Vogel, B. Rosner and B. Tesche, *J. Phys. Chem.* **1993**, *97*, 11611; W. Vogel, D. A. H. Cunningham, K. Tanaka and M. Haruta, *Catal. Lett.* **1996**, *40*, 175.
[25] B. D. Hall, *J. Appl. Phys., submitted.*
[26] B. W. van de Waal, *Phys. Rev. Lett.* **1996**, *76*, 1083.
[27] M. Brust, M. Walker, D. Bethell, D. J. Schiffrin and R. Whyman., *J. Chem. Soc., Chem. Commun.* **1994**, 801.
[28] D. Zanchet et al., *J. Phys. Chem. B., submitted.*
[29] W. H. Press, S. A. Teukolsky, W. T. Vetterling, and B. P. Flannery, *Numerical Recipes in C: The Art of Scientific Computing*, Cambridge, Cambridge UK, **1992**.
[30] B. K. Teo, *EXAFS: Basic Principles and Data Analysis*, Heidelberg, Springer-Verlag, **1986**.
[31] P. A. Montano, G. K. Shenoy, E. E. Alp, W. Schulze and J. Urban, *Phys. Rev. Lett.* **1986**, *56*, 2076.
[32] M. A. Marcus, M. P. Andrews, J. Zegenhagen, A. S. Bommannavar and P. Montano, *Phys. Rev. B* **1990**, *42*, 3312.
[33] A. Balerna et al., *Phys. Rev. B* **1985**, *31*, 5058.
[34] M. Borovisky, *J. Phys. IV* **1997**, *7*, C2-259.
[35] J. Rockenberger et al., *J. Phys. Chem.* **1997**, *101*, 2691.
[36] A. Balerna et al., *Surf. Sci.* **1985**, *156*, 206.
[37] A. Pinto et al., *Phys. Rev. B* **1995**, *51*, 5315.
[38] A. Balerna and S. Mobilio, *Phys. Rev. B* **1986**, *34*, 2293.

[39] C. W. Mays, J. S. Vermaak and D. Kuhlmann-Wilsorf, *Surf. Sci.* **1968**, *12*, 134.
[40] We have used the software packaged "EXAFS pour le MAC" developed by A Michailowitz, Ph.D. Thesis, Université Paris Val de Marne 1990.
[41] A. G. McKale, *J. Amer. Chem. Soc.* **1988**, *110*, 3763.
[42] F. W. Lytle, D. E. Sayers and E. A. Stern, *Phys. B* **1989**, *158*, 701.
[43] R. C. Elder et al., *ACS Sym. Ser.* **1983**, *209*.
[44] F. Sette, T. Hashizume, F. Comin, A. A. MacDowell and P. H. Citrin, *Phys. Rev. Lett.* **1988**, *61*, 1384.
[45] A. Filipponi, A. Di Cicco and C. R. Natoli, *Phys. Rev. B* **1995**, *52*, 15122; A. Filipponi and A. Di Cicco, *Phys. Rev. B* **1995**, *52*, 15135; A. Di Cicco, *Phys. B* **1995**, *208&209*, 125.
[46] P. Eisenberg and G. S. Brown, *Solid State Commun.* **1979**, *29*, 481.
[47] R. E. Benfield et al., *Phys. B* **1995**, *208&209*, 671.
[48] L.B. Hansen, P. Stoltze, J. K. Noskov, B. S. Clausen, W. Niemann, *Phys. Rev. Lett.* **1990**, *64*, 3155.

3 Transmission Electron Microscopy and Spectroscopy of Nanoparticles

Zhong Lin Wang

One of the typical characters of nanophase materials is the small particle sizes. Although some structural features can be revealed by x-ray and neutron diffraction, direct imaging of nanoparticles is only possible using transmission electron microscopy (TEM) and scanning probe microscopy. TEM is unique because it can provide a real space image on the atom distribution in the nanocrystal and on its surface [1]. Today's TEM is a versatile tool that provides not only atomic-resolution lattice images, but also chemical information at a spatial resolution of 1 nm or better, allowing direct identification the chemistry of a single nanocrystal. With a finely focused electron probe, the structural characteristics of a single nanoparticle can be fully characterized. To reveal the capabilities of a modern TEM, this chapter is designed to illustrate the fundamentals of TEM and its applications in characterization of nanophase materials. The fundamentals and applications of scanning transmission electron microscopy (STEM) will be given in Chapter 4.

3.1 A transmission electron microscope

A modern TEM can be schematically shown in Figure 3-1, which is composed of a illumination system, a specimen stage, an objective lens system, the magnification system, the data recording system(s), and the chemical analysis system. The electron gun is the heart of the illumination system, which typically uses LaB_6 thermionic emission source or a field emission source. The LaB_6 gun gives a high illumination current but the current density and the beam coherence are not as high as those of a field emission source. Field emission source is unique for performing high coherence lattice imaging, electron holography and high spatial resolution microanalysis. The illumination system also includes the condenser lenses that are vitally important for forming a fine electron probe. Specimen stage is a key for carrying out structure analysis, because it can be used to perform in-situ observations of phenomena induced by annealing, electric field, or mechanical stress, giving the possibility to characterize the physical properties of individual nanostructures. The objective lens is the heart of an TEM, which determines the limit of image resolution. The magnification system consists of intermediate lenses and projection lenses, and it gives a magnification up to 1.5 million. The data recording system tends to be digital with the use of a charge coupled device (CCD), allowing quantitative data processing and quantification. Finally, the chemical analysis system is the energy dispersive x-ray spectroscopy (EDS) and electron energy-loss spectroscopy (EELS), both can be used complimentary to quantify the chemical composition of the specimen. EELS can also provide information about the electronic structure of the specimen.

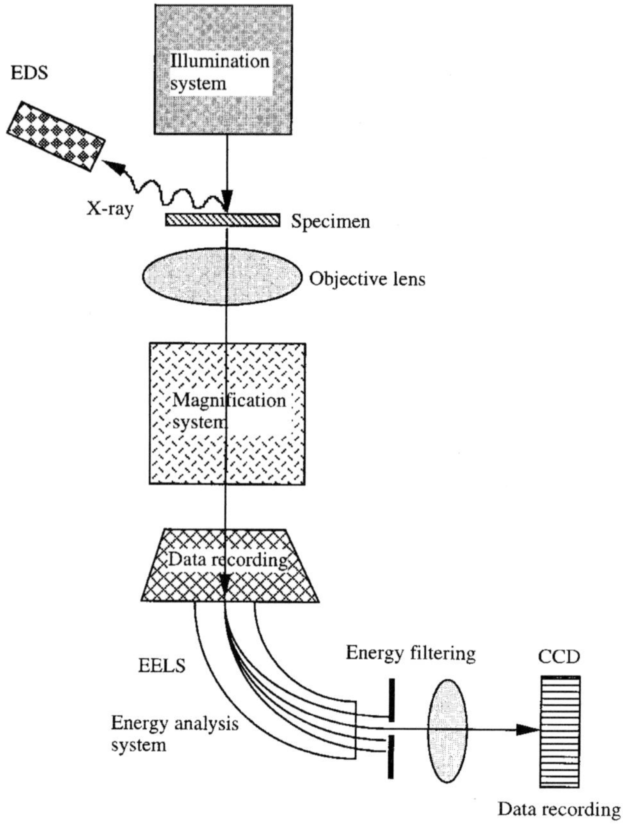

Figure 3-1. Schematic structure of a transmission electron microscope.

3.2 High-resolution TEM lattice imaging

3.2.1 Image formation

As a start, we first illustrate the image formation process in an TEM [2]. For easy illustration, an TEM is simplified into a single lens microscope, as given in Figure 3-2, in which only a single objective lens is considered for imaging and the intermediate lenses and projection lenses are omitted. This is because the resolution of the TEM is mainly determined by the objective lens. The entrance surface of a thin foil specimen is illuminated by a parallel or nearly parallel electron beam. The electron beam is diffracted by the lattices of the crystal, forming the diffracted beams which are propagating along different directions. The electron-specimen interaction results in phase and amplitude changes in the electron wave that are determined by quantum mechanical diffraction theory. For a thin specimen and high-energy electrons, the transmitted wave function $\Psi(x, y)$ at the exit face of the specimen can be assumed to be composed of a forward-scattered wave.

The non-near-axis propagation through the objective lens is the main source of non-linear information transfer in TEM. The diffracted beams will be focused in the back-focal plane, where an objective aperture could be applied. An ideal thin lens brings the parallel transmitted waves to a focus on the axis in the back focal plane.

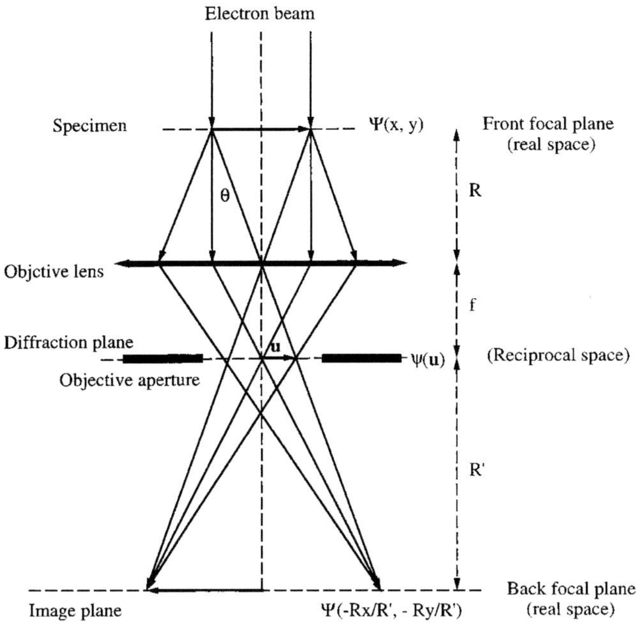

Figure 3-2. Abbe's theory of image formation in an one-lens transmission electron microscope. This theory is for a general optical system in TEM.

Waves leaving the specimen in the same direction (or angle θ with the optic axis) are brought together at a point on the back focal plane, forming a diffraction pattern. The electrons scattered to angle θ experience a phase shift introduced by the chromatic and spherical aberrations of the lens, and this phase shift is a function of the scattering angle, thus, the diffraction amplitude at the back-focal plane is modified by

$$\psi'(\mathbf{u}) = \psi(\mathbf{u}) \exp[i\chi(\mathbf{u})] \tag{3-1}$$

where $\psi(\mathbf{u})$ is the Fourier transform of the wave $\Psi(\mathbf{r})$ at the exit face of the specimen, $\mathbf{u}$ is the reciprocal space vector that is related to the scattering angle by $u = 2\sin\theta/\lambda$, and $\chi(\mathbf{u})$ is determined by the spherical abberation coefficient C_s of the objective lens and the lens defocus Δf [3]

$$\chi(\mathbf{u}) = \frac{\pi}{2} C_s\lambda^3 u^4 - \pi\, \Delta f \lambda\, u^2 \tag{3-2}$$

where λ is the electron wavelength. The aberration and defocus of the lens is to modulate the phases of the Bragg beams distributed in reciprocal space.

The electron image is the interference result of the beams scattered to different angles, and this interference pattern is affected by the phase modulation introduced by the aberration of the objective lens. The image is calculated according to

$$I(x,y) = |\Psi(\mathbf{r}) \otimes t_{\mathrm{obj}}(x,y)|^2 \tag{3-3}$$

where $\otimes$ indicates a convolution calculation of (x, y), $t_{obj}(x,y)$ is the inverse Fourier transform of the phase function $\exp[i\chi(\mathbf{u})]$. The convolution of the lens transfer function introduces the non-linear information transfer characteristics of the objective lens, leading to complexity in image interpretation.

3.2.2 Contrast mechanisms

Images in TEM are usually dominated by three types of contrast. First, diffraction contrast [4], which is produced due to a local distortion in the orientation of the crystal (by dislocations, for example), so that the diffracted intensity of the incident electron beam is perturbed, leading to contrast observed in bright-field image. The nanocrystals oriented with their low-index zone-axis parallel or nearly parallel to the incident beam direction usually exhibit dark contrast in the bright field image, that is formed by selecting the central transmitted beam. Since the diffraction intensities of the Bragg reflected beams are strongly related to the crystal orientations, this type of image is ideally suited for imaging defects and dislocations. For nanocrystals, most of the grains are defect-free in volume, while a high density of defects are localized at the surface or grain boundary, diffraction contrast can be useful for capturing strain distribution in nanocrystals whose sizes are larger than 15 nm. For smaller size nanocrystals, since the resolution of diffraction contrast is in the order of 1–2 nm, its application is limited.

Secondly, phase contrast is produced by the phase modulation of the incident electron wave when transmits through a crystal potential [1]. This type of contrast is sensitive to the atom distribution in the specimen and it is the basis of high-resolution TEM. To illustrate the physics of phase contrast, we consider the modulation of a crystal potential to the electron wavelength. From the *de Brogli relation*, the wavelength λ of an electron is related to its momentum, p, by

$$\lambda = \frac{h}{p} \tag{3-4}$$

When the electron goes through a crystal potential field, its kinetic energy is perturbed by the variation of the potential field, resulting in a phase shift with respect to the electron wave that travels in a space free of potential field. For a specimen of thickness d, the phase shift is

$$\phi \approx \sigma V_p(\mathbf{b}) = \sigma \int_0^d dz\ V(\mathbf{r}) \tag{3-5}$$

where $\sigma = \frac{\pi}{\lambda U_0}$, $\mathbf{b} = (x, y)$, U_0 is the acceleration voltage, and $V_p(\mathbf{b})$ the thickness-projected potential of the crystal. Therefore, from the phase point of view, the electron wave is modulated by a phase factor

$$Q(\mathbf{b}) = \exp[i\sigma V_p(\mathbf{b})] \tag{3-6}$$

This is known to be the *phase object approximation.* (POA), in which the crystal acts as a phase grating filter. If the incident beam travels along a low-index zone-axis, the variation of $V_p(\mathbf{b})$ across atom rows is a sharp function because an atom can be approximated by a narrow potential well and its width is in the order of 0.2–0.3 Å. This sharp phase variation is the basis of *phase contrast*, the fundamental of atomic-resolution imaging in TEM.

Finally, mass-thickness or atomic number produced contrast. Atoms with different atomic numbers exhibit different powers of scattering. If the image is formed by collecting the electrons scattered to high-angles, the image contrast would be sensitive to the average atomic number along the beam direction. This type of imaging is usually performed in STEM (see Chapter 4).

3.2.3 Image interpretation

In high-resolution TEM (HRTEM) images, one usually wonders if the atoms are dark or bright. To answer this question one must examine the imaging conditions. For the clarity of following discussion, the *weak scattering object approximation* (WPOA) is made. If the specimen is so thin that the projected potential satisfies $|\sigma V_p(\mathbf{b})| << 1$, the phase grating function is approximated by

$$\Psi(\mathbf{b}) \approx 1 + i\sigma V_p(\mathbf{b}) \tag{3-7}$$

From Eq. (3-3) and ignoring the σ^2 term, the image intensity is calculated by

$$I(x,y) \approx 1 - 2\,\sigma V_p(\mathbf{b}) \otimes t_s(\mathbf{b}) \tag{3-8}$$

where $t_s(\mathbf{b}) = \mathrm{Im}[t_{obj}(\mathbf{b})]$. The second term in Eq. (3-8) is the interference result of the central transmitted beam with the Bragg reflected beams. Any phase modulation introduced by the lens would result in contrast variation in the observed image. Under the Scherzer defocus, $t_s(\mathbf{b})$ is approximated to be a negative Gaussian-like function with a small oscillating tail, thus, the image contrast, under the WPOA, is directly related to the two dimensional thickness projected potential of the crystal, and the image reflects the projected structure of the crystal. This is the basis of structure analysis using HRTEM. On the other hand, the contrast of the atom rows is determined by the sign and real space distribution of $t_s(\mathbf{b})$. The convolution of $t_s(\mathbf{b})$ with the potential changes the phases of the Bragg reflected beams, which can be explicitly illustrated as following.

A Fourier transform is made to the both sides of Eq. (3-8), yielding

$$\mathbf{FT}[I_p] \approx \delta(\mathbf{u}) - 2\,\sigma\mathbf{FT}[V_p(\mathbf{b})]\;T_s(\mathbf{u}) \tag{3-9}$$

The $\delta(\mathbf{u})$ function represents a strong central transmitted (000) beam. The Fourier transform of the crystal potential, $\mathbf{FT}[V_p(\mathbf{b})]$, is the diffraction amplitude of the Bragg beams under the kinematic scattering approximation. The contribution from each diffracted beam to the image is modified by the function $T_s(\mathbf{u}) = \sin\chi\, E(\mathrm{u})$, and $E(\mathrm{u})$ is the envelope function due to a finite energy spread of the source, the focus spread, beam convergence, the mechanical vibration of the microscope, the specimen drift during the recording of the image, and electric voltage and lens current instability. The envelope function defines the maximum cut-off frequency that can be transferred by the optic system. $T_s(\mathbf{u})$ is known as the *phase-contrast transfer function* (PCTF), characterizing the information transfering property of the objective lens. To illustrate this result, we consider a *fcc* structured crystal which gives {111}, {200}, {220} and {311} reflections. The angular distribution of these beams are schematically shown on the horizontal u axis. The diffraction amplitudes $\phi_{\{111\}}$ and $\phi_{\{220\}}$ of beams {111} and {220}, respectively, are chosen as two representatives to show the characteristics of

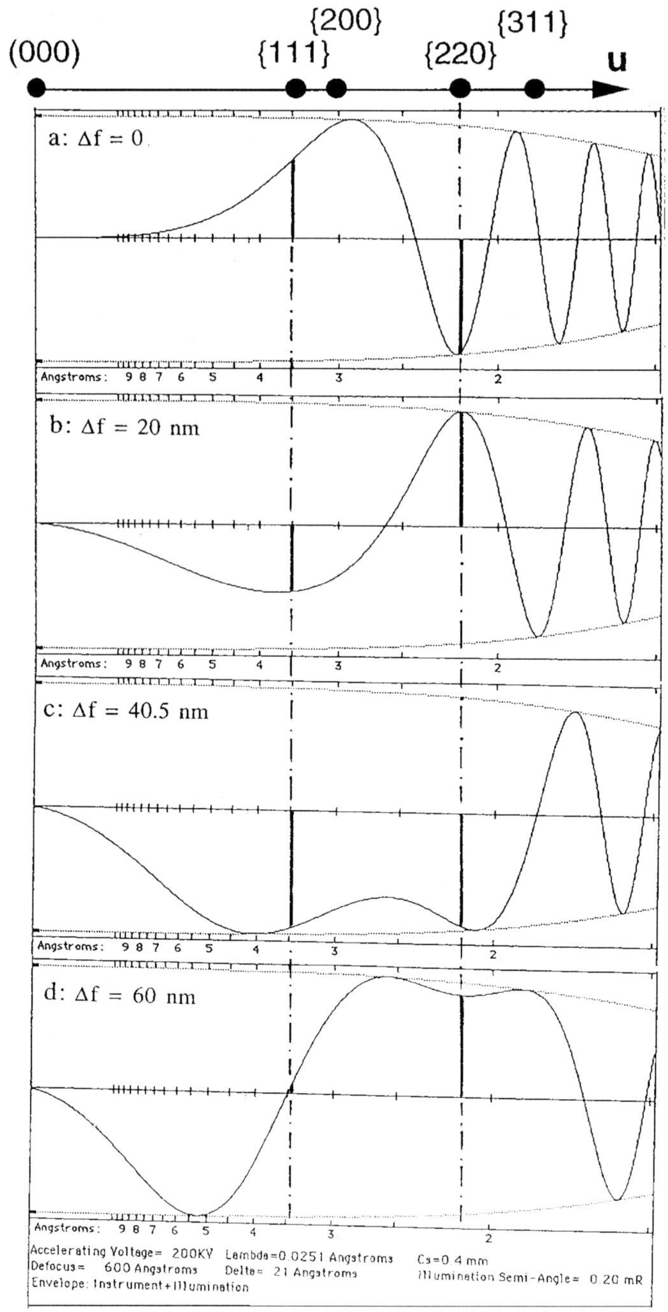

Figure 3-3. Calculated contrast transfer function sinχ for several defocus values showing information transfer under different conditions (see text). The horizontal axis is labeled with the corresponding image resolution R (in Å), corresponding to a reciprocal space vector $\mathbf{u} = 1/R$. The dashed lines indicate the angular positions of the {111} and {220} beams.

phase transfer. When the defocus is zero (Figure 3-3a), $\phi_{\{111\}}$ is transferred with positive sign ($\sin\chi > 0$), while $\phi_{\{220\}}$ is transferred with negative sign ($\sin\chi < 0$). The sign reverse of $\phi_{\{220\}}$ with respect to $\phi_{\{111\}}$ results in a contrast reversal in the interference pattern due to the destructive summation of the two. At $\Delta f = 20$ nm (Figure 3-3b), the two amplitudes are transferred with the same sign except the relative weight factor of the two is changed. At $\Delta f = 40.5$ nm (Figure 3-3c), $\phi_{\{111\}}$ and $\phi_{\{220\}}$ are both transferred with the negative sign, and there is no relative phase change between $\phi_{\{111\}}$ and $\phi_{\{220\}}$, thus, the interferences between $\phi_{\{000\}}$, $\phi_{\{111\}}$ and $\phi_{\{220\}}$ give the image that is directly related to the crystal structure (i.e., the atom rows and planes).

The PCTF depends sensitively on the defocus of the objective lens. The interpretable image resolution that is directly associated with the crystal structure (e.g., the structural resolution) is determined by the width of the information passing band. It was first investigated by Scherzer [5] that the highest structural resolution of $R = 0.66\ \lambda^{3/4}\ C_s^{1/4}$ would be obtained at defocus $\Delta f = (4/3\ C_s\lambda)^{1/2}$. In this focus condition, the $t_s(\mathbf{b})$ function is approximated to be a negative Gaussian function, thus, the atoms tend to be in dark contrast for thin crystals. This is the most desirable imaging condition of HRTEM.

As a summary, the image contrast in HRTEM is critically affected by the defocus value. A slight change in defocus could lead to contrast reversal. A change in signs of the diffraction amplitudes makes it difficult to match the observed image directly with the projection of atom rows in the crystal. This is one of the reasons that image simulation is a key step in quantitative analysis of HRTEM images.

3.2.4 Image simulation

Image simulation needs to include two important processes, the dynamic multiple scattering of the electron in the crystal and the information transfer of the objective lens system. The dynamic diffraction process is to solve the Schrödinger equation under given boundary conditions. There are several approaches for performing dynamic calculations [6]. For a finite size crystal containing defects and surfaces, the multislice theory is most adequate for numerical calculations. The multislice many-beam dynamic diffraction theory was first developed based on the physical optics approach [7]. The crystal is cut into many slices of equal thickness Δz in the direction perpendicular or nearly perpendicular to the incident beam (Figure 3-4). When the slice thickness tends to be very small the scattering of each slice can be approximated as a phase object, the transmission of the electron wave through each slice can be considered separately if the backscattering effect is negligible, which means that the calculation can be made slice-by-slice. The defect and 3-D crystal shape can be easily accounted for in this approach.

The transmission of the electron wave through a slice can be considered as a two step processes – the phase modulation of the wave by the projected atomic potential within the slice and the propagation of the modulated wave in "vacuum" for a distance Δz along the beam direction before striking the next crystal slice. The wave function before and after transmitting a crystal slice is correlated by

$$\Psi(\mathbf{b},z+\Delta z) = [\Psi(\mathbf{b},z)\ Q(\mathbf{b},z+\Delta z)] \otimes P(\mathbf{b},\Delta z) \tag{3-10}$$

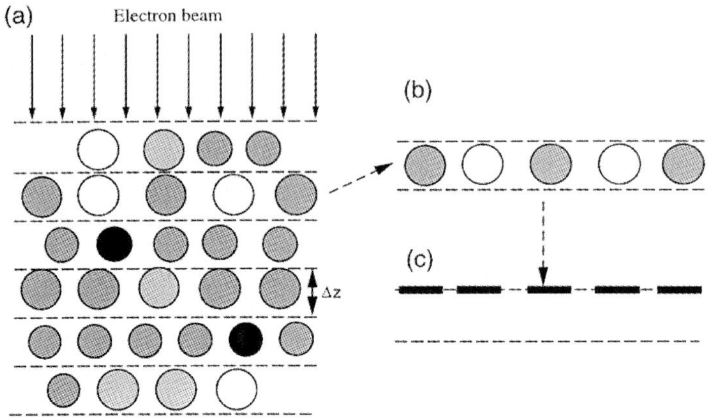

Figure 3-4. (a) A schematic diagram showing the physical approach of the multislice theory for image and diffraction pattern calculations in TEM. (b) Transmission of electron wave through a thin crystal slice. (c) An approximated treatment of the wave transmission through a thin slice.

where $Q(\mathbf{b},z+\Delta z) = \exp[i\sigma\int_z^{z+\Delta z} dz V(\mathbf{b},z)]$ is the phase grating function of the slice, and $P(\mathbf{b},\Delta z) = \frac{\exp(\pi i K|\mathbf{b}|^2/\Delta z)}{i\lambda\Delta z}$ is the propagation function. The most important characteristic of this equation is that no assumption was made regarding the arrangement of atoms in the slices, so that the theory can be applied to calculate the electron scattering in crystals containing defects and dislocations. This is the most powerful approach for nanocrystals.

The next step is the information transfer through the objective lens system. By taking a Fourier transform of the exit wave function, the diffraction amplitude is multiplied by the lens transfer function $\exp(i\chi)\ E(u)$, where the defocus is a variable. The defocus value of the objective lens and the specimen thickness are two important pa-

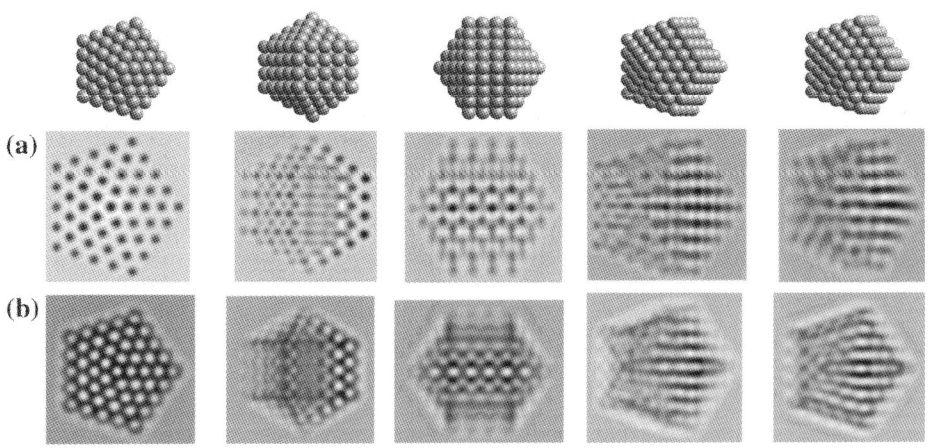

Figure 3-5. Theoretically simulated images for a decahedral Au particle at various orientations and at focuses of (A) Δf = 42 nm and (B) Δf = 70 nm. The Fourier transform of the image is also displayed (Courtsey of Drs. Ascencio and M. José-Yacamán).

rameters which can be adjusted to match the calculated images with the observed ones. Figure 3-5 shows systematic simulations of a decahedral Au particle in different orientations. The particle shape can only be easily identified if the image is recorded along the five-fold axis [8]. The group A and group B images were calculated for two different defocuses, exhibiting contrast reversal from dark atoms to bright atoms. In practice, with consideration the effects from the carbon substrate, it would be difficult to identify the particle shape if the particle orientation is off the five-fold axis.

3.3 Defects in nanophase materials

Nanocrystals exhibiting distinctly different properties from the bulk are mainly due to their large portions of surface atoms and the size effect and possibly the shape effect as well. A particle constituting a finite number of atoms can have a specific geometrical shape, and most of the defects are on the surface, while the volume defects are dominated by twins, stacking faults and point defects. Surface defects and planar defects can be imaged directly using HRTEM, but the analysis of point defects is still a challenge.

3.3.1 Polyhedral shape of nanoparticles

Surface energies associated with different crystallographic planes are usually different, and a general sequence may hold, $\gamma_{\{111\}} < \gamma_{\{100\}} < \gamma_{\{110\}}$. For a spherical single-crystalline particle, its surface must contain high index crystallography planes, which possibly result in a higher surface energy. Facets tend to form on the particle surface to increase the portion of the low index planes. Therefore, for particles smaller than 10-20 nm, the surface is a polyhedron. Figure 3-6a shows a group of cubo-octahedral

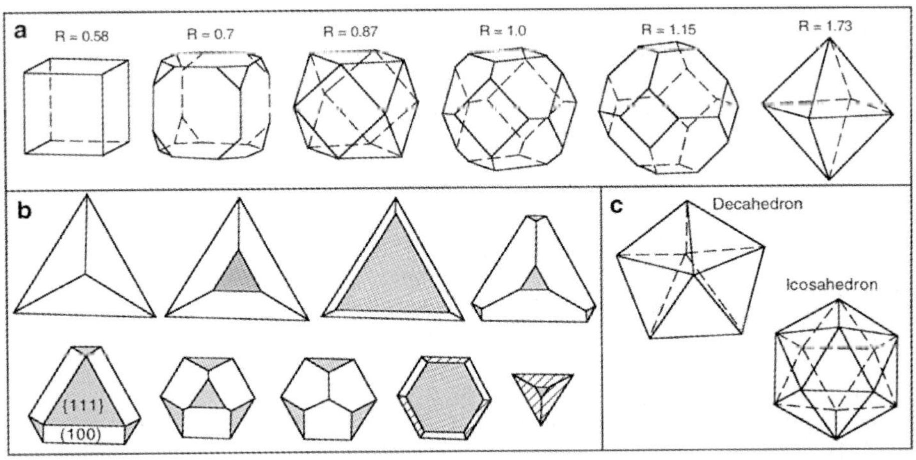

Figure 3-6. (a) Geometrical shapes of cubo-octahedral nanocrystals as a function of the ratio, *R*, of the growth rate along the <100> to that of the <111>. (b) Evolution in shapes of a series of (111) based nanoparticles as the ratio of {111} to {100} increases. The beginning particle is bounded by three {100} facets and a (111) base, while the final one is a {111} bounded tetrahedron. (c) Geometrical shapes of multiply twinned decahedral and icosahedral particles.

shapes as a function of the ratio, R, of the growth rate in the <100> to that of the <111>. The longest direction in a cube is the <111> diagonal, the longest direction in the octahedron is the <100> diagonal, and the longest direction in the cubo-octahedron ($R = 0.87$) is the <110> direction. The particles with $0.87 < R < 1.73$ have the {100} and {111} facets, which are named the truncated octahedral (TO). The other group of particles has a fixed (111) base with exposed {111} and {100} facets (Figure 3-6b). An increase in the area ratio of {111} to {100} results in the evolution of particle shapes from a triangle-based pyramid to a tetrahedron.

If the particle is oriented along a low index zone axes, the distribution of atoms on the surface can be imaged in profile, and the surface structure is directly seen with the full resolution power of an TEM [9, 10]. This is a powerful technique for direct imaging the projected shapes of nanoparticles particularly when the particle size is small. With consideration the symmetry in particle shapes, HRTEM can be used to determine the 3-D shape of small particles although the image is a 2-D projection of a 3-D object.

Figure 3-7a gives a profile HRTEM image of a cubic Pt nanocrystals oriented along [001] [11]. The particle is bounded by {100} facets and there is no defect in the bulk of the particle. The distances between the adjacent lattice fringes is the interplanar distance of Pt {200), which is 0.196 nm, and the bulk structure is face centered cubic. The surface of the particle may have some steps and ledges particularly at the regions near the corners of the cube. To precisely image the defects and facets on the cubic particles, a particle oriented along [110] is given (Figure 3-7b). This is the optimum orientation for imaging cubic structured materials. The {110} facets are rather rough, and the {111} facets are present. These higher energy structural features are present because the particles were prepared at room temperature.

An octahedron has eight {111} facets, four {111} facets are edge-on if viewed along [110] . If the particle is a truncated octahedron, six {100} facets are created by cutting the corners of the octahedron, two of which are edge-on while viewing along [110]. Figures 3-8a and b show the HRTEM images of [110] oriented truncated octahedron and octahedral Pt particles, respectively. A variation in the area ratio of {100} to {111} results in a slight difference in particle shapes.

A tetrahedral particle is defined by four {111} faces and it usually gives a triangular shape in HRTEM. Figure 3-9 gives two HRTEM images of truncated tetrahedral particles oriented along [110]. Two {111} facets and one {001} facet (at the top of the

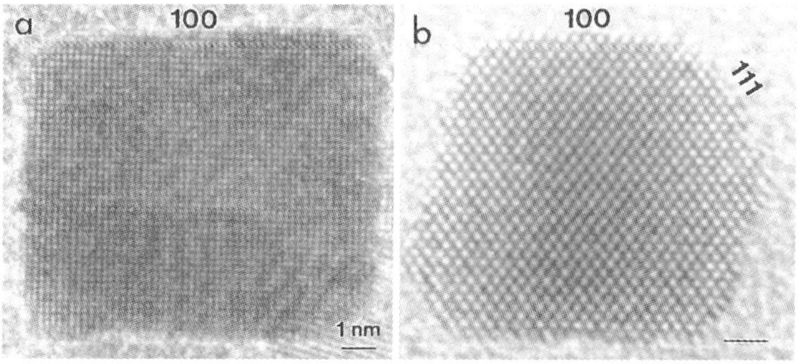

Figure 3-7. HRTEM images of cubic Pt nanocrystals oriented along (a) [001] and (b) [110], showing surface steps/ledges and the thermodynamically unequilibrium shapes.

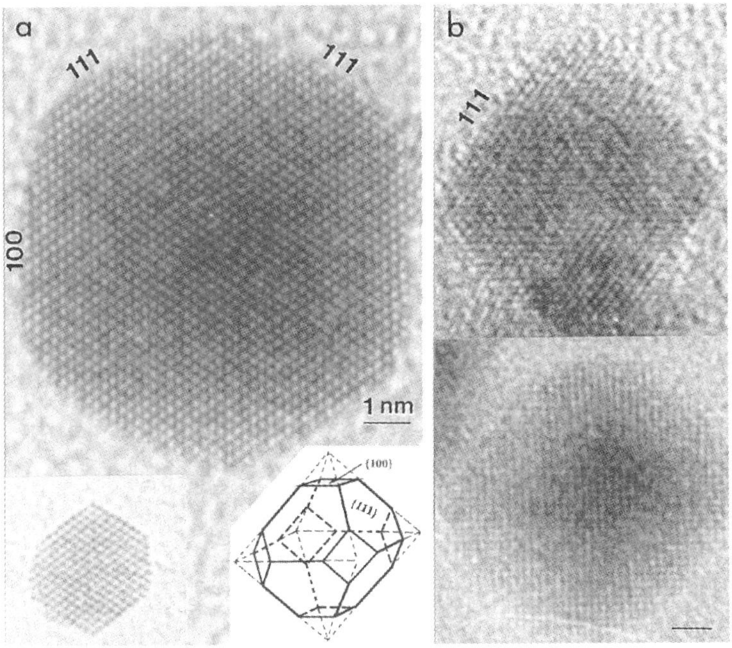

Figure 3-8. HRTEM images of Pt nanocrystals (a) with a truncated octahedral shape and oriented along [110], and (b) with a octahedral shape and oriented along [110] and [001]. The inset in (a) is a model of the particle shape.

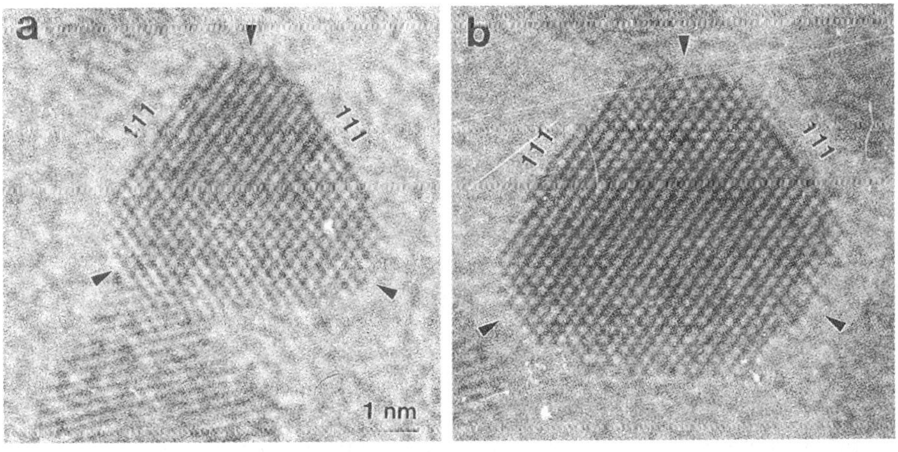

Figure 3-9. HRTEM images of truncated tetrahedral Pt nanocrystals oriented along [110]. The surface steps and ledges at the truncated corners are clearly resolved.

image as a result of truncation) are imaged edge-on. There are some atom-high surface steps on the {111} surfaces and the corners. These atomic-scale structures are likely important for enhancing the catalysis activities of the nanocrystals.

3.3.2 Surface reconstruction

Simply speaking, surface atoms have less bonds in comparison to the atoms in the bulk because of the loss in nearest neighbors. The surface atoms tend to find new equilibrium positions to balance the forces, resulting in surface reconstruction. The most typical example is the Si {111} 7×7 reconstruction [12]. Surface restructure can be the rearrangement of the surface atoms and/or the relaxation of the top few layers. A classical example of surface reconstruction in nanocrystals is the Au (110) 2×1 [13].

Figure 3-10 gives a group of images recorded from Pt nanocrystals. The {111} faces in Figures 3-10a, c and d are relatively flat, while the {100} faces show a large deviation from the perfect lattice structure, suggesting the occurrence of a non-ordered surface reconstruction. Surface relaxation is clearly seen in Figure 3-10b in the layers indicated by arrowheads. Surface stress due to the capillary effect may have significant influence on the lattice constant of nanocrystals [14, 15].

Figure 3-11a gives a bright-field TEM image of a monolayer self-assembled Pt nanocrystals, most of which are single crystalline without twins or stacking faults. The dark-field image is formed using an objective aperture that selects a small section of the {111} and {200} reflection rings, so that the particles whose Bragg reflections falling into the angular range of the aperture would display bright contrast, while those not falling into the angular range are expected to exhibit dark-contrast. The dark-field image, however, displays some ring contrast in some of the particles. The size of the rings corresponds well to the actual size of the particles in the bright-field TEM image and the ring contrast is not the thickness fringes because the particle size is less than

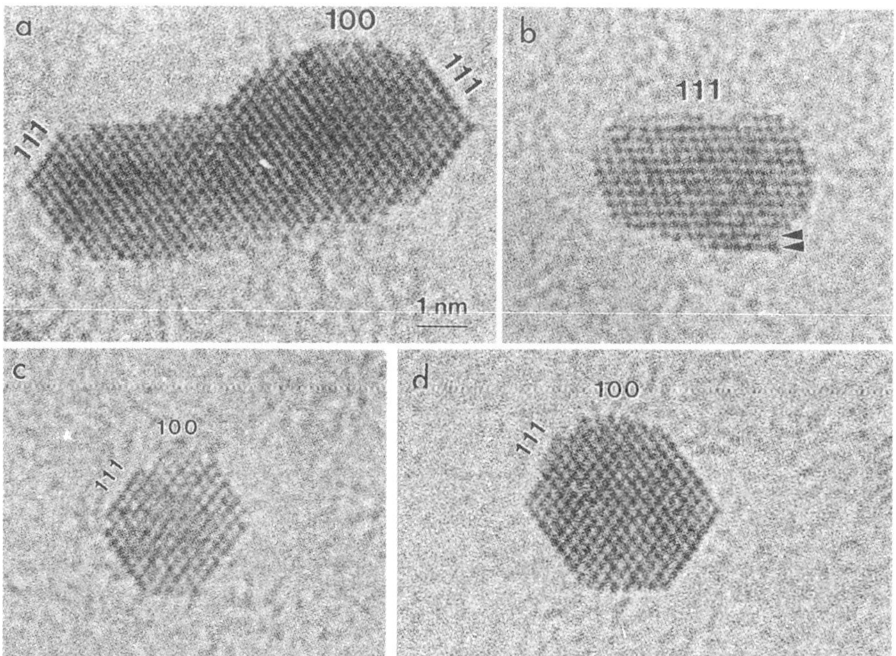

Figure 3-10. HRTEM images of Pt nanocrystals oriented along [110], showing the reconstructed {100} surfaces and surface relaxation (indicated by arrowheads).

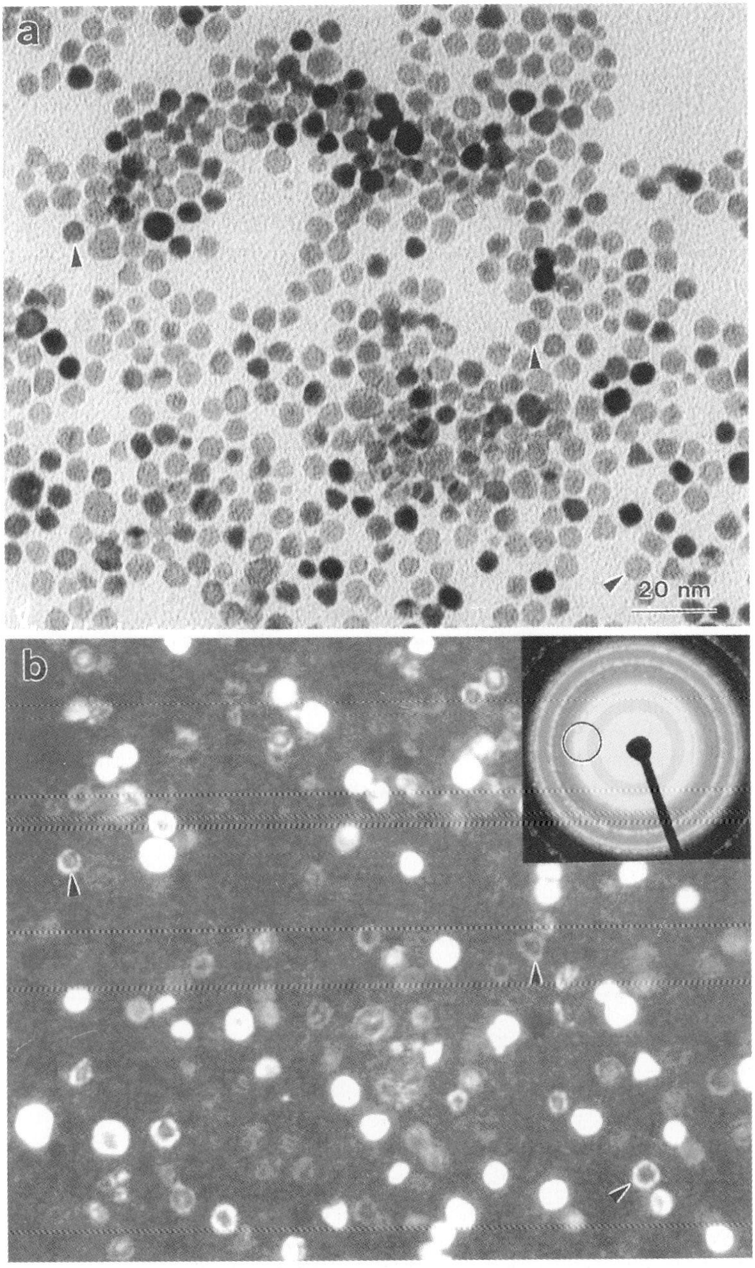

Figure 3-11. A pair of (a) bright-field and (b) dark-field images from Pt nanocrystals, showing a ring shape contrast in the dark field image (see the particles indicated by arrowheads), which is likely due to the diffuse scattering from the surface reconstructed/relaxed layers. The dark field image was recorded using a small objective aperture, as shown in the inserted electron diffraction pattern, positioned on the {111} and {200} reflection rings.

10 nm, smaller than the extinction distances of Au {111) (15.9 nm) and Au (200) (17.9 nm). This type of contrast is due to the diffuse scattering generated from the top few atomic layers of the nanocrystal owing to surface reconstruction/relaxation. A non periodic perturbation to the crystal lattice would produce diffuse scattering that is distributed between the Bragg reflections in the entire diffraction plane. In the angular selection range of the aperture, the particles whose Bragg reflections are not falling into the aperture should be completely dark, but the diffuse scattering from the particles would be partially collected, resulting in a bright ring at the edge of the particle. This is another way to image the reconstruction/relaxation of the surface atoms in nanocrystals.

3.3.3 Twining structure and stacking faults

Twining is one of the most popular planar defects in nanocrystals, and it is frequently observed for fcc structured metallic nanocrystals. Twining is the result of two subgrains by sharing a common crystallographic plane, thus, the structure of one subgrain is the mirror reflection of the other by the twin plane. Figure 3-12a gives a HRTEM image of a Pt particle oriented along [110], which is composed of two grains connected at a twin relation. The twin plan is indicated by a dotted line. The fcc structured metallic nanocrystals usually have {111} twins, which is the fundamental defect mechanism for the growth of spherical-like particles (see Section 3.2.4).

Stacking faults are a typical planar defect. Stacking faults are produced by a distortion on the stacking sequence of atom planes. The (111) plane stacking sequence of a *fcc* structure follows A-B-C-A-B-C-A-B-C-. If the stacking sequence is changed to A-B-C-A-B-A-B-C-, a stacking fault is created. Figure 3-12b shows an Au particle that contains a twin and a stacking fault. It is know that nanocrystals usually contain no dislocation but strain. Twins and stacking faults are created probably due the high strain energy in the volume.

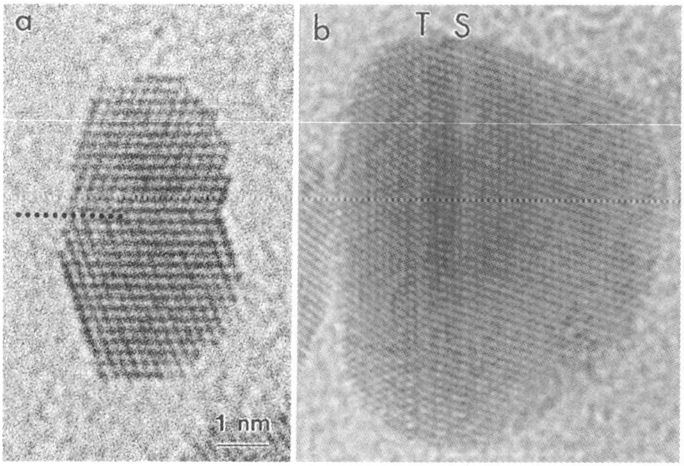

Figure 3-12. HRTEM images of Pt nanocrystals having twin (T) and stacking fault (S) structures. The nanocrystals are oriented along [110] and the twin plane and stacking fault are parallel to the direction of the electron beam.

3.2.4 Decahedral and icosahedral particles

The two most typical examples of multiply twinned particles (MTP) are decahedron and icosahedron [16, 17]. Starting from a *fcc* structured tetrahedron, a decahedron is assembled from five tetrahedral sharing an edge (Figure 3-13a). If the observation direction is along the five-fold axis and in an ideal situation, each tetrahedron shares an angle of 70.5°, five of them can only occupy a total of 352.6°, leaving a 7.4° gap. Therefore, strain must be introduced in the particle to fill the gap [18]. An icosahedron is assembled using 20 tetrahedra via sharing an apex (Figure 3-13b and c). The icosahedral and decahedral particles are the most extensively studied twinned nanocrystals [19–21].

The most easy orientation for identifying the MTPs are along the five fold symmetry axis. Figure 3-14a shows a large decahedral Au particle, the five fold twin structure is clearly seen. The strain inteoduced is also visible. High-resolution TEM image for a smaller one clearly shows the crystal structure of the particle (Figure 3-14b). Identification the shape of small MTPs in TEM is not trivial because the image depends sensitively on the orientation of the particle [8, 21]. With the presence of a thin amorphous carbon substrate, it is even more difficult to identify the particle shape directly.

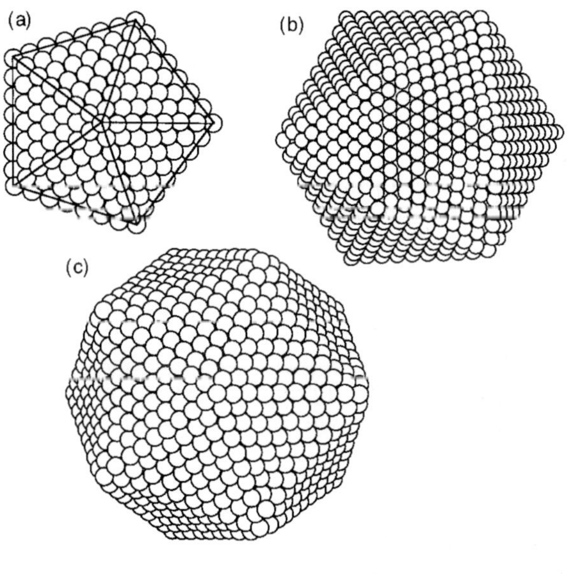

Figure 3-13. Atomic models of (a) decahedral, (b) and (c) icosahedral nanocrystals.

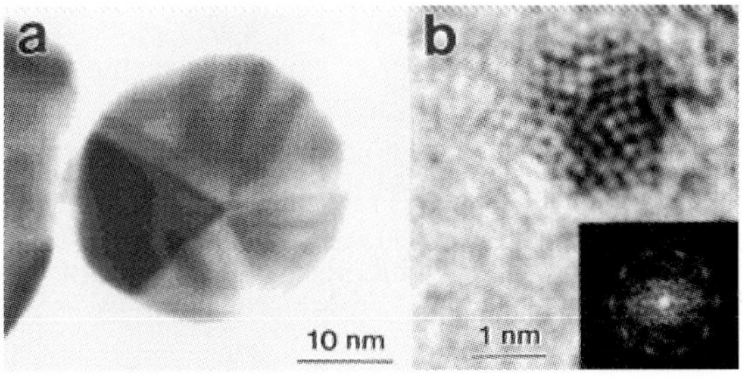

Figure 3-14. TEM images of decahedral Au nanocrystals when the incident electron beam is parallel or nearly parallel to the five-fold symmetry axis. The strain field introduced by the five-fold twins is apparent in (a). A Fourier transform of the image clearly reveals the five-fold symmetry.

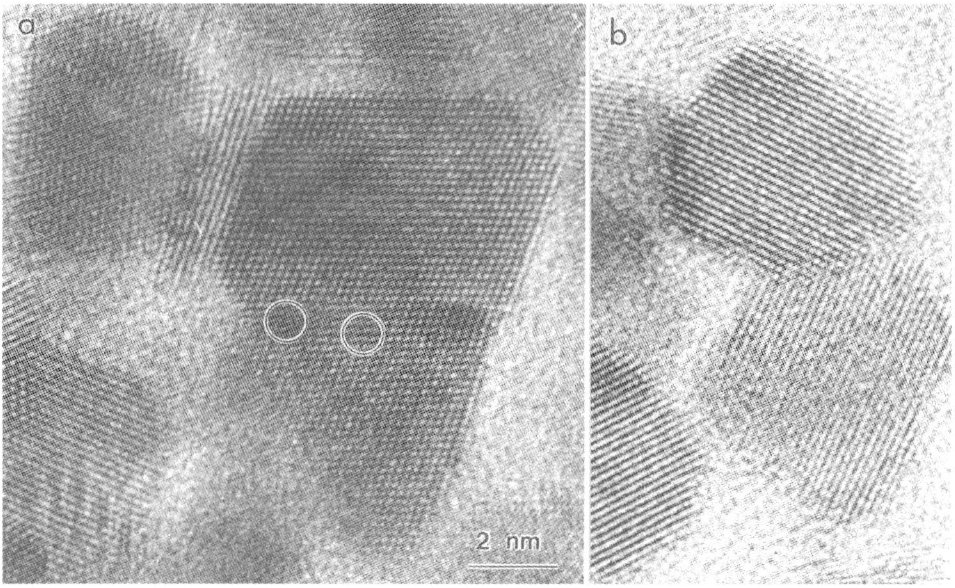

Figure 3-15. HRTEM observations of dislocations at low-angle and high-angle grain boundaries of Pt nanocrystals.

3.3.5 Interface defects and dislocations

Engineering materials made of nanocrystals must involve grain boundaries. The misorientation or lattice mismatch between nanocrystals creates interface dislocations. Figure 3-15a shows a coherent interface formed by two Pt nanocrystals, where the lattices of the two particles are closely aligned and matched fairly well, while edge dislocations are still seen at regions indicated by the circles. Edge dislocations are commonly seen at interfaces to accommodate the accumulated strain created by lattice rotation/mismatch. The interface angle of a coherent interface is usually small. The interface seen in Figure 3-15b is a large angle grain boundary, where the lattices exhibit large incoherence and the grain boundary energy is expected to be higher.

3.4 Electron holography

In conventional HRTEM, the phase and amplitude images of the exit electron wave are mixed non-linearly and convoluted with the phase contrast transfer function. It is usually a rather sophisticated and complex process to retrieve the phase and amplitude images. Electron holography is a technique which can recover the true object image. There are numerous ways of performing electron holography in TEM and STEM [22]. The most popular technique is the off-axis holography. The principle of electron holography was first introduced by Gabor [23] in an attempt to exceed the point-to-point resolution of an electron microscope. Holography is based on the interference and diffraction properties of waves, thereby producing a true image of an object (including amplitude and phase) without any distortion of the lenses. The development of high-brightness high-coherence electron sources has made it possible to obtain holograms using electron waves in TEM [24–26]. The high spatial coherence of the source is the key to perform holography.

3.4.1 Principle of off-axis holography in TEM

Electron holography is designed as a two-step imaging technique [27]. The first step is to form an interference hologram as shown in Figure 3-16a. A high spatial coherent electron wave is emitted from the source, half of which will transmit through a thin specimen and the other half is reserved as a reference wave. The reference wave is the standard to determine the phase shift image of the wave went through the specimen. The interference of the two waves is possible by inserting a positively charged biprism at the intermediate focal plane of the projection lenses. The biprism defects the electrons on the both sides towards with each other and introduce an additional phase shift function due to the Aharonov-Bohm effect, resulting in the formation of an interference hologram.

The second step is to reconstruct the phase image from the hologram (Figure 3-16b). This can be achieved optically or digitally. In the optical reconstruction, a laser beam illuminates the recorded hologram. Due to the presence of interference fringes in the hologram, the transmitted beam will be diffracted along the direction perpendicular to the fringes, forming two side-bands and a center band. The side-band is directly associated with the electron wave function at the exit face of the specimen, and it gives both the amplitude and phase images of the object.

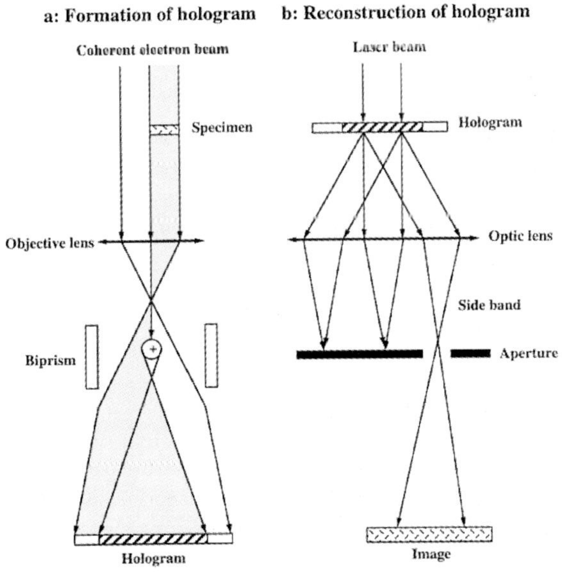

Figure 3-16. (a) Formation of off-axis electron hologram in TEM using an electrostatic biprism, and (b) reconstruction of electron hologram using laser diffraction.

3.4.2 Imaging electrostatic field and charge distribution

The phase image can provide the distribution of electrostatic field around a charged particle as well as the thickness projected potential image of a nanocrystal. The former can be used to extract the charge distribution in the particle [28], and the latter is useful to determine if the particle contains a hard core or an empty cavity [29]. This measurement is based on the electron phase shift after transmission a region with electrostatic potential. The perturbation of the field to the electron wave results in a change in the electron wavelength. From the discussion given in Section 3.2.2, the relative phase shift of the electron wave with respect to the reference wave is given by Eq. (3-5). For a homogeneous specimen with constant thickness, in addition to the average potential introduced by the atoms in the crystal, charge barrier can be created at interfaces and defects. If the former contributes only a background, the latter can be retrieved experimentally.

Figure 3-17 shows a phase image of the MgO cubic crystal oriented along [210]. The MgO cube is bounded by four {100} faces. A thickness projection of the cube along [210] results in rapid variation in the projected mass-thickness towards the edge of the cube, while there is no change in the projected-thickness at the center (see the inset). This result is clearly revealed by the phase image reconstructed from the hologram.

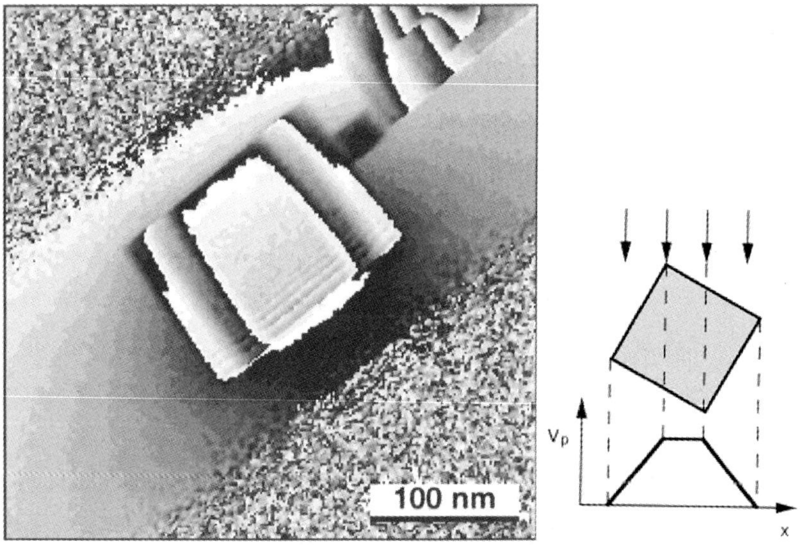

Figure 3-17. A phase image of a MgO cube oriented along [210], showing the phase variation resulted from the change in the projected potential. A schematic model of the projection geometry is given, where the profile of the projected potential across the particle is illustrated (Courtesy of Dr. B. Frost).

3.4.3 Imaging magnetic domains and flux lines

Imaging of magnetic domains and interdomain interaction is vitally important for understanding the behavior and improving the properties of magnetic materials. Although domain structures can be directly seen using Fresnel contrast and Lorentz microscopy, electron holography is probably the only technique which can provide quantitative information about the structure of domains and the field distribution around it [24, 30]. The mechanism of imaging magnetic domains in magnetic materials relies on the Aharonov-Bohm effect [31], which describes the relative phase shift of the two electron waves traveling along two routes enclosing a zone with magnetic field distribution,

$$\Delta\phi = \frac{2\pi e}{h} \int_s \mathbf{B} \cdot d\mathbf{S} \tag{3-11}$$

where $d\mathbf{S}$ is a surface integral over the area enclosed by the two routes. Equation (3-11) means that the phase difference is determined by the magnetic flux enclosed by the two electron trajectories. If a magnetic specimen is inset on the way of the object wave, measurement of the electron phase shift directly gives the magnetic flux enclosed by the two trajectories. Figure 3-18a is a bright-field TEM image of a barium ferrite particle with a truncated octahedral shape. The phase image of the particle

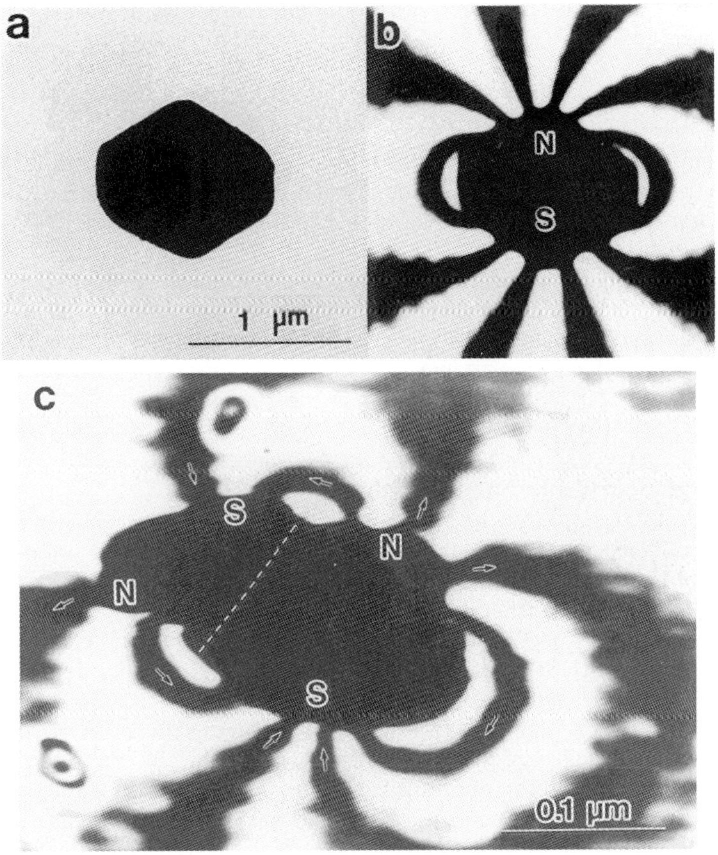

Figure 3-18. (a) A bright-field TEM image of a barium ferrite particle and (b) the corresponding phase image of the particle, reconstructed from a hologram, exhibiting the distribution of magnetic force lines. This particle is single domain. (c) Phase image of a two-domain magnetic barium ferrite particle. The phase is magnified ten time to amplify the effect (Courtesy of Drs. T. Hirayama and A. Tonomura).

(Figure 3-18b) shows the single magnetic domain structure, where the N and S poles of the magnetic domain can be clearly identified [32]. A particle consisting of two domains is given in Figure 3-18c and the domain size can be as small as 80 nm.

A key step in observation of magnetic flux lines is to create a magnetic field-free zone in which the specimen to be examine is placed. This can be achieved either by switching off the objective lens or by using a Lorentz objective lens. In the former, the experiments can be performed using a conventional TEM equipped with an FEG at the expense of image magnification, which is usually reduced by 50–80 times. The image resolution is also greatly reduced. Thus, the technique can only be applied to image magnetic domains of large sizes. On the other hand, the concomitant reduction in image magnification can be compensated by the optics of the electron energy filtering system attached to the microscope, which provides an addition magnification of 25x (at the CCD plane), thus, a magnification as large as 450,000x can be achieved, which is suitable for imaging particles smaller than 10 nm. Unfortunately, the image resolution is still limited by the aberration effects of the intermediate lenses and the projection lenses. With the use of a Lorentz lens the image magnification is unsacrificed and the image resolution is still preserved.

3.5 In-situ microscopy

In-situ studying the temperature induced phase transformation, structural and chemical evolution of nanocrystals are important for understanding the structure and structural stability of nanomaterials. TEM is an ideal system for conducting this type of experiments, in which a specimen can be cooled down to the liquid nitrogen or liquid helium temperature or heated to 1300 °C. The in-situ process can be recorded at TV rate (27 frames per second) for exhibiting the time and temperature dependent phenomena. In-situ observation can also be carried out for the mechanical stress, electric current and magnetic field induced phenomena. This section focuses mainly on the in-situ behavior of nanocrystals due to the effects of temperature and electric field.

3.5.1 Thermodynamic properties of nanocrystals

The large percentage of surface atoms in nanocrystals is the origin of their unique properties. The melting of Pt particles and the surface sublimation of Au nanorods are taken here as examples to illustrate the unusual thermodynamic properties of nanocrystals. The catalytic activity and selectivity of nanosize colloidal platinum (Pt) nanocrystals may depend strongly on their sizes and shapes. We had successfully demonstrated that platinum nanoparticles with a high percentage of cubic-, tetrahedral- and octahedral-like shapes, respectively, can be synthesized by changing the ratio of the concentration of polymer capping material (polyacrylate) to that of Pt^{2+} being reduced by H_2 from K_2PtCl_4 at room temperature [33]. The polymer acts not only as the passivative protection layer of the nanocrystals, but also the key factor in controlling their shapes [34, 35]. To determine the shape dependent catalytic behavior of Pt nanocrystals, the passivation layer on the nanocrystal must be removed. Several questions need to be considered: (1) how high does the annealing temperature need to be to remove the capping polymer while the particle shape is still preserved? (2) to what temperature is the particle shape still stable? (3) is there a temperature induced shape transformation due to surface diffusion? and (4) how high is the melting point of the Pt particles? In-situ TEM observation of the particle shape transformation and melting can provide answers to these questions [36].

Figure 3-19 shows a series of TEM images recorded from the same region when the specimen temperature was increased from 25 to 610 °C. For easy notation, particles are labeled as groups to trace their shape transformation behavior. Most of the parti-

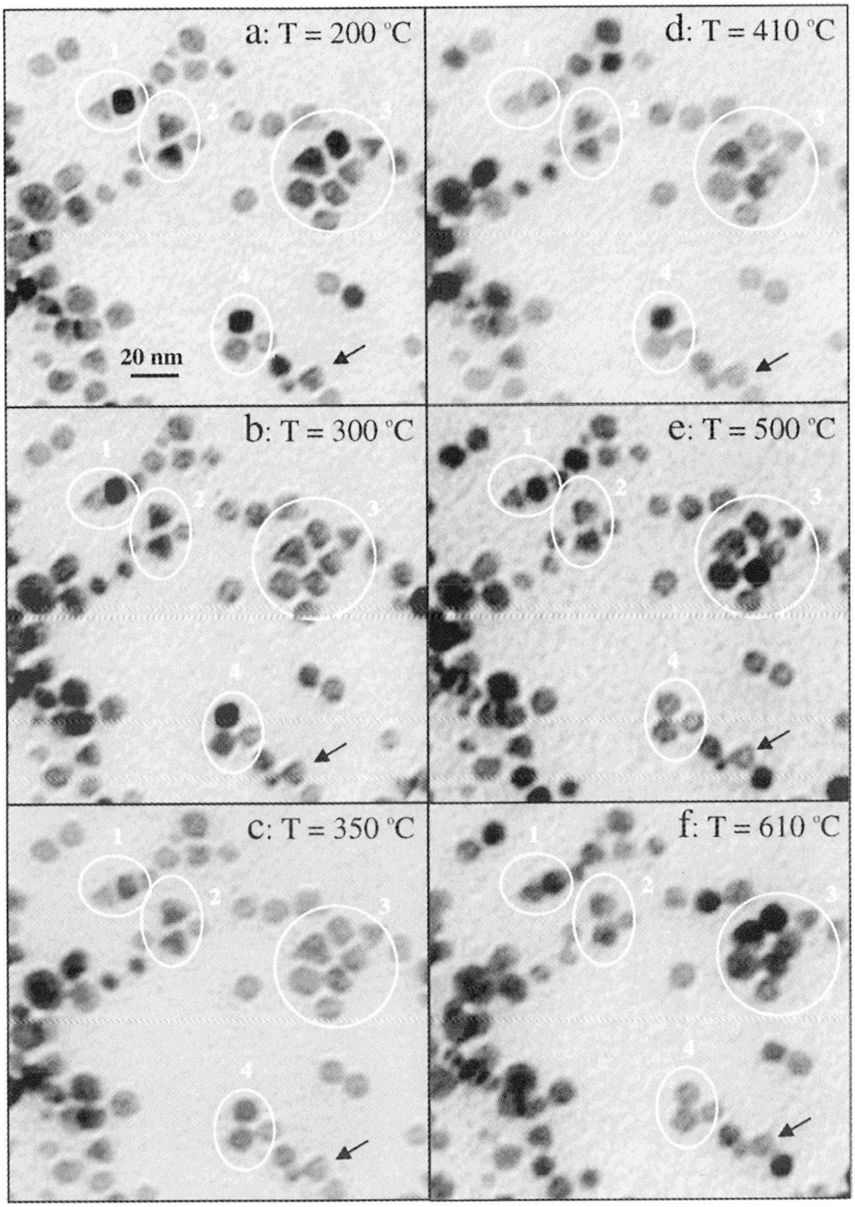

Figure 3-19. A series of in-situ TEM images recorded at different specimen temperatures, displaying the shape and structural transformation of Pt nanocrystals.

cle shapes showed no significant change when the specimen temperature was below 350 °C (Figures 3-19a-c). Truncated cubic and tetrahedral particles were formed when the temperature arrived 410 °C (Figure 3-19d). The corners and edges of the particles were disappearing because the local atoms have higher energy. The tetrahedral particles could still be identified even when the temperature reached 500 °C (Figure 3-19e), while the cubic particles became spherical-like when the temperature was above 500 °C. This indicates that the tetrahedral particles are more stable than cubic ones possibly because the {111} surfaces have lower surface energy than the {100}. A feature observed in Figure 3-19 is that the tetrahedral shape of a particle indicated by an arrowhead was preserved when the specimen temperature was as high as 610 °C. This is possibly due to the contact of the apexes of the particle with the adjacent particles, so that the interparticle diffusion can still sustain the shape of the apexes. Our results indicate that the surface capping polymer is removed by annealing the specimen to a temperature of 180–250 °C, while the particle shape shows no change up to ~ 350 °C. In a temperature range of 350 to 450 °C, a small truncation occurs in the particle shape but no major shape transformation. The particle shape experiences a dramatic transformation into spherical-like shape when the temperature is higher than ~ 500 °C.

The thermodynamic properties of nanocrystals depend strongly on not only the crystal size but also their shapes [37]. The thermodynamic properties of Au nanorods is an example. Gold nanorods were prepared by an electrochemical method [38]. To

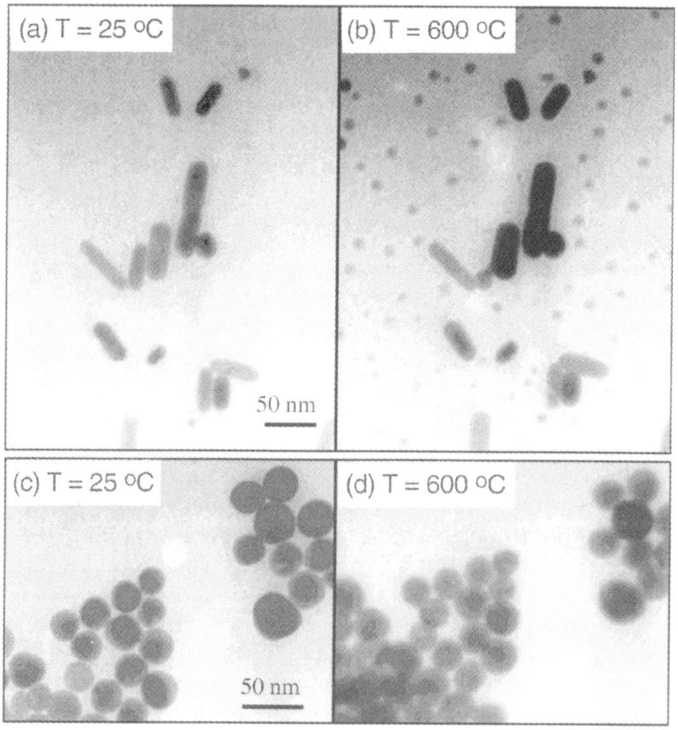

Figure 3-20. TEM images recorded in-situ at T = 25 °C and T = 600 °C, respectively, from the as-prepared Au nanorods (a,b) and from the Au spheres (c,d) (prepared by femtosecond laser photothermal conversion of the Au nanorods), showing the dependence of gold atoms sublimation rates on the shape of gold nanocrystals.

reveal the unique surface processes associated with Au nanorods, they were photoisomerized to form gold nanospheres by the influences of femtosecond laser pulses [39]. The structural evolution in the gold sphere and gold rods at 25 °C and 600 °C, respectively, is compared in Figs. 3-20 (a,b) and (c,d). Both the nanospheres and nanorods have the same average volume and almost equally exposed surface areas (with consideration of the contact of the particle with the substrate), thus, the size effect of the two geometrical shapes should not make a significant difference. After being heated to as low as 220 °C, small Au particles had formed on the substrate in the as-prepared nanorods specimen, while no particles were formed near the gold nanosphere. This clearly shows that the newly formed Au particles are unique characteristics of Au nanorods.

Using HRTEM, it has found that the nanorods have the unstable {110} facets [37], in addition to the {100} and {111} facets present in the spheres. For the face-centered Au nanocrystals, the surface energy follows a relationship of $\gamma_{\{110\}} > \gamma_{\{100\}} > \gamma_{\{111\}}$. The presence of a significant percentage of gold atoms on the {110} surface suggests that the nanorod is relatively thermodynamically unstable. The {111} and {100} faceted Au spheres have lower surface energy. Thus, their atoms may not sublime at relatively low temperatures. Therefore, the sublimation of Au atoms from the nanorods at lower temperatures may be contributed in part by the shape of the nanorods, which has the unstable {110} surfaces.

3.5.2 Electric field driven phenomena

In-situ structural evolution can also be driven by an electric field. Studying of domain boundary movement in ferroelectric crystals is an example. For nanocrystals, TEM provides a powerful approach for directly imaging the nanostructure under the application of an electric field/current.

A key step in this study is to build a specimen holder which can apply a controllable voltage across the specimen. We have demonstrated the study of quantum conductance of multiwalled carbon nanotubes [40, 41]. The nanotubes conduct current ballistically and don't dissipate heat. A current density of $J > 10^7$ A/cm^2 can pass a single nanotube at ambient temperature, setting a new record of room temperature superconductivity. These results strongly suggest that carbon nanotubes are ideal candidate of the interconnects in future molecular electronics.

Figure 3-21a shows an TEM image, in which a carbon fiber is connected with a mercury drop by a single carbon nanotube. The current conducted through the tube can be measured by applying a voltage across the two electrodes, and this measured transport property can be directly linked to the microstructure of the tube revealed by electron imaging and diffraction [42]. On the other hand, at the non-contacting case, an electric potential can be applied across the electrodes, resulting in charging up of the nanotube (Figures 3-21c). This is an ideal case for studying the field emission at individual nanotube. The electric field due to the electrostatic potential causes the charged nanotube to deflect, as can be seen by comparing the images before (Figure 3-21b) and after (Figure 3-21c) applying an electrostatic potential of 40 V. The geometry of the deflected tube can be modeled using the elasticity theory, providing a potential technique for quantifying the charge distribution on the tube if its Young's modules is known (see Figures 1-2 and 1-3 for details).

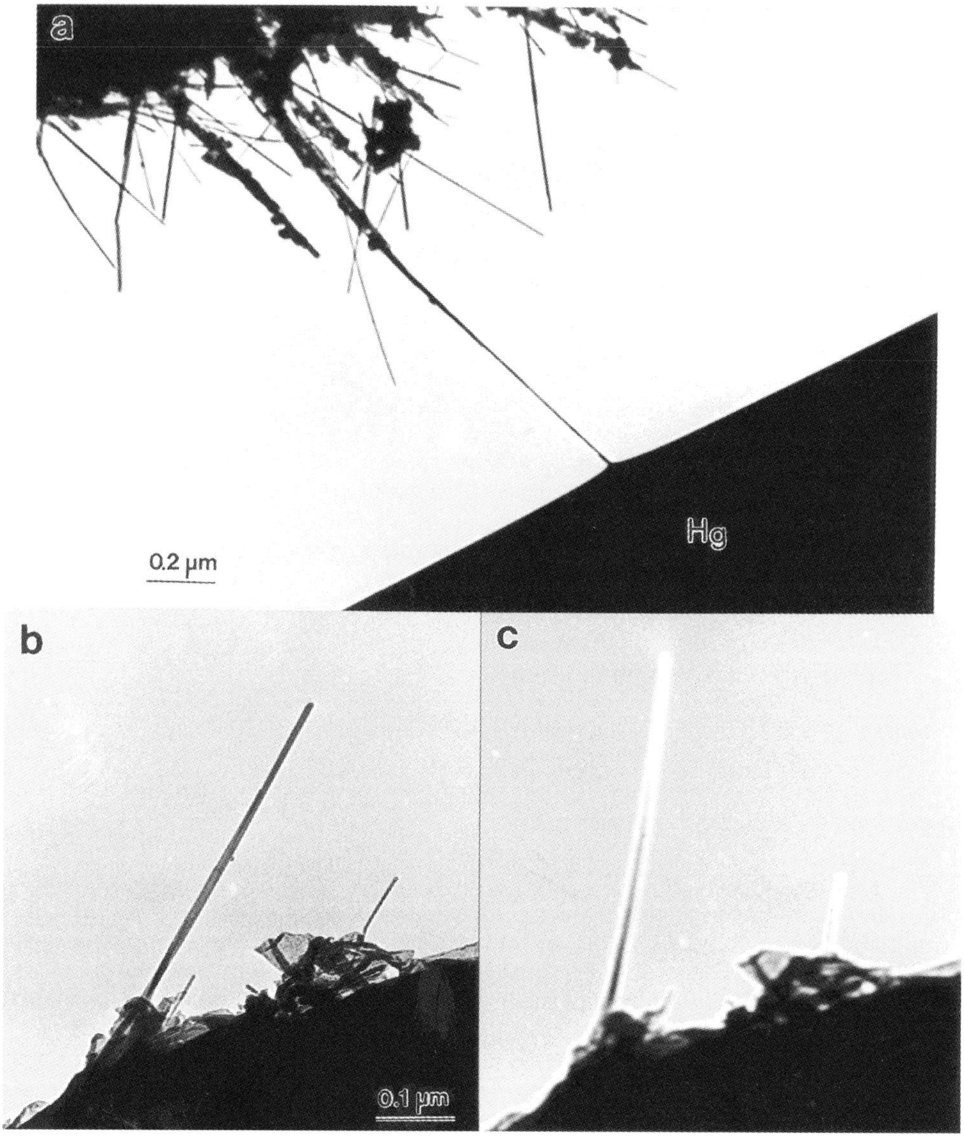

Figure 3-21. (a) TEM image recorded in-situ when a carbon nanotube at the tip of a carbon fiber is in contact with a mercury drop, while an electric current flows through. TEM images of carbon nanotubes recorded at a large defocus condition (b) before and (c) after an electrostatic potential of 40 V is applied across the electrodes. The bright contrast at the tips of the tubes is due to the charging up effect.

3.6 Electron energy-loss spectroscopy of nanoparticles

HRTEM is a powerful technique for identifying the crystal structure of nanophase materials, but it may not be sensitive enough to detect a small change in atomic number. To precisely determine the content of a particular element in a specimen, one must know the characteristics of the atoms. The structure of an atom is usually characterized by the quantized energy levels and the associated electronic states. Under the impact of an incident electron, the electrons bounded to the atoms may be excited either to a free electron state or to a unoccupied energy level with a higher energy. The quantum transitions associated with these excitations will emit photons (or x-rays) and electrons such as secondary electrons, Auger electrons and ionized electrons, these inelastic scattering signals are the finger prints of the elements that can provide quantitative chemical and electronic structural information.

Figure 3-22 shows the main inelastic processes that may be excited in high-energy electron scattering [6, 43]. When a fast electron passes through a thin metal foil, the most noticeable energy-loss is to plasmon oscillations in the sea of conduction electrons. For an ideal case in which the electrons can move "freely" in the sea, the system can be treated as an electron gas. This case is best represented by aluminum, in which the outer-shell electrons can be considered as free electrons. The negatively charged particles are mixed with nuclei, forming a solid state plasmon "gas". The resonance frequency of this *plasmon* is directly related to the density of electrons in the solid. The decay of plasmons results in the emission of *ultraviolet light*. The *cathodulumines-cence* (C.L.) technique is based on the detection of the visible light, which is emitted when an electron in a higher-energy state (usually at an impurity) fills a hole in a lower state that has been created by the fast electron.

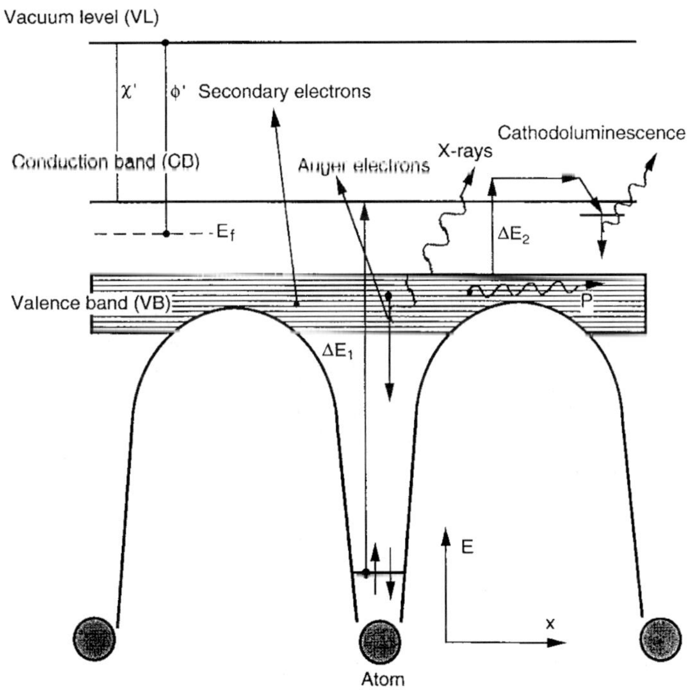

Figure 3-22. Schematic one-electron energy level diagram plotted against the positions of atoms showing the characteristics excitations by an incident electron in a semiconductor material. Here, E_f is the Fermi level, E_b the binding energy, and C.B., V.B. and V.L. are the conduction-band minimum, valance-band maximum, and vacuum level, respectively. ΔE_1 a K shell excitation; ΔE_2 a single-electron excitation; C.L. cathodoluminescence photon; P a plasmon.

Second, *atomic inner-shell ionization* is excited by the energy transfer of the incident electrons, resulting in an ejected electron from a deep-core state (Figure 3-23a). Since only one inner-shell electron is involved in the excitation, this process is also called *single-electron excitation*. The excitation introduces an energy-loss in the range of a few tens to thousands eV which is the finger print of the element. Since the intensity and threshold energy of the inner shell electron transition are determined not only by the binding energy of the atom but also by the density of states in the valence band, the energy-loss near edge structure usually carries some solid state effects, and this is the basis of analyzing the valence states of an element. Analogous to C.L., the holes created at deep core states tend to be filled by the core-shell electrons from higher energy levels, the energy is released in the form of photons (or x-rays) (Figure 3-23b). The energies of the emitted x-rays are quantized and they are fingerprints of the corresponding elements and are used for chemical microanalysis. The holes, created by the ionization process, in deep-core states may alternatively be filled by the electrons from outer shells, the energy released in this process may ionize another outer shell electron (Figure 3-23c), resulting in the emission of *Auger electrons*. Accompanying to these processes, *second electrons* can be emitted from the valence band. The difference between Figures 3-23b and 3-23c is the emission of photon in the former and the Auger electron in the latter, and the two processes are complimentary.

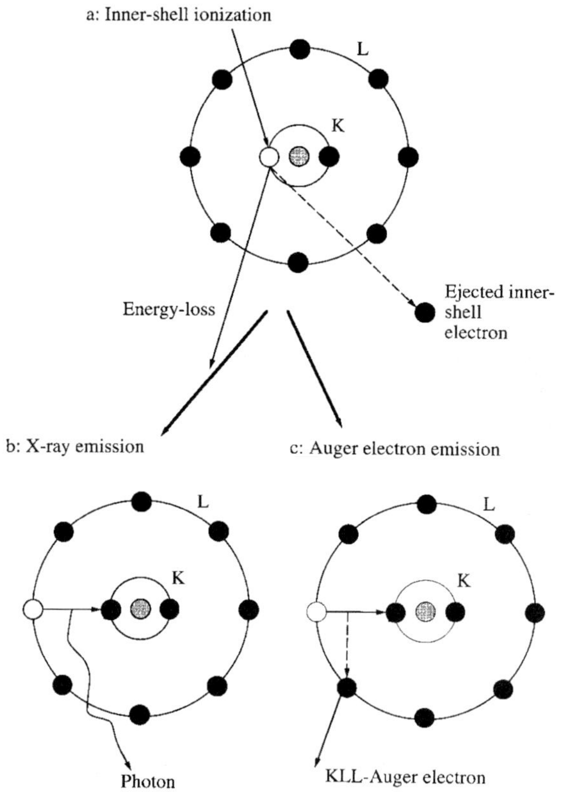

Figure 3-23. (a) Ionization of an atom bounded inner-shell electron by an incident electron, resulting in (b) X-ray emission and (c) Auger electron emission (see text).

In other words, the elements with higher Auger electron yields have lower x-ray emission, and vice versa. Thus, the Auger electron spectroscopy is more sensitive to light elements while EDS is to heavier elements.

The last characteristic inelastic excitation is *phonon scattering* (or *thermal diffuse scattering*, TDS) [44], which is the result of atomic vibrations in crystals. This process does not introduce significant energy-loss but produces large momentum transfers, which can scatter the incident electron into a high angular range. This is a localized inelastic scattering process. As will be shown in Chapter 4 the collection of TDS electrons can produce compositional sensitive images.

3.6.1 Valence excitation spectroscopy

In studying of nanoparticles, it is necessary to probe the electronic structure of a single nanocrystal. This is possible only with the use of a probe that is smaller than the size of the nanocrystal. The valence excitation spectrum of a nanoparticle is most sensitive to its electronic structure and size effects [45, 46]. The spectra can be acquired in TEM and STEM using a fine electron probe. The quantification of the spectra relies on theoretical calculation. The valence band excitation of a nanoparticle is most easily and accurately described using the dielectric response theory. The impact of an incident electron is equivalent to a time-dependent pulse, which causes transitions of valence electrons. In classical dielectric response theory, an incident electron is treated as a particle following a pre-defined trajectory, which is assumed not being affected by the interaction between the incident electron and the dielectric media, as shown in Figure 3-24. Electron energy-loss is a continuous process, in which the electron is decelerated due to the attractive force $F_z = (-\,e)E_z$ owing to the field of the induced charges, resulting in energy-loss. For a general case in which the incident electron is moving along the z-axis and under non-relativistic approximation, if the instantaneous position of the electron is denoted by $\mathbf{r}_0 = (x_0, 0, z'=vt)$, where x_0 is called the impact parameter, the energy-loss spectrum of an incidence electron due to surface plasmon excitation of a finite dielectric medium is calculated by [45, 47]

$$\frac{dP(\omega)}{d\omega} = \frac{e}{\pi\hbar v^2}\int_{-\infty}^{\infty} dz' \int_{-\infty}^{\infty} dz\, \omega \mathrm{Im}\{-\exp[i\omega(z'-z)/v]\, V_i(\mathbf{r},\mathbf{r}_0)\} \Big|_{\mathbf{r}=(x_0,0,z),\ \mathbf{r}_0=(x_0,0,z')} \qquad (3\text{-}12)$$

where $Vi(\mathbf{r},\mathbf{r}_0)$ is the potential due to the induced charge when a "stationary" electron is located at $\mathbf{r}_0 = (x_0, 0, z')$, i.e., it is the homogeneous component of V satisfying

$$\nabla^2 V(\mathbf{r},\mathbf{r}_0) = -\frac{e}{\varepsilon(\omega)\varepsilon_0}\,\delta(\mathbf{r}-\mathbf{r}_0) \qquad (3\text{-}13)$$

for the dielectric media considered. It is important to note that $Vi(\mathbf{r},\mathbf{r}_0)$ is ω-dependent. The potential distribution in space is a quasi-electrostatic potential for each point along the trajectory of the incident electron. The integral over z' is to sum over the contributions made by all of the points along the trajectory of the incident electron. Therefore, the calculation of valence-loss spectra is actually to find the solution of the electrostatic potential for a stationary electron located at $\mathbf{r}_0$ in the dielectric media system. We now use a spherical particle as an example to illustrate the application of this theory.

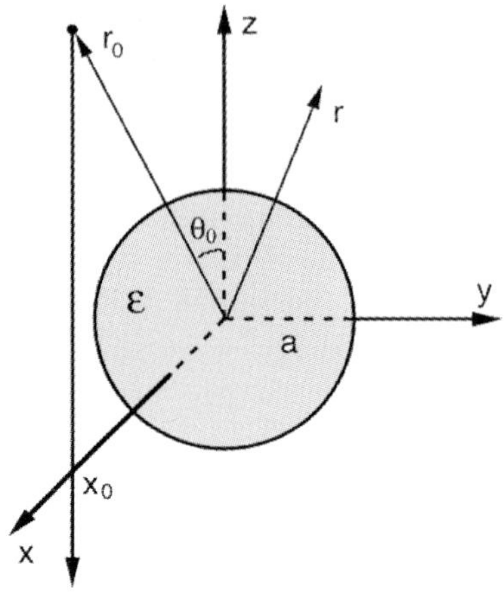

Figure 3-24. Excitation of a spherical particle by an external incident electron with an impact parameter of x_0 ($x_0 > a$). The radius of the sphere is a and its dielectric function is $\varepsilon(\omega)$.

Consider a point electron moving at a constant velocity **v** in vacuum along a trajectory specified by $\mathbf{r}_0 = (x_0,0,z')$, as schematically shown in Figure 3-24. For simplicity one assumes that the electron does not penetrate the particle. In terms of spherical coordinates (r,θ,ϕ), $r = (x_0^2 + z^2)^{1/2}$, $\cos\theta = z/r$, $r_0 = (x_0^2 + z'^2)^{1/2}$ and $\cos\theta_0 = z'/r_0$. The potential due to the incident electron in free-space for $r_0 > r$ is

$$V_e(\mathbf{r},\mathbf{r}_0) = -\frac{e}{4\pi\varepsilon_0|\mathbf{r}-\mathbf{r}_0|} = -\frac{e}{4\pi\varepsilon_0 r_0}\sum_{L=0}^{\infty}\sum_{m=0}^{L} N_{Lm}\,(r/r_0)^L\, P_{Lm}(\cos\theta)\, P_{Lm}(\cos\theta_0)\cos(m\phi) \tag{3-14a}$$

where P_{Lm} is the associated Legendre function and

$$N_{Lm} = \frac{(2-\delta_{0m})(L-m)!}{(L+m)!} \tag{3-14b}$$

where δ_{0m} is unity if $m = 0$ and is zero otherwise. If $r > r_0$, then r and r_0 are exchanged in Eq. (3.149). The solution of Eq. (3-13) consists of the field by the free point charge and the induced charge. The potential outside the sphere has a form of

$$V_{out}(\mathbf{r},\mathbf{r}_0) = V_e(\mathbf{r},\mathbf{r}_0) - \frac{e}{4\pi\varepsilon_0}\sum_{L=0}^{\infty}\sum_{m=0}^{L} A_{Lm}\frac{1}{r}(a/r)^{L+1}\, P_{Lm}(\cos\theta)\, P_{Lm}(\cos\theta_0)\cos(m\phi) \tag{3-15}$$

and inside the sphere the potential is

$$V_{in}(\mathbf{r},\mathbf{r}_0) = -\frac{e}{4\pi\varepsilon_0 a}\sum_{L=0}^{\infty}\sum_{m=0}^{L} B_{Lm}\,(r/a)^L\, P_{Lm}(\cos\theta)\, P_{Lm}(\cos\theta_0)\cos(m\phi) \tag{3-16}$$

Matching boundary conditions:

$$V_{out}(\mathbf{r},\mathbf{r}_0)\big|_{r=a} = V_{in}(\mathbf{r},\mathbf{r}_0)\big|_{r=a}, \text{ and } \frac{\partial V_{out}(\mathbf{r},\mathbf{r}_0)}{\partial r}\Big|_{r=a} = \varepsilon(\omega)\frac{\partial V_{in}(\mathbf{r},\mathbf{r}_0)}{\partial r}\Big|_{r=a} \tag{3-17}$$

the A_{Lm} and B_{Lm} coefficients are determined to be:

$$A_{Lm} = N_{Lm} \frac{L(1-\varepsilon)}{L\varepsilon+L+1} (a/r_0)^{L+1}, \text{ and } B_{Lm} = N_{Lm} \frac{2L+1}{L\varepsilon+L+1} (a/r_0)^{L+1} \tag{3-18}$$

Substituting Eq. (3-15) into (3-14), performing the analytical integral with the use of an identity

$$\int_{-\infty}^{\infty} dt\, r_0^{-(L+1)} P_{Lm}(\cos\theta_0)\exp(i\omega t) = \frac{2i^{L-m} |\omega/v|^L K_m(|\omega x_0/v|)}{v(L-m)!} (\omega/|\omega|)^{L-m} \tag{3-19}$$

the spectrum is given by

$$\frac{dP}{d\omega} = \frac{e^2 a}{\pi^2 \varepsilon_0 \hbar v^2} \sum_{L=0}^{\infty} \sum_{m=0}^{L} M_{Lm} [K_m(\omega x_0/v)]^2 (\omega a/v)^{2L} \operatorname{Im}[\frac{L(\varepsilon(\omega)-1)}{L(\varepsilon(\omega)+L+1}] \tag{3-20a}$$

where

$$M_{Lm} = \frac{(2-\delta_{0m})}{(L-m)!(L+m)!} \tag{3-20b}$$

From the energy-loss function, the resonance free frequencies of the surface plasmons are determined by [50]

$$L\varepsilon(\omega) + L + 1 = 0 \tag{3-21}$$

The homogeneous medium theory has been extended recently for the cases of anisotropic dielectric medium, such as carbon onion structure and carbon nanotubes [48, 49]. Shown in Figure 3-25a are a group of calculated EELS spectra for a carbon onion

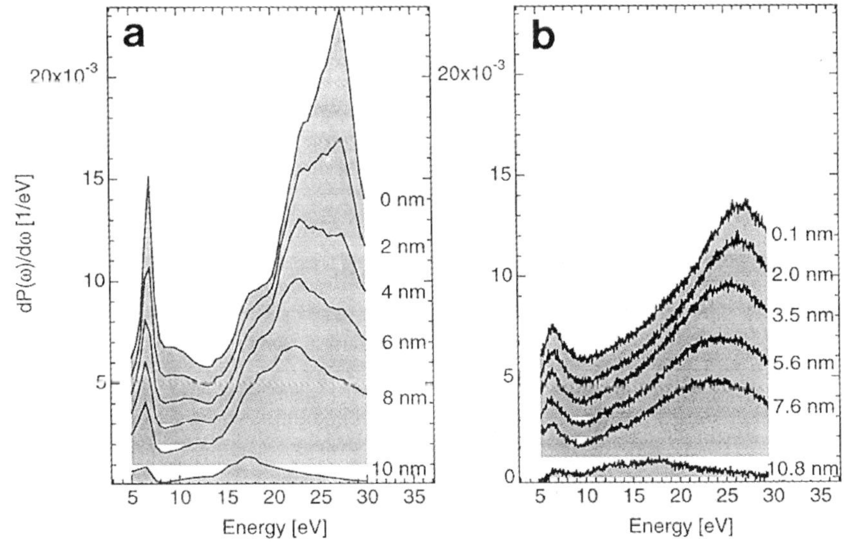

Figure 3-25. (a) Calculated EELS spectra of a carbon sphere (radius = 10 nm) with graphitic onion-like structure as a function of the electron impact parameter x_0. (b) Experimentally observed EELS spectra from a carbon sphere (Courtesy of T. Stöckli).

of radius 10 nm as a function of the electron impact parameter. The calculation has to consider the anisotropic dielectric properties of the graphitic onion-like structure. At small impact parameters, the excitation is dominated by volume plasmon, and at larger impact parameters, the surface excitation becomes dominant. When the electron probe is outside of the sphere, the entire spectrum is the surface excitation. This type of calculation gives quantitative agreement with experimental observations (Figure 3-25b) and it can be used to quantify the dielectric properties of a single nanostructure.

3.6.2 Quantitative nanoanalysis

Energy dispersive x-ray spectroscopy (EDS) and electron energy-loss spectroscopy (EELS) in TEM have been demonstrated as powerful techniques for performing microanalysis and studying the electronic structure of materials [43]. Atomic inner shell excitations are often seen in EELS spectra due to a process in which an atom-bounded electron is excited from an inner shell state to a valence state accompanied by incident electron energy loss and momentum transfer. This is a localized inelastic scattering process, which occurs only when the incident electrons are propagating in the crystal. Figure 3-26 shows an EELS spectrum acquired from $YBa_2Cu_4O_8$. Since the inner-shell energy levels are the unique features of the atom, the intensities of the ionization edges can be used effectively to analyze the chemistry of the specimen.

After subtracting the background, an integration is made to the ionization edge for an energy window of width Δ accounted from the threshold energy. Thus, the intensity oscillation at the near edge region is flattered, and the integrated intensity is dominated by the properties of single atoms. This type of information is most useful in material analysis and the integrated intensity is given by

$$I_A(\Delta) \approx I_0(\Delta)\, \sigma_A(\Delta,\beta)\, n_A\, d \tag{3-22}$$

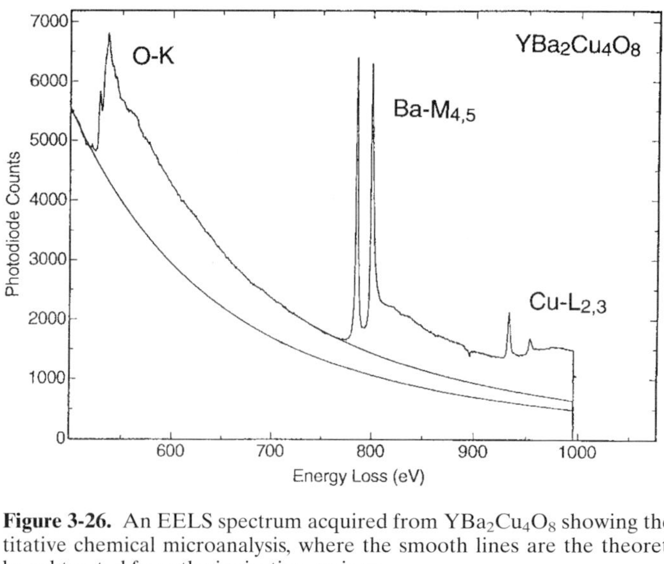

Figure 3-26. An EELS spectrum acquired from $YBa_2Cu_4O_8$ showing the application of EELS for quantitative chemical microanalysis, where the smooth lines are the theoretically simulated background to be subtracted from the ionization regions.

where $I_0(\Delta)$ is the integrated intensity of the low-loss region including the zero-loss peak for an energy window Δ; $\sigma_A(\Delta,\beta)$ is the energy and angular integrated ionization cross-section. In imaging mode, β is mainly determined by the size of the objective aperture or the upper cut-off angle depending on which is smaller. In diffraction mode, the β angle is determined not only by the size of the EELS entrance aperture and the camera length but also by the beam convergence. In general, the width of the energy window is required to be more than 50 eV to ensure the validity of Eq. (3-21), and $\Delta = 100$ eV is an optimum choice. If the ionization edges of two elements are observed in the same spectrum, the chemical composition of the specimen is

$$\frac{n_A}{n_B} = \frac{I_A(\Delta)}{I_B(\Delta)} \frac{\sigma_B(\Delta,\beta)}{\sigma_A(\Delta,\beta)} \tag{3-23}$$

This is the most powerful application of EELS because the spatial resolution is almost entirely determined by the size of the electron probe. The key quantity in this analysis is the ionization cross-section. For elements with atomic numbers smaller than 14, the K edge ionization cross-section can be calculated using the SIGMAK program [43], in which the atomic wave function is approximated by a single-electron hydrogen-like model. The ionization cross-section for elements with $13 < Z < 28$ can be calculated using the SIGMAL program. For a general case, the ionization cross-section may need to be measured experimentally using a standard specimen with known chemical composition.

3.6.3 Near edge fine structure and bonding in transition metal oxides

The energy-loss near edge structure (ELNES) is sensitive to the crystal structure. This is a unique characteristics of EELS and in some cases it can serve as a "fingerprint" to identify a compound. A typical example is the intensity variation in the π^* and σ^* peaks observed in the C-K edge, as shown in Figure 3-27. Diamond is almost completely dominated by the σ^* bonding, while the π bonding appears in graphite and

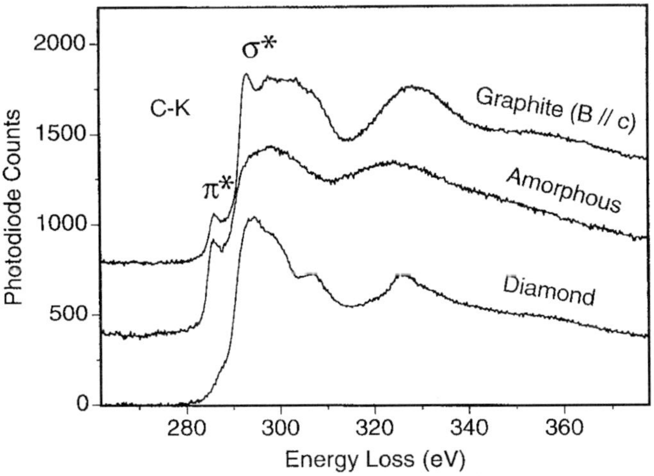

Figure 3-27. EELS C-K edge spectra acquired from diamond, amorphous carbon and graphite, respectively, showing the sensitivity of EELS to bonding in carbon related materials.

amorphous carbon. The disappearance of the π^* peak in C-K edge can be uniquely used to identify the presence of diamond bonding in a carbon compound. It must be pointed out that the spectrum for graphite shown here was acquired when the incident beam parallel to the c-axis. If a small size aperture is used, only a small portion of electrons corresponding to the π^* peak is collected.

In EELS, the L ionization edges of transition-metal and rare-earth elements usually display sharp peaks at the near edge region (Figure 3-28), which are known as *white lines.* For transition metals with unoccupied 3d states, the transition of an electron from 2p state to 3d levels leads to the formation of white lines. The L_3 and L_2 lines are the transitions from $2p^{3/2}$ to $3d^{3/2}3d^{5/2}$ and from $2p^{1/2}$ to $3d^{3/2}$, respectively, and their intensities are related to the unoccupied states in the 3d bands. Numerous EELS experiments have shown that a change in valence state of cations introduces a dramatic change in the ratio of the white lines, leading to the possibility of identifying the occupation number of 3d orbital using EELS.

EELS analysis of valence state is carried out in reference to the spectra acquired from standard specimens with known cation valence states. Since the intensity ratio of L_3/L_2 is sensitive to the valence state of the corresponding element, if a series of EELS spectra are acquired from several standard specimens with known valence states, an empirical plot of these data serves as the reference for determining the valence state of the element present in a new compound [51–59]. The L_3/L_2 ratios for a few standard Co compounds are plotted in Figure 3-29a. EELS spectra of Co-$L_{2,3}$ ionization edges were acquired from $CoSi_2$ (with Co^{4+}), Co_3O_4 (with $Co^{2.67}+$), $CoCO_3$ (with Co^{2+}) and $CoSO_4$ (with Co^{2+}). Figure 3-29b shows a plot of the experimentally measured intensity ratios of white lines L_3/L_2 for Mn. The curves clearly show that the ratio of L_3/L_2 is very sensitive to the valence state of Co and Mn. This is the basis of our experimental approach for measuring the valence states of Co or Mn in a new material.

Determination the crystal structure of nanoparticles is a challenge particularly when the particles are smaller than 5 nm. The intensity maxim observed in the x-ray or electron diffraction patterns of such small particles are broadened due to the crystal shape factor, greatly reduced the accuracy of structure refinement. The quality of the

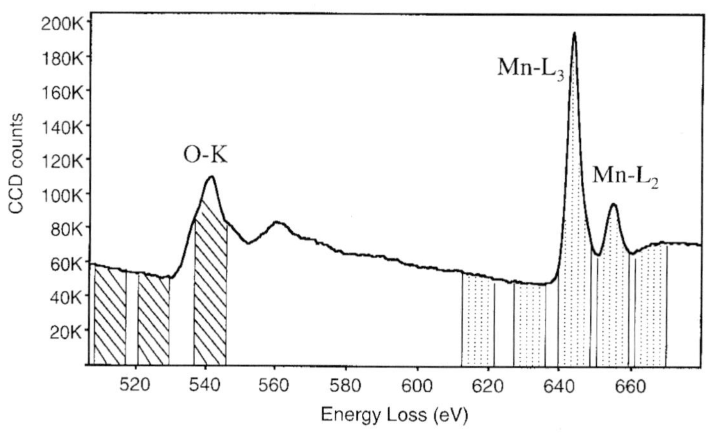

Figure 3-28. EELS spectrum acquired from MnO_2 showing the Mn-L_3 and Mn-L_3 white lines. The five windows pasted in the Mn-L edge are to be used for extracting the image formed by the ratio of white lines.

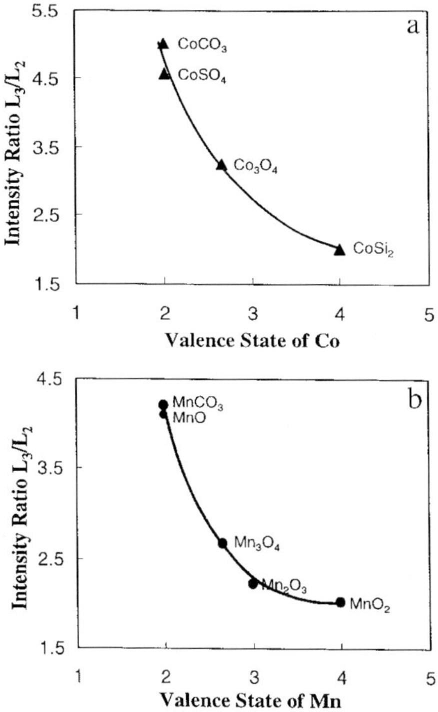

Figure 3-29. Plots of the intensity ratios of L_3/L_2 calculated from the spectra acquired from (a) Co compounds and (b) Mn compounds as a function of the cation valence. A nominal fit of the experimental data is shown by a solid curve.

high-resolution TEM images of the particles is degraded because of the strong effect from the substrate. This difficulty arises in our recent study of CoO nanocrystals whose shape is dominated by tetrahedral of sizes smaller than 5 nm [60]. Electron diffraction indicates the crystal has a fcc-type cubic structure. To confirm the synthesized nanocrystals are CoO, EELS is used to measure the valence state of Co. Figure 3-30

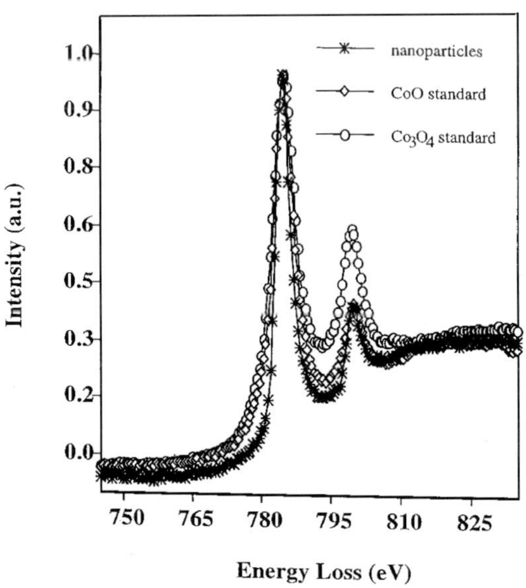

Figure 3-30. A comparison of EELS spectra of Co-$L_{2,3}$ ionization edges acquired from Co_3O_4 and CoO standard specimens and the synthesized nanocrystals, proving that the valence state of Co is 2+ in the nanocrystals. The full width at half maximum of the white lines for the Co_3O_4 and CoO standards is wider than that for the nanocrystals, possibly due to size effect.

shows a comparison of the spectra acquired from Co_3O_4 and CoO standard specimens and the synthesized nanocrystals. The relative intensity of the Co-L_2 to Co-L_3 for the nanocrystals is almost identical to that for CoO standard, while the Co-L_2 line of Co_3O_4 is significantly higher, indicating that the Co valence in the nanocrystals is 2+, confirmed the CoO structure of the nanocrystals.

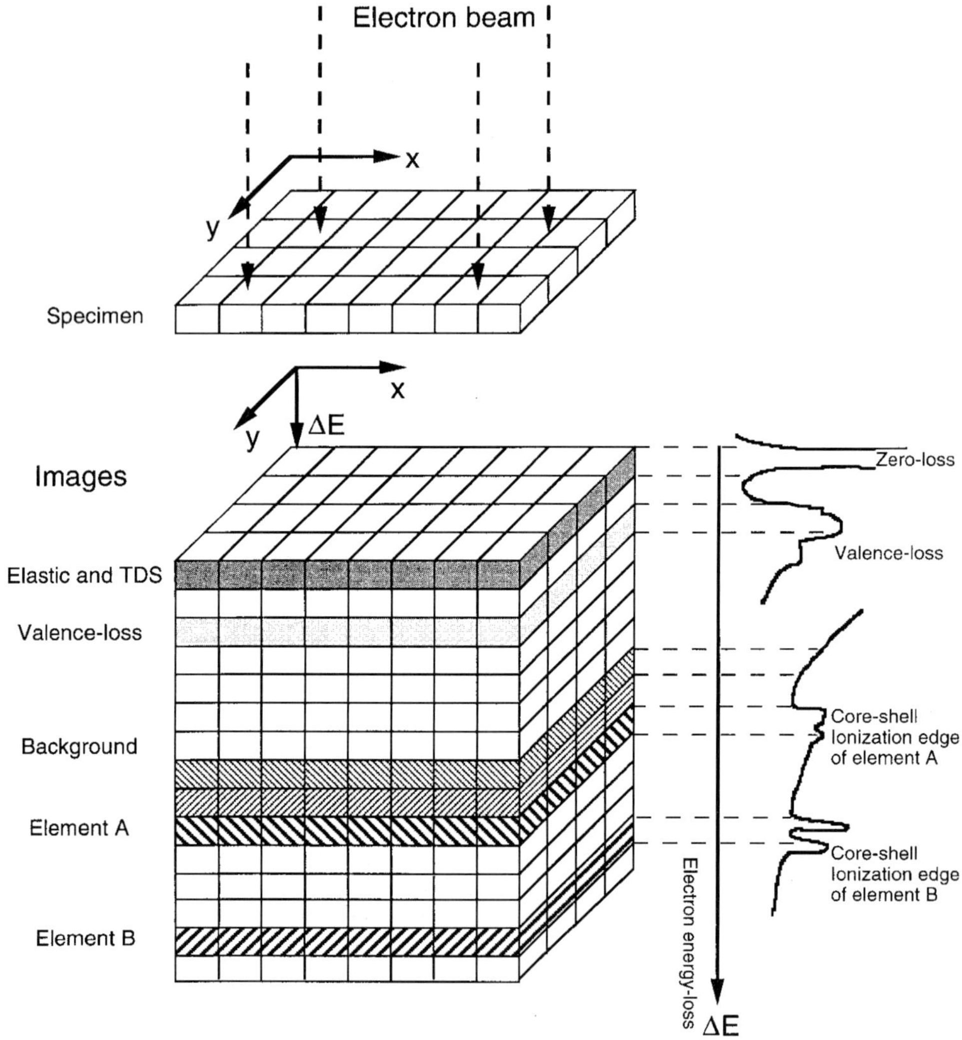

Figure 3-31. A schematic diagram showing energy-filtered electron imaging in TEM. The conventional TEM image is recorded by integrating the electrons with different energy losses. The energy-selected electron images corresponding to different characteristic energy-loss features are shown, which can be used to extract useful structural and chemical information of the specimen.

3.7 Energy-filtered electron imaging

3.7.1 Energy filtering in TEM

Energy-filtered transmission electron microscopy (EF-TEM) is a rapid developing field for high spatial-resolution chemical imaging. With an energy-filter, images (or diffraction patterns) formed by electrons with specific energy-losses can be obtained (see Figure 3-31) [61]. The energy-selected electron images can be simply illustrated using a 3-D data space, in which the z-axis is replaced by the energy-loss of the electrons, and x and y are the real space coordinates of the 2-D image of the specimen. All the inelastic scattering processes introduced above are observed in the EELS spectrum, as schematically shown on the right-hand side of Figure 3-31. The zero-loss (or 0-loss) peak is composed of elastically and thermal diffusely scattered electrons. The images/diffraction patterns produced by the elastically scattered and by thermal diffusely scattered electrons are incoherent. The low-loss region is dominated by valence excitations. However, as was discussed previously, the energy-selected image of valence-loss (or plasmon-loss for metals) electrons is not strongly composition sensitive because their scattering is closely related to the valence-band structure of the material. Atomic inner-shell ionization edges can be employed to form composition-sensitive images. The background observed in the EELS spectrum is produced by scattering processes of multiple valence losses, electromagnetic radiation and electron Compton scattering. In order to extract composition-sensitive images, the contribution made by these processes must be subtracted from the recorded data.

3.7.2 Composition-sensitive imaging using inner-shell ionization edges

Energy-selected electron images of atomic inner shell ionization edges can also be performed in STEM [62, 63]. The image is formed by processing the EELS spectra acquired at each beam position, so that the thickness integrated elemental map is

$$n_A\, d = \frac{1}{\sigma_A(\Delta,\beta)}\frac{I_A(\Delta)}{I_0(\Delta)} \tag{3-24}$$

Figure 3-32 gives a pair of energy-filtered TEM images using the C-K and Ag-L edges, respectively, from a specimen of Ag nanocrystals passivated with $SC_{12}H_{25}$ thiolate chains. The carbon distribution represents the local thickness-projected density of the adsorbed molecules. In the region with less Ag particles, the stronger carbon signal indicates the accumulation of the molecules. The image using Ag-L edge clearly shows the distribution of Ag cores. Energy-filtered image using the O-K edge proves that the particles are not oxidized.

3.7.3 Mapping the bonding and valence state using fine edge structures

If the energy-selected electron image can be formed using the fine features at the near edge region, one can map the spatial distribution of the valence states. The most typical example of mapping the valence state distribution is the case of using the σ^* and π^* peaks observed in C-K edge of diamond specimens [64, 65]. Here we illustrate a new approach that can map the distribution of the valence states of transition metal

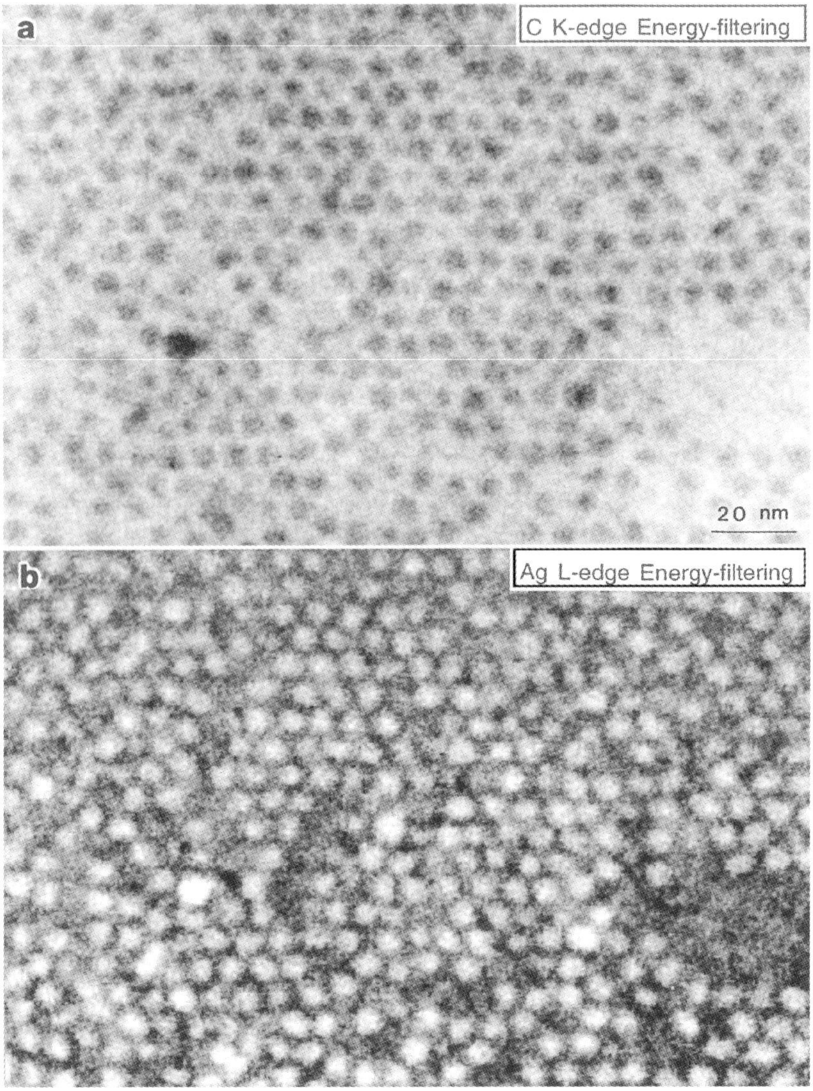

Figure 3-32. Energy-filtered TEM images recorded using (a) the C-K edge and (b) the Ag-L edge from a specimen of Ag nanocrystals passivated with $SC_{12}H_{25}$ molecules and dispersed on an amorphous silica substrate. The energy window width was 20 eV.

elements in oxides [66]. In EF-TEM, an energy window of 12 eV in width is required to isolate the L_3 from L_2 white lines (Figure 3-28). A five-window technique is introduced: two images are acquired at the energy-losses prior to the L ionization edges, and they are to be used to subtract the background for the characteristic L edge signals; two images are acquired from the L_3 and L_2 white lines, respectively, and the fifth image is recorded using the electrons right after the L_2 line that will be used to subtract the continuous background underneath the L_3 and L_2 lines. Then, a L_3/L_2

ratio image will be obtained, which reflects the distribution of valence state across the specimen. It must be pointed out that the thickness effect has been removed in the L_3/L_2 image.

A partially oxidized CoO specimen that contains CoO and Co_3O_4 grain structure was chosen for this study [67]. The CoO and Co_3O_4 phases are separated by clear boundaries and it is an ideal specimen for testing the optimum resolution. Figure 3-33 shows a group of energy-filtered TEM image from a triple point in the CoO-Co_3O_4 specimen. The energy-filtered images using the L_3 and L_2 lines and the post L_2 line region (Figures 3-33a-c) show distinctly difference in contrast distribution due to a difference in the relative white line intensities. From these three images the L_3/L_2 is calculated after subtracting the contribution from the continuous energy-loss region that is due to single atom scattering, and the image clearly displays the distribution of cobalt oxides having different valence states (Figure 3-33e), where the diffraction contrast disappears. The region with lower oxidation state (Co2+) shows stronger contrast, and the ones with high oxidation states show darker contrast (see the L_3/L_2 ratio displayed in Figure 3-29a). The O/Co image was calculated from the images recorded from the O-K edge and the $L_3 + L_2$ white lines for an energy window width of $\Delta = 24$ eV. Although the energy-filtered O-K edge image exhibit some diffraction contrast and the thickness effect, the O/Co compositional ratio image greatly reduces the effect. The high intensity region in the O/Co image indicates the relative high local concentration in oxygen (e.g., higher Co oxidation states), the low intensity region contain relatively less oxygen (e.g., lower Co valence state), entirely consistent with the information provided by the L_3/L_2 image. A spatial resolution of ~ 2 nm has been achieved. This is remarkable in comparison to any existing techniques.

3.7.4 Encrgy dispersive x-ray microanalysis (EDS)

X-rays emitted from atoms represent the characteristics of the elements and their intensity distribution represents the thickness-projected atom densities in the specimen. This is the energy dispersive x-ray spectroscopy (EDS) which has played an important role in microanalysis, particularly for heavier elements [68]. EDS is a key tool for identifying the chemical composition of a specimen. A modern TEM is capable of producing a fine electron probe of smaller than 2 nm, allowing direct identification the local composition of an individual nanocrystal. Details of this type of analysis will be give in Section 3.8.2.

3.8 Structure of self-assembled nanocrystal superlattices

Nanoparticles and the physical and chemical functional specificity and selectivity they possess, suggest them as ideal building blocks for two- and three-dimensional cluster self-assembled superlattice structures, in which the particles behave as well-defined molecular matter and they are arranged with long-range translational and even orientational order. Research has successfully fabricated *self-assembly passivated nanocrystal superlattices (NCS's) or nanocrystal arrays (NCA)* of metal, semiconductor and oxide clusters [35, 60, 69–74], which are a new form of materials with fundamental interests and technological importance. Self-assembled arrays involve self-organization into monolayers, thin films, and superlattices of size-selected nanoclus-

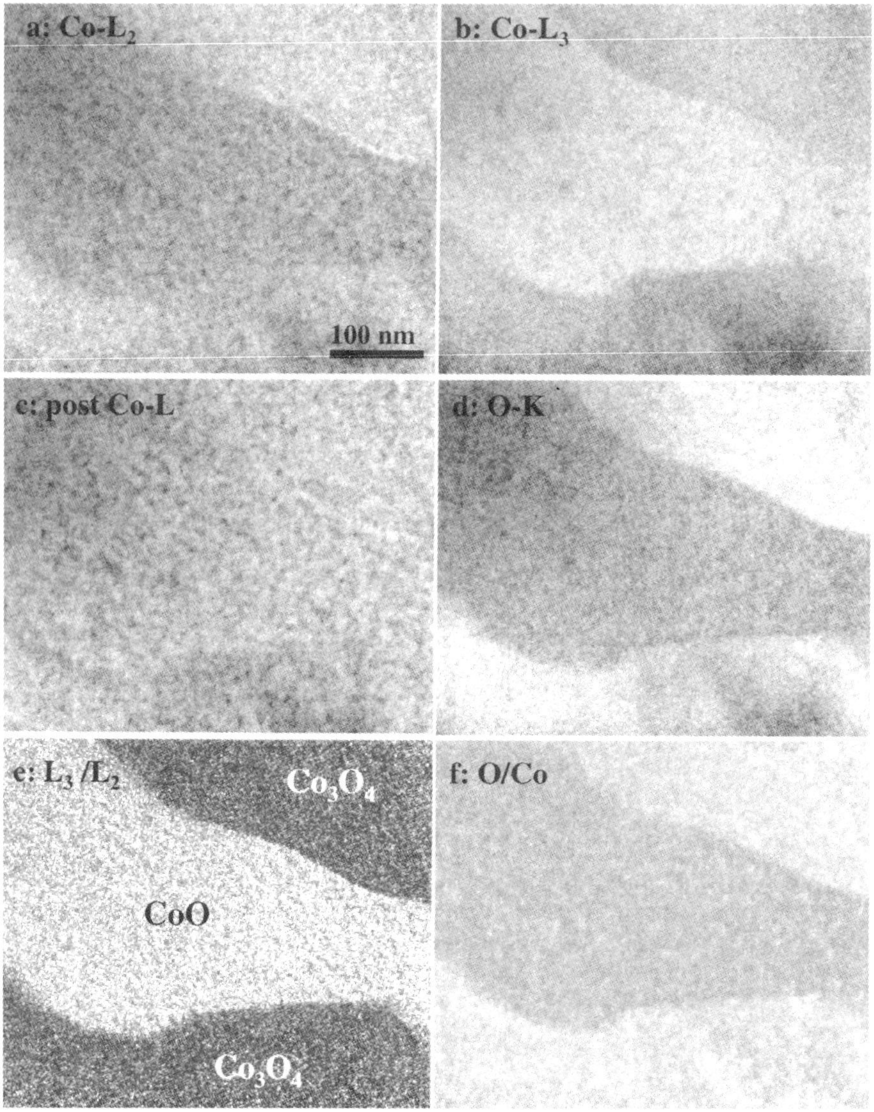

Figure 3-33. A group of images recorded from the same specimen region using signals of (a) the Co-L_3 edge, (b) the Co-L_2 edge, (c) the post Co-L_2 line, and (d) the O-K edge. (e) The processed L_3/L_2 image, displaying the distribution of the valence states. (f) The atomic concentration ratio image of O/Co. The continuous background contributed from the single atom scattering has been removed from the displayed Co-L_3 and Co-L_2 images. The O/Co image is normalized in reference to the standard composition of CoO for the low portion of the image in order to eliminate the strong influence on the ionization cross-section from the white lines. Each raw image was acquired with an energy window width of $\Delta = 12$ eV except for O-K at $\Delta = 24$ eV.

ters encapsulated in protective compact organic coating. By changing the length of the molecular chains, quantum transitions and insulator to conductor transition could be introduced, resulting in tunable electronic, optical and transport properties.

NCS's are characterized by unprecedented size-uniformity and translation and even orientation order, created through a multistage processing strategy involving self-assembly, annealing, etching of defects, reversible passivation by self-assembled monolayers, and macroscopic separation by size of the resulting assemblies. Particles that can be self-assembled usually have sizes smaller than 10 nm, it is in this size range that many exciting and unusual physical properties are enhanced. The nanocrystals are passivated by a monolayer of long-chain organic molecules, such as SR, where R $= n\text{–}C_nH_{2n+1}$, $n = 4, 6, 8, 12...$, called thiolates. The thiolate serves not only as the protection for the particles to avoid direct contact between the particles with a consequence of coalescing, but also the interparticle bonding.

3.8.1 Particle shape and assembling crystallography

If the nanocrystals can be taken as the building blocks, their 3-D assembling is unavoidably affected by the particle shape. In this section, the self-assembling of truncated octahedral Ag nanocrystals is taken as an example to illustrate the roles played by the particle shape in NCSs [2, 10]. Our main goal is to illustrate the formation of directional, interparticle molecular bonds by the thiolates. Figure 3-34a is an TEM image recorded from an Ag NCS deposited on a carbon substrate. The Ag nanocrystals have a truncated octahedral shape and they are oriented along the [110] of the Ag atomic lattice in the image, along which four {111} and two {100} facets are imaged edge-on. The unit cell of the NCS is also oriented along $[110]_s$ of *fcc*, where the subscript refers to the supercrystal. Therefore, the orientational relationship between the Ag particles and the nanocrystal lattice is $[110] \parallel [110]_s$ and $[001] \parallel [1\bar{1}0]_s$ (Figure 3-34c) [35, 69]. Accordingly, a structure model for the NCS is built (Figure 3-34d), in which the nanocrystals are oriented following an assembling principle of *face-to-face*.

Silver nanocrystals can also be dominated by icosahedral and decahedral. The spherical-like of these particles results in hexagonal closing packing [70]. The percentage of multiply twinned particle can be directly estimated from dark-field TEM image recorded using a small size objective aperture. Figure 3-35 shows an TEM image of a monolayer assembly of Ag nanocrystals. Many of the particles show double or even triple split contrast, suggesting these particles are not single crystalline, rather they contain twins. The particle would appear uniform in contrast if it is single crystalline. This is a easy method to estimate the population of the twinned particles. NCS's are rather rubust and they are stable to temperatures as high as 600 °C [75, 76].

3.8.2 Interdigitation of molecular bonds in nanocrystal self-assembly

A careful examination of the image shown in Figure 3-34a indicates there are some white spots in the image, corresponding to open channels formed by the thiolate molecules, suggesting that the thiolates are tethered on the faces of the nanocrystals and they are likely to be erected on the surface. The shortest distance between the face-to-face {100} facets of the two adjacent particles is only ~ 1.5 to 2 nm, almost equal to the 1.5 nm chain length of the thiolate molecules used for passivating the Ag nanocrystals. Therefore, the thiolate molecules tethered on the facets of the nanocrystals are likely to interpenetrate, forming the *interdigitative bonds*.

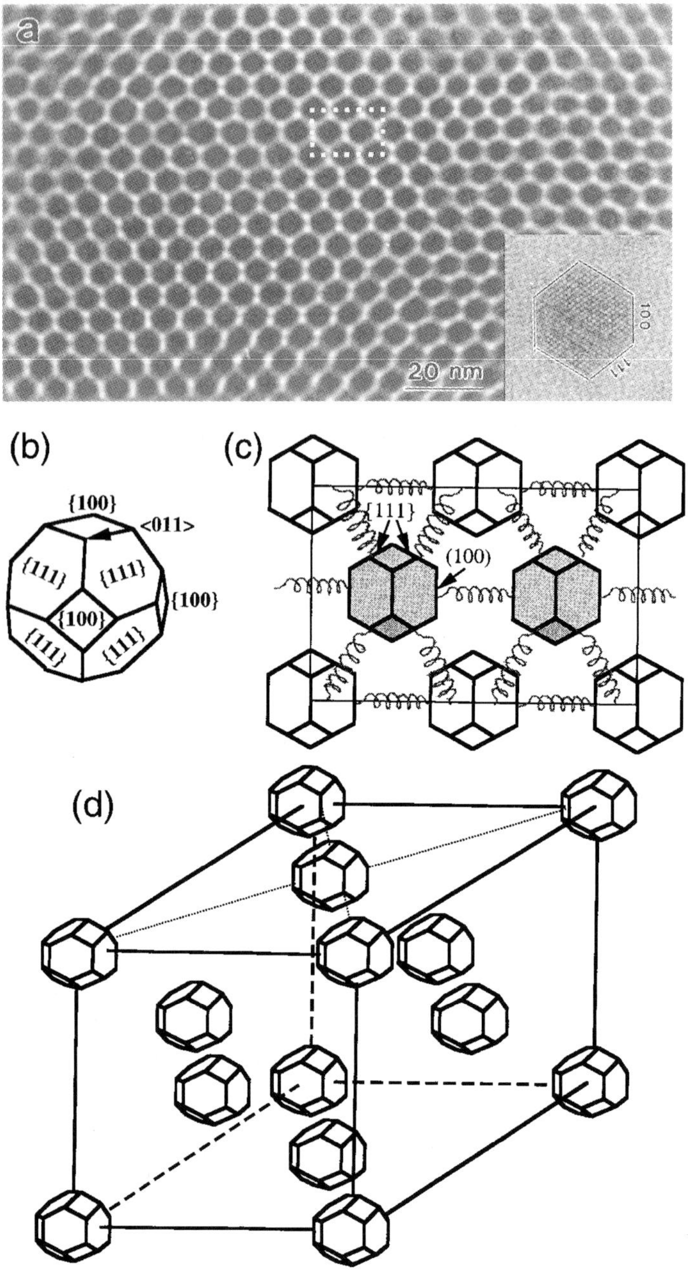

Figure 3-34. (a) TEM image of a face-centered cubic packed Ag nanocrystals whose shape is dominated by truncated octahedral (TO) (see the inset HRTEM image of a single Ag nanocrystal). The image displays the orientational relationship between the nanocrystals and the assembled superlattice. (b) A structural model of the TO particle, (c) the $[110]_s$ projection of the unit cell as indicated in (a), where the grey pasted particles are located at different z heights; and (d) the 3-D structural model of the fcc packing of nanocrystals.

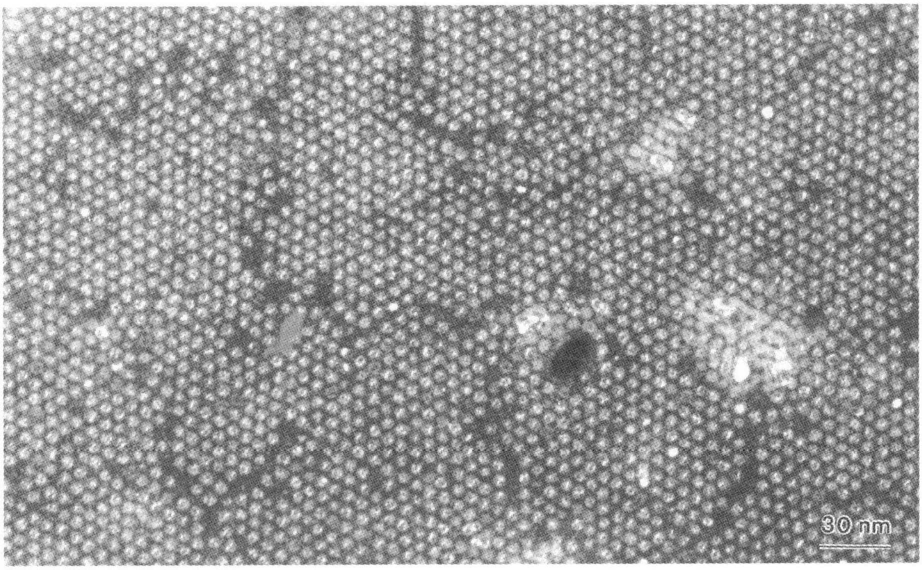

Figure 3-35. A dark-field TEM image recorded from a monolayer self-assembled Ag nanocrystals, showing the multiply split contract across each nanocrystal, and suggesting that the majority of the particles have twin structure.

A direct observation of the thiolates can be provided by the EF-TEM [77]. The thiolate molecules are composed of mainly carbon, thus the EF-TEM of the carbon K edge can give the distribution of the thiolates around the nanocrystals. For this analysis, Ag NCSs are deposited on an amorphous SiO_x substrate and the effects from the substrate can be removed by processing the experimental images acquired pre- and post-edge. The EF-TEM was performed for the Ag NCS oriented along $[110]_s$, which is the optimum orientation for imaging thiolate distribution between the particles. The EF-TEM image acquired using the carbon K edge from a Ag NCS gives an interesting contrast feature (Figure 3-36a). The projected carbon density between the particles shows a contrast pattern that is the strongest between the A and B types of particles, while the contrast is lower between the A and C or B and C types of particles. To interpret this phenomenon, we first construct the $[110]_s$ projection of the NCS based on the 3-D model given in Figure 3-36b, and the result is shown in Figure 3-36b. From the structural point of view, the molecular bonds tend to parallely align the facets on which they are tethered. For the nanocrystals A and B assembled by facing the {100} faces, in addition to the carbon density contributed by the interdigitative thiolates passivated on the {100} facets (which are edge-on while viewed along $[110]_s$), the thiolates passivated on the four {111} planes (not edge-on) also contribute to the projected carbon density although the {111} faces are at an angle with the projection direction. Therefore, the projected density of the thiolate molecules between particles A and B is expected to be higher than that between A and C (or B and C) if the size of {111} faces is the same as the that of {100} as well as the density of the thiolate passivation is the same on both {111} and {100}. With consideration the resolution of the EF-TEM of ~ 2 nm, the channels formed by the bundled thiolates may not be resolved

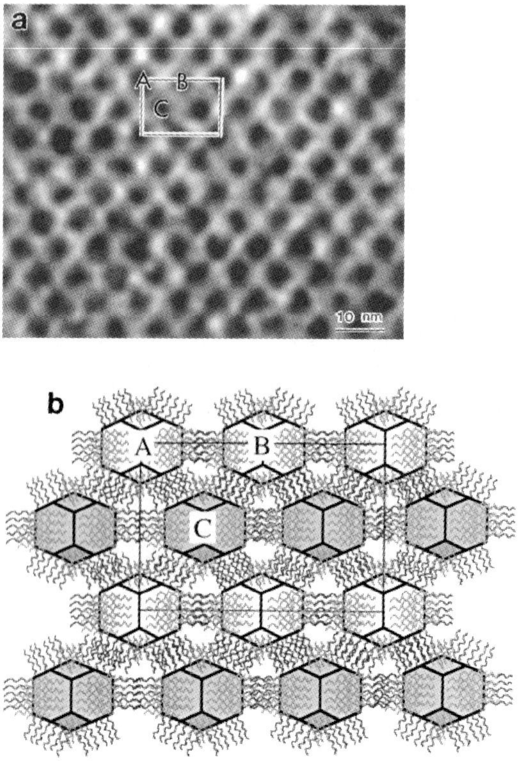

Figure 3-36. (a) Energy filtered TEM image of the Ag nanocrystal assembly using the carbon K ionization edge, showing the projected density of carbon along the $[110]_s$ of the NCS. (b) A model showing the distribution of passivated molecules between the nanocrystals. The projected unit cell is represented by a rectangle.

in this type of images. This result proves that the adsorbed molecules are distributed on the surfaces of the nanocrystals, and they form interdigitative bundling that holds the nanocrystal together.

3.9 Summary

This chapter was concentrated on introduction of TEM and associated techniques for characterization of nanophase materials. The objective was to review the versatility of TEM and the complimentary applications of the techniques. In additional to the high-resolution capability, TEM is also powerful in identifying and quantifying the chemical and electronic structure of nanomaterials. High-spatial resolution analysis is vitally important for solving many of the practical problems of nanomaterials. Spectroscopy analysis of the solid state effects and the valence states mapping are a new direction of quantitative microscopy. Further exploration in the fine structures in EELS is likely to reveal rich information about bonding in crystals and at interfaces. In-situ TEM is anticipated to be important for characterizing and measuring the properties of individual nanostructures, from which the structure-property relationship can be clearly registered to a specific nanoparticle/structure.

Acknowledgment

The author is grateful to his collaborators, M.A. El-Sayed, J.M. Petroski, T. Ahmadi, R.L. Whetten, S.A. Harfenist and I. Vezmar, for kindly provide some of the specimens. Research was partially sponsored by NSF grant DMR-9733160.

References

[1] P. Buseck, J.M. Cowley and L. Eyring eds. *High Resolution Transmission Electron Microscopy and Associated Techniques*, New York, London, Amsterdam, Oxford University Press, **1988**.
[2] J.M. Cowley, *Diffraction Physics*, 3rd revised ed., New York, London, Amsterdam, Elsvier Science B.V., **1995**.
[3] Z.L. Wang, and Z.C. Kang, *Functional and Smart Materials – Structural Evolution and Structure Analysis,* New York, Plenum Press **1998**, Chap. 6.
[4] P.B. Hirsch, A. Howie, R.B. Nicholson, D.W. Pashley and M.J. Whelan, *Electron Microscopy of Thin Crystals*, New York, Roberts E. Krieger Publishing Company, **1977**.
[5] O. Scherzer, *J. Appl. Phys.* **1949**, *20*, 20.
[6] Z.L. Wang, *Elastic and Inelastic Scattering in Electron Diffraction and Imaging,* New York, Plenum Pub. Co., **1995**, Chaps. 2 and 3.
[7] J.M. Cowley and A.F. Moodie, *Acta Cryst.* **1957**, *10*, 609.
[8] J.A. Ascencio, C. Gutiérrez-Wing, M.E. Espinosa, M. Marin, S. Tehuacanero, C. Zorrilla and M. José-Yacamán, *Surf. Sci.* **1998**, *396*, 349.
[9] D.J. Smith and L.D. Marks, *Pil. Mag. A* **1981**, *44*, 735.
[10] M. Yao and D.J. Smith, *J. Microsc.*, **1994**, *175*, 252.
[11] Z.L. Wang, T.S. Ahmadi and M.A. El-Sayed, *Surf. Sci.*, **1996**, *380*, 302.
[12] K. Takayanagi, Y. Tanishiro, M. Takahashi and S. Takahashi, *Surf. Sci.* **1985**, *164*, 367.
[13] L.D. Marks and D.J. Smith, *Nature* **1983**, *303*, 316.
[14] R.G. Linford, *Solid State Surface Science*, ed. M. Green, New York, Dekker **1973**, *2*, 1.
[15] S. Onodera, *J. Phys. Soc. Japan* **1991**, *61*, 2191.
[16] S. Ino, *J. Phys. Soc. Japan,* **1966**, *21*, 346
[17] J.G. Allpress and J.V. Sanders, *Surf. Sci.*, **1967**, *7*, 1
[18] L. D. Marks, *Rep. Prog. Phys.*, **1994**, *57*, 603, and the references there in.
[19] C.Y. Yang, *J. Cryst. Growth*, **1979**, *47*, 274; ibid, **1979**, *47*, 283.
[20] P.-A. Buffat, M. Flüeli, R. Spycher, P. Stadelmann and J.-P. Borel, *Faradat Discuss.*, **1991**, *92*, 173.
[21] S. Ino, *J. Phys. Soc. Japan,* **1969**, *27*, 941.
[22] J.M. Cowley, *Ultramicroscopy* **1992**, *41*, 335.
[23] D. Gabor, *Proc. Roy. Soc. London A* **1949**, *197*, 454.
[24] A. Tonomura, *Electron Holography,* New York, Springer-Verlag **1993**.
[25] H. Lichte, *Adv. Optical and Electron Microsc.* **1991**, *12*, 25.
[26] A. Tonomura, L.F. Allard, G. Pozzi, D.C. Joy and Y.A. Ono eds. *Electron Holography,* Amsterdam, Elsevier Science B.V. **1995**.
[27] G. Möllenstedt, G. *Adv. Optical and Electrons* **1991**, *12*, 1.
[28] B.G. Frost, L.F. Allard, E. Volkl, and D.C. Joy, in *Electron Holography,* eds., A. Tonomura, L.F. Allard, G. Pozzi, D.C. Joy and Y.A. Ono, Amsterdam, Elsevier Science B.V., **1995**, 169.
[29] A.K. Datye, D.S. Kalakkad, E. Vöikl and L.F. Allard, in *Electron Holography,* A. Tonomura, L.F. Allard, G. Pozzi, D.C. Joy and Y.A. Ono eds., Amsterdam, Elsevier, **1995**, 199.
[30] M. Mankos, J. M. Cowley and M.R. Scheinfein, *Mater. Res. Soc. Bulletin,* XX (October) **1995**, 45.
[31] Y. Aharonov and D. Bohm, *Phys. Rev.* **1959**, *115*, 485.
[32] T. Hirayama, J. Chen, Q. Ru, K. Ishizuka, T. Tanji and A. Tonomura, in *Electron Holography,* A. Tonomura, L.F. Allard, G. Pozzi, D.C. Joy and Y.A. Ono eds., Amsterdam, Elsevier, **1995**, 145.
[33] T.S. Ahmadi, Z.L. Wang, T.C. Green, A. Henglein and M.A. El-Sayed, *Science*, **1996**, *28*, 1924.
[34] R.L. Whetten, J.T. Khoury, M.M. Alvarez, S. Murthy, I. Vezmar, Z.L. Wang, C.C. Cleveland, W.D. Luedtke, U. Landman, *Adv. Materials*, **1996**, *8*, 428.
[35] Z.L. Wang, *Adv. Materials*, **1998**, *10*, 13.
[36] Z.L. Wang, J. Petroski, T. Green and M.A. El-Sayed, *J. Phys. Chem. B*, **1998**, *102*, 6145
[37] Z.L. Wang, M. Mohamed, S. Link and M.A. El-Sayed, *surf. Sci.* **1999**, in press.
[38] Y.-Y. Yu, S.-S. Chang, C.-L. Lee and C.R. Wang, *J. Phys. Chem. B,* **1997**, *101*, 6661.
[39] M.B. Mohamed, K.Z. Ismail, S. Link and M.A. El-Sayed, *J. Phys. Chem. B* **1998**, *102*, 9370.
[40] S. Frank, P. Poncharal, Z.L. Wang, and W.A. de Heer, *Science*, **1998**, *280*, 1744.

[41] P. Poncharal, S. Frank, Z.L. Wang, and W.A. de Heer, Proceeding of the International Conference on Small Clusters **1998**, in press.
[42] P. Poncharal, Z.L. Wang, D. Ugarte and W.A. de Heer, *Science*, **1999**, *283*, 1513.
[43] R.F. Egerton, *Electron Energy-Loss Spectroscopy in the Electron Microscope*, 2nd ed., New York, Plenum Pub. Co., **1996**.
[44] Z.L. Wang, *Philos. Mag. B* **1992**, *65*, 559.
[45] Z.L. Wang, *Micron* **1996**, *27*, 265; and **1997**, *28*, 505.
[46] P. Schattschneider, and B. Jouffrey, B., in *Energy Filtering Transmission Electron Microscopy* , ed. L. Reimer, Springer Series in Optical Science, New York, Springer Verlag **1995**, *71*, 151.
[47] P.M. Echenique, J. Bausells, A. and Rivacoba, *Phys. Rev. B* **1987**, *35*, 1521.
[48] T. Stöckli, J.-M. Bonard, A. Châtelain, Z.L. Wang and P. Stadelmann, *Phys. Rev. B* **1998**, *57*, 15599.
[49] T. Stöckli, Z.L. Wang, J.-M. Bonard, A. Châtelain and P. Stadelmann, *Phil. Mag. B*, in press, **1999**.
[50] C. Colliex, Adv. Optical and Electron Microscopy, ed. V.E. Cosslett and R. Barer, London, Academic Press, **1984**, 9, 65.
[51] H. Kurata and C. Colliex, *Phys. Rev. B*, **1993**, *48*, 2102.
[52] D.H. Pearson, C.C. Ahn, and B. Fultz, *Phys. Rev. B*, **1993**, *47*, 8471.
[53] J.L. Mansot, P. Leone, P. Euzen, and P. Palvadeau, *Microsc. Microanal. Microstruct*, **1994**, *5*, 79.
[54] J.A. Fortner, and E.C. Buck, *Appl. Phys. Lett.*, **1996**, *68*, 3817.
[55] S.J. Lloyd, G.A. Botton, and W.M. Stobbs, *J. Microsc.* **1995**, *180*, 288.
[56] J. Yuan, E. Gu, M. Gester, J.A.C. Bland, and L.M. Brown, *J. Appl. Phys.* **1994**, *75*, 6501.
[57] Z.L. Wang, J.S. Yin, J.Z. Zhang, and W.D. Mo, *J. Phys. Chem. B*, **1997**, *101*, 6793.
[58] Z.L. Wang, J.S. Yin, Y.D. Jiang, and J. Zhang, *Appl. Phys. Lett.*, **1997**, *70*, 3362.
[59] Z.L. Wang and J.S. Yin *Philos. Mag. B*, **1998**, *77*, 49.
[60] J.S. Yin and Z.L. Wang, *Phys. Rev. Lett.*, **1997**, *79*, 2570.
[61] L. Reimer ed., *Energy Filtering Transmission Electron Microscopy* , Springer Series in Optical Science, New York, Springer Verlag, **1995**.
[62] A.J. Craven, and C. Colliex, *J. Microsc. Spectrosc. Electron*, **1977**, *2*, 511.
[63] R.D. Leapman and J.A. Hunt, *Microscopy: The Key Research Tool*, published by The Electron Microsc. Soc. of America, **1992**, *22*, 39.
[64] J. Mayer, J. M. Plitzko, *J. Microsc.* **1996**, *183*, 2.
[65] P.E. Batson, N.D. Browning and D.A. Muller, *MSA Bulletin*, published by the Microscoy Society of America, **1994**, *24*, 371.
[66] Z.L. Wang, J. Bentley and N.D. Evans, *J, Phys. Chem. B*, **1999**, *103*, 751.
[67] J. Bentley, S. McKernan, C.B. Carter, and A. Revcolevschi, *Microbeam Analysis*, edited by J.T. Armstrong and J.R. Porter, **1993**, *2* (Suppl.), S286.
[68] J.I. Goldstein, D.E. Newbury, P. Echlin, D.C. Joy, A.D. Romig, C.E. Lyman, C. Fiori, and E. Lifshin, *Scanning Electron Microscopy and X-ray Microanalysis, a text for biologists, materials scientists and geologists,* New York, Plenum Press, **1992**.
[69] S.A. Harfenist, Z. L. Wang, M. M. Alvarez, I. Vezmar, R. L. Whetten, *J. Phys. Chem.*, **1996**, *100*, 13904.
[70] S.A. Harfenist, Z. L. Wang, M. M. Alvarez, I. Vezmar and R. L. Whetten, *Adv. Mater.*, **1997**, *9*, 817.
[71] L. Motte, F. Billoudet, E. Lacaze, J. Douin, M.-P. Pileni, *J. Phys. Chem B*, **1997**, *101*, 138.
[72] R.P. Andres, T. Bein, M. Dorogi, S. Feng, J.I. Henderson, C.P. Kubiak, W. Mahoney, R.G. Osifchin, R. Reifenberger, *Science*, **1996**, *272*, 1323.
[73] C.B. Murray, C.R. Kagan and M.G. Bawendi, *Science*, **1995**, *270*, 1335.
[74] C.P. Collier, R.J. Saykally, J.J. Shiang, S.E. Henrichs and J.R. Heath, *Science*, **1997**, *277*, 1978.
[75] J.S. Yin and Z.L. Wang, *J. Phys. Chem.*, **1997**, *101*, 8979.
[76] S.A. Harfenist and Z. L. Wang, *J. Phys. Chem B*, **1999**, *103*, 4342.
[77] Z.L. Wang, S.A. Harfenist, R. L. Whetten, J. Bentley and N.D. Evans, *J. Phys. Chem. B*, **1998**, *102*, 3068.

4 Scanning Transmission Electron Microscopy of Nanoparticles

Jingyue Liu

Ultimately, all materials and living organisms have their origins in the collective assemblies of a small number of atoms or molecules. Human beings have been fascinated by the interior world: the secrets of cells and the building blocks of matter. Ever since various types of microscopes were discovered, they have been the primary instruments for helping us to directly observe and understand matter or cells on an ever-decreasing scale. By understanding what the building blocks are and how they are arranged together to possess unique properties or to display specific functions, we hope to modify, manipulate, and sculpt matter at nanoscopic dimensions for desired purposes.

To understand the fundamental properties of nanoparticles or other nanosystems, it is necessary to characterize their structures at a nanometer or atomic level. This, in turn, allows us to understand the synthesis-structure-property relationships of nanoparticle systems, thus achieving the ultimate goal of the molecular engineering of nanoparticles for important industrial applications.

In Chapter 3, we discussed transmission electron microscopy (TEM) techniques and the use of these powerful techniques for characterizing nanoparticles. In this chapter, we will discuss scanning transmission electron microscopy (STEM) and associated imaging, spectroscopy, and diffraction techniques, as well as the applications of these STEM techniques to the characterization of nanoparticles.

4.1 Introduction to STEM and associated techniques

STEM instruments are tailor-made to provide structural, chemical, and morphological information about individual nanoparticles on a nanometer or atomic level. A remarkable capability of STEM instruments is the formation of high-brightness electron probes with diameters < 0.2 nm at 100 keV and as small as 0.13 nm at 300 keV. To form such an electron nanoprobe, the use of a field-emission gun (FEG) is necessary to provide sufficient signal strength for viewing and recording STEM images, spectra, and diffraction patterns. The FEG source has a brightness of at least 10^3 greater than that of a W or a LaB_6 filament, which is typically used in conventional TEM instruments, but with an effective electron source size of only 4–5 nm in diameter. Another attractive feature of STEM is its great flexibility in the detection system. Different signals generated from the same sample area can be collected either independently or simultaneously and can be analyzed in parallel to yield complementary information about the sample. Figure 4-1 shows a schematic diagram illustrating the available signals that can be used in a STEM instrument to give useful information about the sample.

Unlike in TEM where a stationary, parallel electron beam is used to form images, the STEM is a mapping device. In a STEM instrument, a fine electron probe, formed by using a strong objective lens to de-magnify a small electron source, is scanned over a specimen in a two-dimensional raster. Signals generated from the specimen are

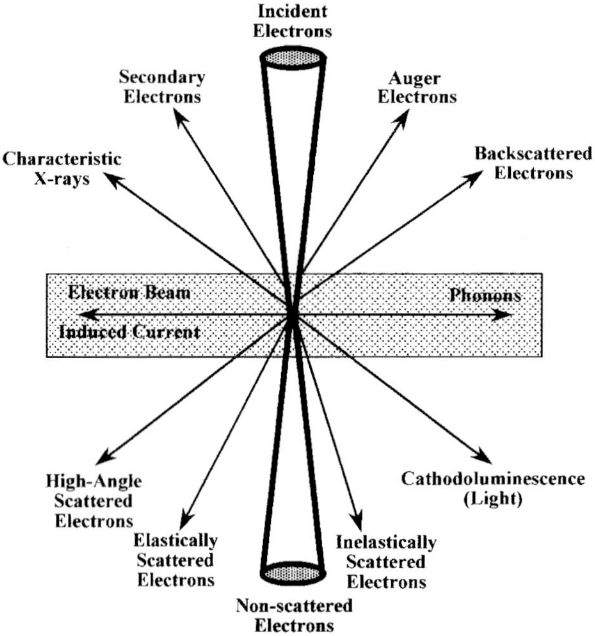

Figure 4-1. Signals generated inside a STEM microscope when a high-energy electron probe interacts with a thin specimen. These signals can be collected, individually or simultaneously, and analyzed to give useful information about the structure and chemistry of the sample.

detected, amplified, and used to modulate the brightness of a second electron beam that is scanned synchronously with the STEM electron probe across a cathode-ray-tube (CRT) display. Therefore, a specimen image is mapped onto the CRT display for observation (Fig. 4-2). If the area scanned on the sample is A_s and the corresponding area on the CRT display is A_d, then the magnification (M) of a STEM image is given by $M = A_d/A_s$. The STEM magnification is purely geometric in origin and can be easily changed by varying the scanned area on the sample.

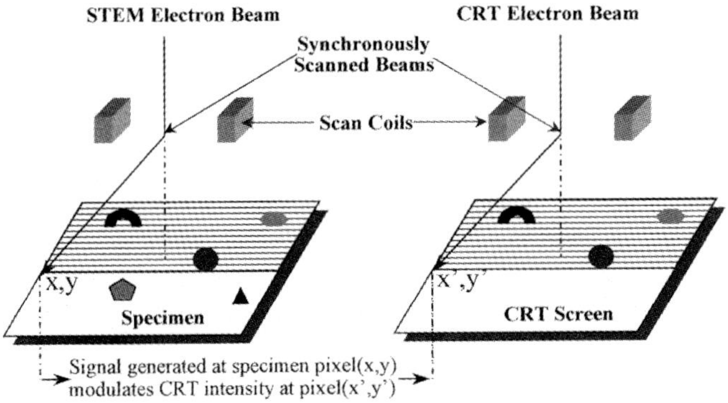

Figure 4-2. Schematic diagram illustrates the formation of a STEM image: signals generated from a thin specimen are detected, amplified, and used to modulate the brightness of a second electron beam scanned synchronously with the STEM electron probe across a cathode-ray-tube (CRT) display. A specimen image is "mapped" to the CRT display for observation.

Since STEM is a serial recording instead of a parallel recording system, the whole process of generating a STEM image is much slower than that in TEM. A STEM image usually builds up over several seconds to several minutes, depending on the types of signals used to produce the image. Unlike in TEM, there is no rotation between the object and image planes, and the microscope magnification can be changed without refocusing the electron beam to obtain a focused image. The resolution of STEM images is primarily determined by the incident probe size, the stability of the microscope, and the inherent properties of the signal generation processes.

Multiple imaging and analytical detectors have been developed to simultaneously collect several signals that can be displayed individually or combined in perfect register with each other. This unique capability makes the STEM a powerful nanoanalytical tool since multiple views of a sample, in different imaging, diffraction, and spectroscopy modes, can be collected, analyzed, and compared in a single pass of the incident electron beam. Figure 4-3 illustrates the available detectors commonly used for collecting imaging and analytical signals in dedicated STEM instruments.

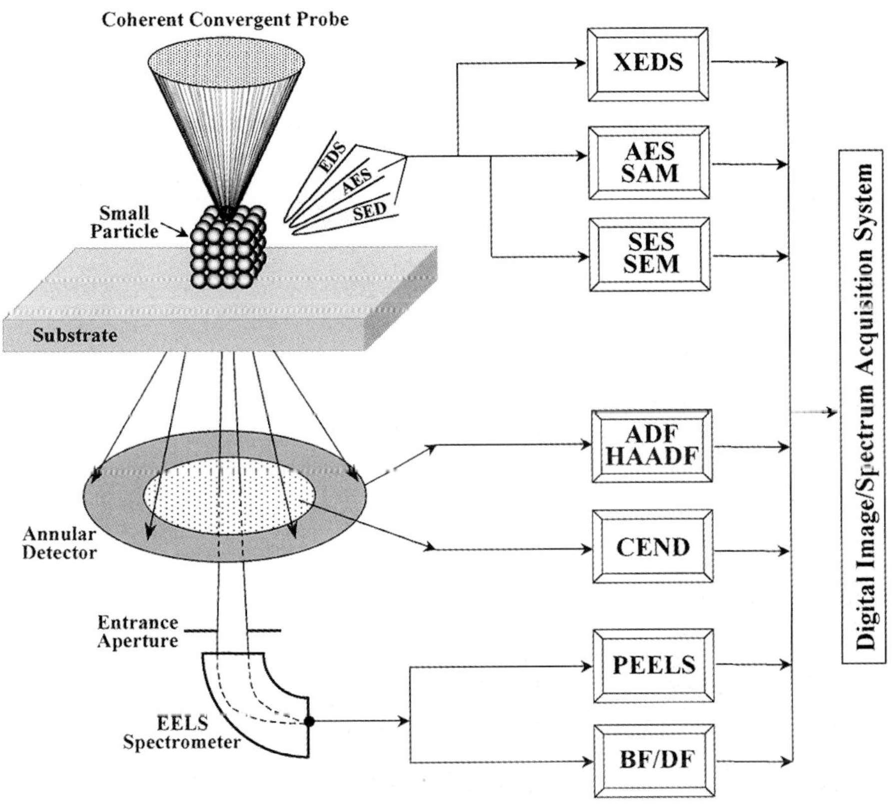

Figure 4-3. Schematic diagram illustrates imaging, diffraction, and spectroscopy modes commonly used in a dedicated STEM instrument: X-ray Energy Dispersive Spectroscopy (XEDS); Auger Electron Spectroscopy (AES) and Scanning Auger Microscopy (SAM); Secondary Electron Spectroscopy (SES) and Secondary Electron Microscopy (SEM); Annular Dark-Field (ADF) and High-Angle Annular Dark-Field (HAADF) microscopy; Coherent Electron Nano-Diffraction (CEND); Parallel Electron Energy-Loss Spectroscopy (PEELS); and Bright-Field (BF) and Dark-Field (DF) microscopy.

By collecting high-angle scattered electrons with an annular detector, high-angle annular dark-field (HAADF) images (or Z-contrast images) can be formed to provide information about structural variations across the sample at an atomic level. Electron energy-loss spectroscopy (EELS) and X-ray energy-dispersive spectroscopy (XEDS) can give quantitative data describing changes of elemental composition, electronic structure, or state of oxidation associated with inhomogeneous structures of the sample. The combination of XEDS and EELS with HAADF imaging technique can provide detailed information on the composition, chemistry, and electronic and crystal structure of nanoparticles with atomic resolution and sensitivity.

Information about surface topography or surface chemistry of nanoparticles can be obtained by collecting and analyzing secondary electron (SE) and Auger electron (AE) signals. By positioning an electron nanoprobe at the area of interest, diffraction patterns from individual nanoparticles can be acquired to provide information about the crystallographic structure of nanoparticles. The structural relationship of nanoparticles to the surrounding materials and, in some cases, the shape of nanoparticles can also be extracted from nanodiffraction patterns. The powerful combination of atomic resolution HAADF imaging with nanospectroscopy and nanodiffraction techniques has proved invaluable in solving a wide range of materials problems.

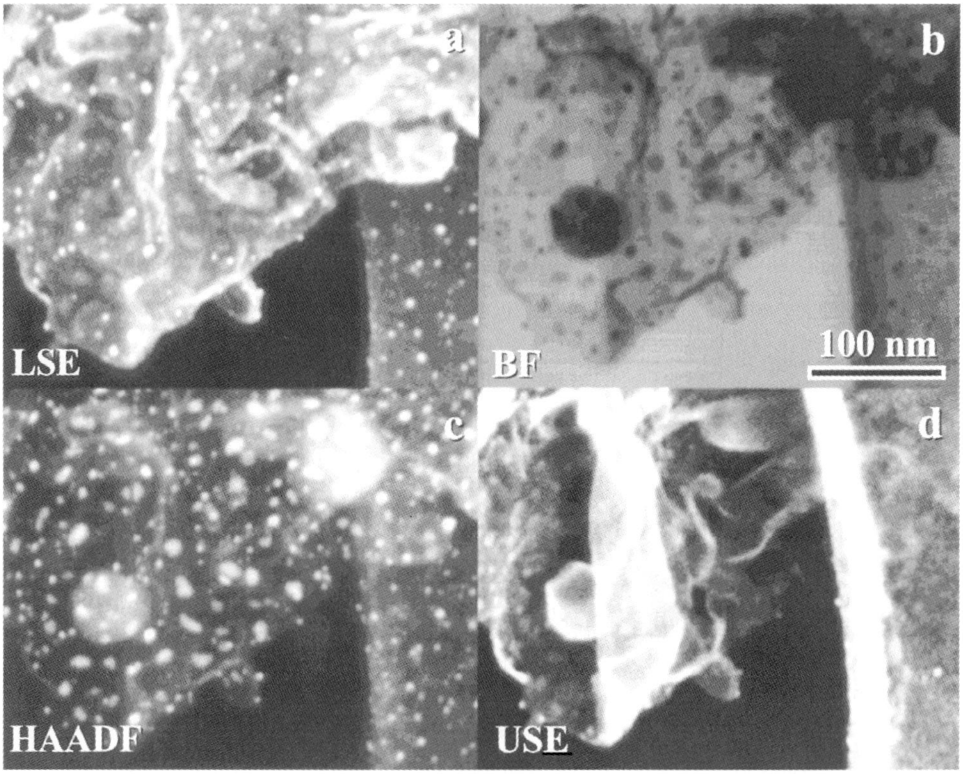

Figure 4-4. Simultaneously acquired STEM images of a supported catalyst: a) entrance-surface SE image; b) BF image; c) HAADF image; d) exit-surface SE image. The size distribution and the relative locations of the metal particles, as well as the morphology of the catalyst support, are clearly revealed in this set of STEM images.

Figure 4-4 shows a set of images, acquired simultaneously, of a supported metal catalyst. Figure 4-4a is an SE image of the entrance surface of the catalyst showing small particles and a detailed surface morphology of the catalyst support. Figure 4-4b is the corresponding bright-field (BF) STEM image, revealing small particles with a dark contrast. Metal particles of various shapes and sizes, dispersed in or on the catalyst support, are clearly revealed in the HAADF image (Fig. 4-4c). The exit-surface SE image (Fig. 4-4d) reveals an unusual surface topography of the catalyst support. Detailed analyses of such sets of images can provide important information about the catalyst: distribution of sizes and relative locations of various types of metal particles present in the supported catalyst. The SE images clearly indicate that the small, spherical particles are located on the entrance side of the support. The larger metal particles with irregular shapes are located in the interior of the catalyst support since these particles are clearly revealed in the HAADF image but are not shown in either the entrance- or the exit-surface SE images. XEDS analysis of these particles can give information about the composition of the individual nanoparticles. The correlation of HAADF images with SE images and XEDS spectra is very effective in identifying which type of the metal particles is exposed to the reacting molecules during a catalytic reaction.

4.2 STEM instrumentation

4.2.1 Dedicated STEM and STEM attachment in TEM

There are two types of STEM instruments commonly used by electron microscopists: dedicated STEM (DSTEM) and STEM attachment in TEM. The DSTEM instruments, which were exclusively made by VG Microscopes (UK), have unique designs and capabilities. A DSTEM microscope uses a cold field-emission gun to generate high-brightness electron probes with sub-nanometer sizes. The sizes of electron probes used in TEM/STEM instruments are generally much larger than those in DSTEM instruments. Furthermore, the operation of STEM attachments is not as flexible as that of DSTEM.

However, recent advances in field-emission technology and the increasing use of field-emission guns in high-resolution 200 kV TEM instruments have made the STEM attachment and associated STEM techniques dominant features in field-emission TEM (FE-TEM) microscopes. When operated in the STEM mode, the recent generation of FE-TEM instruments can provide high-brightness electron probes with sizes as small as those obtainable in DSTEM instruments. HAADF images can now be obtained in FE-TEM microscopes with a resolution similar to, or, in some instances, better than, that obtainable in DSTEM instruments. The distinction between DSTEM and STEM attachments in FE-TEM is rapidly diminishing. In the following, we will focus our discussion exclusively on DSTEM instruments. However, the fundamental principles of DSTEM and associated techniques discussed in this chapter equally apply to STEM operation in FE-TEM. Unless stated otherwise, we will use the acronym STEM to represent DSTEM for the rest of this chapter.

4.2.2 Principal features of a STEM instrument

In a STEM instrument, the cold field-emission gun is generally operated at 100 kV or 300 kV and is placed at the bottom of the microscope column to minimize mechanical disturbances of the field-emission tip. Two condenser lenses and a strong objective lens are usually used to form a small electron probe on the specimen. A virtual objective aperture (VOA) is usually placed before the condenser lenses to control the convergent angle of the incident electron probe. The use of VOA is desirable for reducing stray X-rays, generated inside the microscope column, and hence simplifying the interpretation of XEDS spectra and X-ray elemental maps. A real objective aperture situated just before the specimen is sometimes used to define the angle of incident beam convergence. The size and the intensity of the high-energy electron probe can be manipulated by adjusting the two condenser lenses and by selecting the proper size of the VOA or the real objective aperture.

In some STEM instruments, the electrons, passing through the specimen, directly reach the detector plane without the use of any post-specimen lenses. It is, however, desirable to have post-specimen lenses to offer great flexibility for effectively utilizing various detector configurations and for conveniently observing and recording nanodiffraction patterns.

The easy access to the top of the STEM column provides necessary versatility for modification or construction of various STEM detectors. Interchangeable annular detectors, for example, can be installed to provide flexibility for ADF imaging or for special imaging modes using configured detectors. The attachment of a series EELS or a parallel EELS (PEELS) detector to the top of the microscope column makes it possible to analyze the chemistry of the sample at an atomic resolution. It also allows BF or dark-field (DF) imaging with only elastically scattered electrons or with other selected energy-loss electrons.

A scintillator screen and a low-light level TV camera, together with a VCR, can be used to view or record nanodiffraction patterns at a TV rate. A charge coupled device (CCD) system can also be used for quantitative recording of nanodiffraction patterns, shadow images, or electron holograms.

For effectively collecting characteristic X-rays, a retractable, windowless XEDS spectrometer is usually attached to the column of a STEM instrument. Because of the small volume probed by the electron nanoprobe, the XEDS detector has to be placed close to the sample region to increase the strength of the collected X-ray signal. Two XEDS detectors can be interfaced to the microscope column to significantly increase the strength of the collected X-ray signal.

A stable operation of a cold field-emission gun requires the vacuum in the gun chamber to be better than 10^{-10} torr. The column vacuum is generally better than 10^{-8} torr to prevent significant back streaming of gas molecules into the gun chamber and to reduce the effects of contamination on the specimen surface. Most of the STEM instruments can be baked at moderate temperatures for extended periods to obtain a high vacuum.

Some STEM instruments were specifically designed and constructed to have a true UHV system for surface studies. These instruments have a vacuum better than 10^{-10} torr throughout the whole microscope system and have elaborate auxiliary chambers for preparing, testing, and manipulating specimens in situ. One of these special STEM instruments is the microscope code-named MIDAS (a Microscope for Imaging, Diffraction, and Analysis of Surfaces) which was specifically designed for the study of nanostructures of surfaces and small particles with high spatial resolution. The various

STEM imaging, analytical, and diffraction techniques can be performed on clean nanoparticles or thin films, prepared inside the specimen chamber and treated at different temperatures or with various gases. In addition, secondary and Auger electrons, emitted from both the entrance and the exit surfaces of a sample, can be collected to form surface images. Figure 4-5 shows a photograph of the essential components of the MIDAS system.

Because of the intrinsic nature of a mapping device, the STEM is ideal for digital imaging and for on-line or off-line processing of images, spectra, and nanodiffraction patterns. Signals from several detectors can be digitally acquired either simultaneously or independently; different signals can also be combined together by addition, subtraction, multiplication, or other mathematical manipulations to gain insight about the structure of the sample. These digital images or spectra can be electronically transferred to remote locations through the Internet or the World Wide Web for fast dissemination of vital information.

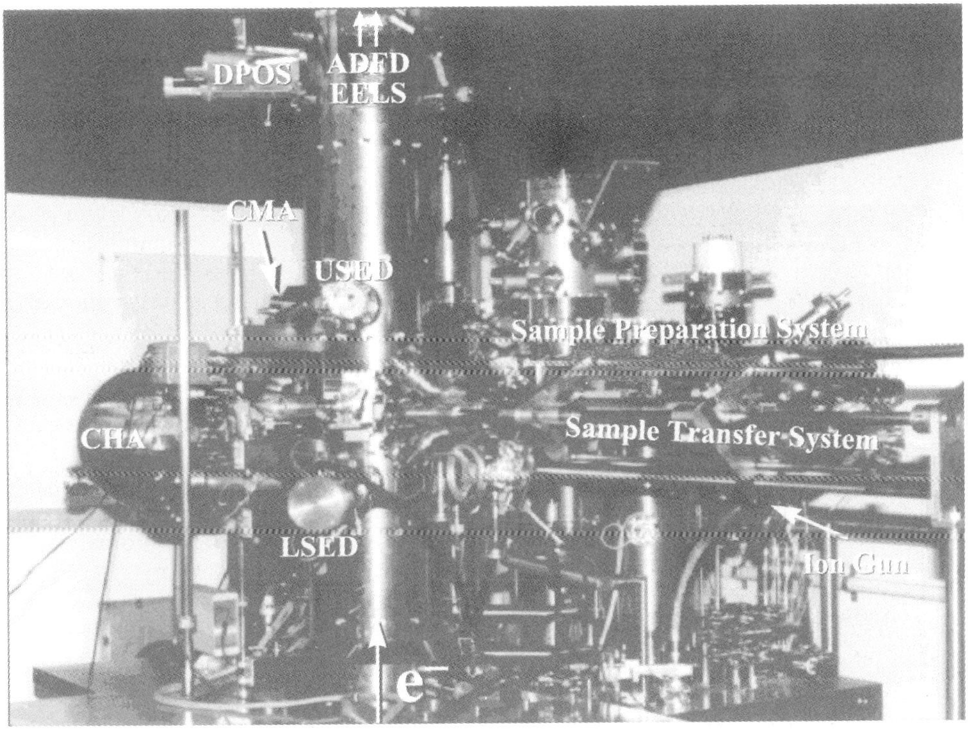

Figure 4-5. A photograph shows the main features of a specially designed UHV STEM HB-501S housed at Arizona State University (code-named MIDAS—Microscope for Imaging, Diffraction, and Analysis of Surfaces). The electron beam travels from bottom to top in dedicated STEM instruments. Annular Dark-Field Detector (ADFD); Electron Energy-Loss Spectrometer (EELS); Diffraction Pattern Observation Screen (DPOS); Cylindrical Mirror Analyzer (CMA); Concentric Hemispherical Analyzer (CHA); Upper Secondary Electron Detector (USED); Lower Secondary Electron Detector (LSED).

4.3 Imaging with high-energy electrons

4.3.1 The principle of reciprocity

The principle of reciprocity developed in the light optics can be equally applied to electron optical systems [1–2]. The wave amplitude at a point P due to a point source at Q is identical to the wave amplitude at Q due to a point source at P (Fig. 4-6). The essential components of a STEM imaging system are similar to those of a TEM microscope: the ray diagram of STEM is the reciprocal of that of TEM. This is demonstrated with the aid of the schematic ray-diagram of Fig. 4-6. The STEM detector replaces the TEM electron source; the STEM gun is placed in the detector plane of the TEM; and the scanning system translates the STEM source to cover the TEM recording plate. Therefore, for a particular detector configuration, the contrast of STEM images can often be inferred by finding the equivalent TEM geometry. For example, BF STEM images obtained by collecting electrons scattered into a small-angle θ can be interpreted the same way as those of BF TEM images obtained with an illumination angle of θ. Images obtained with a large BF STEM detector are similar to those obtained in TEM with a large illumination angle of the incident beam. The principle of reciprocity makes it possible to apply imaging theories developed in TEM to the interpretation of the corresponding STEM images.

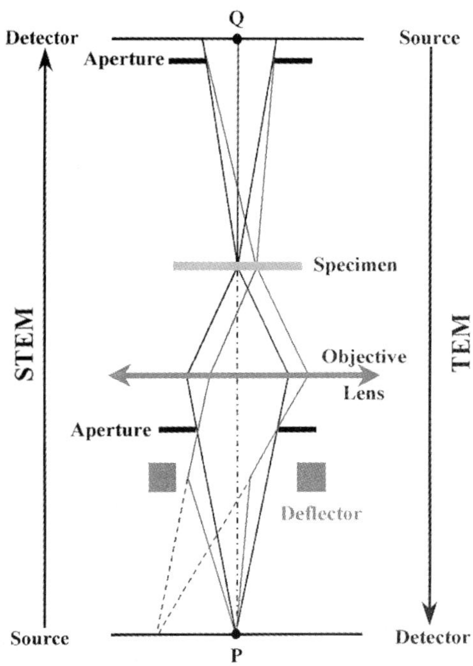

Figure 4-6. Schematic ray-diagram illustrates the Principle of Reciprocity in electron optics: the ray diagram of STEM is the reciprocal of that of TEM. The STEM detector replaces the TEM electron source; the STEM gun is placed in the detector plane of the TEM; and the scanning system translates the STEM source to cover the TEM recording plate.

4.3.2 Theoretical background

When a finely focused electron probe interacts with a thin specimen, the high-energy incident electrons are scattered by the sample electrons and nuclei. The amplitude distribution of the incident electrons at the exit surface of the sample can be described by a wave function $\Psi(\mathbf{K})$. The variable **K** is a two-dimensional vector in the reciprocal space with $|\mathbf{K}| = 2\sin(\theta/2)/\lambda$ (θ is the scattering angle and λ is the wavelength of the incident electrons). When the electron probe is scanned across the sample, variations of $\Psi(\mathbf{K})$ carry information about the electron beam-specimen interactions. If the exit wave function $\Psi(\mathbf{K})$ of the scattered high-energy electrons can be determined, we can extract structural information about the sample of interest. It is, however, not possible to directly measure $\Psi(\mathbf{K})$. What can be obtained experimentally are images of the sample formed by collecting the directly transmitted or scattered electrons. The observed image intensity, $I(\mathbf{X})$, as a function of the beam position **X** is related to $\Psi(\mathbf{K})$ by:

$$I(\mathbf{X}) = \int D(\mathbf{K})\, |\Psi(\mathbf{K}, \mathbf{X})|^2\, \mathrm{d}\mathbf{K} \tag{4-1}$$

where $D(\mathbf{K})$ is the transmission function of the detector. The amplitude function $\Psi(\mathbf{K}, \mathbf{X})$, to first approximation [3, 4], can be expressed as:

$$\Psi(\mathbf{K}, \mathbf{X}) = Q(\mathbf{K}) * T(\mathbf{K}) \exp\{2\pi i \mathbf{K} \cdot \mathbf{X}\} \tag{4-2}$$

where $Q(\mathbf{K})$ is the Fourier transform of the object transmission function, $q(\mathbf{x})$, of the sample and the * symbol represents convolution. The transfer function of the microscope, $T(\mathbf{K})$, is given by:

$$T(\mathbf{K}) = A(\mathbf{K}) \exp\{i\chi(\mathbf{K})\} \tag{4-3}$$

The aperture function, $A(\mathbf{K})$, is given by:

$$A(\mathbf{K}) = \begin{cases} 1 & \text{for } \mathrm{K} < \mathrm{K}_0 \\ 0 & \text{for } \mathrm{K} > \mathrm{K}_0 \end{cases} \tag{4-4}$$

The aberration function of the objective lens, $\chi(\mathbf{K})$, is given by:

$$\chi(\mathbf{K}) = \pi \Delta \lambda K^2 + \pi C_s \lambda^3 K^4/2 \tag{4-5}$$

where Δ is the defocus value, C_s is the spherical aberration coefficient of the objective lens, and K_0 is the cut-off wave-vector determined by the size of the objective aperture.

The amplitude distribution of the incident probe, $t(\mathbf{x})$, is determined by the Fourier transform of $T(\mathbf{K})$ which is determined by the aperture function $A(\mathbf{K})$ and the aberration function $\chi(\mathbf{K})$ of the objective lens. The probe size, therefore, depends on the spherical aberration coefficient of the objective lens, the wavelength of the incident electrons, the size of the objective aperture, and the defocus of the electron beam.

To determine the transmission function of the object, $q(\mathbf{x})$, requires a reconstruction of the exit wave function and a solution to the inverse dynamic diffraction problem which is beyond the scope of this book. In order to gain insight about the image resolution and contrast, however, various types of approximations can be made to simplify $q(\mathbf{x})$. In the phase object approximation [3, 4], for example, $q(\mathbf{x})$ can be expressed as:

$$q(\mathbf{x}) = \exp(-i\sigma\phi(\mathbf{x})) \tag{4-6}$$

where $\sigma = \pi/(\lambda E_0)$ = interaction constant, E_0 = accelerating voltage, and $\phi(\mathbf{x})$ = projected specimen potential along the incident beam direction. For most practical studies, however, rigorous dynamical calculations have to be performed to determine the image of specific specimen structures.

The detector function $D(\mathbf{K})$ plays a significant role in determining the final form of STEM images. For example, if $D(\mathbf{K}) \equiv 1$ for all scattering angles, the STEM image is formed by collecting all the high-energy electrons penetrating through a thin sample. Then, the image intensity $I(\mathbf{X})$ does not vary with the beam position $\mathbf{X}$ because of the conservation of the total number of high-energy electrons (we neglect here the effect of backscattered electrons which is negligible for very thin specimens). Therefore, no contrast will be observed in the STEM image, and no information about the specimen can be inferred.

If $D(\mathbf{K}) = \delta(\mathbf{K})$ or $\delta(\mathbf{K}-\mathbf{G})$ where $\mathbf{G}$ is a reciprocal lattice vector, the Eq. (4-1) reduces to: $I(\mathbf{X}) = |\Psi(\mathbf{0}, \mathbf{X})|^2$ or $|\Psi(\mathbf{G}, \mathbf{X})|^2$. This is exactly same as for BF or DF TEM imaging with parallel illumination. For other detector configurations, the evaluation of the Eq. (4-1) is not straightforward. Various types of approximations, however, can be made to gain insight into the characteristics of the images formed by collecting the corresponding signals [4].

With configured STEM detectors we can, nevertheless, rewrite the Eq. (4-1) as:

$$I_i(\mathbf{X}) = \int_{K_i}^{K_{i+1}} D_i(\mathbf{K})|\Psi(\mathbf{K}, \mathbf{X})|^2 \, d\mathbf{K} \tag{4-7}$$

$$\sum_i \int_{K_i}^{K_{i+1}} D_i(\mathbf{K})|\Psi(\mathbf{K}, \mathbf{X})|^2 \, d\mathbf{K} = 1 \tag{4-8}$$

where the summation is over the whole diffraction plane.

The resolution and contrast of STEM images depend on the configuration of the STEM detector used to form these images. By selecting the shape and the size of the STEM detector, a variety of imaging modes can be used to extract complementary information about the sample.

When a small electron probe is positioned on the sample, all information carried by high-energy electrons is contained in the whole diffraction pattern. The intensity distribution in the diffraction plane varies with the change of the incident beam position provided the electron probe is small enough to resolve the lattice spacings of a crystal. The total integrated intensity across the whole diffraction plane does not vary with the change of the probe position.

By digitally recording the whole diffraction pattern with energy (E) discrimination for each pixel (probe position $\mathbf{X}$) on the sample, a five-dimensional function $I(\mathbf{K}, \mathbf{X}, E)$ can be generated. All information about the specimen can be extracted by off-line processing of these digitally stored, energy-selected diffraction patterns. By selecting certain portion(s) of the scattered electrons as an input signal, various types of images can be formed to give information about the structure and the chemistry of the sample with atomic resolution. This process, however, needs a tremendous amount of computer work, fast image-acquisition systems, a large collection of data, and a high stability of the microscope parameters. Impressive progress has been made in the last few years [5–7]. In the following, we will focus our discussion on a few simple, but very powerful, imaging modes which can provide high-resolution structural information about nanoparticles.

4.3.3 High resolution BF and DF STEM imaging

When a finely focused STEM probe interacts with a thin crystal oriented along a principal zone-axis, an electron diffraction pattern consisting of a set of convergent beam discs is obtained. Each disc subtends the same semi-angle α, determined by the size of the objective aperture, at the specimen (Fig. 4-7a). If $\alpha > \theta_B$ (θ_B is the Bragg diffraction angle of the diffracting planes), then the convergent beam diffraction discs overlap. For thin, perfect crystals, the electron intensity within non-overlapping regions (e.g., the region indicated by the numeral 1 in Fig. 4-7b) is independent of the probe position and the aberrations of the probe forming lens [8]. The electron intensity within regions where discs do overlap depends on the probe position, the lens aberrations, and the defocus values of the objective lens. The intensity modulations in the

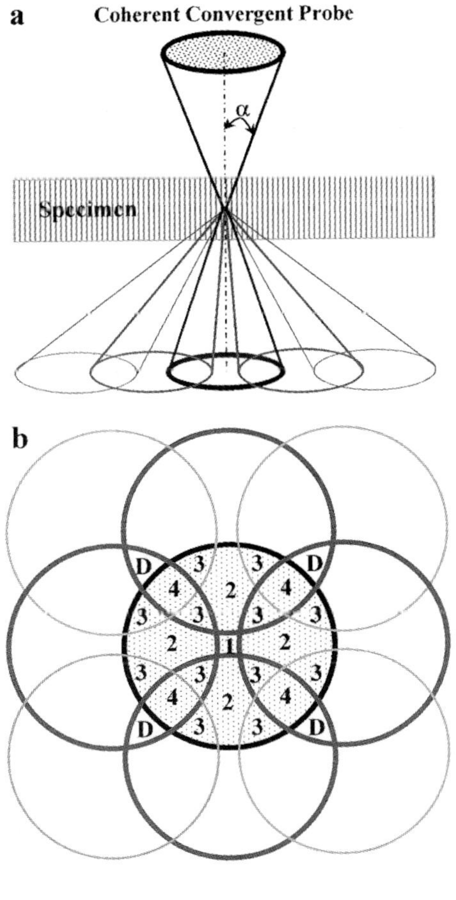

Figure 4-7. Schematic diagrams illustrate the overlap of diffraction discs in a convergent beam, electron diffraction pattern: (a) side view, (b) plane view. Lattice fringes can be obtained by positioning a small STEM detector at any point in the regions of overlapping discs. The numeral 2 stands for two-beam interference, 3 for three-beam interference, 4 for four-beam interference, and the letter D indicates detector positions for dark-field lattice imaging.

overlapping regions are caused by coherent interference of high-energy electrons with different incident directions. The intensity at any point in the overlapping regions varies, in simple cases, sinusoidally with the periodicity ($\lambda/2\theta_B$) of the crystal lattice [8].

If a STEM detector is positioned at any point in the overlapping regions, lattice fringes can be acquired by scanning the electron probe across the sample. Two-dimensional lattice fringes can be obtained by positioning the STEM detector at a point where three or more non-systematic diffraction discs overlap. These multiple-beam regions are labeled as numeral 3 (three-beam interference) and numeral 4 (four-beam interference) in Fig. 4-7b. High-resolution BF STEM images are, by invoking the principle of reciprocity, similar to high-resolution TEM images discussed in Chapter 3. Figure 4-8 shows such a BF STEM lattice image of oxide nanoparticles dispersed on a thin carbon film.

High-resolution DF STEM images can be easily obtained by moving the detector to a point outside the directly transmitted disk. For example, a two-dimensional DF STEM lattice image can be obtained by shifting the STEM detector to the position D labeled in Fig. 4-7b. DF STEM imaging technique is useful for identifying small particles in supported metal catalysts, defects in extended crystals, and different phases in polycrystalline nanophase materials.

The contrast of high-resolution STEM images varies with the displacement of the STEM detector. The movement of the STEM detector corresponds to beam tilt in TEM. In STEM, however, the relative shift of the BF detector is easily accomplished by deflecting the whole diffraction pattern with the use of scanning coils. Unlike beam tilt in TEM, the movement of scanning coils does not disturb the optical alignment of STEM microscopes. Thus, the contrast of specific features of a sample can be conveniently enhanced or reduced by shifting the position of the STEM detector without the

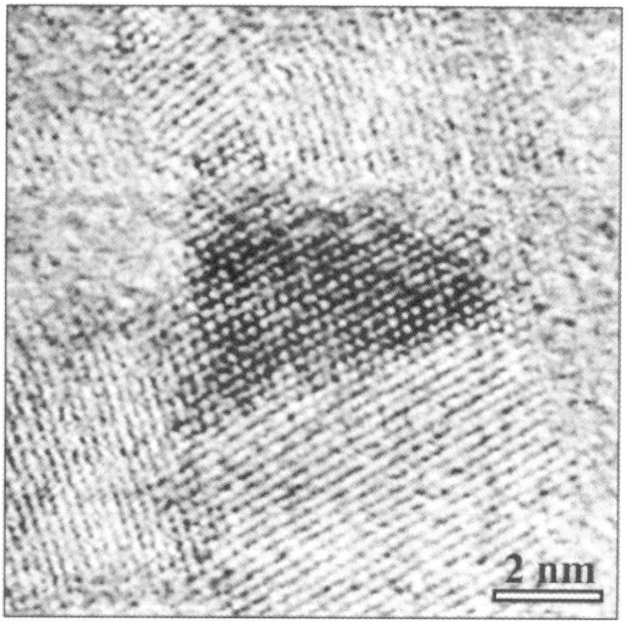

Figure 4-8. Atomic resolution BF STEM image of oxide nanoparticles supported on an amorphous carbon film.

complication of disturbing the microscope alignment. This method is useful for imaging highly inhomogeneous samples, especially for identifying small particles and for imaging interphase interfaces with enhanced chemical sensitivity.

There are some advantages of using BF STEM over TEM for examining thick specimens. Since inelastic scattering produces a spread of energies of the transmitted electrons, the chromatic aberration of the objective lens degrades the image resolution in TEM. In STEM, on the other hand, because the objective lens comes before the specimen, such energy spread does not affect the image resolution. Traditionally, the use of an electron energy-loss spectrometer gives more flexibility in STEM imaging. For example, STEM images can be formed with only elastically scattered electrons, or with plasmon-loss electrons, or any other energy-loss electrons. With the increasing use of energy filters in modern TEM instruments [9], however, the distinction between energy-filtered TEM and STEM imaging is rapidly abating.

4.3.4 Large-angle bright-field imaging

The phase contrast of STEM images rapidly decreases with the increase of the detector collection angle. Due to the principle of reciprocity, the increase of the collection angle in STEM is equivalent to the increase in the illumination convergent angle in TEM. The use of large convergent angle of illumination in TEM pushes the first crossover of the contrast transfer function (CTF) to higher values and causes a rapid damping of high frequency oscillations in the CTF obtained with parallel-beam illumination [10]. Thus, interpretable image resolution can be improved at the expense of image contrast.

For a STEM detector large enough to coincide with the disc of directly transmitted electrons, i.e. $D(\mathbf{K}) = A(\mathbf{K})$, imaging theory suggests that, with a phase object approximation, the image intensity can be approximated by [4, 11]:

$$I_{BF}(\mathbf{X}) = 1-2[1-\cos(\sigma\phi(\mathbf{X}))]*|t(\mathbf{X})|^2 \tag{4-9}$$

In a weak phase object approximation $\cos(\sigma\phi(\mathbf{X})) \sim 1-\frac{1}{2}(\sigma\phi(\mathbf{X}))^2$, thus:

$$I_{BF}(\mathbf{X}) = 1-(\sigma\phi(\mathbf{X})^{2}*|t(\mathbf{X})|^2 \tag{4-10}$$

This is a form of incoherent imaging: the phase contrast is washed out and the image resolution is determined by the probe current distribution inside the sample.

By changing the strength of the post-specimen lenses, the collection angle of the STEM detector can be easily varied. If we collect all the directly transmitted electrons, plus a large portion of the scattered electrons, a large-angle BF (LABF) image of the sample can be obtained. The dominant phase contrast visible in BF STEM images (Fig. 4-9a) is mostly suppressed in LABF images (Fig. 4-9b). The contrast of LABF images is predominantly due to absorption effects, weak diffraction contrast, plus an electron channeling effect.

For crystals with principal zone-axes aligned in the incident beam direction, the diffraction and phase contrast are significantly reduced in LABF images; but the image resolution is improved. Figure 4-10a shows a high-resolution BF STEM lattice image of a GaAs crystal oriented in the [110] zone-axis. Figure 4-10b shows a corresponding image obtained with a LABF detector (semi-collection angle of about 30 mrad). The LABF image clearly shows a better image resolution and an enhanced image contrast.

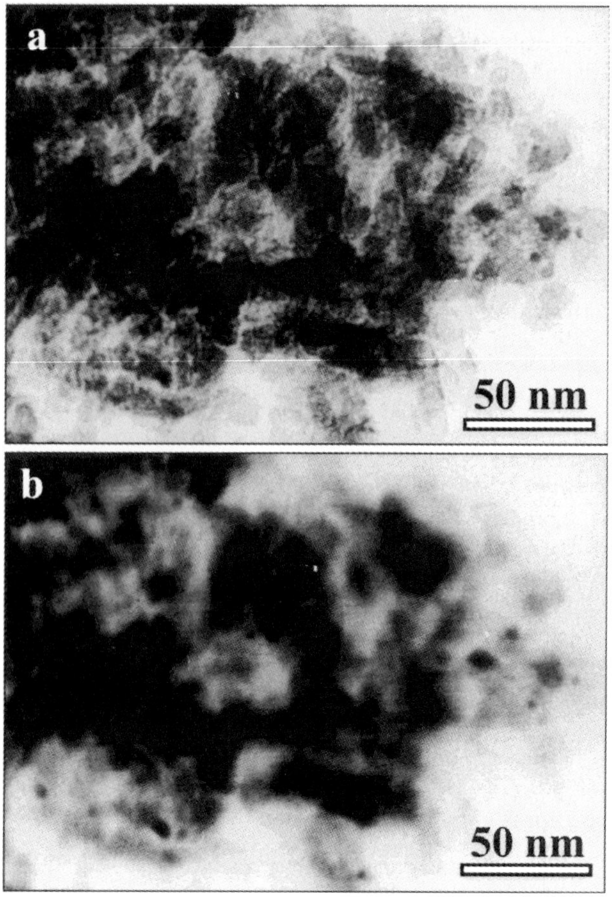

Figure 4-9. BF (a) and large-angle BF (b) STEM images of Pt nanoparticles supported on γ-alumina crystallites. The strong phase and diffraction contrast present in the BF image is suppressed in the large-angle BF image.

The contrast characteristics of LABF lattice images include less dependence on the beam defocus and sample thickness, but strong dependence on the channeling condition of the crystal. These are similar to the characteristics of annular dark-field (ADF) images (see detailed discussions in section 4.3.5).

To understand the characteristics of LABF images, we rewrite the Eq. (4-8) as:

$$\int [D_{\mathrm{LABF}}(\mathbf{K}) + D_{\mathrm{ADF}}(\mathbf{K})] |\Psi(\mathbf{K},\mathbf{X})|^2 d\mathbf{K} = I_{\mathrm{LABF}}(\mathbf{X}) + I_{\mathrm{ADF}}(\mathbf{X}) \equiv 1 \qquad (4\text{-}11)$$

$$I_{\mathrm{LABF}}(\mathbf{X}) = 1 - I_{\mathrm{ADF}}(\mathbf{X}) \qquad (4\text{-}12)$$

Thus, the contrast of LABF images is complementary to the contrast of ADF images obtained with an inner collection angle as large as that of the LABF detector. LABF images can be interpreted in the same way as low-angle ADF images: improvement in image resolution and increased atomic-number sensitivity.

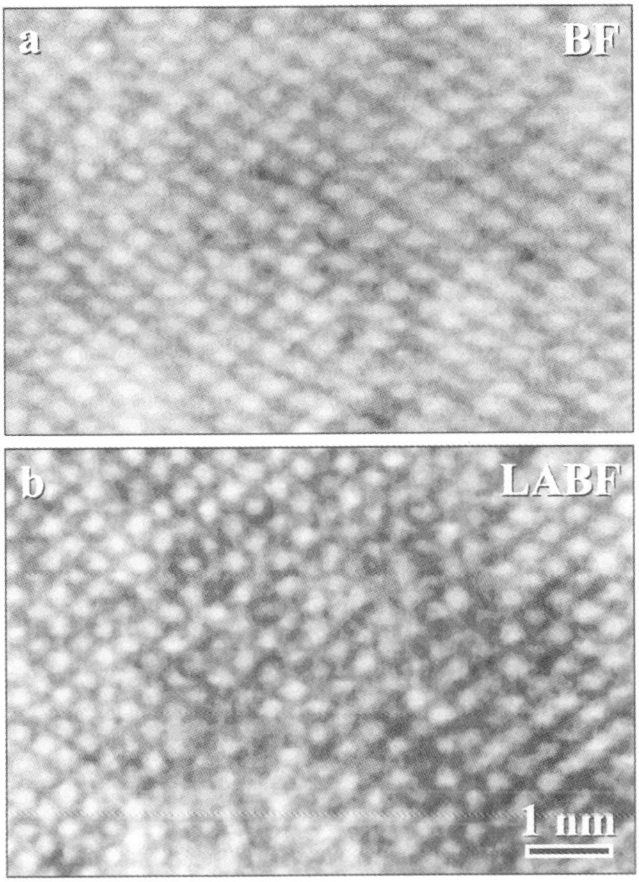

Figure 4-10. Atomic resolution BF (a) and large-angle BF (b) STEM images of the same area of a GaAs crystal oriented along the [110] zone axis. The large-angle BF STEM image clearly shows a better resolution and a higher contrast.

4.3.5 Annular dark-field and high-angle annular dark-field imaging

By collecting electrons scattered outside the central beam in the diffraction pattern, a high-resolution ADF image of the sample can be obtained. In fact, atomic resolution imaging was first achieved in STEM by using an ADF detector to collect electrons scattered from heavy atoms supported on ultra-thin, light-element substrate [12–14]. The high efficiency of the ADF signal made it possible to image individual atoms of heavy-elements, and to study the motion and diffusion of atom clusters [15]. Atomic rows in small crystallites and different layers of atoms were also observed in ADF STEM images, and for very small clusters, quantized levels of integrated intensity of atoms were observed [16–17].

However, the ADF imaging mode has its drawbacks. Because of the low collection angle of the ADF detector, strong dynamical diffraction effects may obscure the compositional sensitivity in ADF images of polycrystalline materials or of small metal particles that are supported on polycrystalline substrates.

The contrast of ADF images critically depends on the inner and outer collection angles of the ADF detector. The diffraction effects in ADF images of crystalline materials can be greatly suppressed by increasing the inner collection angle of the ADF detector beyond the Bragg reflections so that only high-angle scattered electrons contribute to the collected signal [18]. This new imaging mode is called HAADF or Z-contrast imaging. In HAADF images, the diffraction and phase contrast is significantly suppressed and the compositional sensitivity is recovered.

The development of HAADF imaging technique has proved very successful for characterizing small particles and supported metal catalysts with sub-nanometer or atomic resolution and high compositional sensitivity [19–24]. Small metal particles supported on polycrystalline materials cannot be reliably detected in BF or low-angle ADF images because of the strong diffraction and phase effects (Fig. 4-11a), but these nanoparticles can be easily recognized in HAADF images (Fig. 4-11b).

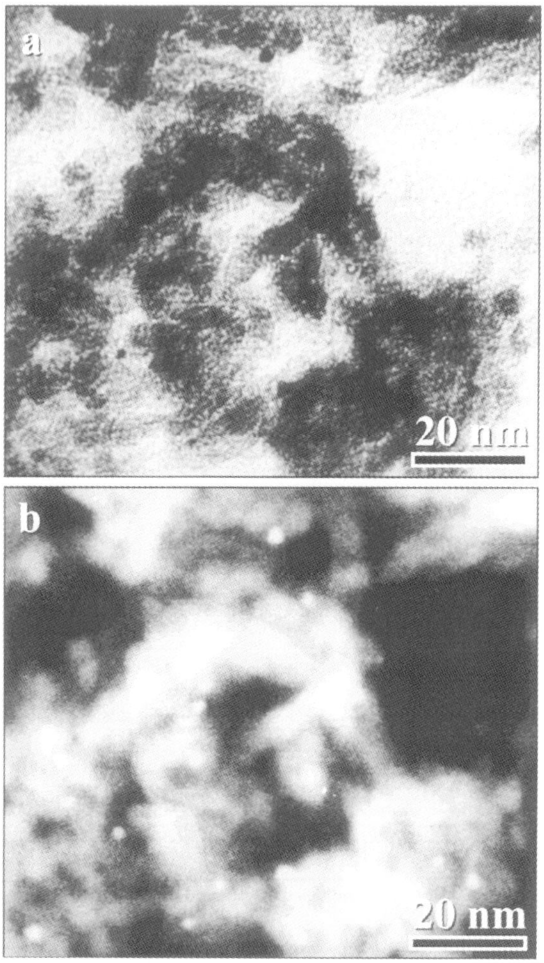

Figure 4-11. BF (a) and HAADF (b) STEM images of Pt nanoparticles supported on γ-alumina crystallites. The phase and diffraction contrast of the γ-alumina crystallites is significantly reduced and the Pt particles are clearly revealed in the HAADF image.

By using post-specimen lenses, we can conveniently switch from low-angle ADF to HAADF imaging mode to study the contrast variations in ADF images. With the increase of the inner collection angle, the phase and diffraction contrast slowly decreases, the atomic-number contrast becomes dominant; but the signal-to-noise ratio of the ADF images decreases. The inner collection angle (α_1) for low-angle ADF images is usually greater than 10 mrad and the outer collection angle (α_2) is generally set at about 50 mrad (Fig. 4-12). For HAADF imaging, however, the appropriate inner collection angle of the ADF detector depends on the nature of the sample and the orientation of crystalline materials. For imaging polycrystalline materials or small particles, the inner collection angle (β_1) is usually greater than 100 mrad and the outer collection angle (β_2) can be as high as a few hundred mrad. For imaging zone-axis crystals, however, β_2 is usually set at an angle smaller than the first-order Laue pattern to reduce contributions of high order Bragg scattering, and β_1 is about 50–60 mrad.

The resolution attainable in HAADF images is better than the resolution obtainable in BF STEM or TEM images using the same lenses [4, 15, 25–27]. Observation of high-resolution details of crystalline specimens on a scale < 0.2 nm can be achieved by using 100 keV electrons [28–32], and a resolution of less than 0.13 nm has been achieved by using 300 keV electrons [33, 34]. Because of the high atomic-number sensitivity and high spatial resolution, the HAADF technique is undoubtedly one of the most useful imaging techniques for studying nanoparticles, supported catalysts, and interfaces in semiconductors, ceramics, and superconducting materials.

The strength of the high-angle scattering, which gives the HAADF imaging signal, depends on several parameters including: 1) large-angle elastically scattered electrons, 2) phonon scattered electrons, and 3) inelastically scattered electrons. Various simulations using dynamical diffraction theory have been performed to understand the nature of HAADF imaging signal [35–43].

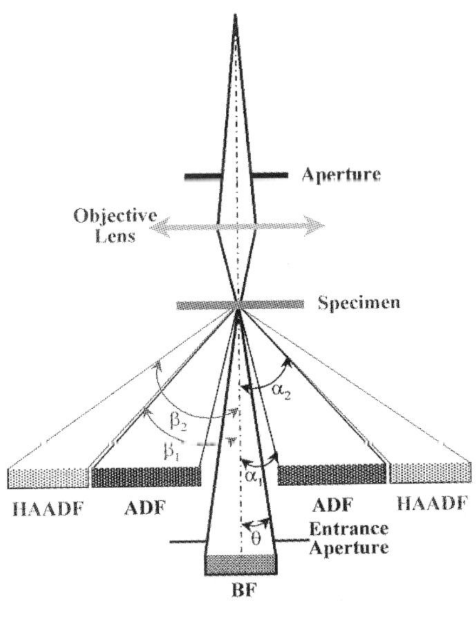

Figure 4-12. Schematic diagram illustrates the geometric arrangement of BF, ADF, and HAADF detectors. The parameter θ represents the collection (semi-) angle of the BF detector; α_1 and α_2 are the inner and outer collection (semi-) angle of the ADF detector, respectively; β_1 and β_2 are the inner and outer collection (semi-) angle of the HAADF detector, respectively.

The imaging theory of HAADF microscopy follows that of incoherent imaging: the high-angle scattered electrons can be treated incoherently. The lateral coherence of the scattered electrons is almost completely suppressed because of detector geometry (averaging effect) and thermal diffuse (phonon) scattering of high-energy electrons. The columnar coherence of the scattered electrons is significantly reduced because of phonon scattering although a small residue of the coherence still exists along the incident beam direction [42, 43]. The contrast characteristics of incoherent HAADF imaging include: 1) high atomic-number sensitivity, 2) less dependence on beam defocus and sample thickness, and 3) absence of proximity effects at interfaces.

For zone-axis crystals, high-energy electrons may preferentially travel along paths of low-energy potentials in the sample. This phenomenon is called the electron channeling in crystalline materials [44]. The propagation of a coherent-convergent electron probe inside a perfect crystal in the zone-axis channeling condition has been widely investigated [36, 44–49]. Remarkable electron focusing effects can occur under channeling conditions. The penetration of the incident electrons is different for probes focused onto atomic columns of different species. The channeling effect of a convergent probe is important in interpreting high-resolution HAADF images of crystals oriented in principal zone-axes [36]. Thermal diffusing scattering, plus the channeling effect, forms the basis of atomic resolution HAADF imaging of crystalline materials. The channeling effect, however, may not significantly affect the contrast of small particles since small particles are usually randomly oriented.

In the incoherent imaging limit, the image contrast becomes a pure "number contrast": the total number of high-angle scattered electrons determines the image intensity at that pixel. Thus, HAADF images can be viewed as the convolution of the intensity distribution of the incident probe with appropriate cross-sections for high-angle scattering processes. Since high-angle scattering processes are highly localized, the resolution of HAADF images is necessarily determined by the size of the incident electron probe.

With coherent, convergent beam illumination, the intensity distribution of the incident probe, $I_0(\mathbf{x})$, rather than the vaguely defined probe size, is usually used to describe the performance of a STEM instrument. The form of $I_0(\mathbf{x})$ strongly depends on the size of the objective aperture, the spherical aberration coefficient of the objective lens, the beam defocus, the energy of the incident electrons, and the instability of the microscope. The "optimum" probe sizes in a STEM instrument depend on the selected imaging and analytical modes [11, 50]. With the use of small objective apertures, usually used for nanodiffraction and nanoanalysis, the intensity distribution of the electron probe does not vary appreciably with the change of beam defocus. On the other hand, with the use of large objective apertures, such as those used in high-resolution STEM imaging, the intensity distribution within the electron probe becomes increasingly sensitive to the change of beam defocus. Figure 4-13 shows calculated intensity distributions of a small electron nanoprobe at different defoci. The central peak of the nanoprobe narrows with increasing under-focus, but the side lobes also become significant.

For high-resolution imaging of zone-axis crystals, it is desirable to use a large objective aperture and to work at an under-focus value slightly larger than the Scherzer focus to improve image resolution without introducing complications in image interpretation [32]. For imaging small particles, however, an analytical probe (top-hat shape) may give shaper images.

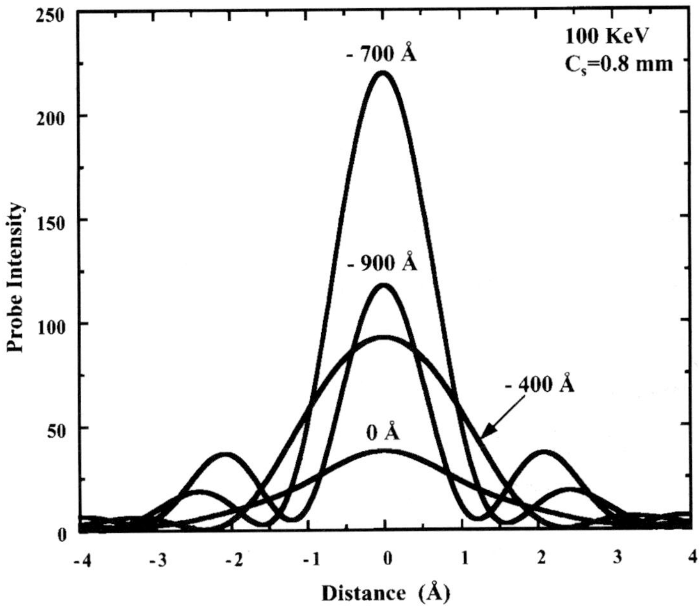

Figure 4-13. Calculated intensity distributions of an electron nanoprobe at different values of defocus: $C_s = 0.8$ mm; $E = 100$ keV; incident beam semi-angle of convergence $\alpha = 12$ mrad.

HAADF images of nanoparticles can be intuitively interpreted as the convolution of the intensity distribution of the probe with the scattering power of the particles (Fig. 4-14a). For an infinitely small electron probe, the image intensity profile should give information about the shape of nanoparticles with an atomic resolution. In practice, however, the finite probe complicates the interpretation of HAADF images. Therefore, it is difficult to unambiguously determine the shape of nanoparticles in HAADF images. When the under-focus value is too large, the intensity in the central peak of the probe can be much less than that in the side lobes; thus an annular ring-like probe may be formed. The corresponding HAADF images of small particles should show annular rings with an intensity modulation weighted by the scattering power of the particles (Fig. 4-14b).

The high atomic-number sensitivity, the incoherent imaging characteristics, and the intuitive relationship to the object make HAADF imaging the most powerful STEM technique for characterizing small particles, interfaces, and superstructures. For example, supported nanoparticles with sizes < 1 nm are difficult to be identified in high-resolution TEM (HRTEM) images. Image distortion also poses a severe problem for correctly interpreting HRTEM images of small particles without extensive image simulations [51]. However, sub-nanometer sized particles and even single atoms can be detected in HAADF images with high contrast [20–24].

The effective penetration depth of high-energy incident electrons increases with their scattering angle. This general argument is true for both elastic and inelastic scattering processes. Thus, it is an advantage to examine relatively thick samples by collecting high-angle scattered electrons. Depending on the inner and outer collection angle of the ADF detector, plural and multiple scattered electrons may significantly contribute to the collected signal. Multiple scattering events redistribute the angular

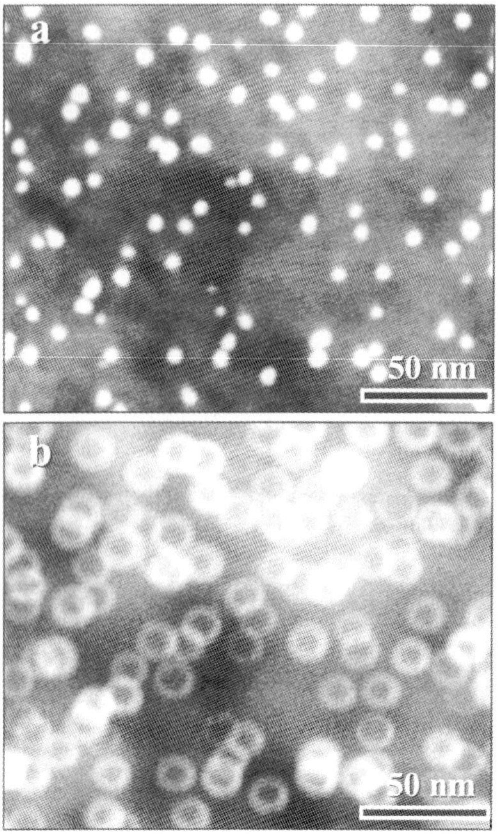

Figure 4-14. In-focus (a) and under-focus (b) HAADF images of the same sample area of Pt nanoparticles dispersed on a highly activated carbon support. The annular rings shown in image (b) represent the shape of the under-focused electron nanoprobe. All the rings have a similar size, but their intensities are weighted by the scattering power of the individual Pt particles.

distribution of the scattered electrons in the diffraction plane; they broaden the width of the angular distribution of the scattered electrons so that the high-angle ADF signal is enhanced for thick specimens.

For imaging thick samples consisting of metal particles dispersed on a light-element support, BF TEM or STEM images can not give much information about the sample because of plural and multiple scattering of incident electrons (Fig. 4-15a). ADF images, however, can still give useful information about the size and spatial distribution of nanoparticles [52]. At low collection angle, high atomic-number particles show white contrast in thin regions of the sample; but these particles show dark contrast in thick regions of the sample because of the plural and multiple scattering effects (Fig. 4-15b). With the increase of the inner and outer collection angle, the contrast of particles in thick regions diminishes first (Fig. 4-15c), and then changes to bright (Fig. 4-15d). This technique can be effectively utilized for correlating the features observed in the thinner regions of a specimen to those in thicker regions.

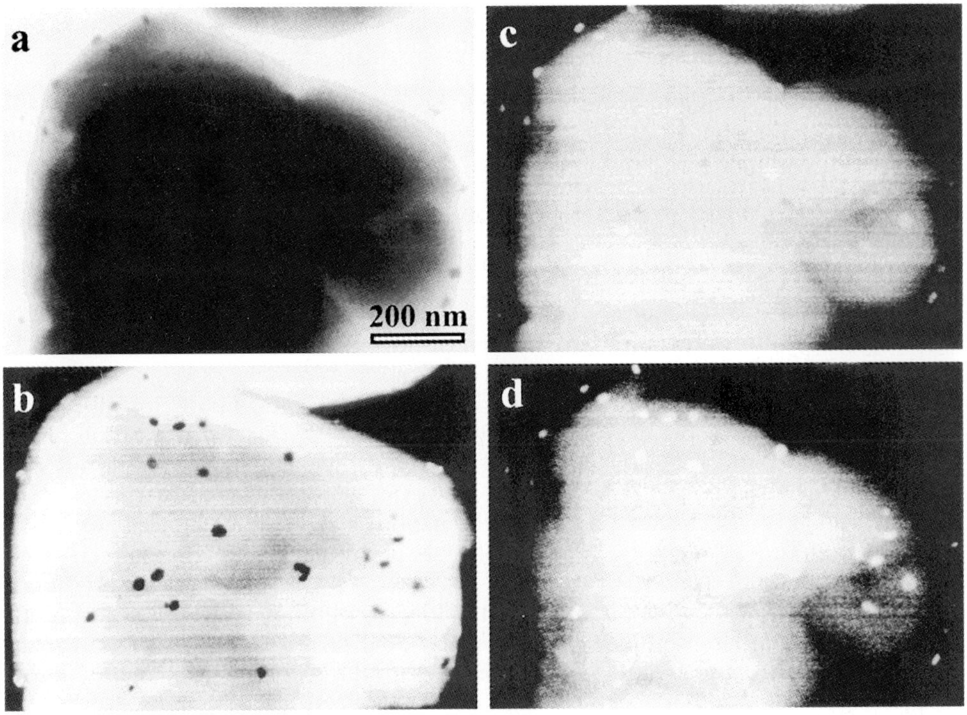

Figure 4-15. A set of STEM images of metal particles supported on zeolite crystals illustrates the variations of image contrast of nanoparticles with the change of the scattering angle. The BF image (a) shows a few metal particles near the edge of the large zeolite particle. The low-angle ADF image (b) shows some metal particles near the edge of the support with a slightly bright contrast and metal particles inside the support with a dark contrast. The ADF image (c) obtained with an intermediate inner collection angle does not show much contrast of the metal particles at all. The HAADF image (d) shows all metal particles with a bright image contrast.

4.3.6 Thin annular detector and other configured detectors

The central beam of the diffraction pattern can be expanded, by use of post-specimen lenses, to overlap the inner edge of the ADF detector. A thin ring at the outer edge of the directly transmitted beam, plus a small portion of the scattered beams, can be collected to form an image of the specimen. A specially designed thin annular detector (TAD) with only about 10% difference between the inner and outer collection angle can also be used to obtain TAD images. The TAD can be used to form bright-field (TADBF) or dark-field (TADDF) images [53–54]. By applying the principal of reciprocity, this imaging mode is equivalent to hollow-cone illumination in TEM. Detailed treatment of the imaging process suggested that the resolution of TADBF images can be improved [53–54].

The TAD imaging modes take advantage of selecting the range of frequencies that give higher image resolution and excluding the lower frequencies that contribute to background signal. The TAD may also be used to collect signals of small-angle scattering to produce images of amorphous materials or light-element particles. For example, carbon nanoparticles supported on amorphous silica can be detected with good contrast [55]. The combination of TAD with HAADF imaging technique can be very effective in examining both heavy-element and light-element nanoparticles with atomic resolution.

Other specially configured detectors can be constructed to increase image resolution, to enhance image contrast, or to extract unique information about certain features of the sample [56, 57]. For example, circular detectors splitting into halves or quadrants have been used to study magnetic field and magnetic domain structures of thin films and small particles [58, 59]. Complicated multiple detectors have also been proposed for special purposes.

4.3.7 Size measurement and distribution of nanoparticles

It is often a challenge to accurately determine the exact sizes of particles with diameters < 1 nm, highly dispersed on supporting materials. Because of the strong phase and diffraction contrast of the supports, it is even difficult to recognize particles of sub-nanometer sizes in HRTEM images. However, the knowledge of this class of nanoparticles is important for understanding nanoparticle systems. For example, the catalytic properties of supported metal catalysts are directly related to the size and spatial distribution of the metal particles.

Although nanoparticles can be observed in high-resolution HAADF images, particles with sizes comparable to the size of the incident probe cannot be reliably measured. Because of the convolution of the probe with the real dimensions of the particles, particles with sizes smaller than the size of the electron probe may give similar sizes in HAADF images.

Because of the incoherent nature of HAADF imaging, the signal strength of high-angle scattered electrons from an isolated small particle is proportional to the total number of atoms in that particle. In principle, the integrated intensity of a particle in a HAADF image is independent of the probe size, the defocus value, and the substrate thickness. It only depends on the type of atoms, the total number of atoms in the particle, and the total current of the incident probe. In practice, however, the presence of noise in HAADF images affects the accurate measurement of the total integrated image intensity of a particle.

Quantification of the intensity distribution of HAADF images of metal particles can be accomplished by image processing techniques [20]. By measuring the integrated intensity, I_1, associated with the N_1 pixels of a particle and I_2 associated with an expanded area of N_2 pixels, the net integrated intensity I_{total} due to only the particle is given by:

$$I_{total} = \frac{I_1 N_2 - I_2 N_1}{N_2 - N_1} \qquad (4\text{-}13)$$

In the incoherent imaging limit, the total integrated image intensity of a particle in a HAADF image is proportional to the probe current and the total scattering power of the particle. The signal strength of high-angle scattered electrons is linearly dependent on the number of atoms of each type within the probed volume, weighted by their scattering cross-sections. Assuming the radius of a spherical particle is R, then we can write $I^{1/3} = \alpha R + \beta$. The parameter α is proportional to $(\sigma/\Omega)^{1/3}$; β is a parameter related to the microscope and the noise level introduced in the image acquisition and analysis processes; σ is the atomic scattering cross-section; and Ω is the atomic volume [20].

In the absence of image noise, the total scattering strength of a particle should be independent of the size of the integration area and is not sensitive to lens aberrations or beam defocus. This unique property makes the intensity integration method a much more powerful technique than the conventional sizing methods that rely on diameter or area measurements. Diameters measured in HRTEM images can change in a complicated way with defocus and substrate thickness, especially for particles with a sub-nanometer size [51]. In practice, however, particle visibility in HAADF images change with beam focus. Low signal-to-noise ratio is another limiting factor for measuring small particles. The intensity measurement technique is sensitive to noise, either intrinsic to the specimen or generated from the image-acquisition system [20].

With clean metal particles supported on thin carbon films, the measurement of the total integrated intensity of individual metal particles can be easily performed through a simple intensity-threshold process. Figure 4-16 shows an intensity$^{1/3}$-radius plot from the HAADF signal of silver nanoparticles. The plot lies on a straight line with a slope of α which is related to the intensity increment per atom and the particle morphology. The number of atoms per cluster can be estimated from the intensity$^{1/3}$-radius plots [20].

Because of the extremely complex structures of supports and nanoparticles in practical systems, it is a formidable task to obtain the true integrated intensity of a nanoparticle. Higher beam current densities and smaller probe sizes are preferred to increase the signal-to-noise ratio of particles with sub-nanometer sizes. Sample movement, decomposition, and damage due to intense electron beam irradiation, will all affect the accuracy of the intensity measurement. Sample contamination may also render complications in determining the true signal strength of individual nanoparticles. Sophisticated image analyses are usually needed to subtract uneven background signals before performing particle size and intensity measurement (Fig. 4-17).

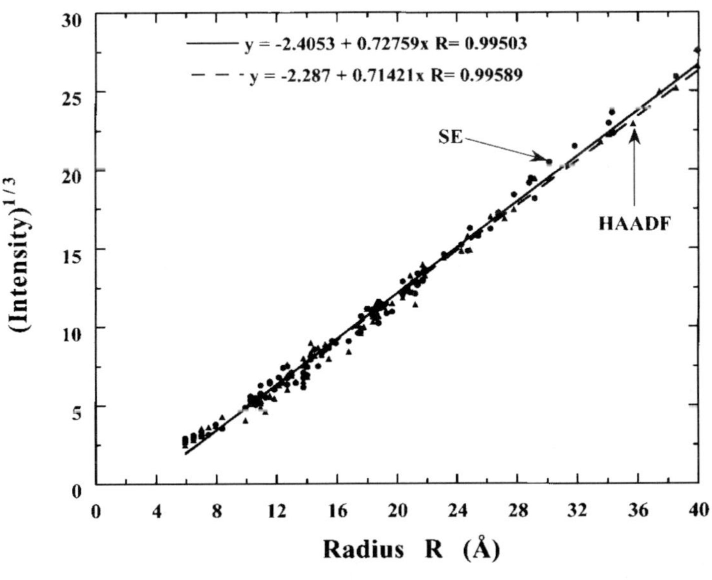

Figure 4-16. Intensity and radius data from particles in HAADF and SE images of silver nanoparticles plotted as (intensity)$^{1/3}$ against particle radius R. The HAADF and SE images were simultaneously collected.

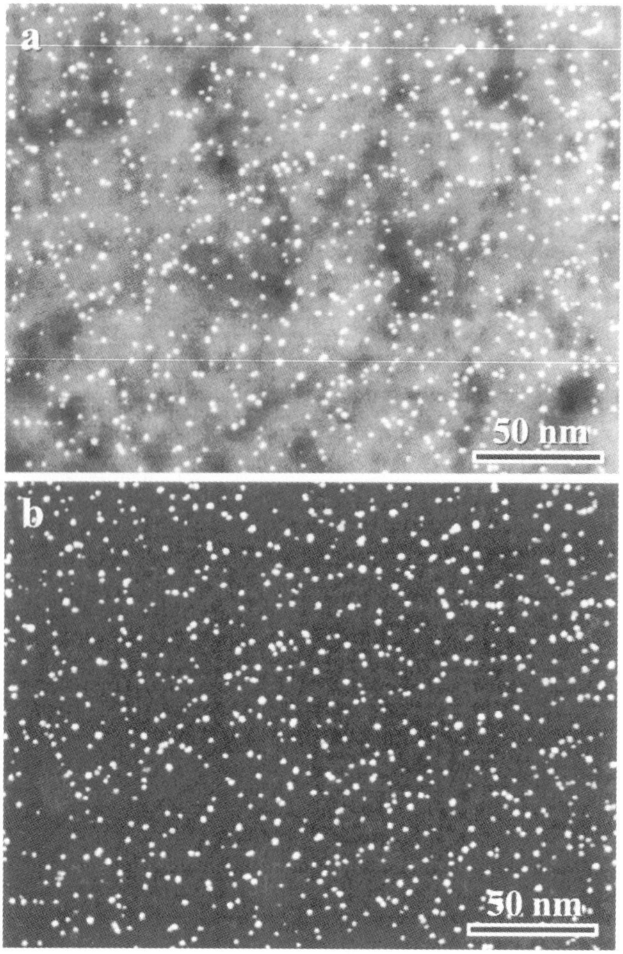

Figure 4-17. HAADF image (a) of Pt nanoparticles dispersed on highly activated carbon support shows small Pt particles with a bright contrast, and intensity variations of the carbon support. The background-subtracted image (b) shows a much "cleaner" image for size and intensity measurement of individual Pt nanoparticles.

In principle, the intensity-size plot could be used to give accurate size distributions of particles, to distinguish between raft-like and spherical particles, and to give information about alloys of different compositions. Although this technique has been used to assess compositional variations in bimetallic nanoparticles, the interpretation of intensity-size plots is much more complicated [60].

4.4 Coherent electron nanodiffraction

When an electron nanoprobe is stopped at any point of interest on a sample, a convergent beam electron-diffraction (CBED) pattern is formed on the detection plane of a STEM instrument. The size of the diffraction discs is determined by the convergent angle of the incident electron probe. With a two-dimensional detector such as a phosphor screen or a CCD system, the diffraction patterns can be observed, recorded, and analyzed in the same away as CBED patterns obtained in TEM instruments [53, 61–62].

The use of a field-emission gun in STEM, however, introduces important new features in STEM CBED patterns [61]. First, the sizes of the electron probes are usually 1 nm or less in diameter, much smaller than those used in TEM. Thus, the diffraction patterns obtained in STEM are usually called micro- or nano-diffraction patterns. Secondly, the use of a field-emission gun warrants the coherent nature of a convergent nanoprobe: the illuminating aperture is filled with completely coherent radiation, and the final probe entering the specimen can be treated as perfectly coherent. In contrast, the illuminating aperture in conventional TEM is considered incoherently filled and the illumination is treated as completely incoherent.

Coherent electron nanodiffraction (CEND) is the only technique that gives full diffraction information about individual nanoparticles. Figure 4-18 shows a schematic diagram illustrating a coherent, convergent electron probe with a size smaller than the size of a multi-faceted nanoparticle. Diffraction patterns from the various parts of the nanoparticle can be obtained to give information about the structure as well as the morphology of the nanoparticle.

Although CEND patterns obtained in a STEM instrument are necessarily CBED patterns, different operating modes of STEM require different incident beam convergent angles. The spatial distribution of the electron probe depends on the convergent angle of the incident beam. For example, high-resolution lattice imaging requires the smallest electron probe and overlapping diffraction discs; thus, a large convergent angle of the incident probe is used to satisfy these conditions. On the other hand, it is necessary to use a small convergent angle to obtain sharp diffraction spots from small particles or localized crystal defects. If we denote the convergent semi-angle of the incident probe as α and the Bragg angle of a crystal as θ_B (Fig. 4-19), then three distinctive types of CEND patterns need to be discussed: 1) $\alpha \leq \theta_B$; 2) $\alpha > \theta_B$; and 3) $\alpha >> \theta_B$. In the following, we will discuss each type of the diffraction pattern in detail.

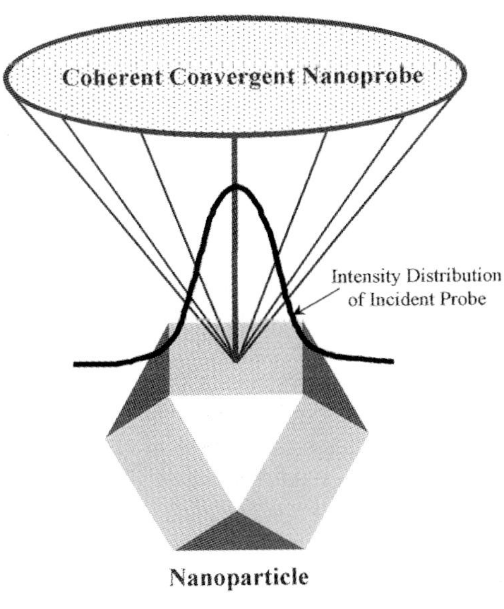

Figure 4-18. Schematic diagram illustrates a coherent, convergent electron probe with a size smaller than the size of a multifaceted nanoparticle. Nanodiffraction patterns from the various facets of the nanoparticle can be obtained to give information about the structure as well as the morphology of the nanoparticle.

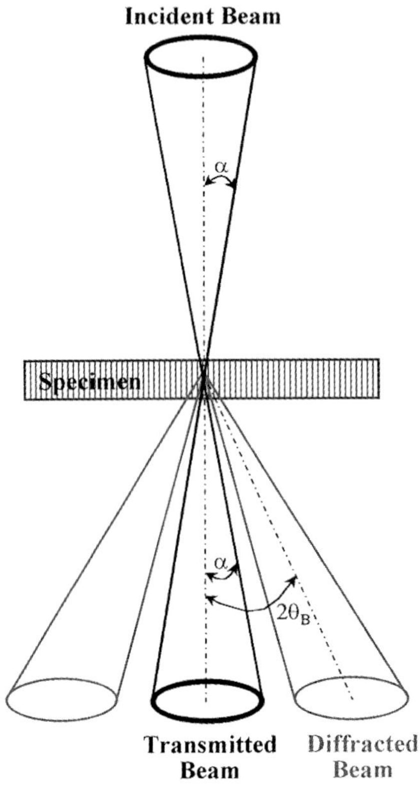

Figure 4-19. Schematic diagram illustrates coherent electron nanodiffraction from crystalline materials: α is the convergent semi-angle of the incident beam; θ_B is the Bragg angle of a diffracting plane.

4.4.1 Coherent electron nanodiffraction with $\alpha \leq \theta_B$

4.4.1.1 Perfect crystals

For an ideally perfect, thin crystal (no thickness variation, no defects, no bending), there are no differences in the diffraction patterns obtained with a coherent or an incoherent electron beam if the diffraction discs do not overlap ($\alpha \leq \theta_B$ in Fig. 4-19). This is a consequence of the Bragg law: for each incident direction of the electron beam only scattering through mutiples of the Bragg angle is allowed. Thus, electrons with different incident beam directions cannot interfere with each other although the incident electron probe is completely coherent. If the crystal is thicker, the intensity distribution within the diffraction discs may become non-uniform, with sets of lines, bands, or more complicated shapes [62]. This is mostly due to dynamical diffraction effects giving a variation of the incident and diffracted beam intensities as a function of the angle of incident beam. If we ignore the fine-details present in CBED patterns, CEND patterns can be treated the same way as those generated by an incoherent electron beam with a nanometer-size probe. CEND patterns from perfect crystals can be interpreted the same way as CBED patterns obtained in conventional TEM [62].

4.4.1.2 Imperfect crystals

For crystals containing defects (crystal edges and bending, stacking faults and dislocations, thickness variations, and other defects), elastic diffuse scattering from these defects can coherently interfere with each other or with Bragg diffracted electrons. Different patterns, characteristic of the nature of the defects, can be observed. For thicker or strongly scattering samples, any discontinuity in the sample can have some observable effect on the CEND patterns.

For example, when a nanoprobe approaches a straight edge of a MgO cube aligned along the [001] zone-aixs, first the central transmitted disc shows strong streaking towards the crystal; then diffraction spots appear at non-Bragg positions (Fig. 4-20). Fine structures in these CEND patterns change rapidly with the movement of the probe position. The streaking of the central transmitted spot is attributable to the influence of the crystal inner potential. The interference among waves arriving from different incident beam directions gives rise to perturbations of the Bragg diffraction spots. When part of the incident probe is positioned inside the MgO crystal, electrons with different incident directions interact with different parts of the crystal. The scattered electrons interfere with each other to give a characteristic diffraction pattern related to the position of the electron probe. Dynamical diffraction simulations show that the intensity distributions in CEND patterns are sensitive to the surface or the defect structures of the sample [63]. A surface channeling effect may also be responsible for the fine features observed in CEND patterns from straight edges of small crystals [63].

When the incident probe is positioned near the edge of a crystal, CEND discs may show annular rings or splitting of diffracted spots [64]. Internal discontinuities, such as fault planes, out-of-phase boundaries, and thin layer precipitates, may give character-

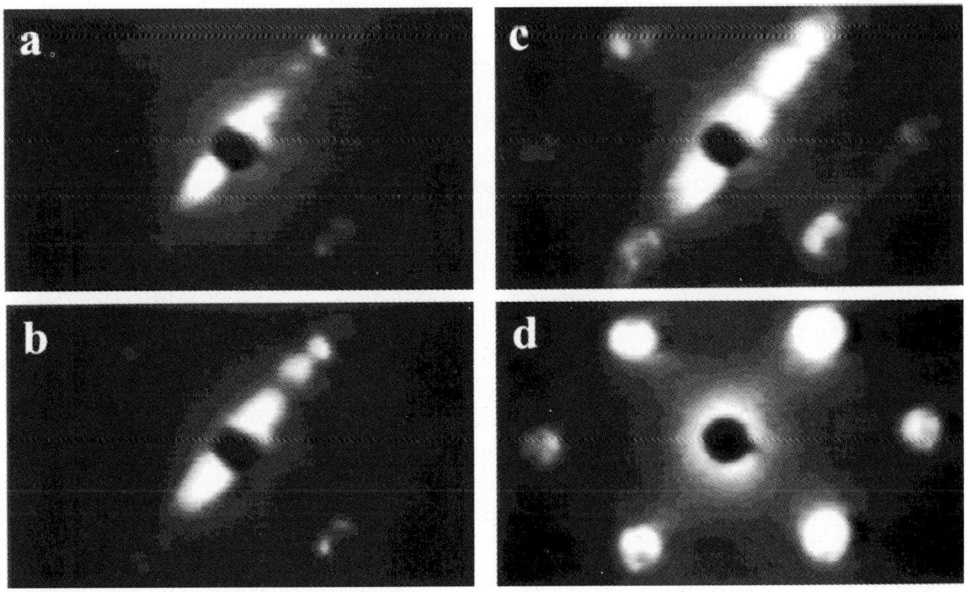

Figure 4-20. Nanodiffraction patterns obtained with an electron nanoprobe (~ 0.8 nm in diameter) moving toward a large MgO cube oriented along the [001] zone axis. The electron nanoprobe was right outside (a) and inside (d) the MgO cube. The black spot represents the center of the directly transmitted electron beam.

istic structures in their corresponding CEND patterns. For example, CEND patterns from antiphase domain boundaries in ordered alloys show spot splitting of superlattice reflections [65].

One of the important applications of CEND is the structural study of metal particles in supported catalysts [21, 66–69]. Other applications of CEND include study of empty and filled carbon nano-tubes [70], precipitates in metals and alloys [71], phase separation in magnetic materials [72], and local ordering in thin films.

4.4.1.3 Small particles

Small particles have peculiar chemical and physical properties compared to their corresponding bulk materials. Metallic nanoparticles are especially important in heterogeneous catalysis; they are primarily the active components of supported metal catalysts. A precise knowledge of the size, structure, and composition of metal particles is of primary importance in understanding the synthesis-structure-activity relationships and is essential for a qualitative and quantitative understanding of the performance of supported metal catalysts.

One of the successful applications of STEM techniques is the ability to analyze the structure and composition of individual nanoparticles. Figure 4-21, for instance, shows a set of CEND patterns, obtained with an electron beam of approximately 1 nm in diameter, of a bimetallic catalyst containing noble metals supported on zeolite crystals which are connected by alumina binders. These CEND patterns provide information about the crystallographic structure of the bimetallic particles and supports, and information about the structural relationship between the metal particles and their supports or binder materials.

It is, however, impossible to make accurate measurements of lattice parameters in CEND patterns because of the large sizes of the diffraction spots. An error of 5% or higher is common in determining lattice constants of small particles, and much larger errors can frequently occur because of coherent interference effects.

It is important to correlate the characteristic features of CEND patterns to particle properties, such as the structure of the particle, the nature of defects within the particle, and the shape and size of the particle. A frequently observed characteristic feature is the splitting of diffraction spots along certain crystallographic directions (Fig. 4-21). The spot splitting in non-overlapping CEND patterns is attributable to the coherent nature of electrons diffracting from an abrupt discontinuity of the scattering potential at particle edges [64]. It is also observed that the spot splitting is related to the geometric forms of the diffracting particles; some splitting occurs in a well-defined crystallographic direction. Depending on the probe position relative to the center of a particle, annular rings may be observed in CEND patterns of small particles [64].

Dynamical simulations reveal that for a particle which has facets smaller than the size of the incident probe, the incident electrons may interact with several facets of the small particle [73]. The thickness of the particle may vary rapidly even within a region of only about 1 nm in diameter. The electron probe effectively interacts with the "particle morphology" under illumination. The intensity variations of the splitting spots are related to the probe positions with respect to the particle facets and are related to the length of the facets along the incident beam direction.

The direction of spot streaking or splitting is directly related to specific edges or facets of a small particle [73]. Furthermore, the intensity profiles across the splitting spots vary with the types of particle wedges. In principle, it is possible to deduce the 3-dimensional structure of nanoparticles by quantitatively analyzing the intensity distri-

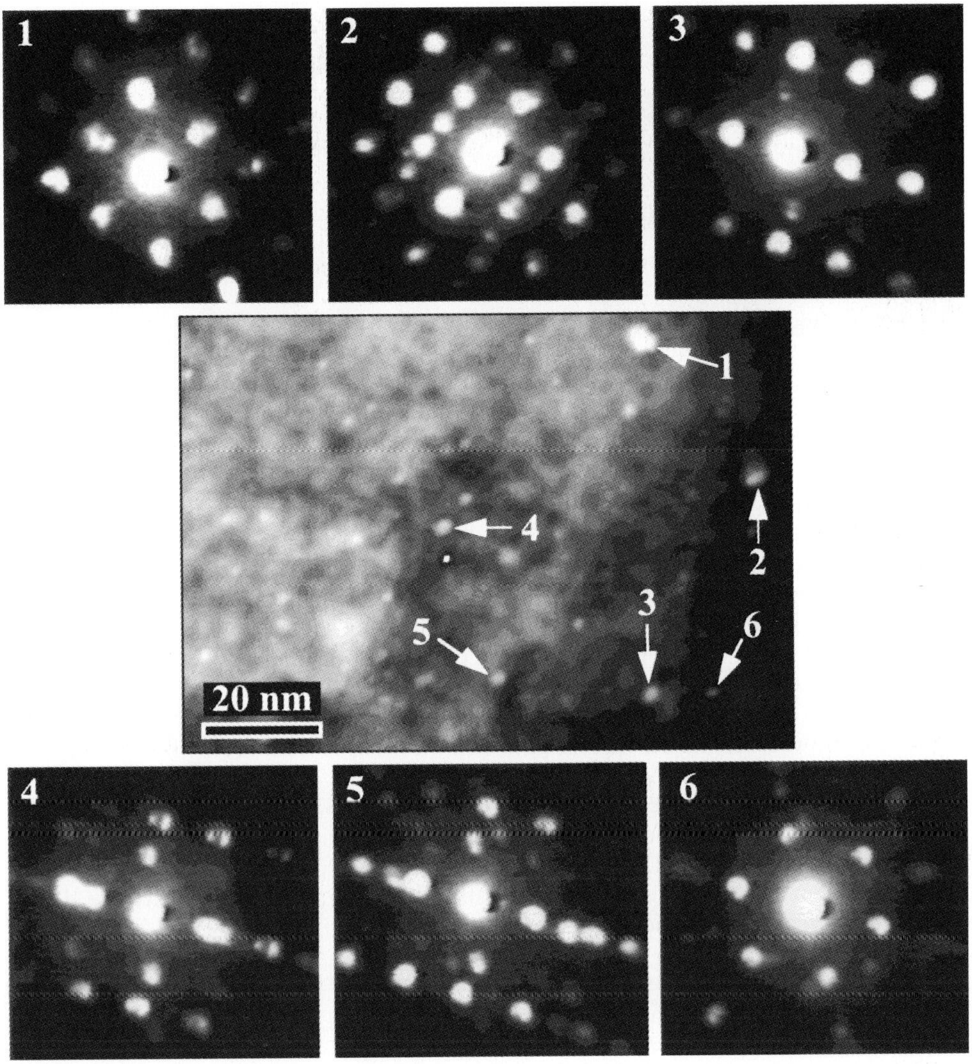

Figure 4-21. HAADF image of bimetallic particles supported on zeolite crystallites connected by alumina binders and electron nanodiffraction patterns obtained from the six nanoparticles labeled in the HAADF image. Some of the diffraction spots are clearly split along certain crystallographic orientations (spot splitting in the diffraction patterns # 3 and # 4).

butions of their CEND patterns. Before this technique can be effectively and reliably utilized to extract the sample information coded in CEND patterns of small particles, many experimental difficulties, such as particle stability, contamination, and accurate control of beam defocus, have to be overcome.

When metal atoms aggregate from the vapor phase or in a liquid, they usually form a crystal, having shapes of regular pentagonal bi-prisms or icosahedra (see Chapter 3 for details). Their internal structure is a complex arrangement of five or twenty twinned components. Large metal particles with shapes of cuboctahedron, decahe-

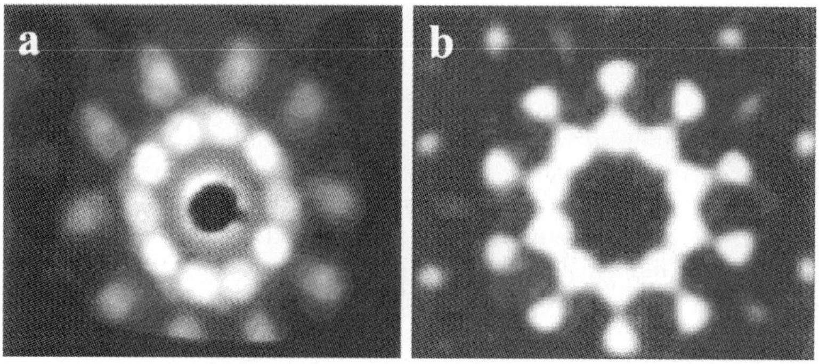

Figure 4-22. Experimental (a) and simulated (b) coherent electron nanodiffraction patterns of a silver nanoparticle with an icosahedral shape. The incident beam was along the five-fold symmetry axis of the icosahedral particle.

dron, icosahedron, and other multiple-twinned structures have been observed [74–76]. For particles with sizes smaller than 2 nm in diameter, however, it is difficult to unambiguously determine their shape by imaging techniques.

CEND technique can provide information about the shape of clean, metallic nanoparticles. For example, a large portion of clean silver nanoparticles with a size smaller than 3 nm in diameter was observed to give unique CEND patterns exhibiting almost five-fold symmetry. These clean silver nanoparticles were formed by in situ deposition in a UHV STEM instrument (MIDAS). Figure 4-22a shows such a CEND pattern and Fig. 4-22b shows a simulated CEND pattern of a small icosahedron with the incident beam direction along the five-fold symmetry axis of the icosahedron. The simulated CEND pattern closely matches the experimental one. These small particles are not stable under intense electron beam irradiation. They rapidly change their orientation and crystal structure [77]. Detailed analyses of CEND patterns recorded on videotapes, however, can provide information about the shape of, as well as the defect structure in, clean nanoparticles.

4.4.2 Coherent electron nanodiffraction with $\alpha > \theta_B$

When the incident electron convergent semi-angle α is larger than the Bragg angle, θ_B, of the diffracting planes, interference effects occur in regions of overlapping discs (see schematic diagrams of Fig. 4-7). In these overlapping regions, the amplitudes of waves from different incident beam directions are added. The phase differences among these waves depend on the values of the transfer function of the probe-forming lens and the relative phases of the specific reflections of the crystal [8]. The phase differences vary with the incident beam position relative to the origin of the unit cell of the crystal. The intensity distribution in overlapping regions becomes sensitive to the probe position and the microscope parameters, such as beam defocus, spherical aberration coefficient of the objective lens, and stability of the microscope. Lattice images can be obtained by placing a STEM detector in the overlapping regions (see Section 4.3.3).

It is possible to determine the detailed atomic arrangement at the core of a defect by coherent overlapping CBED technique [78, 79]. These CBED patterns can also be used to determine the local symmetry of a crystal [80]. In principle, analysis of the intensity distributions in all overlapping regions of the diffraction pattern should provide information about the relative phases of all diffracted beams, the basis for unambiguous structure analysis of crystals.

A new technique for determining atomic positions using a series of coherent CBED patterns from overlapping regions has been developed; auto-correlation functions are computed for each probe position, giving atomic coordinates with an accuracy of 0.02 nm [5]. However, the most frequent use of these overlapping diffraction patterns is for the final stages of lens alignment and a stigmatism correction of the objective lens for high-resolution STEM imaging [81].

When the convergent angle of the incident beam is made even larger, the diffraction discs overlap at any point of the CEND pattern so that individual discs are no longer visible. The coherent incident beam may have a central peak smaller than the projection of the unit cell of a crystal. The CEND pattern depends on the symmetry and the structure of only that small portion of the projected potential. The intensity distribution of the whole diffraction pattern changes as the beam is moved across the unit cell of the crystal. The changes of intensity and symmetry in CEND patterns can also be observed when an incident beam is scanned across localized defect [79]. By comparing the intensity variations of CEND patterns across crystal defects to those simulated by using many-beam dynamical theory, it is possible to deduce the nature of localized defects in crystalline materials [78, 79].

Recent experiments performed in FE-TEM instruments have demonstrated distortion-free interference fringes in coherent CBED patterns with a lattice spacing < 0.13 nm [82]. The phases of crystal structure factors can be determined from the relative positions of the interference fringes [83]. This coherent interference technique has been applied to the investigation of glide planes, stacking faults, and other defects in crystalline materials [82, 84].

4.4.3 Coherent electron nanodiffraction with $\alpha >> \theta_B$

When the convergent semi-angle of the incident beam is much larger than the Bragg angle of the crystal, the diffraction discs are no longer identifiable, and a point projection image of the specimen is formed on the diffraction plane. In practice, this condition corresponds to using a very large objective aperture or no objective aperture at all. In the absence of spherical aberration, this projection image would be a direct representation of the object transmission function with a resolution determined by the effective probe size. The magnification of projection images is inversely proportional to the beam defocus [61]. The effect of the spherical aberration of the objective lens is to introduce distortions in projection images [81, 85]. These distortions become severe when the in-focus position is approached [85].

For a thin crystal, a special type of projection image, called electron Ronchigram, can be observed in the diffraction plane (Fig. 4-23) [81, 85]. These patterns are called electron Ronchigrams because the geometry used to obtain them is identical to that used in testing optical lenses and mirrors [86]. Electron Ronchigrams can be conveniently used to measure the spherical aberration coefficient and the defocus value of the objective lens [85]. At large defocus values, projection images are similar to low magnification TEM images.

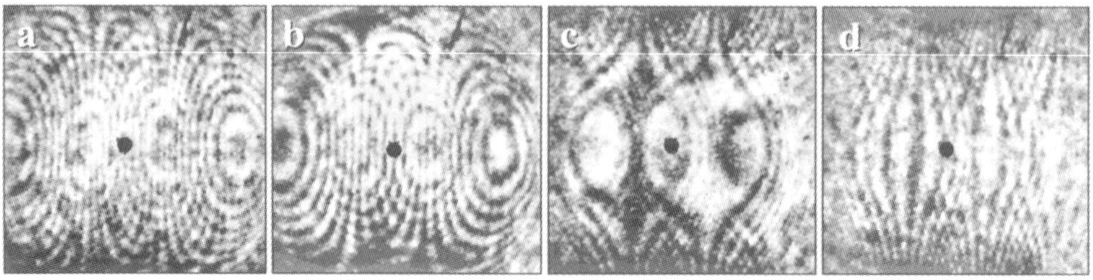

Figure 4-23. A through-focus series of electron Ronchigrams from under-focus (a) to over-focus (d), obtained from a thin GaAs crystal.

A projection image formed by a point source distance Z from an object produces an identical image to that produced by a plane-wave illumination on a plane at a distance Z beyond the object. When a thin crystal is oriented along a principle-zone axis, a slightly defocused projection image can give two-dimensional lattice fringes representative of the projected crystal structure.

With the correction of lens aberrations, increased stability of the microscope parameters, and smaller incident probe sizes, we should be able to study the local structure of crystals in projection images without using any lenses or scanning systems.

Point projection images formed in STEM are exactly the in-line electron holograms proposed by Gabor: in-line holograms could be recorded and used to reconstruct the object wave [87–88]. Gabor suggested that if the reconstruction could be done with an optical system with appropriate aberrations, the effect of the aberrations on the original electron optical images could be corrected and the resolution of the electron microscope could be enhanced. Initial attempt to reconstruct the object function using digitally recorded holograms in a computer has shown limited success [89].

For greatly defocused projection images, an off-line electron holography technique can be effectively utilized to extract quantitative information about the magnetization in small magnetic particles with a nanometer resolution [90–91].

4.5 Imaging with secondary electrons

High-energy electrons impinging on a solid sample experience elastic scattering by atomic nuclei and inelastic scattering by sample electrons. Inelastic scattering results in the transferring of energy from the high-energy electrons to the sample electrons. Thus, specimen-specific electrons can be excited to high-energy states. Some of the excited electrons travelling to the sample surface can be emitted out of the sample as secondary or Auger electrons. By collecting these low-energy electrons, high-resolution surface images can be obtained to give morphological information about the sample (Fig. 4-4). In the following, we will discuss the emission and collection of secondary electrons, the resolution and contrast of SE images, and the application of the SE imaging technique to the characterization of nanoparticles.

4.5.1 Emission of secondary electrons

Electron-induced SE emission is a complicated process and is not well understood. Nevertheless, the emission of secondary electrons from a solid sample can be described as a three-step process: generation, transportation, and emission.

The generation of internal secondary electrons is directly related to the electron energy-loss processes of high-energy incident electrons that undergo a series of inelastic scattering events with an average energy loss of about 20 eV per event. The main energy-loss peaks are usually associated with the generation of plasmons, the direct excitation of outer-shell electrons, and the excitation of core electrons [92]. Secondary electrons can be generated via plasmon-decay or electron-electron scattering processes. The relative contribution of each generation process to the total number of internal secondary electrons depends on the properties of the material under study as well as the energy of the primary incident electrons. The number of secondary electrons generated inside a solid is often regarded as proportional to the stopping power (total energy loss per unit path length) of the incident electrons [93–94]. The relative importance of plasmon-decay processes in the SE production has also been emphasized for conducting materials [95–96].

The internal secondary electrons thus generated interact with different components of the sample such as electrons, phonons, etc. During their diffusion through the solid, the internal secondary electrons can be scattered both elastically and inelastically. Elastic scattering modifies their angular distribution while inelastic scattering changes their energy distribution. The scattering processes of secondary electrons can be best described by a multiple scattering or diffusion theory of low-energy electrons in solids [96–99]. This diffusion process of internal secondary electrons is also called the cascade process of SE transportation.

Only a small fraction of the internal secondary electrons may reach the sample surface because of elastic and inelastic scattering processes. Those secondary electrons that can surmount the energy barrier (work function) at the sample surface can be emitted out of the sample to become detectable secondary electrons. The final emission process is sensitive to the surface properties of the sample such as work function, surface adsorbates, thin layer deposition, sample contamination, surface charging, etc.

The current density of emitted secondary electrons depends on the initial inelastic scattering events, the decay processes of the initial excitations, the transport of low-energy electrons through the sample, and the work function and modifications of the sample surface.

4.5.2 Detection of secondary electrons

In a STEM instrument, a specimen is usually placed inside the pole pieces of a highly excited objective lens. The emitted secondary electrons first experience a strong magnetic field before being collected by an SE detector. Due to the effect of the magnetic field, an emitted SE spirals in a cyclotron orbit with a radius R that depends on the energy and the emission angle of the SE, and the strength of the magnetic field. When secondary electrons travel up or down the optic axis of the microscope, their emission angles are compressed. After spiraling out of the bores of the objective lens, secondary electrons are collected by an SE detector through a transverse electric field. Because of the effect of the magnetic field on the trajectory of the emitted secondary electrons, the collection efficiency of secondary electrons in a STEM instrument is much higher than that in a conventional scanning electron microscope.

The collection efficiency of low-energy electrons can be further enhanced by the use of electron "parallelizers" located inside the bores of the objective lens [100–101]. The "parallelizers" provide a magnetic field that is strong enough to keep low-energy electrons moving in a small spiral trajectory and traveling out of the bores of the objective lens. The detection efficiency of low-energy electrons in such systems can be significantly increased [100].

The energy distribution of the collected secondary electrons can be analyzed by a low-energy electron spectrometer. Secondary electron spectroscopy (SES) can be used to investigate the energy distribution of secondary electrons from different materials; to measure work function of solid samples; and to study the charging effects of non-conducting materials [102–103]. Localized SES spectra also give information about the nature of high-resolution SE imaging. High-resolution SES has been used to investigate SE emission processes from metals, semiconductors, and insulators [52, 104–106].

Figure 4-24 shows a SES spectrum obtained from a small MgO cube. Since the onset energy of the secondary electrons does not shift from the zero value, we can conclude that there is no observable charge-up of this particular crystal although MgO is an insulator. Furthermore, the SES spectrum shows that more than 90% of the emitted secondary electrons have energies < 5 eV. Secondary electrons emitted from clean, non-charging MgO crystals have a maximum intensity at 1.5 eV with full-width-at-half-maximum (FWHM) of about 2 eV. Energy-selected SE imaging can be performed by selecting certain portion(s) of an SES spectrum as an imaging signal to enhance the contrast of specific features or to improve image resolution and surface sensitivity.

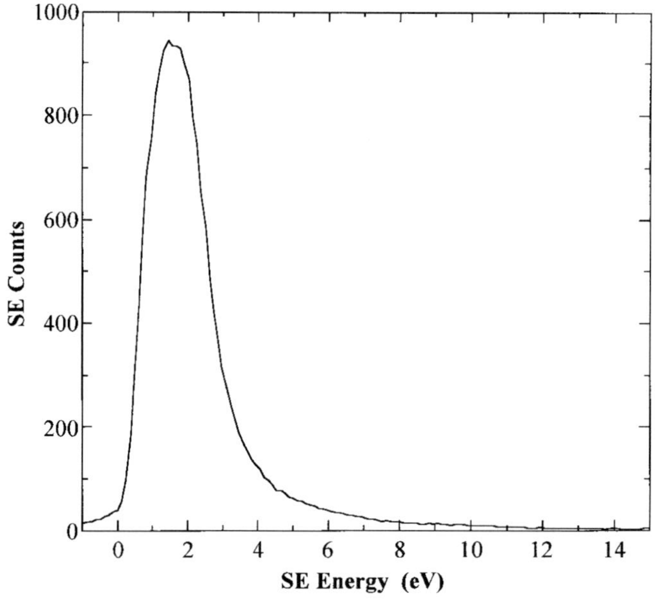

Figure 4-24. High-energy resolution secondary electron spectrum obtained from a small MgO cubic crystal shows that more than 90% of the collected secondary electrons have energies less than 5 eV.

4.5.3 Resolution and contrast of secondary electron images

There are two types of secondary electrons emitted at the surface of a sample. Secondary electrons that are directly generated by the incident beam are termed SE1 and those that are generated by backscattered electrons are termed SE2. Signals of SE1 and SE2 cannot be distinguished from their energy or angular distributions. Depending on the incident beam energy and the type of samples under investigation, the total signal strength of SE2 can be stronger than that of SE1.

The resolution of SE images depends on the rate at which the signal changes as the probe is scanned across a sample. The contrast of SE images depends on the relative intensity variations among different probe positions. Therefore, both the resolution and the contrast of SE images depend on the local current density of the emitted secondary electrons. With very small probe sizes, the spatial distributions of SE1 and SE2 determine the image contrast and resolution.

In a STEM or a field-emission SEM instrument, the current density of SE1 is much higher than that of SE2, although the total signal strength of SE2 could be stronger than that of SE1 for thick or bulk samples. At high magnifications, the SE2 signal varies much more slowly with the movement of the incident probe than that of SE1, thus SE1 and SE2 signals can be spatially separated. For thin specimens, the contribution of SE2 is usually negligible. The contrast and the resolution of SE images of thin specimens are entirely determined by the spatial distribution of SE1 signals.

Sub-nanometer surface details can be observed in high-resolution SE images [107–111]. This implies that the generation processes of secondary electrons are localized to within 1 nm or less. It was first pointed out and later experimentally proved that the generation of secondary electrons is directly related to large-angle inelastic scattering of the high-energy incident electrons [112–113]. There exist large momentum transfer mechanisms during inelastic interaction processes such as Umklapp or phonon-assisted electron excitation processes. Inelastic scattering events involving these processes are highly localized. The resolution obtainable in SE images is currently limited by the incident probe size to about 0.5 nm.

The contrast in SE images is primarily due to topographic effect, although other contrast mechanisms (material contrast, work function contrast, etc.) may play a role in determining the contrast of SE images of specific samples [93, 112, 114]. Figure 4-25 shows an SE image of a MgO particle revealing surface steps, facets, and the var-

Figure 4-25. High-resolution secondary electron image of a large MgO cube clearly shows flat {001} faces, terraced {011} faces, and the faceted (111) face. The {011} and {111} surfaces are composed of {001} terraces or facets.

ious faces of the incomplete MgO cube. Individual atomic steps can be observed in high-resolution SE images with characteristic black and white contrast for up- and down-steps, respectively [106]. For normal incidence of the electron beam, all steps are shown bright but sharper and fainter. The observed contrast of SE images of surface steps can be explained in terms of SE emission from the steps. The amount of material that generates and emits secondary electrons is greater for down-steps than for a flat surface and is less for up-steps [106].

4.5.4 Image contrast of small particles

When the radius of a small particle becomes comparable to, or even smaller than, the escape depths of secondary electrons, most of the secondary electrons, generated inside the particle, with energies higher than the surface barrier may escape. In contrast, because of the effect of total internal reflection of low-energy secondary electrons, only about 10% of the total internal secondary electrons which have energies higher than the surface barrier can escape from a flat surface. Because of this geometric effect in SE emission, small particles are often observed with a bright contrast in high-resolution SE images (Fig. 4-26) [109, 112].

The SE signal strength of a small particle increases with the size of the particle. Furthermore, for metal particles with a radius much smaller than the average escape depth of the collected secondary electrons, the SE image intensities of these particles follow exactly those of HAADF images. Figure 4-27 shows SE and HAADF intensity line-scans, acquired simultaneously, across small, clean silver nanoparticles. These intensity line-scans clearly show that the SE signal is as localized as the HAADF signal within the resolution limit which is about 0.6 nm in these images. The SE intensity profiles almost overlap those of HAADF signal. In fact, the total integrated SE intensity (I_{SE}) from a small particle is proportional to the volume of that particle. If we plot $I_{SE}^{1/3}$ against the particle radius R, a straight line is obtained (see Fig. 4-16), similar to that of the HAADF signal. Thus, the total integrated SE intensities of small particles are proportional to the volume of the particles.

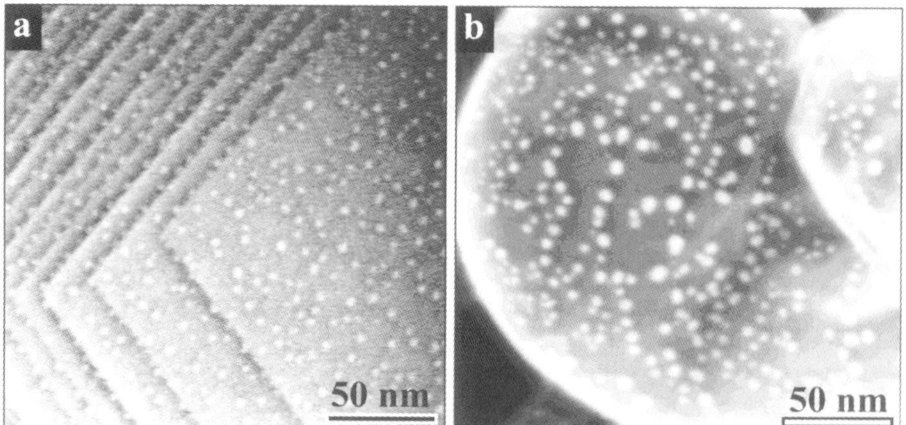

Figure 4-26. High-resolution secondary electron images of silver nanoparticles deposited onto clean surfaces of (a) MgO and (b) α-alumina crystals. Small silver particles are clearly shown with a bright particle contrast.

The particle contrast in SE images can be parameterized by the ratio of the particle radius (R) to the average escape-depth (L) of the collected secondary electrons (Fig. 4-28). If $R/L < 1$, the brightness of a particle increases with the size of the particle and the image intensity has a maximum at the center of the particle. If $R/L > 1$, the particle intensity slowly increases with the size of the particle and the highest image intensity is approximately at a distance $d = (R - L)$ from the center of the particle. For very large particles, the particle contrast evolves into the edge-brightness contrast commonly observed in SE images.

Although the resolution of SE images is comparable to the size of the incident probe, it is impossible to extract information about the shape of nanoparticles with sizes less than the escape depth of the collected secondary electrons. Therefore, we cannot extract information about detailed surface morphology of very small particles. However, we can obtain useful information about the relative locations of nanoparticles with respect to the surface topography of supports (Fig. 4-4).

Because of the use of strong probe-forming lenses and field-emission electron sources, it is possible to obtain SE images with a resolution of about 0.5 nm and 2 nm at 30 kV and 1 kV, respectively. The collection efficiency of secondary electrons in a field-emission SEM is also significantly improved due to the utilization of novel SE detection configurations. It is especially attractive to operate at low electron energies: the increase in signal strength, the reduced volume of electron-specimen interactions, and the neutralization of charging effects for non-conducting materials [115–116]. Other advantages of low-voltage SEM include enhanced surface sensitivity at medium image resolution and reduced radiation damage for delicate samples [115–116]. Metal nanoparticles, as well as detailed surface morphology of supports, can be clearly revealed in low-voltage high-resolution SE images with high contrast and high surface sensitivity (Fig. 4-29).

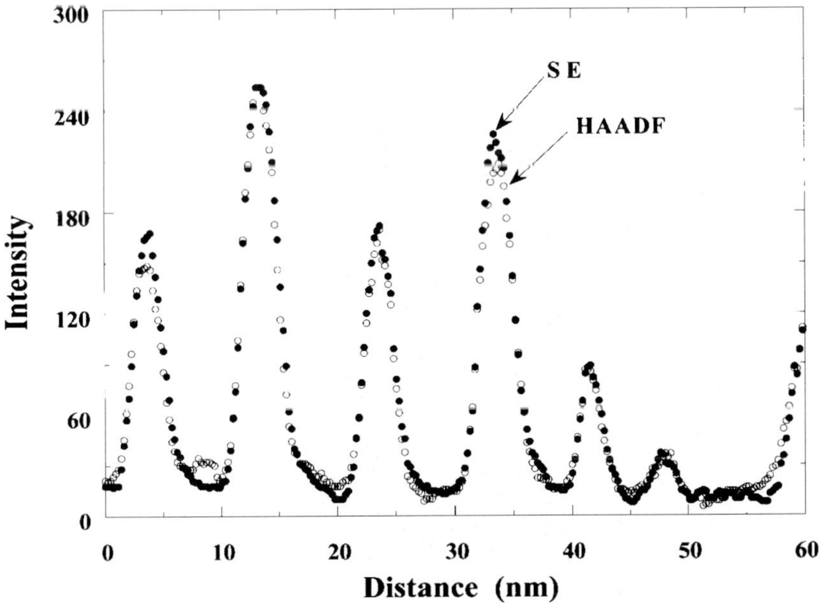

Figure 4-27. HAADF and SE intensity line-scans across silver nanoparticles. The SE intensity line-scan closely follows that of the HAADF signal.

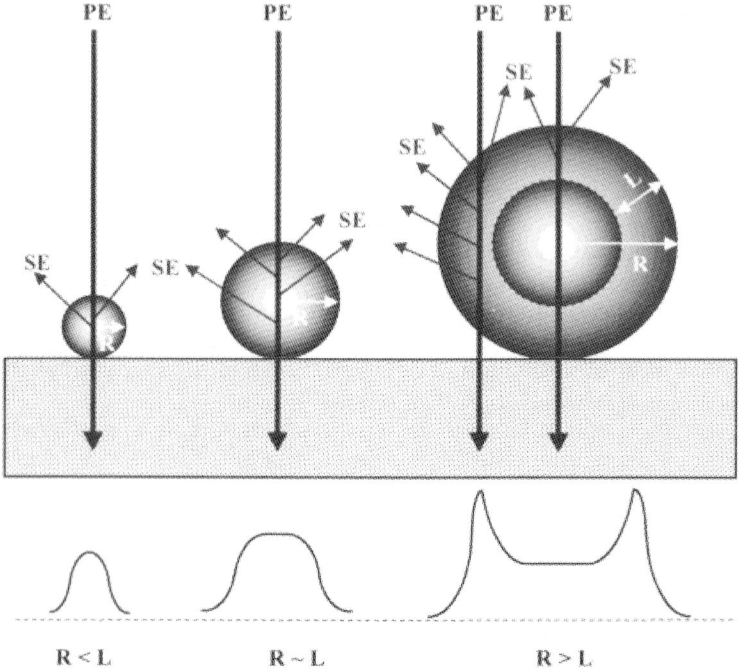

Figure 4-28. Schematic diagrams illustrate the emission of secondary electrons from spherical particles of different sizes and the corresponding SE intensity line-scans across the particles. The parameters *R* and *L* represent the particle radius and the average escape depth of the secondary electrons, respectively.

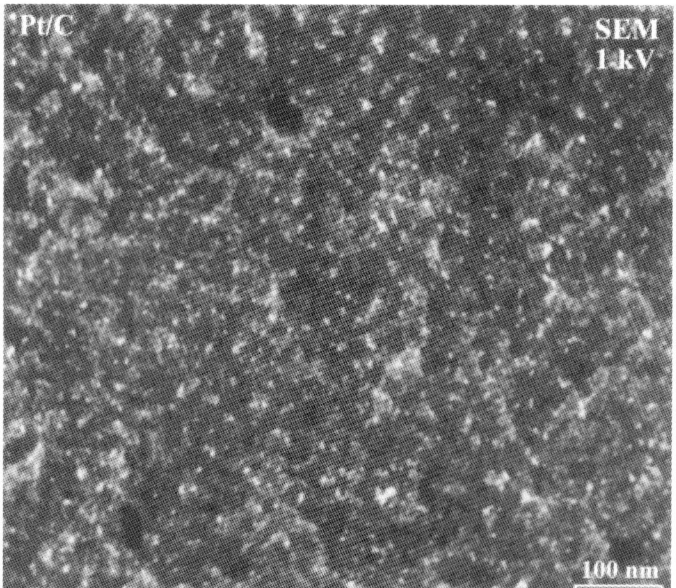

Figure 4-29. Low voltage, high-resolution secondary electron image of Pt nanoparticles dispersed in a highly activated carbon. The Pt particles, as well as the nano-pores of the carbon support, are clearly revealed.

4.6 Imaging with Auger electrons

An atomic inner-shell vacancy produced by the incident electrons can be relaxed through a two-electron process: one electron fills the inner-shell vacancy and the other is emitted from the atom (Fig. 4-30). The emitted electron is called an Auger electron. The energies of the primary excitation and of the emitted Auger electrons are both characteristic of the element(s) involved [117]. The production of Auger electrons is essentially similar to that of low-energy secondary electrons; the initial excitation produced by the inelastic scattering of the incident electrons decays to generate a low-energy electron that can escape into the vacuum. In contrast to the diffusion of secondary electrons, Auger electrons must escape from the specimen surface without losing any energy in order to be registered as Auger peak signals. The reason that Auger electron spectroscopy is a surface-sensitive technique lies in the intense inelastic scattering that occurs for electrons in this energy range. Thus, only Auger electrons generated from the outmost atom layers of a solid can survive to be ejected and measured in the Auger electron spectrum. Most of the emitted Auger electrons are produced within a very short distance from the sample surface, typically 0.3 to 3 nm [118].

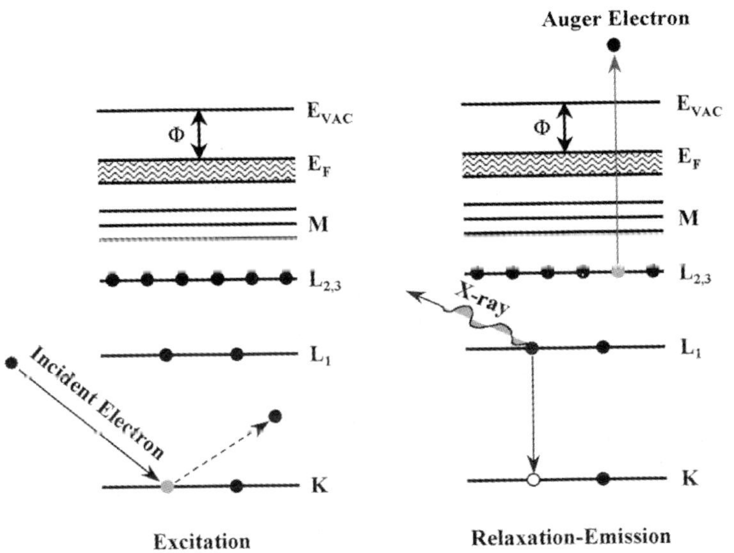

Figure 4-30. Schematic diagram illustrates the emission of characteristic X-rays and Auger electrons from a solid sample. The excitation of a core-level electron can be relaxed either by emission of a photon (characteristic X-ray) or by emission of another electron (Auger electron).

4.6.1 High spatial resolution Auger electron spectroscopy

In a STEM instrument, Auger electrons, emitted from either the entrance or the exit surface of a specimen, can be collected and analyzed using a CMA (cylindrical mirror analyzer) or a CHA (concentric hemispherical analyzer) electron spectrometer. Because of the high-energy and high-brightness of the incident electrons, the employment of magnetic "parallizers", and the use of thin specimens in STEM instruments, high-quality Auger electron spectra can be acquired with extremely high peak-to-background ratios [119–121]. Figure 4-31a shows a high-energy resolution AES spectrum of clean silver nanoparticles supported on a thin carbon film; the silver MNN doublet is clearly resolved. The sizes of the silver nanoparticles range from 1 to 5 nm in diameter.

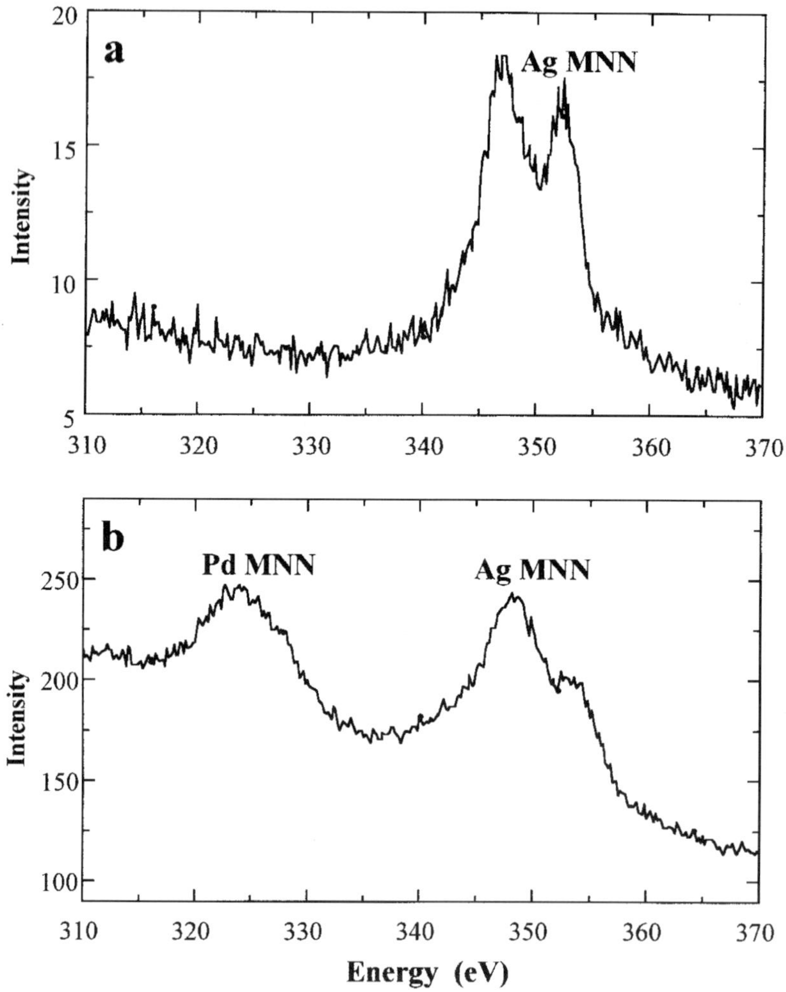

Figure 4-31. Auger electron spectra obtained in a UHV STEM instrument (MIDAS): (a) high-energy resolution MNN Auger electron spectrum of silver nanoparticles; (b) Auger electron spectrum obtained from Ag/Pd bemetallic particles.

Surface compositional analysis of individual nanoparticles is essential for understanding the activity and selectivity of industrial bimetallic or multi-component catalysts used in a variety of chemical reactions. The overall composition of these individual nanoparticles can usually be obtained by the use of high spatial resolution XEDS (see section 4.7). It is, however, extremely difficult to extract information about preferential surface segregation or aggregation of individual components in nanoparticles of different sizes. Because of the high-surface sensitivity of Auger electrons, it is possible to determine qualitatively and, in some cases, quantitatively, the surface composition of nanoparticles consisting of multiple components. Figure 4-31b shows an Auger electron spectrum obtained from a sample containing palladium, silver, or palladium-silver nanoparticles highly dispersed on a high-surface area support. Both the silver and the palladium MNN Auger peaks are clearly revealed. Quantitative analyses of this type of spectrum can provide information about the surface enrichment of specific elements and information about how this enrichment varies with the size of the particles.

4.6.2 High spatial resolution scanning Auger microscopy

Considerable efforts have been devoted to the development of scanning Auger microscopy (SAM) and the improvement of its imaging resolution since the pioneering work of MacDonald and Waldrop [122]. The driving force for developing high spatial resolution SAM is related to the characterization of nanoparticles and other nanostructured materials. The best resolution of SAM images obtained in conventional SEM geometry is approximately 30 nm under favorable conditions [123–124]. A spatial resolution of about 10 nm can be achieved using the high current density and the small probes of STEM instruments [125]. A similar or better image resolution may also be achievable in the new generation of field-emission SEM instruments. Further improvement in resolution by using smaller electron probes significantly reduces the strength of the collected Auger signal.

With the improvement in the detection efficiency of low-energy electrons in the MIDAS system by employing magnetic "parallelizers" [100], a spatial resolution of $<$ 3 nm in Auger peak images can be obtained on bulk samples [120]. For electron transparent specimens, typically used in STEM instruments, an image resolution $<$ 1 nm can be achieved [121]. Silver nanoparticles $<$ 1 nm in diameter and containing as few as 15 silver atoms can be detected [119, 121].

Metal nanoparticles highly dispersed on supporting materials represent a significant fraction of heterogeneous catalyst systems. Surface chemical information about such systems can be obtained by collecting Auger electrons to form high spatial resolution, elementally specific, and surface sensitive images. Figure 4-32 shows a set of images of a model catalyst consisting of clean silver nanoparticles deposited onto a thin carbon film. Both the silver nanoparticles and the surface details of the support are revealed in the SE image. However, the nano-scale pore structure of the carbon film is clearly revealed in the C KLL Auger peak image with high contrast. The high intensity of the silver particles in the C KLL Auger peak image could occur due to electron scattering processes or because of the high SE background from the silver nanoparticles [121].

The silver nanoparticles are shown with very high contrast in the silver MNN Auger peak image. The silver Auger map, obtained by subtracting the background image from the silver MNN Auger peak image, shows silver clusters as small as 1–2 nm in diameter. The average background pixel value in the silver Auger map is $<$ 3 while the

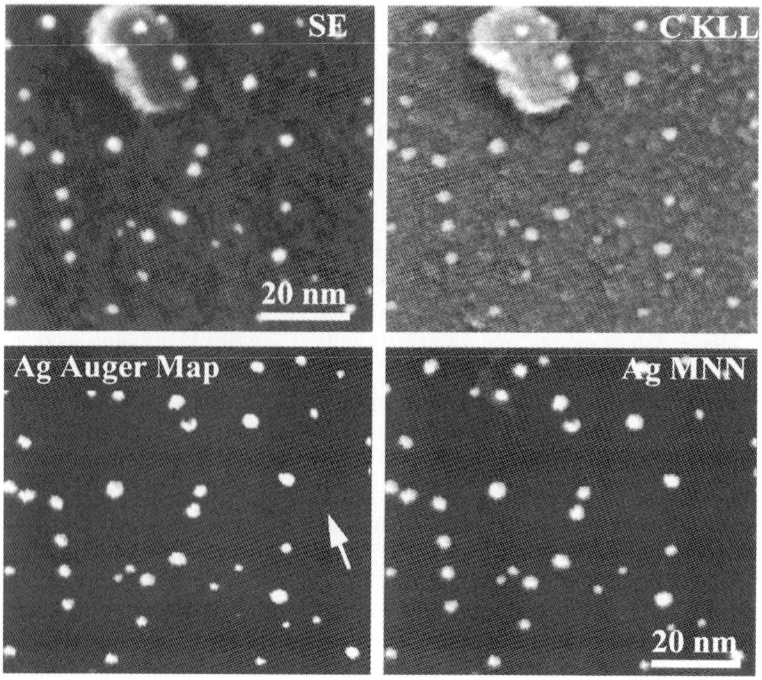

Figure 4-32. A set of images of silver nanoparticles supported on a carbon film: SE image; carbon KLL Auger peak image; silver MNN Auger peak image; and silver Auger map obtained by subtracting a background image from the silver MNN Auger peak image.

smallest silver particles have a pixel value of about 30 at the center of the particle (indicated by the arrow in the silver Auger map). Larger particles have pixel values of about 340 at the center of the particles. Further particle analyses can be performed on this type of "clean" image, yielding information about the size distribution and the composition of the particles. The intensity variations of Auger signals with the sizes of silver nanoparticles can also be investigated.

The digitally acquired Auger peak and background signal images contain quantitative data on the Auger peak height and SE background intensity. Data sets obtained by processing the digital images contain quantitative information about the contrast of the Auger peak and background signal images [121].

The topographic contrast in SAM images of bulk samples can be, in general, minimized by using the algorithm (P–B)/(P+B). Here P represents the signal at the Auger peak and B represents a background signal taken at an energy level just above the Auger peak [126]. Despite the successful use of this algorithm in low resolution SAM images, it cannot be directly applied to nanometer-resolution SAM images. In fact, the (P–B)/(P+B) or (P–B)/B display only enhances the edge contrast since the escape depth of the background signal is much greater than the attenuation length of the Auger electrons [121, 127]. For nanometer-resolution SAM images, the P–B display may best represent the elemental map of the sample surface.

4.6.3 Resolution limit

The resolution in SAM images depends on several sample- and instrument-related effects. The sample-related effects include: 1) surface topography, 2) escape depth of the collected Auger electrons, 3) contribution from backscattered electrons, and 4) localization of the Auger electron production processes. The last factor sets the ultimate resolution limit that will be achievable in SAM images. Since the primary inelastic scattering processes involve excitation of inner-shell electrons, the generation of Auger electrons is highly localized. With thin specimens and high-energy incident electrons, the contribution from backscattered electrons should be negligible. It may, however, degrade the image resolution and affect the image contrast of bulk samples.

The instrument-related effects include: 1) the intensity distribution of high-energy electron probes, 2) the collection efficiency of the emitted Auger electrons, and 3) the instability of the STEM microscopes. The capabilities and limitations of high spatial resolution AES and SAM are discussed in a recent review article [128]. At present, the instrument-related factors set the limits of obtainable resolution to about 1 nm in Auger peak images of thin specimens [121].

4.6.4 Detection sensitivity

Because the Auger electron intensity profile of a particle is a complicated convolution of the electron probe with the real morphology of the particle, the true size of the particle may be smaller than that measured in a SAM image. A de-convolution process might be used to extract the real sizes of the observed nanoparticles. However, knowledge of the intensity distribution of the probe, the radial and angular distributions of the emitted Auger electrons, and the detailed transmission function of the electron analyzer would have to be known.

The emission of Auger electrons from small particles is different from that of flat surfaces. Electrons generated inside a small particle may all escape if the radius of the particle is much smaller than the inelastic mean free path (λ) of the collected Auger electrons. In this limit, the total number of the collected Auger electrons is proportional to the volume of the particle; and the distinction between "bulk" and surface signals is no longer valid. A quantitative estimate of the minimum particle size detectable in SAM images can be developed from a simple relationship between the signal strength and the size of the particle. By using an intensity-ratio method, it is possible to estimate the total number of atoms detectable in high spatial resolution Auger maps [119].

The total integrated signal strength of a small particle, in the absence of image noise, should be independent of the size of the incident electron probes. The integrated intensity (I) of a particle in SAM images is proportional to the total number of atoms (N) which contribute to the collected Auger signal, $I = \alpha N$. The parameter α is dependent on the Auger electron cross-section, detection efficiency, incident beam current, and other instrument-related parameters. However, for each SAM image α can reasonably be assumed a constant.

Thus, the number of atoms contained in a small particle N_1 with an integrated intensity I_1 can be estimated from a large particle with N_2 atoms and an integrated intensity I_2: $N_1 = N_2(I_1/I_2)$. This relationship indicates that a large particle could be used as an internal calibration standard to determine the total number of atoms contained in smaller particles. On this basis, it is estimated that particles containing as few as 15 silver atoms can be detected in high-resolution SAM images (see Fig. 4-32). At present, the minimum detectable mass in high spatial resolution SAM images is less than 3×10^{-21} g [119].

The minimum detectable dimension is different from the resolution of SAM images [128–129]. While the latter is currently limited by the sizes of the incident electron probe, the former is related to, but not limited by, the dimension of the incident electron probe. The minimum detectable size can be much smaller than the incident probe diameter. The minimum particle size detectable in a SAM image is directly related to the signal-to-noise and signal-to-background ratios, as well as the radiation damage of the sample. For example, the minimum detectable dimension would be the size of a single atom if the signal strength were not a limiting factor, and that the atom were stable under intense electron beam irradiation.

4.7 Nanoanalysis with energy-loss electrons and X-rays

The remarkable capability of forming an electron probe with sub-nanometer sizes in STEM instruments makes it practical to perform chemical analysis with a nanometer or sub-nanometer resolution. When high-energy incident electrons interact with a specimen, a variety of signals are generated from within the interaction volume (see Fig. 4-1). Some of these signals such as characteristic X-ray photons, energy-loss electrons, Auger electrons, and light photons, carry information about the composition or the chemistry of the sample. These analytical signals can be collected and analyzed by appropriate spectrometers to give useful information about the sample.

Unlike microscopy that records the location and distribution of atoms, particles, and phases, spectroscopy is concerned with the measurement of energies or wavelengths of the collected analytical signals. We can transform various spectra into useful data that give information about elemental composition or distribution of different phases in the sample at a nanometer and, in some cases, an atomic resolution. Among these spectroscopic techniques, XEDS and EELS are the two most frequently used analytical techniques in STEM instruments.

EELS is the analysis of the energy distribution of high-energy incident electrons that have been inelastically scattered by the specimen. These inelastically scattered electrons carry important information about the electronic structure of the specimen atoms. Thus, EELS spectra contain essential details of the nature of these atoms, their bonding and nearest-neighbor coordination, charge transfer, and dielectric response (see Chapter 3 for detailed discussions).

XEDS is the analysis of the energy distribution of emitted characteristic X-ray photons, which are related to the excitation of inner-shell electrons of the specimen atoms (Fig. 4-30). The dependence of the emitted characteristic X-rays on the nature of the excited atoms present in a sample provides the basis of nanoanalysis by XEDS. The fundamental principles of XEDS and EELS have been discussed in many textbooks [9, 92, 130–132]. We discuss here only those features that are unique to STEM configurations.

4.7.1 Electron energy-loss spectroscopy

By performing parallel EELS (PEELS) in conjunction with the HAADF imaging technique in STEM, detailed information about the composition, chemistry, and structure of the specimen can be obtained with atomic resolution and sensitivity. This strategy has recently been applied to the study of a variety of materials problems with great success [133–140]. For example, in the study of iron nanoparticles used as a fuel additive to reduce or enhance soot formation during combustion, PEELS gave information about the spatial variation of the oxidation states of iron species within individual nanoparticles 141]. This type of information is extremely important for understanding the role of nanoparticles during catalytic reactions.

Electron energy-loss images, formed by selecting characteristic energy-loss peaks in EELS spectra as input signals, provide information about the spatial distribution of elements or electronic and oxidation states in the specimen [135, 142–143]. For example, energy-loss electrons were used to obtain maps of sp^2 and sp^3 states of carbon at sub-nanometer spatial resolution across the silicon-diamond interface [135]. By analyzing the electron energy-loss images, it was concluded that diamond nucleates on an amorphous carbon layer, with the transition from sp^2 to sp^3 carbon occurring over a distance less than 1 nm. This type of valuable information can only be obtained by using a high-brightness, sub-nanometer electron probe available in STEM instruments.

A recent development in quantitative EELS mapping is the spectrum-imaging technique: acquire and store a spectrum at each pixel in a two-dimensional image [142–144]. A spectrum-image in its most common form is similar to a normal STEM image; but a spectrum-image contains an entire spectrum at each pixel. A spectrum-image can be viewed as a three–dimensional image with two spatial coordinates and one energy coordinate. One of the greatest advantages of the spectrum-image technique is that it can be processed a posterior, thus, offering access to the exact relationship between the spatial coordinates and the spectral feature. Elements in unexpected locations may be found by off-line processing the digitally stored spectra. The major limitations of this technique include acquisition and manipulation of a tremendous amount of spectral data and the time-consuming process of analyzing millions of spectra for each sample. Quantitative EELS imaging is, however, a powerful method for spatial interpretation of elemental, chemical, and other important information related to the physical properties of nanophase materials.

Another unique application of ELLS in STEM is the study of valence excitation processes of high-energy electrons passing through a thin film or a small particle, or passing close to a surface or interface of a solid material [145–149]. Surface or bulk plasmon excitations, inter-band or intra-band single-electron transitions, and defect-induced electronic states can be investigated with a sub-nanometer spatial resolution (see Chapter 3 for detailed discussion on the excitation processes).

EELS technique is ideal for analysis of low atomic-number materials of thin specimens. When a specimen is thicker, plural and multiple inelastic scattering may become dominant and thus obscure the characteristic features of the EELS spectra.

4.7.2 X-ray energy-dispersive spectroscopy

XEDS is now routinely used, in TEM, SEM, or STEM instruments, to identify unknown phases or to obtain information about the spatial distribution of certain phases of interest. In a STEM instrument, XEDS can be conveniently used to analyze the features revealed in HAADF images by stopping the incident probe at any point of interest. For thin specimens, qualitative interpretation of XEDS spectra can be performed on-line.

A consequence of high spatial resolution in STEM is that the X-ray signal originates from a much smaller volume of the specimen; a weaker signal means long acquisition time in order to obtain statistically meaningful results. Therefore, specimen-drift correction, either manually or automatically, is usually used to collect enough X-ray signals when very small particles are analyzed. Nevertheless, XEDS can detect the presence of just a few atoms if the analyzed volume is small enough [150].

XEDS technique has been widely used to solve materials problems including identifying precipitates in alloys, active phases in supported catalysts, elemental segregation at grain boundaries, and quantitative composition of multi-component phases.

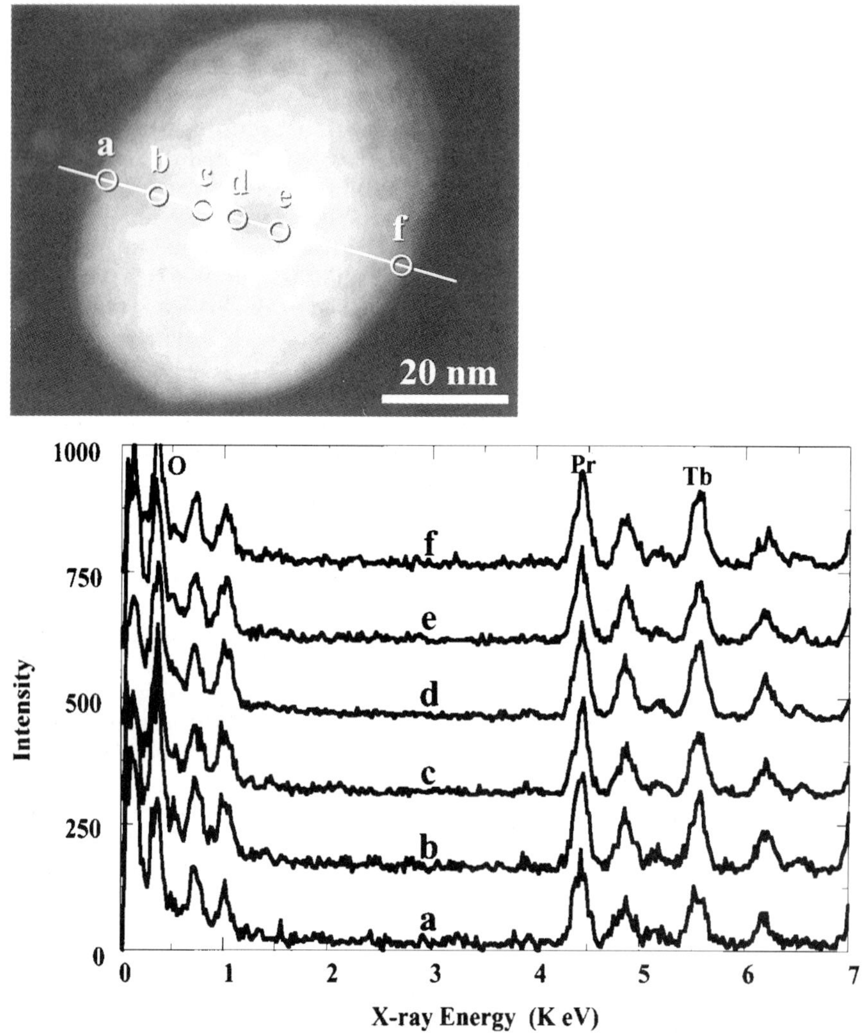

Figure 4-33. Nanoanalysis of a colloidal $Pr_{0.5}Tb_{0.5}O_x$ nanoparticle. The XEDS spectra were obtained from the corresponding locations labeled in the HAADF image.

The combination of qualitative XEDS with HAADF imaging technique is now routinely used in STEM instruments for fast analysis of unknown phases and the spatial distribution of these phases.

In the study of in situ decomposition of hydroxycarbonate ($Pr_{0.5}Tb_{0.5}OHCO_3$) colloidal nanoparticles, it is necessary to understand the chemical homogeneity of the decomposition products within individual nanoparticles. Figure 4-33a shows a HAADF image of one of these nanoparticles after approximately 20 minutes intense electron beam irradiation. To determine the variations in elemental composition across the small particle, a series of XEDS spectra were obtained at different probe positions (Fig. 4-33b). By quantitatively analyzing the XEDS spectra obtained from different particles and comparing these nanoanalytical results with those of HRTEM, a model of the decomposition process and the structure of the final decomposition product were proposed [151].

Similar to EELS mapping, X-ray maps can provide valuable information about two-dimensional elemental distributions in the sample. Particles as small as 2 nm in diameter can be detected in digitally acquired X-ray maps [152]. The ultimate limitation for X-ray mapping of nanoparticles is the extremely low X-ray signal strength although high-brightness field-emission guns are used. The conditions for optimum X-ray mapping include: 1) high beam current with small probe size, 2) high X-ray collection efficiency, and 3) long acquisition times per pixel with specimen-drift correction.

4.7.3 Comparison of nanoanalysis by EELS and XEDS

Both EELS and XEDS are important analytical techniques available in STEM instruments. These two techniques are complementary to each other. EELS is preferable for analysis of light-elements of thin specimens with a sub-nanometer or atomic resolution; XEDS is mostly used for identifying, and determining the composition of, unknown phases composed of heavy-elements. Furthermore, EELS can provide useful information other than just the composition of the sample. Although EELS has a much higher sensitivity than XEDS has for detecting light-elements [153], little useful information can be extracted in the acquired EELS spectra unless the specimen is very thin. Because of the complexity of the experimental procedures and because of the large number of variables related to the acquisition and the subsequent analysis, EELS, however, is not as widely used as XEDS in solving industrial problems.

Qualitative analysis of both XEDS and EELS spectra can be performed on-line. Quantification of these spectra, however, is not straightforward, especially for analysis of nanoparticles. Low signal-to-noise ratio often poses a severe problem for quantitative analysis of XEDS spectra while low peak-to-background ratio in EELS spectra presents difficulties for quantification.

The achievable resolution in XEDS is limited by the size of the incident probe, the spreading of the propagating probe, the signal-to-noise ratio in the spectra, the specimen drift, and the electron beam damage of the sample [154]. The achievable resolution in EELS is limited by the size of the incident probe, the delocalization of the primary excitation processes, the signal-to-background ratio, instrument stability, and electron beam damage of the sample [155]. Under highly favorable conditions, single atoms can be detected in EELS spectra [156]. With special precautions and under optimized conditions, about 10 atoms can be detected in XEDS spectra [150].

4.8 Summary

STEM can be used in various forms to provide useful information about nanoparticles or nanoparticle systems. Surface morphology of the sample can be examined on a nanometer scale by collecting secondary electrons. Nanoparticles can be identified

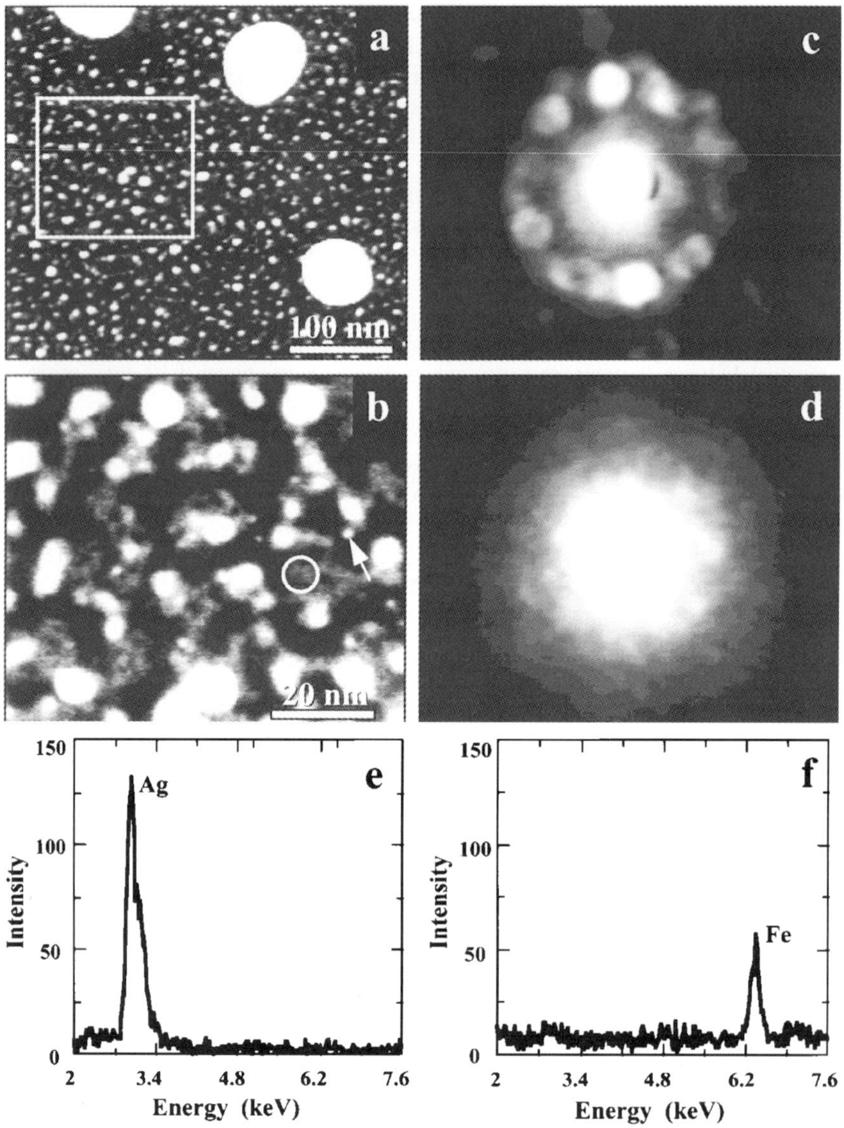

Figure 4-34. Nanoanalysis of Fe_XAg_{1-X} giant magnetoresistance films. HAADF images clearly show two phases with distinctive contrast. Nanodiffraction patterns show that some of the small particles have an icosahedral shape (c) and the matrix film is highly disordered (d). XEDS spectra (e) and (f) were obtained from the small particle (indicated by the arrow in (b)) and the matrix film (the region inside the circle in (b)), respectively. The XEDS spectra show that there is a complete phase separation in the Fe_XAg_{1-X} film, and the bright particles are silver and the highly disordered matrix consists of only Fe.

in images formed by collecting high-angle scattered electrons; and thus, the spatial and the size distributions of nanoparticles can be obtained. Nanodiffraction patterns, formed by positioning an electron nanoprobe at the area of interest, can provide crystallographic information about nanoparticles and their supports. XEDS and EELS can be used to give complementary information about the composition, electronic structure, and state of oxidation of nanoparticles as well as their supports. Elementally specific and surface sensitive techniques such as AES and SAM can be used to extract information about surface chemistry of nanoparticles or other nano-structured materials.

The combination of all these high spatial resolution imaging, diffraction, and analytical techniques in a single microscope makes the STEM the most powerful tool for characterizing nanoparticles (see Figs. 4-4 and 4-22). For example, in the study of giant magnetoresistance (GMR) in granular Fe_XAg_{100-X} magnetic films, it was found that the GMR is sensitive to the Fe volume fraction, and the size distribution and the inter-particle distances of the ferromagnetic particles. STEM and associated techniques were used to characterize the nano-structure of the magnetic granular films. Figure 4-34 summarizes the major results of the STEM analysis [72]. Based on these results preliminary models of the observed GMR of granular Ag-Fe magnetic films were proposed [157].

With the recent development of electron nanoprobes with a size < 0.2 nm in commercially available 200 kV field-emission TEM instruments, the unique STEM capabilities described in this chapter will soon be available to most researchers. STEM and its associated techniques will become more widespread and useful in solving materials problems in the 21st century.

Acknowledgements

The author is deeply grateful to many collaborators and friends for valuable and insightful discussions on characterization of nanoparticles. The author wishes to express his gratitude to Drs. J. M. Cowley, P. A. Crozier, G. E. Spinnler, J. R. Ebner, J. A. Venables, G. G. Hembree, M. Pan, and Z. L. Wang for giving advice and encouragement.

References

[1] J. M. Cowley, *Appl. Phys. Lett.* **1969**, *15*, 58.
[2] E. Zeitler, M. G. R. Thomson, *Optik* **1970**, *31*, 258, 359.
[3] J. M. Cowley, *Diffraction Physics*, 2nd ed., New York, North-Holland, **1981**.
[4] J. M. Cowley, *Ultramicroscopy* **1993**, *49*, 4.
[5] J. Konnert, P. D'Antonio, J. M. Cowley, A. Higgs, H. J. Ou, *Ultramicroscopy* **1989**, *30*, 371.
[6] J. M. Rodenburg, R. H. T. Bates, *Philos. Trans. R. Soc. Lond.* **1992**, *A339*, 521.
[7] J. M. Rodenburg, B. C. McCallum, P. D. Nellist, *Ultramicroscopy* **1993**, *48*, 304.
[8] J. C. H. Spence, J. M. Cowley, *Optik* **1978**, *50*,129.
[9] L. Reimer, *Energy-Filtering Transmission Electron Microscopy*, New York, Springer, **1995**.
[10] M. Mitome, K. Takayanagi, Y. Tanishiro, *Ultramicroscopy* **1990**, *33*, 255.
[11] C. Mory, C. Colliex, J. M. Cowley, *Ultramicroscopy* **1987**, *21*, 171.
[12] A. V. Crewe, J. Wall, J. Langmore, *Science* **1970** *168*,1338.
[13] A. V. Crewe, J. Wall, *J. Mol. Biol.* **1970**, *48*, 373.
[14] A. V. Crewe, *Rep. Progr. Phys.* **1980**, *43*, 621.
[15] M. Isaacson, D. Kopf, M. Utlaut, N. W. Parker, A. V. Crewe, *Proc. Natl. Acad. Sci.*, **1977**, *74*, 1802.

[16] M. Isaacson, D. Kopf, M. Ohtsuki, M. Utlaut, *Ultramicroscopy* **1979**, *4*, 101.
[17] M. Ohtsuki, *Ultramicroscopy* **1980**, *5*, 325.
[18] A. Howie, *J. Microscopy* **1979**, *117*, 11.
[19] M. M. J. Treacy, A. Howie, C. J. Wilson, *Philos. Mag.* **1978**, *A38*, 569.
[20] M. M. J. Treacy, S. B. Rice, *J. Microsc.* **1989**, *156*, 211.
[21] J. Liu, J. M. Cowley, *Ultramicroscopy* **1990**, *34*, 119.
[22] S. B. Rice, J. Y. Koo, M. M. Disco, M. M. J. Treacy, *Ultramicroscopy* **1990**, *34*, 108.
[23] S. A. Bradley, M. J. Cohn, S. J. Pennycook, *Microsc. Res. Tech.* **1994**, *28*, 427.
[24] P. D. Nellist, S, J, Pennycook, *Science* **1996**, *274*, 413.
[25] K. J. Hanssen, G. Ade, *Optik* **1976**, *44*, 237.
[26] H. Rose, *Ultramicroscopy* **1977**, *2*, 251.
[27] J. Silcox, P. Xu, R. F. Loane, *Ultramicroscopy* **1992**, *47*, 173.
[28] S. J. Pennycook, L. A. Boatner, *Nature* **1988**, *336*, 565.
[29] D. H. Shin, E. J. Kirkland, J. Silcox, *Appl. Phys. Lett.* **1989**, *55*, 2456.
[30] P. Xu, E. J. Kirkland, J. Silcox, R. Keyse, *Ultramicroscopy* **1990**, *32*, 93.
[31] D. E. Jesson, S. J. Pennycook, S. J. Baribeau, *Phys. Rev. Lett.* **1991**, *66*, 750.
[32] J. Liu, J. M. Cowley, *Ultramicroscopy* **1993**, *52*, 335.
[33] A. J. McGibbon, S. J. Pennycook, J. E. Angelo, *Science* **1995**, *269*, 519.
[34] S. J. Pennycook, D. E. Jesson, A. J. McGibbon, P. D. Nellist, *J. Electron Microsc.* **1996**, *45*, 36.
[35] S. J. Pennycook, D. E. Jesson, *Phys. Rev. Lett.* **1990**, *64*, 938.
[36] S. J. Pennycook, D. E. Jesson, *Ultramicroscopy* **1991**, *37*, 14.
[37] R. F. Loane, P. Xu, J. Silcox, *Acta Crystallogr.* **1991**, *A47*, 267.
[38] S. Hillyard, R. F. Loane, J. Silcox, *Ultramicroscopy* **1993**, *49*, 14.
[39] S. Hillyard, J. Silcox, *Ultramicroscopy* **1993**, *52*, 325.
[40] Z. L. Wang, J. M. Cowley, *Ultramicroscopy* **1989**, *31*, 437.
[41] Z. L. Wang, J. M. Cowley, *Ultramicroscopy* **1990**, *32*, 275.
[42] M. M. J. Treacy, J. M. Gibson, *Ultramicroscopy* **1993**, *52*, 31.
[43] D. E. Jesson, S. J. Pennycook, *Proc. R. Soc. Lond.* **1993**, *A441*, 261.
[44] J. Fertig, H. Rose, *Optik* **1981**, *59*, 407.
[45] R. F. Loane, E. J. Kirkland, J. Silcox, *Acta Crystallogr.* **1988**, *A44*, 912.
[46] J. M. Zuo, J. C. H. Spence, *Philos. Mag.* **1993**, *A68*, 1055.
[47] J. Broeckx, M. Op de Beeck, D. Van Dyck, *Ultramicroscopy* **1995**, *60*, 71.
[48] J. T. Fourie, *Optik* **1994**, *95*, 128.
[49] J. M. Cowley, J. C. H. Spence, V. V. Smirnov, *Ultramicroscopy* **1997**, *68*, 135.
[50] J. K. Weiss, R. W. Carpenter, A. A. Higgs, *Ultramicroscopy* **1991**, *36*, 319.
[51] J. O. Malm, M. A. O'Keefe, *Ultramicroscopy* **1997**, *68*, 13.
[52] J. Liu, J. M. Cowley, *Ultramicroscopy* **1991**, *37,* 50.
[53] J. M. Cowley, *Ultramicroscopy* **1993**, *49*, 4.
[54] J. M. Cowley, M. S. Hansen, S. Y. Wang, *Ultramicroscopy* **1995**, *58*, 18.
[55] J. M. Cowley, V. I. Merkulov, J. S. Lannin, *Ultramicroscopy* **1996**, *65*, 61.
[56] N. H. Dekkers, H. de Lang, *Optik* **1974**, *41*, 452.
[57] L. H. Veneklasen, *Optik* **1975**, *44*, 447.
[58] J. M. Chapman, P. E. Batson, E. M. Waddell, R. P. Ferrier, *Ultramicroscopy* **1978**, *3*, 203.
[59] I. R. McFayden, J. M. Chapman, *Electron Microsc. Soc. Amer. Bull.* **1992**, *22*, 64.
[60] R. Darji, A. Howie, *Micron* **1997**, *28*, 95.
[61] J. M. Cowley, *J. Electron Microsc. Tech.* **1986**, *3*, 25.
[62] J. C. H. Spence, J. M. Zuo, *Electron Microdiffraction*, New York, Plenum Press, **1992**.
[63] M. Pan, J. M. Cowley, *Ultramicroscopy* **1988**, *26*, 205.
[64] J. M. Cowley, J. C. H. Spence, *Ultramicroscopy* **1981**, *6*, 359.
[65] J. Zhu, J. M. Cowley, *Acta Cryst.* **1982**, *A38*, 718.
[66] J. M. Cowley, R. A. Roy, *Scanning Electron Microsc.* **1981**, 143.
[67] A. Howie, L. D. Marks, S. J. Pennycook, *Ultramicroscopy* **1982**, *8*, 163.
[68] M. Pan, J. M. Cowley, I. Y. Chan, *J. Appl. Cryst.* **1987**, *20*, 300.
[69] J. M. Cowley, R. J. Plano, *J. Catal.* **1987**, *108*, 199.
[70] J. M. Cowley, P. Nikolaev, A. Thess, R. E. Smalley, *Chem. Phys. Lett.* **1997**, *265*, 379.
[71] J. Zhu, J. M. Cowley, *Ultramicroscopy* **1985**, *18*, 419.
[72] J. Liu, Z. G. Li, H. Wan, A. Tsoukatos, G. C. Hadjipanayis, *Mat. Res. Soc. Symp. Proc.* **1994**, *332*, 303.
[73] M. Pan, J. M. Cowley, J. C. Berry, *Ultramicroscopy* **1989**, *30*, 385.
[74] J. G. Allpress, J. V. Sanders, *Surf. Sci.* **1967**, *7*, 1.
[75] M. J. Yacaman, K. Heinemann, C. Y. Yang, H. Poppa, *J. Cryst. Growth* **1979**, *47*, 187.
[76] M. J. Yacaman, M. A. Borja, *Catal. Rev.-Sci. Eng.* **1992**, *34*, 55.

[77] S. Iijima, T. Ichahasi, *Phys. Rev. Lett.* **1986**, *50*, 616.
[78] J. M. Cowley, J. C. H. Spence, *Ultramicroscopy* **1979**, *3*, 433.
[79] J. M. Cowley, M. A, Osman, P. Humble, *Ultramicroscopy* **1984**, *15*, 311.
[80] H. J. Ou, A. A. Higgs, J. M. Cowley, *Mater. Res. Soc. Symp. Proc.* **1989**, *139*, 223.
[81] J. M. Cowley, *Ultramicroscopy* **1979**, *4*, 413.
[82] K. Tsuda, M. Tanaka, *J. Electron Microsc.* **1996**, *45*, 59.
[83] W. J. Vine, R. Vincent, P. Spellward, J. W. Steeds, *Ultramicroscopy* **1992**, *41*, 423.
[84] S. Yamada, M. Tanaka, *J. Electron Microsc.* **1995**, *44*, 213.
[85] J. A. Lin, J. M. Cowley, *Ultramicroscopy* **1986**, *19*, 31.
[86] V. Ronchi, *Appl. Optics* **1964**, *3*, 437.
[87] D. Gabor, *Nature* **1948**, *161*, 777.
[88] D. Gabor, *Proc. R. Soc. Lond.* **1949**, *A197*, 454.
[89] J. A. Lin, J. M. Cowley, *Ultramicroscopy* **1986**, *19*, 179.
[90] M. Mankos, M. R. Scheinfein, J. M. Cowley, *J. Appl. Phys.* **1994**, *75*, 7418.
[91] M. Mankos, P. de Haan, V. Kambersky, G. Matteucci, M. R. McCartney, Z. Yang, M. R. Scheinfein, J. M. Cowley in *Electron Holography* (Eds.: A. Tonomura, L. F. Allard, G. Pozzi, D. C. Joy, Y. A. Ono), New York, Elsevier Science B. V. **1995**, p329.
[92] R. F. Egerton, *Electron Energy-Loss Spectroscopy in the Electron Microscope*, 2nd ed., New York, Plenum Press, **1996**.
[93] L. Reimer, *Scanning Electron Microscopy*, Berlin, Springer, **1985**.
[94] D. C. Joy, *J. Microscopy* **1987**, *147*, 51.
[95] M. S. Chung, T. E. Everhart, *Phys. Rev.* **1977**, *B15*, 4699.
[96] M. Rosler, W. Brauer, *Phys. Status Solidi* **1981**, *B104*, 161.
[97] R. Bindi, H. Lantari, P. Rostaing, *J. Phys.* **1980**, *D13*, 461.
[98] M. Rosler, W. Brauer, *Phys. Status Solidi* **1988**, *B148*, 213.
[99] J. Schou, *Scanning Microscopy* **1988**, *2*, 607.
[100] P. Kruit, J. A. Venables, *Ultramicroscopy* **1988**, *25*, 183.
[101] G. G. Hembree, P. A. Crozier, J. S. Drucker, M. Krishnamurthy, J. A. Venables, J. M. Cowley, *Ultramicroscopy* **1989**, *31*, 111.
[102] H. Seiler, *Scanning Microscopy* **1988**, *2*, 1885.
[103] J. A. Venables, G. D. T. Spiller, D. J. Fathers, C. J. Harland, M. Hambucken, *Ultramicroscopy* **1983**, *11*, 149.
[104] A. L. Bleloch, A. Howie, R. H. Milne, *Ultramicroscopy* **1989**, *31*, 99.
[105] J. Liu, J. M. Cowley, *Ultramicroscopy* **1993**, *52*, 335.
[106] A. Howie, *J. Microsc.* **1995**, *180*, 192.
[107] D. Imeson, R. H. Milne, S. D. Berger, D. McMullan, *Ultramicroscopy* **1985**, *17*, 243.
[108] A. Howie, R. H. Milne, *Ultramicroscopy* **1985**, *18*, 427.
[109] J. Liu, J. M. Cowley, *Ultramicroscopy* **1987**, *23*, 463.
[110] T. Nagatani, S. Saito, *Scanning Microscopy Suppl.* **1987**, *1*, 901.
[111] J. Liu, J. M. Cowley, *Scanning Microscopy* **1988**, *2*, 65.
[112] J. Liu, J. M. Cowley, *Scanning Microscopy* **1988**, *2*, 1957.
[113] J. Drucker, M. R. Scheinfein, J. Liu, J. K. Weiss, *J. Appl. Phys.* **1993**, *74*, 7329.
[114] H. Seiler, *J. Appl. Phys.* **1983**, *54*, R1.
[115] D. C. Joy, J. P. Pawley, *Ultramicroscopy* **1992**, *47*, 80.
[116] D. C. Joy, C. S. Joy, *Micron* **1996**, *27*, 247.
[117] D. Briggs, J. C. Riviere in *Practical Surface Analysis: Auger and X-ray Photoelectron Spectroscopy* (Eds.: D. Briggs, M. P. Seah), 2nd ed., New York, John Wiley & Sons, **1990**, Chapter 3.
[118] M. P. Seah, W. A. Dench, *Surf. Interface Anal.* **1979**, *1*, 2.
[119] J. Liu, G. Hembree, G. Spinnler, J. Venables, *Surf. Sci.* **1992**, *256*, L111; *Catal. Lett.* **1992**, *15*, 133.
[120] G. G. Hembree, J. A. Venables, *Ultramicroscopy* **1992**, *47*, 109.
[121] J. Liu, G. G. Hembree, G. E. Spinnler, J. A. Venables, *Ultramicroscopy* **1993**, 52, 369.
[122] N. C. MacDonald, J. R. Waldrop, *Appl. Phys. Lett.* **1971**, *19*, 315.
[123] A. P. Janssen, J. A. Venables, *Surf. Sci.* **1978**, *77*, 351.
[124] M. M. El Gomati, A. P. Janssen, M. Prutton, J. A. Venables, *Surf. Sci.* **1979**, *85*, 309.
[125] J. Cazaux, J. Chazelas, M. N. Charasse, J. P. Hirtz, *Ultramicroscopy* **1988**, *25*, 31.
[126] C. J. Harland, J. A. Venables, *Ultramicroscopy* **1985**, *17*, 9.
[127] J. Ferron, E. C. Goldberg, *Surf. Sci.* **1992**, *275*, 114.
[128] J. Cazaux, *J. Surf. Anal.* **1997**, *2*, 286.
[129] J. Cazaux, *Surf. Interface Anal.* **1989**, *14*, 354.
[130] D. B. Williams, G. I. Goldstein, D. E. Newbury (Eds.), *X-ray Spectrometry in Electron Beam Instruments,* New York, Plenum Press, **1995**.

[131] M. M. Disko, C. C. Ahn, B. Fultz (Eds.), *Transmission Electron Energy-Loss Spectrometry in Materials Science,* Pennsylvania, TMS, Warrendale, **1992**.
[132] J. J. Hren, J. I. Goldstein, D. C. Joy (Eds.), *Introduction to Analytical Electron Microscopy*, New York, Plenum Press, **1979**.
[133] N. D. Browining, M. F. Chisholm, S. J. Pennycook, *Nature* **1993**, *366*, 143.
[134] S. J. Pennycook, N. D. Browning, D. E. Jesson, M. F. Chisholm, A. J. McGibbson, *Appl. Phys.* **1993**, *A57*, 385.
[135] D. A. Muller, Y. Tzou, R. Raj, J. Silcox, *Nature* **1993**, *366*, 725.
[136] N. D. Browining, S. J. Pennycook, *J. Microsc.* **1995**, *180*, 230.
[137] P. E. Batson, *J. Electron Microsc.* **1996**, *45*, 51
[138] C. Colliex, *J. Electron Microsc.* **1996**, *45*, 44.
[139] D. A. Muller, D. A. Shashkov, R. Benedek, L. H. Yang, J. Silcox, D. N. Seidman, *Phys. Rev. Lett.* **1998**, *80*, 4741.
[140] G. Duscher, N. D. Browning, S. J. Pennycook, *Phys. Status Solidi* **1998**, *A166*, 327.
[141] D. J. Wallis, N. D. Browning, C. M. Megaridis, P. D. Nellist, *J. Microsc.* **1996**, *184*, 185.
[142] J. A. Hunt, D. B. Williams, *Ultramicroscopy* **1991**, *38*, 47.
[143] S. Q. Sun, S. L. Shi, J. A. Hunt, R. D. Leapman, *J. Microsc.* **1995**, *177*, 18.
[144] C. Jeanguillaume, C. Colliex, *Ultramicroscopy* **1989**, *28*, 252.
[145] A. Howie, *Ultramicroscopy* **1983**, *11*, 141.
[146] R. Garcia-Molina, A. Gras-Marti, A. Howie, R. H. Ritchie, *J. Phys. C: Solid State Phys.* **1985**, *18*, 5335.
[147] R. H. Richie, A. Howie, *Philos. Mag.* **1988**, *A58*, 753.
[148] Z. L. Wang, J. M. Cowley, *Ultramicroscopy* **1987**, *21*, 77, 335.
[149] M. G. Walls, A. Howie, *Ultramicroscopy* **1989**, *28*, 40.
[150] C. E. Lyman, J. I. Goldstein, D. B. Williams, D. W. Ackland, S. Von Harrach, A. W. Nicholls, P. J. Statham, *J. Microsc.* **1994**, *176*, 85.
[151] J. Tong, J. Liu, Z. C. Kang, L. Eying, *J. Alloys and Compounds* **1994**, *207*, 416.
[152] C. E. Lyman, *Ultramicroscopy* **1986**, *20*, 119.
[153] R. D. Leapman, J. A. Hunt, *Microsc. Microanal. Microstruct.* **1991**, *2*, 231.
[154] C. E. Lyman in *Physical Aspects of Microscopic Characterization of Materials* (Eds.: J. Kirschner, K. Murata, J. A. Venables) Scanning Microscopy International, AMF O'Hare, Illinois, **1987**, p123.
[155] C. Mory, H. Kohl, M. Tence, C. Colliex, *Ultramicroscopy* **1991**, *37*, 191.
[156] O. L. krivanek, C. Mory, M. Tence, C. Colliex, *Microsc. Microanal. Microstruct.* **1991**, *2*, 257.
[157] Z. G. Li, H. Wan, J. Liu, A. Tsoukatos, G. C. Hadjipanayis, L. Liang, *Appl. Phys. Lett.* **1993**, *63*, 3011.

5 Scanning Probe Microscopy of Nanoclusters

Lifeng Chi and Christian Röthig

5.1 Introduction

The invention of the *scanning tunneling microscopy* (STM) [1] presents the beginning of a novel class of near-field microscopy with nanometer till atomic resolution in real space. Originally the STM was designed for the surface analysis under ultrahigh vacuum (UHV) conditions on conducting surfaces. It was verified soon that this method could also be used under a variety of environmental conditions such as ambient air [2], water [3], oil [4] and electrolytes [5]. These exciting discoveries motivated the further development of a whole family of other *scanning probe microscopes* (SPM) [6]. Examples include *scanning force microscopy* (SFM) (or *atomic force microscopy*, AFM) [7] and its extensions [8] and *scanning near-field optical microscopy* (SNOM) [9].

The applications of SPM increased rapidly in the last decade. Since the requirement of a conductive surface for STM is not the necessary condition for SFM, the objects for SPM are extended to insulating materials and cover a wide range of surfaces of bulk systems such as metals [10], semiconductors [11], adsorbates on solid surfaces [12], polymers [13], biological materials [14] and organic layered systems [15]. Besides the fundamental research of surface structure, surface morphology and surface reconstruction, the SPM can also be used to reveal local physical properties of the sample, e.g. electronic energy spectra by using *local tunneling spectroscopy* (LTS, or *scanning tunneling spectroscopy*, STS), local tribological properties by using *frictional force microscopy* (FFM), local elasticity by using *force modulation microscopy* (FMM), local magnetic properties by using *magnetic force microscopy* (MFM), and local electrostatic properties by using *electric force microscopy* (EFM). Another important approach of SPM techniques is in the field of nano-lithography [16].

In recent years, nanometer-sized clusters of metals and semiconductors have received increasing scientific and technological interest [17]. Introducing chemically synthesized or physically evaporated nanoclusters on a variety of surfaces is a recent approach to generate nanostructured surfaces, which are promising for novel sensoric, electronic and photonic devices and are therefore inducing increased research activities. The SPM, especially STM and SFM, can yield valuable information on nanoclusters as they may reveal the electronic and geometric structure of the cluster's. Further more, they can serve as ideal tools to characterize cluster covered surfaces since a wide range of materials can be chosen as substrates for cluster deposition.

In the following, we will discuss the fundamental principle of SPM (Section 5.2) and different operating modes used experimentally (Section 5.3). In Section 5.4, selected applications of SPM on nanoclusters will be presented, including:

- imaging of individual nanoclusters and cluster covered surfaces
- local physical properties
- SPM-based local deposition and modification
- tunneling spectra of nanoclusters

Besides the typical nanocrystalline materials prepared with gas-phase synthesis, and organic ligand stabilized monocrystalline clusters, systems described here will also include the nano-scale metallic dots or islands generated by STM/SFM tips, by nanosphere lithography and by electroless deposition. The limitation of SPM methods to study the nanophase materials and the prospects will be given in Section 5.5.

5.2 Fundamental of the techniques

The basic idea of *scanning probe microscopy* (SPM) is relatively simple : take a probe susceptible to the property one wants to measure, bring it into the vicinity of a surface and measure the reactions of the probe. As one is interested in microscopic information the probe has to be sufficiently small and its movements have to be controlled on a length scale comparable to its size. By taking measurements at different locations close to the surface, for instance by scanning the probe at constant height, a topographic representation of the surface can be gained. Depending on the specific property measured, the type of probe and the way in which the reaction of the probe is amplified, all existing SPM methods can be differentiated.

The probe in SPM is usually connected to or consists of the very end of a rigid tip. The movements of this tip relative to the surface are controlled by a scanner constructed from piezoceramics in most setups which scans either the tip or the surface. The distance between tip and surface is controlled by a similar transducer, quite often all three transducers are combined in a singular unit providing movement in all three spatial directions. If one is only interested in recording a special surface property as a function of the spatial coordinates it is sufficient to amplify and record the reactions of the probe in a proper way. If the interest is on constant-property-maps or if the property changes very drastically with slight changes in tip-sample distance, a feedback circuitry of some kind is necessary. With it the interesting property is held constant and the distance is changed until the set-point is reached. By recording the distance as a function of the two other spatial dimensions a surface plot can be produced. Obviously such a scheme is relatively easy to implement if the interesting property is a monotonous function of the distance (interatomic forces being a prominent example of a more complicate case). Figure 5-1 shows a typical setup for SPM schematically.

The oldest SPM method which lead to resolutions comparable to the best electron microscopes was *scanning tunneling microscopy* (STM). With it the local electronic density of the surface is measured by applying a small voltage between the probe and surface. The probe is usually formed by the foremost atom of a thinned rigid piece of wire and is connected via a sensitive amperemeter to the surface. If the probe approaches the surface, tunneling of electrons between surface and tip can be observed. This tunneling current I is a function of easily controllable independent parameters like the coordinates of the probe relative to the surface and the applied voltage and dependent parameters like the work function of the surface or the electric conductivity. The general form of the tunneling current is given by

$$I = I(x, y, z, U) \tag{5-1}$$

Even from very basic quantum mechanical calculations, i.e. one-dimensional tunneling through a rectangular barrier, it is evident that the tunneling current exhibits an exponential dependence from the distance between tip and surface. One finds [18]:

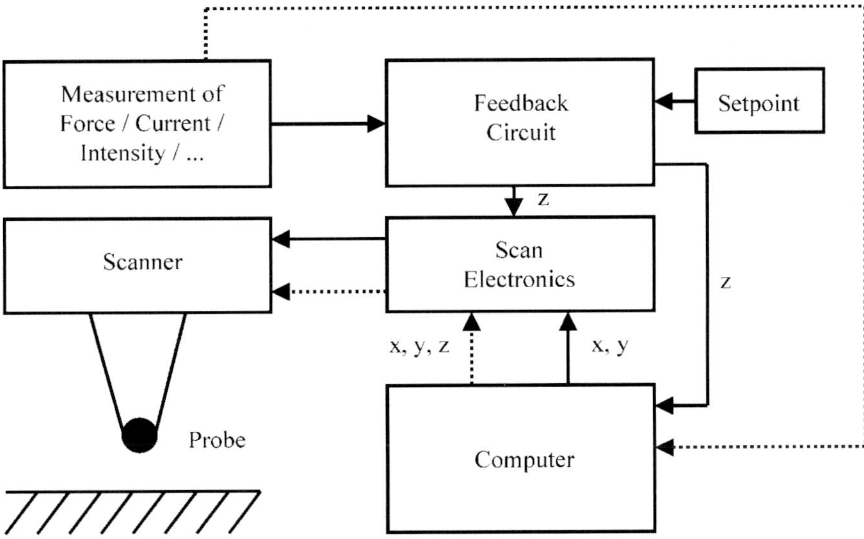

Figure 5-1. Schematic SPM setup. The solid lines represent a setup with feedback circuitry, the dotted lines a setup without.

$$I \propto U\rho_s(0, E_F)\mathrm{e}^{-1.025\sqrt{\Phi W}} \tag{5-2}$$

local density of states of the sample:

$$\rho_s(z, E) = \frac{1}{\varepsilon} \sum_{E_n - E\ \varepsilon}^{E} |\psi_n(z)|^2 \tag{5-3}$$

where:

U tunneling voltage
Φ work function
W distance between tip and sample

Typical values for the work function are in the range of 4 to 5 eV which leads to typical decay constants of about 10 times per Å, i.e. the tunneling current is reduced by an order of magnitude if the distance between tip and sample is increased by only 1 Å.

Scanning force microscopy (SFM) does not record surface properties directly like STM but measures the forces that are caused by the interaction between probe and surface. These minute forces are recorded by measuring the bending of a small cantilever to which the probe is connected and which compensates the forces acting on the probe. The movements of the cantilever are therefore a very good representation of the tip-surface interaction. One finds that for ratio of the displacement of the cantilever and the piezo [18]:

$$\frac{\Delta z_c}{\Delta z_p} = \frac{1}{\frac{k}{F'(d)} - 1} \tag{5-4}$$

with Δz being the displacement of the cantilever and piezo, respectively, k the spring constant and $F'(d)$ the force gradient, a function of the distance d between probe and sample.

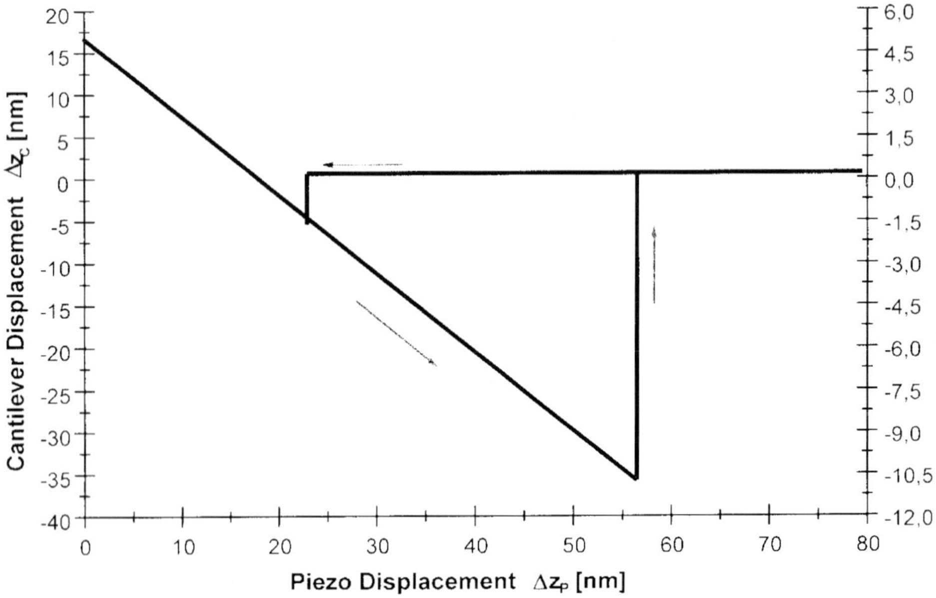

Figure 5-2. Typical force-distance curve as seen in SFM experiments.

Another local probe method uses the optical properties of the surface: *scanning near-field optical microscopy* (SNOM). In contrast to the optical far-field which is being used in common microscopes the optical near-field is capable to carry information about objects with lateral dimensions much smaller than the wavelength of light used. To exploit this approach the probe has to be susceptible to light intensity, so usually tips made from glass fibers or small apertures are used. If one probes the optical near-field intensity in this way one does not only get distance information by measuring the intensity, but one can also measure the spectra of the emitted light to get additional information about the chemical composition or other optical properties of the surface.

5.3 Experimental approaches and data interpretation

In this section, we will focus on different operation modes of *scanning tunneling microscopy, scanning force microscopy* and *scanning near-field microscopy.* We will not describe all the modes but only those included in this chapter for the characterization of nanoclusters.

5.3.1 Scanning tunneling microscopy (STM)/ Scanning tunneling spectroscopy (STS)

For *scanning tunneling microscopy*, a number of measurement modes have been developed which represent special surfaces of the function $I = I(x, y, z, U)$:

$$I = I(x, y,)|_{U,z\,=\,\text{const.}} \tag{5-5}$$

Constant Height Mode (CHM)

$$I = I(x, y, z)|_{U = \text{const.}} = \text{const.} \tag{5-6}$$

Constant Current Mode (CCM), z being the feedback parameter

$$I = I(U)|_{x,y,z = \text{const.}} \tag{5-7}$$

(Scanning) Tunneling Spectroscopy and

$$I = I(z)|_{x,y,U = \text{const.}} \tag{5-8}$$

(Scanning) Tunneling Spectroscopy

One can deduce from Eq. 5-2 that the first spectroscopy mode $I = I(U)$ will yield information about the local density of states ρ_S while $I = I(z)$ will provide information about the local barrier height. By scanning the tip over the surface at a constant current I a surface of constant local density of states of the sample at the location of the tip is imaged according to this simple model. The assumptions of this model are valid if small tunneling voltages are used and an arbitrarily fast method is employed to correct the distance W.

For larger tunneling voltages, especially for the interpretation of STS measurements, more sophisticated models have to be employed. One of these approaches is the first order perturbation approach by Bardeen [19], later refined by Tersoff and Hamann [20, 21]. If $k_B T$ is much smaller than the feature sizes in the energy spectrum of interest one can show that

$$I = \frac{4\pi c}{\hbar} \int_0^{eU} \rho_s(E_F - eU + \varepsilon)\rho_T(E_F + \varepsilon)|M|^2 \, d\varepsilon \tag{5-9}$$

$$M_{\mu\nu} = -\frac{\hbar^2}{2m} \int_\Sigma (\Phi_\nu^* \nabla\Psi_\mu - \Psi_\mu \nabla\Phi_\nu^*) dS \tag{5-10}$$

with ρ_s and ρ_T being the density of states of sample and tip, respectively, and Ψ_μ and Φ_ν their wave functions. The tunneling matrix element $M_{\mu\nu}$ has the dimension of energy and describes the lowering of energy due to the overlap of the two states μ and ν. It is important to note that the expression is symmetric, i.e. exchanging tip and sample does not change I. That means for practical purposes that there is no easy way to decide whether the tip has measured the surface structure or the surface has measured the tip structure (for instance via a sharp protrusion). The resulting image will always represent a convolution of tip and surface structure.

Apart from the topographic structure and position on a given surface the electronic structure is of major interest as well from the viewpoint of basic understanding of clusters as from a technological point of view. For basic understanding clusters are expected to exhibit a number of electronic properties which position them right in the middle between molecules and the solid state. If we consider the mode of tunneling spectroscopy with respect to the measurement of the gap resistance one would expect to measure the distinct electronic energy levels present in a small cluster system right below the Fermi level. In fact we will see that these attempts have been made in the past, a thorough interpretation of the obtained data, however, is extremely difficult due to the vicinity of the substrate's electronic system and temperature effects. More obvious is the charging effect in insulated clusters, i.e. clusters which have a ligand-shell or which are insulated by some other kind of layer from the underlying substrate. Provided these clusters are located not to far away from the conducting surface, tunneling from the tip to the cluster and from the cluster to the surface will take place. If the tunneling frequency from the cluster to the surface is sufficiently low electrons tunneling from the tip to the cluster will experience the electric field from the excess electrons in the cluster resulting in a quantised tunneling behavior which becomes visible as distinct steps in the I/U-curves.

The second spectroscopic mode, i.e. the measurement of $I(z)$ to determine the work function, is very interesting from a basic point of view. Small free sodium clusters for instance show a very strong dependence of their ionization potential from the number of atoms present in the cluster [22, 23]. However, data interpretation of this kind of spectroscopy, i.e. the determination of work function as a function of cluster size, is quite difficult as the observed clusters will usually not be free but more or less strongly attached to a surface and the number of atoms such clusters consist of can only be determined from a measurement of the cluster volume which in itself is subject to severe uncertainties due to the folding between cluster and tip structure.

5.3.2 Scanning force microscopy (SFM)

Scanning force microscopy senses the overall forces between the probe and the sample surface, where the probe is attached to a cantilever-type spring. The images are related to the total density of (valence-) electron states up to the Fermi level at the surface. It is therefore independent of the electrical conductivity of the sample. The force exerted on the tip deforms the cantilever elastically. Since the spring constant c of the cantilever is known, the net force F can be derived directly from the deflection Δz according to the equation $F = c\Delta z$. Measurement of the deflection of the cantilever is a main issue in performing SFM.

Different methods were used at the beginning stage of instrument development [24]. Now most of the commercial instruments use the laser beam deflection technique proposed first by Meyer et al [25] and Alexander et al [26]. A laser beam is reflected from the backside of the cantilever. The deflection is monitored with a position-sensitive detector (PSD). This signal is used to drive the feedback loop. The experimental setup is schematically shown in Fig. 5-3. Depending on the motion of the cantilever, SFM can be divided into *static SFM* and *dynamic SFM*. The typical

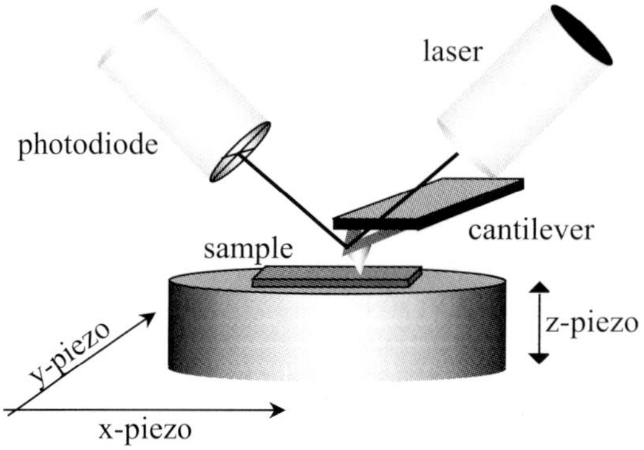

Figure 5-3. Schematic illustration of the scanning force microscope using the laser beam deflection method for measuring the cantilever bend and the torsion due to the forces between the tip and the sample surface.

interaction force between the tip and the sample is between 10^{-6} to 10^{-12}N, depending on the spring constant of the cantilever and the environmental medium. Most of the commercially available SFM tips are made of microfabricated silicon nitride (Si_3N_4) or silicon with a tip radius of 2–10 nm.

5.3.2.1 Contact mode SFM

Contact SFM runs in quasistatic mode. The probing tip is brought into a static mechanical contact with the sample. In analogy to the constant current and constant height mode of STM, the common contact mode SFM operates in *constant force mode* or *constant height mode.* In *constant force mode*, the force is controlled by keeping the deflection of the cantilever constant by means of the feedback loop. The output of the feedback yields the motion of the z piezo thus maps the surface contour. In the *constant height mode* the feedback loop is not used and the deflection of the cantilever is measured. In this way higher scan rates can be achieved which minimizes the thermal drift effects.

Detecting the torsion mode of the cantilever instead of the bending mode during scanning allows the study of lateral (friction) forces with the SFM. By using a quadrant PSD, the normal and the lateral (frictional) force can be monitored simultaneously [27]. The development of *friction force microscopy* (FFM) allows the study of local tribological properties of samples.

5.3.2.2 Dynamic mode SFM

A significant increase in possibilities of the force microscopes was achieved by the use of dynamic modes. Dynamic modes are usually more sensitive leading to less (damaging) interactions between probe and surface and they provide information about dynamic properties such as energy dissipation or local elasticity.

In *dynamic SFM* the cantilever is driven to vibrate near its resonant frequency by means of a) a piezoelectric element, or b) an external force. Instead of measuring the quasi-static cantilever deflections, changes in the resonant frequency as a result of tip-sample interaction are detected. This a.c.-detection method is sensitive to the force gradient rather than to the force itself. Two different methods are commonly used to measure the frequency shift. One is to measure the amplitude changes or the phase shift of the cantilever vibration with the deflection sensor and a lock-in amplifier [28]. The feedback loop adjusts the tip-sample separation by keeping the force gradient constant. The other one is to measure the change in oscillation frequency with a frequency counter or a frequency modulation (FM) discriminator. The feedback loop is adjusted to keep the frequency constant. Some physical properties such as magnetic forces or surface potential depending on the field gradients can be separated from the surface topography by *dynamic SFM* methods.

Small amplitude working in the repulsive regime. In this mode, the tip is oscillated in a range of 1 nm in the repulsive contact region, named *force modulation mode* (FMM). It provides an elastic or stiffness contrast of the sample. By measuring the response to the tip oscillation amplitudes, areas with different viscoelasticities can be detected.

High amplitude mode in the intermittent regime. In this mode the SFM tip oscillates with high amplitudes (10–100 nm) so that the tip propagates through large parts of the interaction potential in a single oscillation cycle, i.e. through attractive and repulsive regimes. Thus it is called *intermittent contact mode* (IC-mode) or *tapping mode*™ (Digital Instruments). Approaching the sample surface the oscillation amplitude and the frequency of the cantilever decreases with the average tip-sample distance. The feedback loop is controlled by keeping the amplitude or the frequency constant. In this mode, the contact time between the tip and the sample is reduced by two orders of magnitude as compared to the contact mode and is thus touching the surface only "gently".

High amplitude mode in the attractive regime. The high amplitude mode can also work in the effective attractive regime if the tip is set above the sample surface in a distance of 10–100 nm ("non-contact" mode). Another reliable way is to introduce an active feedback circuit instead of the standard feedback, Fig. 5-4 [29]. In this design, the effective quality factor (Q-factor, $Q = f/\Delta f$) of the oscillating system is increased by at least one order of magnitude. As a consequence the sensitivity is enhanced allowing to prevent the onset of tapping. This concept provides a means to control and enhance the size of the attractive interaction regime and thereby makes this operation mode more stable and usable in air and in liquids. This method is very useful for measuring soft materials which stick weakly to the substrate.

Magnetic force microscope (MFM) Long range forces such as magnetic or electrostatic forces dominate the tip-sample interaction if the tip-sample separation exceeds 10 nm. The long range magnetic or electrostatic forces are usually probed by using the a.c.-detection method. For magnetic force detection magnetic probing tips are necessary and it is realized by magnetic coating of the Si_3N_4 tips which are commercially available. To achieve the separation of the topographic and the magnetic structures, the strength of the magnetic interaction can be modulated and measured by different methods [30]. Alternatively, a modulated bias voltage can be applied. The resulting electrostatic forces cause an oscillation of the cantilever at the second harmonic. The amplitude of this oscillation can be used to drive the feedback loop [31]. Simultaneously, the d.c. force induced by the magnetic interaction can be detected.

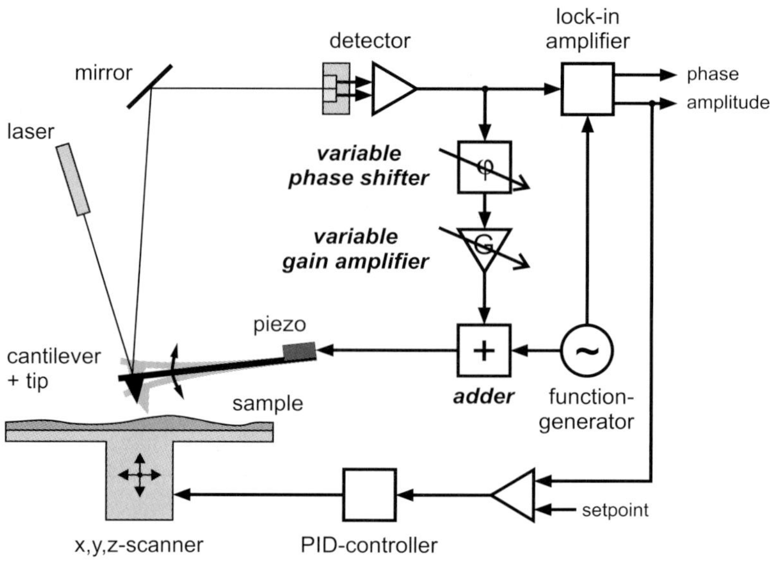

Figure 5-4. Schematic diagram of the SFM with additional feedback circuit for active control of the system response function. The setup allows the effective quality factor of the dynamic system to be changed by more than an order of magnitude (adapted from Anczykowski et al, [29]).

5.3.3 Scanning near-field microscopy (SNOM)

In a transmission-type *scanning near-field optical microscope* (SNOM), as shown schematically in Fig. 5-5, a tip of dimensions smaller than the wavelength of light serves as a light emitting antenna. It is excited by light from an external source by an optical path and a link which serves to transmit light from the optical path to the nanoscopic dimensions of the tip where optical components fail. Light emitted from the tip and transmitted through the object is converted by a detector into an electrical signal which is used for imaging. The probe is mounted on a piezoelectric scanner which scans the probe relatively to the object. In order to obtain a high resolution in the near field image, the distance between tip and object has to be kept smaller than the desired resolution. Therefore an auxiliary signal such as a SFM or STM signal is generally used to control the distance between tip and object. The tip and the link have specific near-field optical functions and therefore are the characteristic parts of a SNOM. Usually, a glass fiber is used as the optical path and a conically tapered end coated with metal serves as a link. At the end of this link an aperture of a diameter in the range of 50 to 100 nm is formed at the tip which serves as a SNOM probe. One alternative to such an aperture probe is the tetrahedral tip which consists of the corner of a glass fragment which is coated with 50 nm of gold. For a general description of SNOM see ref. [32].

The most important advantage of SNOM, named by some authors *near-field scanning optical microscopy* (NSOM), in comparison to other forms of optical microscopy and to other forms of SPM is the potential of a high lateral resolution below the diffraction limit in conjunction with optical spectroscopic contrast.

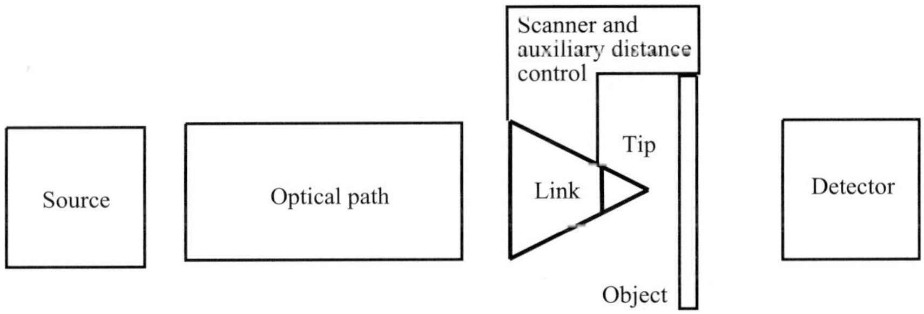

Figure 5-5. The characteristic components of a transmission-type scanning near-field microscope (adapted from U.Ch. Fischer, [32]).

5.4 Applications for characterizing nanophase materials

Applications of SPM methods for characterizing nanophased materials address the following questions: 1) geometry of individual clusters and their size distribution on the surface of supporting substrates; 2) binding and growth behavior of clusters on substrates; 3) local physical properties; 4) two-dimensional arrangement of clusters on substrates and 5) generation and fabrication of nanoclusters with SPM techniques.

5.4.1 Individual nanoclusters

5.4.1.1 Nanoclusters without protective coating

Shape and size distribution. A knowledge of the three-dimensional shape and size-distribution of supported clusters deposited on substrates could provide insight into the microscopic forces acting on surfaces and is important for a detailed understanding of many experimental results. Although research on free standing and supported clusters has clearly demonstrated the size-dependent structural and electronic features differing from bulk systems, it is not easy to measure the three-dimensional shape of supported cluster. SPMs offer alternative methods supplementing *transmission electron microscopy* (TEM). With TEM intriguing questions arise sometimes whether clusters have their stable shapes in a high energy electron beam [33].

The early SPM studies on metal clusters were carried out with STM on highly oriented pyrolytic graphite (HOPG). Abraham et al [34] demonstrated the use of STM as a tool for the study of metal aggregates formed by metal vapor deposition. They were able to image isolated Ag clusters and the motion of clusters on the support. Baró et al [35] obtained constant current images in air of nanometer-sized Au clusters deposited on HOPG by using the multiple expansion cluster source (MECS), Fig. 5-6. The width of the objects was determined by measuring the width at the half-maximum from the contour map. A mean size of 12 Å with a standard deviation of 2.7 Å was obtained, which was in good agreement with the 13 Å diameter predicted by deceleration cell measurements on the cluster beam.

In these studies, the STM scans indicated substantial cluster mobility which hindered the reliable imaging and the following analysis. Suitable substrates have to be chosen for stronger binding of the clusters. Castro et al [36] extended the preliminary experiments by exploring other substrates, such as Au and Pt. They inserted a piece of clean Au or Pt wire, 1mm in diameter, into a gas torch flame. A molten sphere with a diameter of 2 mm was formed at the end of the wire, giving atomically flat facets. Stable STM images of large clusters (10–20 nm in diameter) were obtained on both substrates. Small clusters with diameters < 2 nm were still difficult to image due to the tip-cluster interaction. Hövel et al introduced a method to fix clusters on

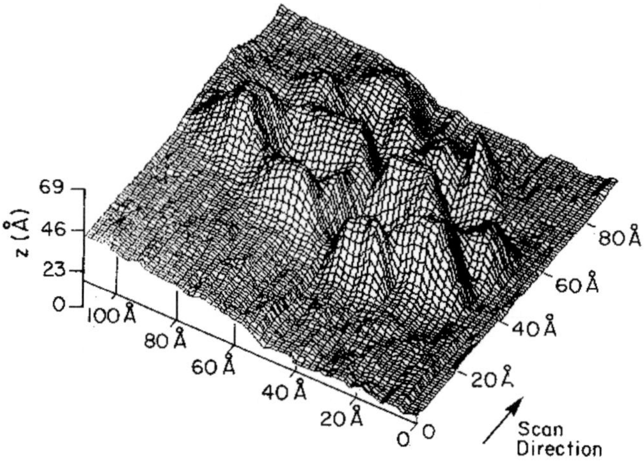

Figure 5-6. Three-dimensional STM image of eight Au clusters (Baró et al, [35]).

HOPG [37]. They produced nanometer-sized pits in the first layer of HOPG by sputtering (with Ne ions) and following oxidation. On such a substrate small clusters, e.g. thermally evaporated Ag clusters with 3.7 nm in height, were stably imaged by UHV-STM. In addition, the pits acted as well-defined condensation centers, which led to a narrow size distribution and controllable cluster coverage.

Schaefer et al demonstrated that with dynamic mode SFM working in attractive regime clusters thermally evaporated on substrates such as HOPG, mica, Si and PbS could be readily imaged [38]. Typical interaction forces between the tip and the sample in this operating mode were in the range of 10^{-10}N. They found, however, that clusters with heights less than 5 nm and more than 18 nm were rarely seen with SFM, even though parallel TEM studies indicated that a detectable number of these clusters were present. It seemed likely that the small clusters were less bound to the surface and could be moved by the SFM tip even in this operating mode. The large clusters could be caused by a flattening of the clusters on the carbon film due to the surface force, or they could possibly be a result of a modification by the electron beam during TEM studies. Differences in the size distribution evaluated from STM and TEM measurements were also reported by other groups, e.g. by Granjaud et al [39].

Li et al [40] studied systematically the 3D shape of preformed Au clusters with radii varying from 1 to 6 nm on flame annealed Au substrates by using STM. Based on the analysis of apparent cluster diameter D and cluster height H (D/H was larger than 1), they found that the shape of the supported clusters resembled spherical caps rather than spheres. A characteristic radius of curvature between 10 and 30 nm was found, which was greater than and not correlated with the original free space radius of the clusters. The flatness of the supported clusters was considered to be attributed to plastic deformation induced by surface forces. Clusters whose radius of curvature was less than a critical value r^* were subjected to internal stress in excess of their yield stress. On deposition this stress was relieved by cluster deformation when the cluster wetted the substrate. They believed that this behavior might be very common and might be useful for explaining other cluster experiments. Wurster et al indicated that the interpretation of the shape of clusters in the size range of 5–50 nm based on the SFM and FFM images had to be careful since the shape of the probing tip could strongly influence imaging [41]. They observed a clear difference in imaging hemispherical shaped indium clusters with an ordinary pyramidal tip (aspect ratio 1:1) and a super tip on the top of the pyramid (aspect ratio > 10:1). The hemispherical shape could only be observed with the super tip.

Sarid et al [42] demonstrated the possibility of applying STM on semiconductor clusters, e.g. BiI_3 deposited from colloidal suspensions on HOPG. Later Jing et al presented STM images of Bi_2S_3 clusters deposited on HOPG and gold surfaces [43]. They observed disk-like structures independent of the nature of the substrate, as expected for clusters having layered symmetry. Again, larger clusters with a lateral dimension of 10 nm containing 600 atoms were reliably imaged, while smaller clusters were never stable enough to allow high quality images.

Growth of nanoclusters on substrates. Ganz et al extended the work of Baró et al to ultrahigh vacuum (UHV) [44]. They determined adsorption sites of single metallic atoms (Ag and Au) and atomic spacing of small two-dimensional islands or rafts of metallic atoms on graphite. The islands contained ordered regions of roughly 50 atoms in rectangular lattices, incommensurate with the substrate lattice, Fig. 5-7. Later studies have been carried out on different cluster and substrate combinations [45–51].

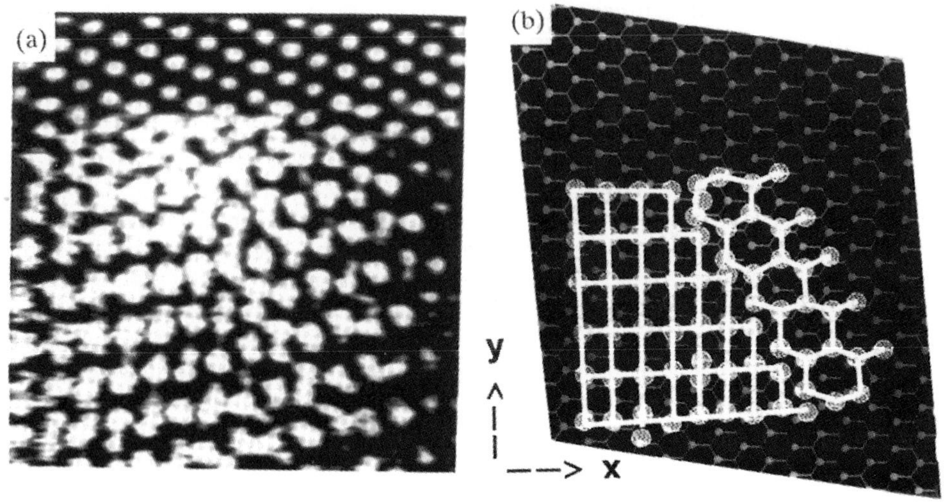

Figure 5-7. STM image of a monolayer Au island on graphite (a) and the computer model showing a rectangular lattice on the left and a honeycomb lattice on the right (Ganz et al, [44]).

Humbert et al [48] indicated with STM images that in the Pd 2D clusters on HOPG the Pd atoms packed non-closed hexagonal with a lattice parameter of 4.26 Å which is definitively larger than the nearest neighbor distance of bulk Pd (2.74 Å). Nie et al [50] proposed that the initial Pd atoms might occupy centers of the hexagonal units of HOPG and leave the six nearest-neighbor centers empty. They showed FFM high resolution images giving the lattice distance of 2.8 Å (on mica substrate), which is comparable to the bulk system. This means that during the growth of the Pd clusters, the lattice distance relaxed to that of the crystal.

Kern and coworkers reported recently the growth of nano-scale metallic islands by position selective *electroless deposition* (ELD) [52]. ELD is an autocatalytic redox process in which metal ions are chemically reduced to metal at a surface in absence of any external current source. They used mixed amino-/alkanethiolate self-assembled monolayers (SAMs) on a gold electrode to provide special binding positions for palladium (Pd binds preferably to amino groups). The Pd^{2+} activated electrode was then used to grow Co islands. STM was used to observe *in situ* the growth of metal islands. It was revealed by STM that the density of the islands with a diameter between 1.5 to 6 nm for Pd and 1 to 10 nm for Co could be adjusted by changing the amine concentration in the aminothiolate/alkanethiolate SAMs. By using such a concept of metal deposition, and combining it with microcontact printing introduced by Whiteside and coworkers [53], it should be possible to fabricate diverse metal cluster arrays in desirable patterns.

Mechanical properties of nanoclusters. Few works have addressed the mechanical properties of nanocluster. That may at least partially be related to the fact that the clusters can be easily swept away by applying contact SFM or FFM measurements [54]. A systematic study of elastic properties of individual gold clusters preformed in the gas phase and deposited on diverse substrates (α-Al_2O_3, mica and HOPG) was reported by Schaefer et al with SFM [55]. They scanned the sample with dynamic SFM first in order to position the tip on a single cluster. The elastic modulus of a cluster was then measured by nanoindentation described elsewhere [56] and compared

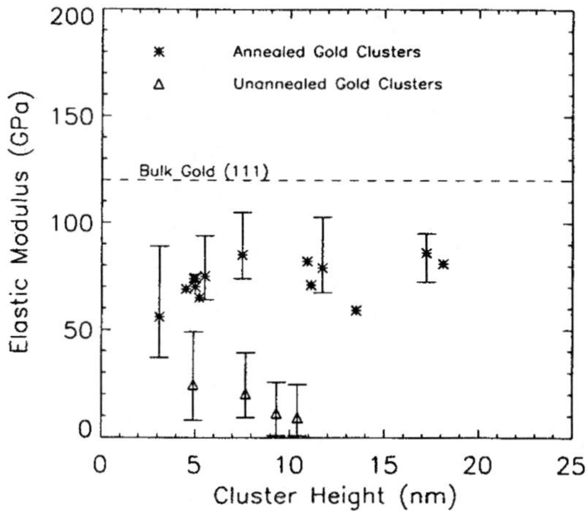

Figure 5-8. Plot comparing the measured elastic modulus of unannealed gold clusters with the measured values on annealed gold clusters by using nanoindentation. For comparison, the bulk elastic modulus of Au(111) is also indicated (Schaefer et al, [55]).

with the bulk gold surface. The results were summarized as shown in Fig. 5-8. For applied force loads less than 20 nN, the annealed Au clusters showed an elastic modulus roughly 2/3 of bulk Au, while the unannealed Au showed an elastic modulus roughly 1/6 of bulk Au. The difference was likely due to the crystal defects in the unannealed cluster rather than to a size effect.

5.4.1.2 Passivated nanoclusters

Clusters stabilized by organic ligand shells (also named *passivated nanocluster*) have received increasing scientific interest in the recent years [57]. The core of the cluster can be made of metals or semiconductors. The metallic clusters have a fixed number of atoms and thus a well defined size. They provide excellent model systems for monodispersed metal clusters, embedded in a dielectric matrix for the investigation of physical properties related to nano-scale particles. The possible electronic applications of these clusters as "quantum dots" are discussed elsewhere [58].

The early works using STM to image passivated metallic clusters were done by van Kempen [59] and Becker et al [60]. They applied STM, under atmospheric conditions, to gold and palladium clusters [$Au_{55}(PPh_3)_{12}Cl$ (a core of 55 gold atoms stabilized by $P(C_6H_5)_3$ and Cl) and Pd_{561} $(Phen)_{38+2}O_{\approx 200}$ (561 palladium atoms stabilized by phenanthrolin and O_2)], synthesized by G. Schmid [61]. The clusters were deposited onto gold and graphite substrates from a droplet of aqueous suspension solutions. They found that the clusters could be easily dragged along with the scanning tip and even be picked up by the tip due to the very loose binding to the substrate. These effects hindered the imaging and could be minimized by increasing the distance between tip and substrate and decreasing the scanning speed to obtain stable imaging. On the other hand, once the clusters were picked up, they stayed at the outer end of the tip and thus provided a well defined tip, with a reproducible geometrical and electronic structure. They called the cluster-attached tip "known tip" and the original PtIr tip cut with a pair of scissors "unknown tip". They were able to measure single free-lying clusters with both kinds of tips. To distinguish single clusters with the STM, nevertheless, became easier if the clusters were stacked densely. The STM images supported the idea that the clusters are more or less in a spherical form. The measured size distribution of the clusters, evaluated from the measured heights of the clusters and their full width at half-maximum (FWHM) with a known tip was much smaller than that measured with an unknown tip. For all the cases, the width of the protrusion was a few times larger than the expected diameter of the clusters, while the height was significantly smaller. The deviation in width was obviously induced by the convolution of the tip, whereas the lack in the thickness was assumed to be induced from the deformation of the cluster, the deformation of the tip, or the fragmentation of the clusters on the substrates.

Reetz et al [62] determined the thickness of the protective surfactant layer on nanosized palladium (Pd) clusters by a combination of STM and high resolution TEM. They used Pd clusters with an average diameter between 2 and 4 nanometers stabilized by tetraalkylammonium salts. The size of the ammonium ions was adjusted in the series $^{+}N(n\text{–}C_4H_9)_4 < {}^{+}N(n\text{–}C_8H_{17})_4 < {}^{+}N(n\text{–}C_{18}H_{37})_4$. STM images showed the size difference between a 4.1 nm Pd cluster stabilized with $N(n\text{–}C_8H_{17})_4\text{–}Br$ and a 2 nm Pd cluster stabilized with the same ammonium ion, Fig. 5-9a and Fig. 5-9b. The difference between the mean diameter (averaged over 100 particles) determined by STM (d_{STM}) and TEM (d_{TEM}) provided fairly detailed information of the interaction between the tetraalkyl ammonium ions and the metal clusters. The experimental results are summarized in Fig. 5-9c and Fig. 5-9d. In the case of different metal core and the same stabilizer, the geometric difference ($d_{STM}–d_{TEM}$) was nearly independent of the metal core diameter d_{TEM} (Fig. 5-9c); while in the case of the same metal core and different stabilizer, $d_{STM}–d_{TEM}$ was directly dependent on the size of the ammonium ion (the length of the alkyl groups on $^{+}NR_4$), Fig. 5-9d. The geometric difference ($d_{STM}–d_{TEM}$) correlated well with the monolayer thickness calculated with a standard MM2 force field. Thus, a monolayer protective coat was involved. The

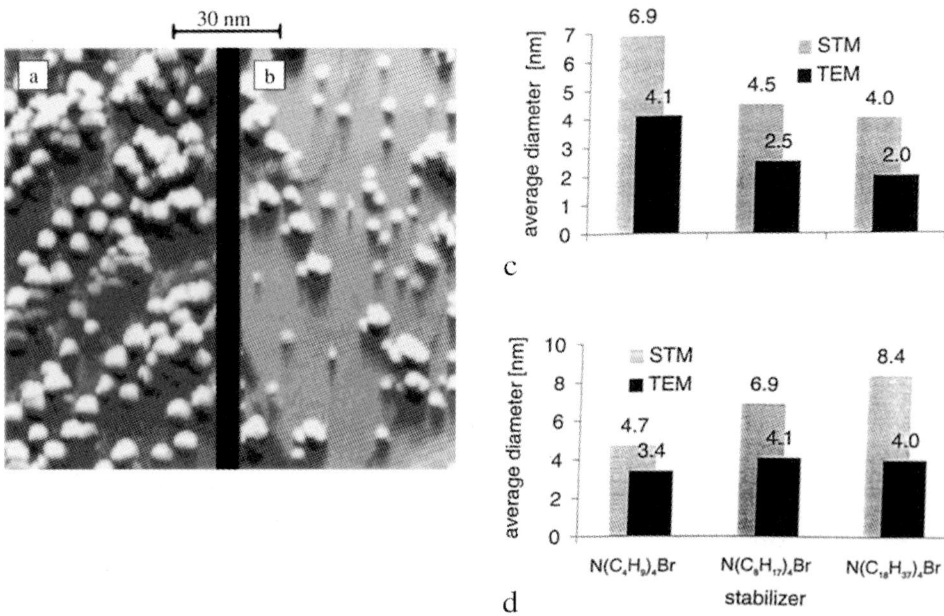

Figure 5-9. STM image of (a) a 4.1 nm Pd cluster stabilized by $N(n\text{–}C_8H_{17})_4\text{–}Br$ and (b) a 2 nm Pd cluster stabilized by $N(n\text{–}C_8H_{17})_4\text{–}Br$. The influence of the metal core size on the STM image is summarized in (c) [stabilizer: $N(n\text{–}C_8H_{17})_4\text{–}Br$] and the influence of the stabilizer on the STM image is summarized in (d) (Reetz et al, [62]).

results indicate that the combination of STM and high resolution TEM leads to valuable information regarding the approximate geometric relation between the metal core and the stabilizing ligand shell.

5.4.2 Structure of two-dimensionally arranged nanoclusters

5.4.2.1 Monolayer of passivated metal clusters

The achievement of ordered arrays of metal dots is the essential step for developing nano-scale electronic devices [63]. With this background, close-packed passivated metal clusters have been intensively studied by different research groups. Andres et al [64] reported the preparation of ordered arrays of 4 nm gold clusters encapsulated by a monolayer of alkyl thiol molecules by so called *colloid self-assembly.* The visualization of ordered cluster arrays was carried out with TEM. The *colloid self-assembly* was successfully applied to some other systems, see [65] as review. There are, however, rarely SPM studies on this kind of samples. This may result from the fact, that the solutions were normally dropped onto carbon film on copper grids and ready for TEM studies.

Two other methods to prepare 2-dimensional arrangements of passivated nanoclusters were reported: two-step *self-assembly* (SA)- and *Langmuir-Blodgett* (LB)-technique [66]. The monolayer formation of ligand stabilized Au clusters or Au colloids on insulating and conductive substrates using two-step SA procedure was reported by several groups [67, 68]. The general idea is: generate a spacer layer carrying functional groups (e.g. $-NH_2$) through self-assembly on the surface as the first step, and deposit the Au clusters with suitable groups in the ligand shell (e.g. $-SO_3H$) as the second step. The surfaces were characterized by using SFM or STM. It turned out that this procedure lead to a high coverage of the surfaces (Fig. 5-10a), but the packing was not as good (Fig. 5-10b) as the colloid preparation. The advantage of this method is the wide selectivity of the substrates.

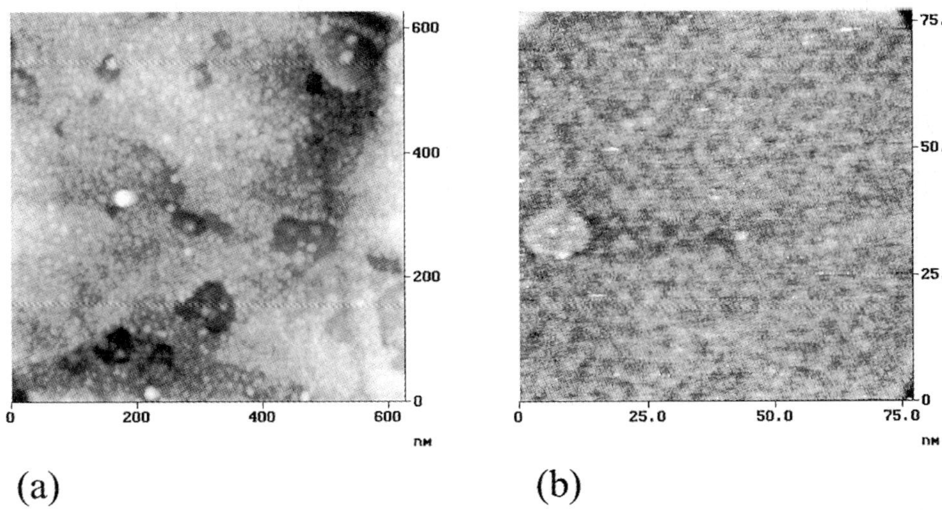

Figure 5-10. Monolayer of Au_{55} cluster on gold substrate prepared by two-step self-assembly. The surface coverage is more than 90% (a, SFM) and the clusters are densely packed (b, STM) (Chi et al, [68]).

Fendler et al applied the LB technique to prepare monolayers of various colloid nanoparticles [69]. Later, Chi et al [70] and Bourgoin et al [71] reported the preparation of passivated cluster monolayers by using the same method and the SFM investigation on these monolayers. Water-insoluble metal clusters with hydrophobic ligand shells, e.g. Au_{55} stabilized with phosphine $(PPh_3)Cl_6$ or with oligo silsesquioxane derivative (T_8–OSS), are not typical amphiphils used for LB preparation, but they are able to form stable monolayers at the air/water interface and can be transferred onto different solid substrates. One monolayer thickness was directly measured with SFM across the dipping line [70].

The transfer did not depend on the nature of the substrates. At a surface pressure of 20 mN/m, about 1–5% uncovered areas or holes were observed by SFM [72]. Some of the clusters were located on the top of the first monolayer. By transferring the monolayers at different surface pressures from the air/water interface, the surface coverage could be adjusted [70]. While "Schmid clusters" could experience higher lateral surface pressures up to 20 mN/m, the dodecanethiol capped gold clusters (C_{12}–Au) form multilayers at 12 mN/m [71]. From SFM investigations, a pressure of 8 mN/m was found to be the optimum transfer pressure to minimize defects (either holes or multilayers) for this system. Such cluster layers were very fragile. In particular, they could be destroyed by STM and contact mode SFM. Bourgoin et al were successful to displace the dodecanethiol by 2,5‴-bis(acetylthio)-5,2′,5′,2″-terthienyl (T3), by emerging the LB monolayer of dodecanethiol capped Au clusters into T3 solution overnight. After the displacement, the samples could be scanned stably with STM, Fig. 5-11. STM images indicated that despite the relative polydispersity of the dodecanethiol Au particles, local hexagonal order was often observed in the film. Square order was also sometimes observed in very small areas.

T_8-OSS-stabilized Au_{55} clusters show increased stability in hydrocarbon solutions such as pentane and could be irradiated with an intensive electron beam in the electron microscope without any aggregation or coalescence [72]. They were, however, very loosely bound to the substrates with LB preparation. It was even difficult to image the monolayer of T_8-OSS-stabilized Au_{55} clusters with tapping mode. The damage to the sample is clearly seen in Fig. 5-12a: the center part ($100 \times 200\ nm^2$) was scanned only once in tapping mode. This effect strongly decreased the resolution since

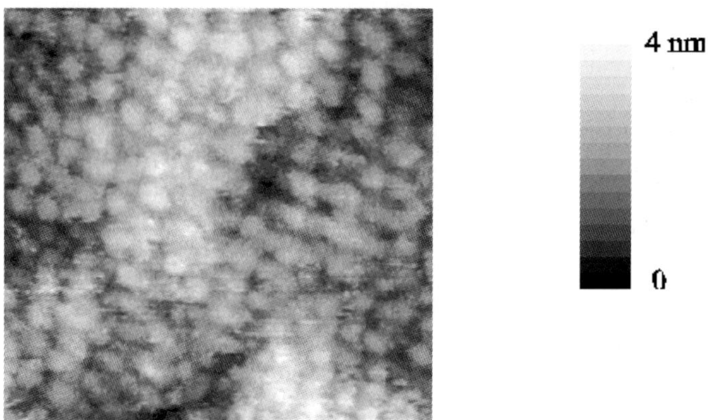

Figure 5-11. STM image taken on an interconnected arrays of gold particles prepared from a C_{12}–Au monolayer deposited at 8 mN/m on epitaxial gold on mica (Bourgoin et al. [71]).

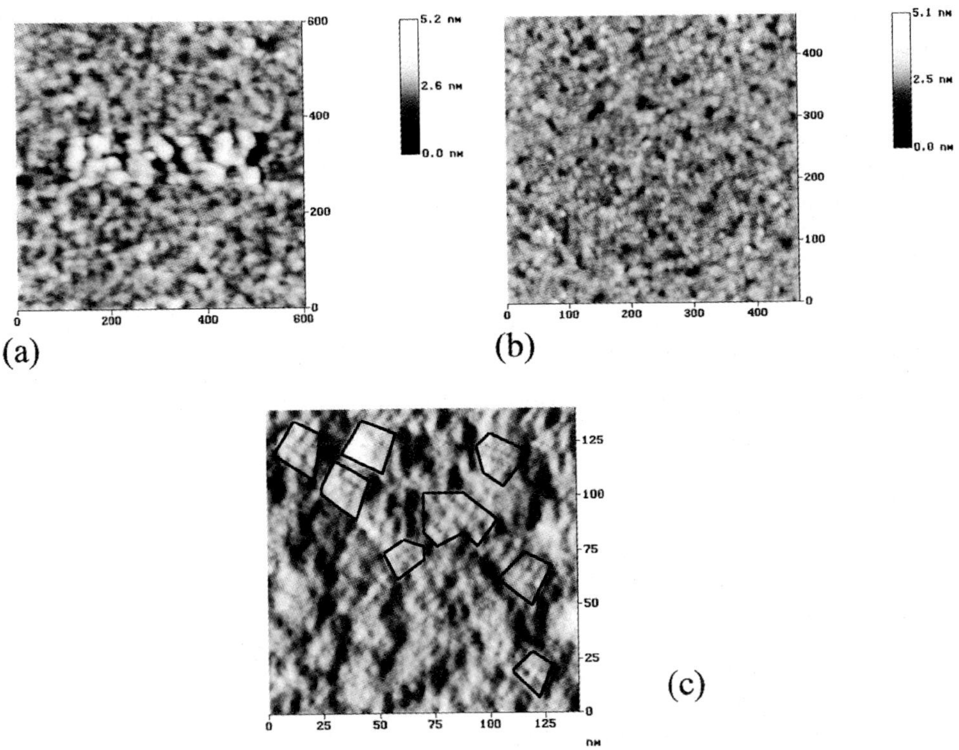

Figure 5-12. (a) Monolayers of T_8–OSS-stabilized Au_{55} clusters undergo an easy deformation during scanning with SFM operated in the intermittent contact regime (middle part). (b) No deformation is observed when SFM is operated in the attractive regime assisted with the active feedback circuit. (c) With SFM operated in the attractive regime, individual Au_{55}–T_8–OSS clusters can be resolved (a and b: unpublished results from L.F. Chi, c: Schmid and Chi, [72]).

the damage became more pronounced by a smaller scanning area. This problem was solved by operating the SFM in the attractive regime using an active feedback circuit as mentioned in the experimental section. An SFM image of a monolayer without damage by operating SFM in this mode is shown in Fig. 5-12b. The center part of the image (200×200 nm^2) was scanned previously for more than one hour and no sample damage was observed. Finally, individual T_8–OSS-stabilized Au_{55} clusters could be visualized as shown in Fig. 5-12c, which was only possible by operating high amplitude SFM in the attractive regime assisted by the active feedback controller. The measured cluster size is 5 nm in diameter in good agreement with the expected value (4.2 nm in diameter). The clusters formed short range ordered arrays, as indicated by black frames in Fig. 5-12c.

5.4.2.2 Nanosized metal/semiconductor dot arrays

Well ordered monolayers of colloid particles can be formed by drying the suspension solutions on substrates and have been examined in detail [73]. This approach offers another attractive method to prepare well ordered nano-dot arrays.

Möller and coworkers [74] used amphiphilic diblock copolymers to form reverse micelles in a selective non-polar solvent, e.g. polystrene-block-poly(2-vinylpyridine) dissolved in toluene forms a core of poly(2-vinylpyridine)(P2VP) and a corona of polystyrene (PS). The polar block solubilizes metal salts such as $HAuCl_4$, H_2PtCl_6, $Pd(Ac)_2$ and $Ti(OR)_4$. Within the core chemical manipulations can be performed, e.g. confined salts can be reduced by UV light, electron beam or by suitable chemical reagents. Ultimately, singular metal or semiconductor particles of similar size are formed. The size of the particles is controlled by the micellar compartment between 1 and 15 nm. Such a micellar solution can be cast on suitable substrates to form mono-micellar films. Finally, the polymer shell can be removed by plasma etching, while ordered arrays of single naked nanoparticles remain.

The formation of the nano-dot arrays was studied by SFM as well as TEM [75]. The ordered structure could extend up to large macroscopic areas (3×3 cm^2). Figure 5-13a shows a SFM image of a layer of PS(800)–b–P[2VP($HAuCl_4$)$_{0.5}$(860)] after being treated with an oxygen plasma operating at 200W for 20 min. The polymer shell has been removed, exposing the single Au particles. The white dots display Au particles with 12 nm in height and 15 nm in lateral dimension. By adjusting the polymer block, they succeeded in controlling the size of the naked nano-dots and the distance between them. Figure 5-13b shows SFM images of organized Au cluster arrays loaded by PS(325)–b–P2VP(75) on mica. The size of the Au clusters is 3 nm in diameter with a periodicity of 30 nm.

Another promising technique to create ordered metallic nano-dot arrays is *nanosphere lithography* [76] based on *natural lithography* [77]. Hereby monolayers of closed packed monodisperse latex spheres with diameters from micrometer down to several ten nanometers are used as lithographic masks through which metals are deposited onto flat surfaces. The formation of metal dot arrays and the SFM observation has been reported by several groups [78–81]. The size of metallic dots prepared with this method can be reduced by annealing. For example, by using latex spheres

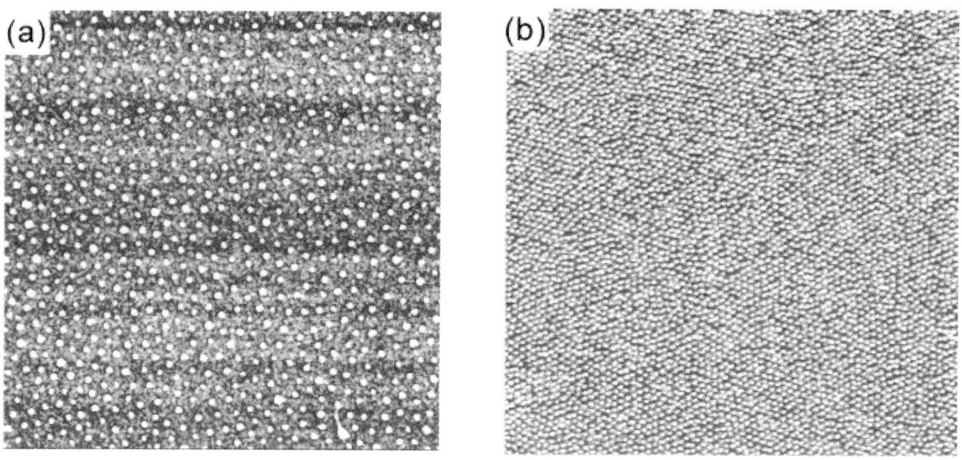

Figure 5-13. SFM topography images of Au cluster arrays on mica substrates after the oxygen plasma treatment. The different interparticle distances are obtained by varying the lengths of the polymer blocks in the following way: (a) PS(800)–b–P[2VP(HAuCl4)0.5(860)], (b) PS(325)–b–P[2VP(HAuCl4)0.5(75)] and (c) PS(1700)–b–P[2VP(HAuCl4)0.1(450)]. The size of each image corresponds to 3μm × 3μm (Spatz et al, [75]).

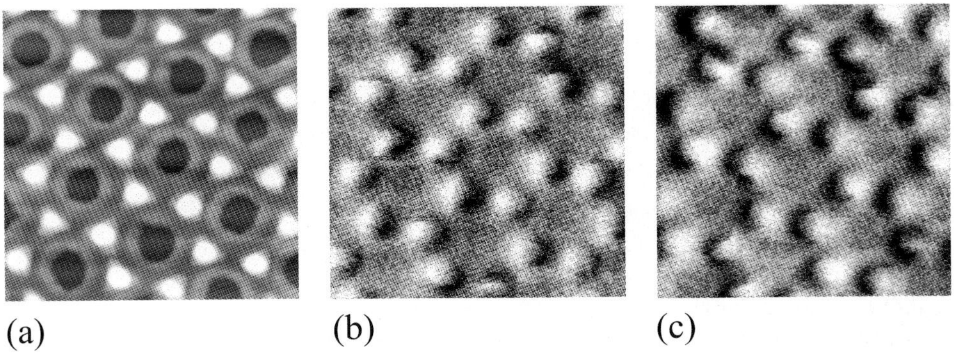

Figure 5-14. (a) Topography SFM-image (1μm × 1μm) of a nanodot array consisting of cobalt particles. (b) MFM image showing the single domain state of the Co particles. The black and white contrast corresponds to the north and south poles of the Co-dots. An external magnetic field of 400 Oe has been applied. (c) Corresponding MFM image of the same Co-dot-array upon reversal of the magnetization direction and the same magnitude (Winzer et al, [81]).

with a diameter of 220 nm, one edge of the evaporated metallic triangular dots is about 70 nm. After the flame annealing treatment, the dots change to a circular shape with a diameter of 20 nm, while the packing of the dots remains unaffected [80].

By evaporating cobalt instead of gold, magnetic dot arrays were formed as shown in Fig. 5-14a. [81]. Application of *magnetic force microscopy* (MFM) revealed the single-domain state of the individual nanometer-sized cobalt dots which were magnetized in an external field of 400 Oe. The black-and white contrast in Fig. 5-14b corresponds to the north and south poles of the Co-dots. Upon switching the magnetization direction, the black and white contrast above the Co-dots reversed (Fig. 5-14c), which clearly proved the magnetic origin of the observed signal.

5.4.2.3 Surface of nanostructured solids

Nanostructured solids are prepared by compacting the clusters synthesized by gas condensation [82] or compacting the passivated clusters. The STM/SFM studies on such solid surfaces concentrate on the nanostructures and the grain boundaries after compacting. Houbertz et al resolved individual passiviated clusters with STM in compact pellets indicating only few clusters agglomerated [83]. Wang and Ying et al imaged nanostructured palladium surfaces with STM and SFM [84, 85], suggesting the equiaxed crystals of 10 nm diameter joined together by grain boundaries. Migration of the grain boundaries was stimulated by STM (but not SFM), resulting in a preferential alignment of nanosized grains. In another report of Ying et al [86] the nanosized titanium oxide (TiO_{2-X}) was found to adopt a prior preferential alignment by compacting. The TiO_{2-X} clusters with diameters of 10–20 nm close packed in rows, as shown in Fig. 5-15a, underwent structure change by heating. The individual clusters linked together forming chain-like structures by heating the compact pellets. With further heating the structures developed into smooth tubular structures, Fig. 5-15b. The alignment phenomenon demonstrated step by step with the SFM could be of importance to mechanical and catalytic applications.

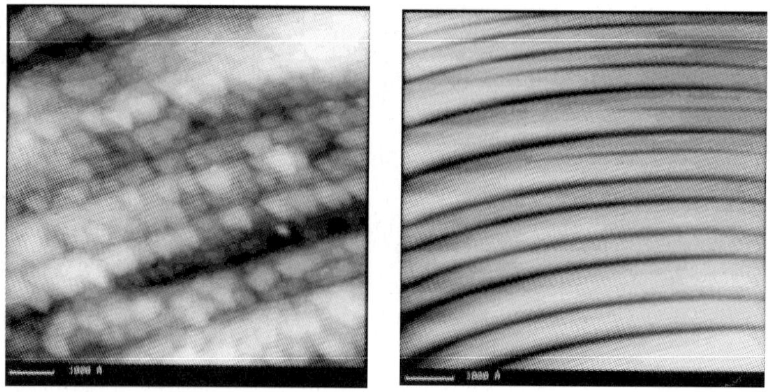

a) b)

Figure 5-15. (a) SFM image of nanocrystalline TiO_{2-X} compact. (b) SFM image of the same compact after being heated to 1000°C revealed the formation of tubular structures (adapted from Ying et al [86]).

5.4.3 SPM-based deposition and manipulation of metallic nanoclusters

Various SPM-based techniques have been successfully employed for nanofabrication, e.g. atom manipulation [87], local oxidation [88], field-induced charge trapping [89]. In this section, we will concentrate on the SPM-based deposition and manipulation of metallic nanoclusters. The general experimental approach is to apply a voltage pulse between the STM tip or metal coated SFM tip and the selected sample surfaces. Beyond a certain voltage threshold, nanosized clusters can be generated. Several mechanisms have been proposed for the cluster deposition, e.g. field evaporation of the tip material [90], point contact of the tip and the sample [91], and tip melting due to high electrical current [92]. The advantage of SPM-based generation of nanoclusters lies in the subatomic accuracy of tip positioning which allows the deposition of clusters at the desired site. STM serves as well as the tool for the *in situ* characterization of the induced structures.

A number of pioneer works were carried out in the late 1980's. Staufer et al [93] created small hillocks of 35 nm in diameter by locally melting glassy substrates. Silver et al [94] deposited metallic structures down to 20 nm in size by decomposing organometallic gas with a STM. Later, McCord et al [95] reduced the size of metallic dots as small as 10 nm by decomposing organometallic gases containing tungsten and gold. Schneir et al [96] and Li et al [97] demonstrated the possibility to deposit Au clusters on flat gold substrates with STM in air. Nanometer sized pits could be created as well with the same scanning tip. At that time, it was not possible to choose between mounts and pits forming under the tip.

A distinguished step to reliably deposit Au clusters on gold substrates was achieved by Mamin et al in 1990 [98]. They believed that the Au mounts deposited with the STM tip in the earlier experiments were induced by field evaporation as suggested by Gomer [90]. To cause gold emission, they applied voltage pulse between the tip made of gold and the gold substrate while the tip was within the tunneling range. The pulses were sufficiently short (a few hundred nanoseconds or less) so that the feedback loop did not need to be disengaged during the pulsing. Figure 5-16a shows the gold mound

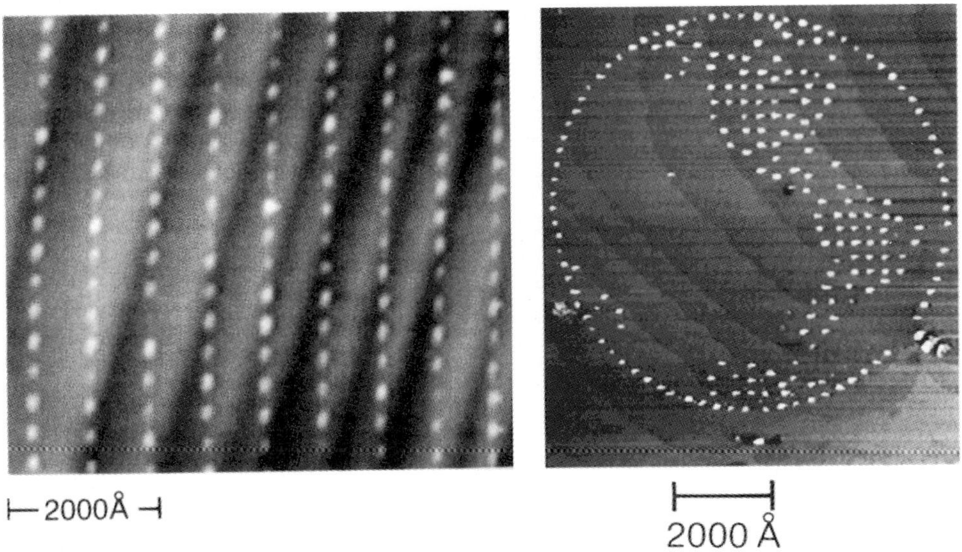

Figure 5-16. (a) Gold mounds made by applying voltage pulses between the gold tip and the gold substrate (3.6 V, tip grounded and 600 ns). (Mamin et al, 1990). (b) Simplified map of the world fabricated by STM (Au dots on Au (111) substrate (Mamin et al [98, 99]).

arrays made with 4.2 V, 300 nsec pulses on a stepped gold surface. More complicated patterns could be created by controlling the movement of the tip [99, 100], Fig. 5-16b. The mounts were typically 10–20 nm across at the base and 2–3 nm high, although they could be as small as 5 nm across. The emission process was reproducible and fast. This technique can be used to write several thousand features without apparent degradation of the tip's ability to emit atoms. The first experiment was carried out in air, and later in UHV [101]. Both positive and negative voltage pulses created mounds, but the mounds were typically larger with the tip positive, and the threshold voltage was slightly lower. These gold dots were stable over a period of weeks, even after heating at 65°C over hours. With the same method, Hsiao et al [102] deposited gold and copper clusters onto hydrogen-terminated Si (111) surfaces which were in contact either with air or dry N_2. The metallic composition of the mounds observed with STM were confirmed by *ex situ* elemental analysis using the *field emission scanning Auger microscope* (FE-SAM). They described the different behavior of structures generated by tip-positive and tip-negative pulses, although they were apparently similar in STM topographic images. The tip-negative mounds were stable during the repeated imaging in STM for hours and during the exposure to an electron beam in the FE-SAM instrument, whereas tip-positive mounds were not stable in both cases. Based on these observations, they speculate that the mounds produced by tip-positive pulses were not metallic.

Fujita et al demonstrated the deposition of Au dots on Si (111) in UHV with gold-coated tungsten STM tips [103]. They tended to believe the model of point contact presented by Tsong [104] and Pascual [91] for the dot deposition. Atomically resolved images of the Si (111) (7×7) structure were observed even after many cycles of atom transfer using the same tip (Fig. 5-17), which indicated the stability of the tip apex.

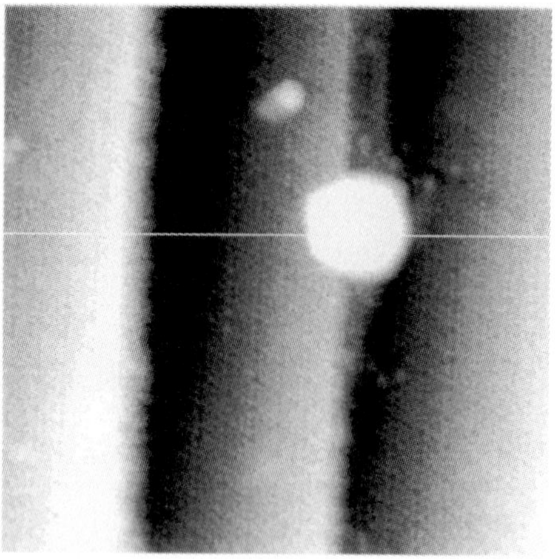

Figure 5-17. Atomically resolved 50 × 50 nm STM image (V_t = –1.5 V, I_t = 0.5 nA) of a gold dot produced on the Si(111) surface using voltage pulses (–5.5 V, tip negative and 5 ms). The Si (111) (7 × 7) structure was observed in the same image (Fujita, [103]).

Again the deposition with tip-positive and tip-negative showed different behaviors: the former one created large overlapped mounds up to a few hundred nanometers in diameter, whereas the later one created 3 to 20 nm gold mounds with a deposition probability up to 60%. They improved the deposition probability later by using a piezo z-pulse instead of the voltage pulse [105]. With this mode, they fabricated gold dots with a diameter of about 5 nm and a height of about 1 nm on the Si surface more reliably so that a continuous gold nanoline of 5 nm in width and 50 nm in length could be produced.

Chang et al reported the creation of gold dots on gold surfaces in a non-conducting liquid [106]. They found that the shape of the created structures depended on materials used as scanning tips, e.g. gold, tungsten and PtIr. They proposed a contact mechanism with the shape of the created structures affected by the adhesion property of the tip and sample materials. In consequence, the nanostructures of different shapes could be formed efficiently in a non-conducting liquid by the tips made of different materials.

Nanosized metallic deposition was also performed with SFM carrying metal coated tips, e.g. by Hosaka [107] and Liu [108]. With SFM, non-conductive substrates such as mica can be used as well as conductive substrates. Gold is the most widely used metal for such studies up to now due to its stability against oxidation and relatively low working function. Some other materials such as aluminum were deposited in the same way, reported by Hu et al [109].

5.4.4 Scanning tunneling spectroscopy on nanocluster systems

Most of the work concerning STS investigations of deposited or produced clusters on surfaces concentrated on the measurement of single-electron-tunneling (SET) effects rather than on effects due to the discrete nature of the electronic system in small clusters. In fact SET effects are much more pronounced in clusters insulated from the surface by a small gap, i.e an oxide layer, an organic molecular layer or any other kind of insulating spacer, and they are at least potentially much more rewarding in terms of technological applications.

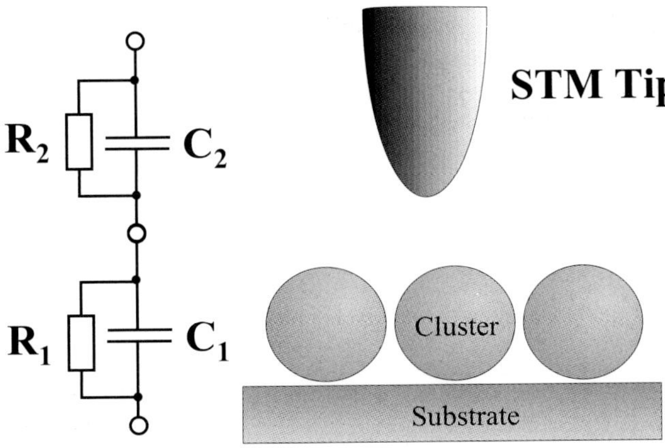

Figure 5-18. Schematic sketch of the two tunneling junctions between STM tip and cluster and between cluster and surface. Both junctions are represented by a DC-resistance R and a capacitance C.

STS measurements on small insulated clusters are usually interpreted in the framework of a two-junction model : the cluster is connected via two tunneling junctions to the tip and the surface, respectively. Both junctions can be represented in electronical terms by a resistance R for the DC tunneling resistance and a capacitance C parallel to the resistance (Fig. 5-18).

If the cluster between current source and drain is sufficiently small one will experience step-like features in the I/U-spectra which are interpreted as single-electron-tunneling effects. They originate from the fact that even one single excess electron will have a significant effect on the electric potential of the cluster. The energy ΔE to add an electron to such a cluster is

$$\Delta E_{n+1} = \frac{e^2}{C_1 + C_2}\left(n + \frac{1}{2}\right), \qquad (5\text{-}11)$$

$C_{1,2}$ being the capacities of the tip-cluster and the cluster-surface junction, respectively, and n the number of excess electrons in the cluster [110]. This increase in energy has to be taken care of by larger tunneling voltages thereby creating the step-like shape of the I/U-curves. Obviously two conditions have to be met to observe such a behavior. Firstly the cluster has to be small enough so that the charging energy of the cluster is large compared to the thermal energy $\frac{3}{2}kT$ and secondly the resistance has to be large compared to the resistance of a one-dimensional conductor $R_1 >> \frac{h}{e^2} \cong 26k\Omega$. Figure 5-19 shows a simulated STS measurement for such a system. One can clearly see that the SET effect creating the step-like structure of the I/U curve is almost completely lost at room temperature.

Measurements in this respect, either by measuring the I/U-curves or the $\frac{\partial I}{\partial U}$-curves using Lock-in techniques have been reported on numerous systems. The very first observation of SET effects was made by Zeller and Giaever in 1969 in a planar setup [112]. It consisted of small Sn particles sandwiched between two planar electrodes inside the oxide barrier separating both. Although a thorough interpretation was not possible in this experiment due to numerous parallel/serial connections, SET effects could be seen easily. Consequently local probe microscopy and well-characterized particles and barriers were used for the investigation of SET effects. The first observation

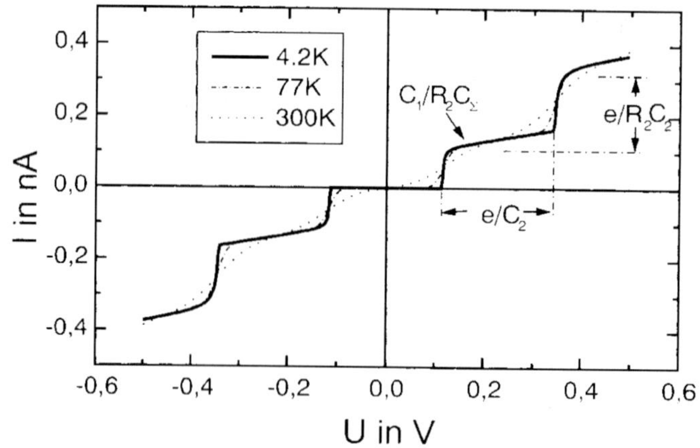

Figure 5-19. Simulated STS measurement of the double junction in Fig. 5-18. Indicated are the effects of temperature broadening (adapted from T. Drechsler [111]).

of such SET effects by STS measurements was reported by van Kempen and coworkers [113, 114]. They investigated the properties of granular Al films deposited onto an oxidized continous Al film, as well as Al on a single crystal of $YBa_2Cu_3O_7$, a high-T_C superconductor. In both cases they observed step-like structures in their experimental curves which they interpreted in the framework of the semiclassical approach by Likharev et al. [115, 116] combined with a Monte-Carlo simulation.

Other authors reported the observation of Coulomb blockades and staircases on Au particles on ZrO_2 covered Gold [117–120], In droplets on oxide covered Al [121], Pb grains on oxide covered Pb [122, 123] and Au droplets on hydrogen terminated Si(111) [124]. These experiments were mostly performed at low temperatures, however, Schönenberger et al. demonstrated in 1992 that it is also possible to observe SET effects at 300 K if the capacitance of the cluster is sufficiently small.

In all the mentioned experiments the production of a sufficiently small cluster and its insulation to the supporting surface was not very reliable. This problem can be solved if the insulating layer is made up from an organic monolayer or if the cluster itself is already supplied with an insulating spacer. For both systems the observation of SET effects in STS measurements was reported by several groups [70, 83, 125]. Van Kempen et al. reported the observation of Coulomb staircases on $Pt_{309}Phen^*_{36}O_{30}$ [125]. Deviations from the expected shape of the staircase they interpreted as quantum size effects. Cluster Pellets made up from Au_{55} clusters were investigated by Hartmann et al [83].

If an organic monolayer is used the surface is usually covered by a self assembly monolayer (SAM) as a first step. Reifenberger and coworkers used a dithiole monolayer on top of a gold(111) surface to attach gold clusters from a multiple expansion cluster source (MECS) [126, 127]. The thickness of the insulating dithiole layer was about 1 nm, the thickness of the gold clusters varied between 1 and 2 nm. The authors could observe SET effects at room temperature as well as give an estimation of the resistance of the gold cluster of about 18 kΩ. Grummt et al reported the investigation of a mixed SAM system on Gold(111) consisting of a monothiole (n-Decanthiole) with small amounts of BHQ1 [128]. Subsequent oxidation in silver nitrate solution lead to a decoration of the oxidized BHQ1 molecules with Ag_4 clusters. STS measurements at room temperature and 100 K in UHV showed Coulomb staircases as expected from simulations.

As mentioned at the beginning of this chapter only very few attempts exist to measure effects arising from the discreteness of electronic states due to electron confinement in small clusters deposited or assembled on surfaces with the STM. Meiwes-Broer and coworkers [129] have performed such a study on small platinum clusters deposited on HOPG with a low-temperature STM (LT-STM). They observed distinct peaks in the $\frac{\partial I}{\partial U}$-curves which they attributed either to a standing electron wave between the top and the bottom of the cluster or to discrete electronic states of the cluster by performing plots of the energetic difference between the first and second peak observed and the inverse thickness of the cluster. So far the authors do not differentiate between the two possible interpretations of their observations.

To illustrate the onset of solid state behavior in clusters Stroscio and coworkers [130] investigated the shape of STS spectra of Fe clusters grown spontaneously from metallic vapor on GaAs. By evaluating $\left(\frac{\partial I}{\partial U}\right)_{U=0}$ representing the local density of states near the Fermi energy (FLDOS) they found that clusters of 150 Å^3 volume, corresponding to roughly 13 Fe atoms, exhibited non-metallic behavior while clusters of about 1000 Å^3 volume, corresponding to about 85 Fe atoms, showed fully metallic behavior. They even succeeded in observing metallic behavior in clusters consisting of only 35 Fe atoms.

A very similar approach was pursued by Bifone et al. [131]. They investigated the FLDOS of palladium clusters on HOPG. The clusters were produced by evaporation onto the cleaned HOPG, the FLDOS of the clusters was determined by evaluating $\left(\frac{\partial I}{\partial U}\right)_{U=0}$ and plotted vs. the apparent size of the cluster. They found that the FLDOS increased linearly with the cluster volume between 10 and 350 nm^3 and leveled further on to a constant value indicating the onset of solid state behavior.

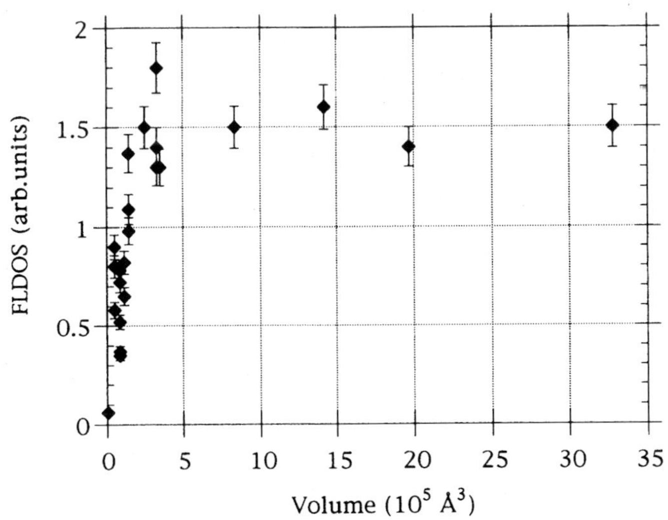

Figure 5-20. Dependence of the FLDOS of palladium clusters as a function of their apparent size on HOPG (Bifone et al [131].)

STS studies of semiconductor clusters were performed by Kuk et al. [51]. They deposited Si_{10} clusters from a pulsed laser vaporization source onto Au(001)(5 × 20). By analyzing $\frac{\partial I}{\partial U}$ they could easily differentiate between the gold substrate, the silicon clusters and gold silicide existing in islands on the surface.

5.4.5 Scanning near-field microscopy and near-field spectroscopy of nanoclusters

A reduced dimension of electron motion in nanometer-sized structures gives rise to new electronic and optic phenomena. One of the most attractive properties of semiconductor nanostructures is the narrow and strong optical transition due to the discrete electronic density of state of zero-dimensional structures, which is favorable for optoelectronic devices [132]. In metallic nanoparticles, collective electron oscillations known as surface plasmons (SPs) excited by light show pronounced optical resonance [133]. The applications of near-field spectroscopy, based on the essential design of *scanning near-field microscopy* (SNOM), allow local spectroscopic measurements of individual structures. The inhomogenous broadening of the spectrum induced by the variation of structure size and the local environment often observed in far-field spectroscopy can thus be minimized. Although only few SNOM and near-field spectroscopy studies of nanometer-scale structures have been reported, the great promise in optical probing is demonstrated.

The high spatial resolution of SNOM has been reported elsewhere [32]. By using a tetrahedral tip, Koglin et al demonstrated a material contrast of nanometer size particles of Ag and Au in a mixed film with a lateral resolution in the range of 1–10 nm [134]. Klar et al performed near-field spectroscopy by using a SNOM setup on gold particles with a radius of 20 nm embedded in a dielectric sol-gel TiO_2 matrix [135]. The homogenous line shape of the surface-plasmon resonance was measured, and the

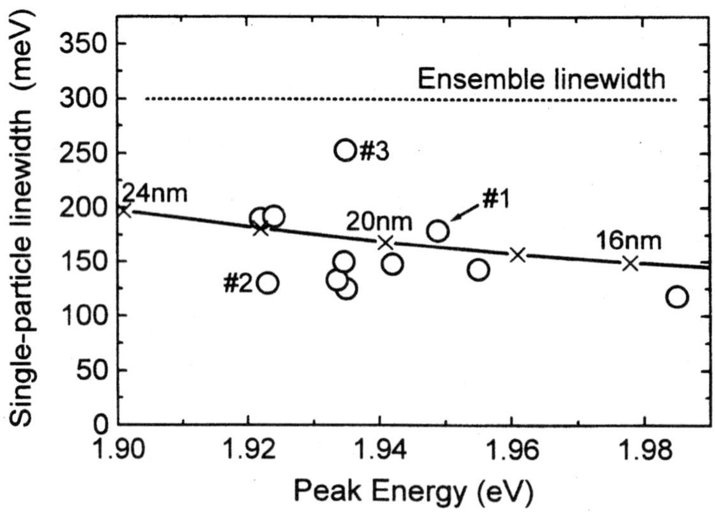

Figure 5-21. Linewidth vs. peak energy determined from the near-field spectra of Au nanoparticles (circles). Solid line: theoretical results from gold spheres of different radii. Dashed line: ensemble linewidth as determined from the far-field extinction spectrum (Klar et al, [135]).

SP linewidth was substantially smaller than that of an ensemble measured with far-field spectroscopy, as shown in Fig. 5-21. Deviations in the spectra of individual particles were interpreted as being due to the variations in the local environment.

Near-field spectroscopy of semiconductor quantum constitutions were first reported by Hess et al [136]. By using an external reflection mode aperture-NSOM suitable for cryogenic temperatures (down to 2K), they were able to identify luminescent centers with sharp spectrally distinct emission lines in GaAs/AlGaAs quantum wells. Meanwhile more detailed investigations of single quantum dot spectra were performed by Gammon et al in layered semiconductor structures [137] using a non scanning near-field method similar to the one previously described by Fischer [138].

3D nanometer-scale semiconductor dots of uniform size were formed by self-organization of deposits during strained growth on a flat substrate and characterized with SFM [139, 140]. Just recently, Saiki et al reported the near-field photoluminescence (PL) spectroscopy of single InGaAs dots at low temperature (down to 5K) [141]. By employing a high sensitive near-field probe with a large aperture of 500 nm ($\lambda/2$) and an optimized shape at the apex of the optical fiber, they measured the PL spectra of individual quantum dots (QDs) and imaged single QDs with a resolution of $\lambda/6$. Based on the evolution of single dot PL spectra, they were able to identify the emission origins of individual dots. In addition to the ground-state emission, excited-state and biexciton emission were observed. In another work reported by Toda et al, the Zeeman spin splitting in PL spectra from single InAs/GaAs self-assembled QDs was observed [142]. In this work, a low temperature NSOM was used incorporating a superconducting magnetic field up to 10 T.

5.5 Limitations and Prospects

SPM methods are in general sensitive and powerful to image surface structures. The topographic data obtained can be quantitatively transformed to 3D images of the sample surfaces as long as the radius of probing tips is much smaller than the object features. In the case of nanoclusters with a size range of 1–20 nm, the images have to be carefully interpreted since the clusters are in the same size range as the probing tip or even smaller. In most cases, the measured lateral size of clusters is larger than that measured with TEM, an effect caused by the convolution of the tip and the cluster. Many authors used the measured height instead of lateral size to deduce the shape of the clusters. It was indicated by some authors that the measured height strongly depended on the adhesion force [143]. The adhesion between the tip and the sample is determined by the material properties and can be influenced by environment conditions (e.g. humidity) and the tip geometry. Knowledge about adhesion forces between tip and cluster and that between tip and substrate, as well as the control of environment conditions, are necessary to evaluate the measured data.

There are no intrinsic limitations of applying SFM to clusters on substrates. The binding of clusters on substrates, however, has to be strong enough to allow reliable measurements. Reducing the force exerted by the probing tip is another way to minimize the influence of the scanning procedure, but it is limited by the dimension of the probing tip and the sensitivity of the detection methods. Suitable combinations of substrates and clusters are also important for measuring local physical properties with SPM methods on individual clusters.

The application of SPM methods on nanophased materials can yield valuable information of geometric and physical properties of the materials. The local spectroscopic studies based on the STM and SNOM are expected to be further applied to a variety of preformed clusters in the future. With the extended use of the instruments, some other physical properties of nanophased materials may be possible measured locally, e.g. the dielectric properties of clusters by *scanning capacitance microscopy* (SCAM) [89]. The deposition of nanoclusters with probing tips in combination with the lithographic use of the SPM was demonstrated. This offers the possibility to deposit or move the preformed clusters to desirable positions. The study of self-assembled passivated nanocrystal superlattices (NCSs) with SFM, which was not systematically done yet, should provide information of the growth of these NCSs on a variety of substrates.

Acknowledgment

We would like to thank all the colleagues who provided illustrations for this chapter. Particular thanks to B. Anczykowski, U. Ch. Fischer, M. Gleiche, S. Gao and S. Höppener for the assistance during the preparation of the manuscript.

References

[1] G. Binnig, H. Rohrer, Ch. Gerber, E. Weibel, *Phys. Rev. Lett.* **1982**, *49*, 57.
[2] S.I. Park and C.F, Quate, *Appl. Phys. Lett.* **1986**, *49*, 1172.
[3] R. Sonnenfeld and P. Hansma, *Science*, **1986**, *232*, 211.
[4] S. Akari, M. Stachel, H. Birk, E. Schreck, M. Lux, K. Dransfeld, *J. Microsc.* **1988**, *152*, 521.
[5] S. Manne, J. Massie, V. B. Elings, P.K. Hansma, A. A. Gewirth, *J. Vac. Sci. Technol.* **1991**, *B9*, 950.
[6] H.K. Wickramasinghe, Scanning Probe Microscopy, *AIP Conference Proceedings*, **1992**, *241*, 9.
[7] G. Binnig, C.F. Quate, Ch. Gerber, *Phys. Rev. Lett.* **1986**, *56*, 930.
[8] D. Sarid, *Scanning Force Microscopy* (Ed.: M. Lapp, H. Stark), Oxford Series in optical and imaging sciences, Oxford, , Oxford University Press **1991**.
[9] D. W. Pohl, W. Denk, M. Lanz, *Appl. Phys. Lett.* **1984**, *44*, 651.
[10] Y. Kuk in *Scanning Tunneling Microscopy I* (Eds.: H.-J. Güntherodt and R. Wiesendanger), Heidelberg, Springer, **1992**, Chapter 3.
[11] R. T. Hamers in *Scanning Tunneling Microscopy I* (Eds.: H.-J. Güntherodt and R. Wiesendanger), Heidelberg, Springer, **1992**, Chapter 5.
[12] J. Winterlin, R.J. Behm in *Scanning Tunneling Microscopy I* (Eds.: H.-J. Güntherodt and R. Wiesendanger), Heidelberg, Springer, **1992**, Chapter 4.
[13] S. Magonov and M.-H. Whangbo, *Surface Analysis with STM and AFM,* Weinheim, VCH, **1996**.
[14] A. Ikai, *Surf. Sci. Rep.*, **1996**, *26*, 261.
[15] J. A. DeRose and R. M. Leblanc, *Surf. Sci. Rep.* **1995**, *22*, 73.
[16] R. Wiesendanger, *Scanning Probe Micrsocopy and Spectroscopy*, Cambridge, Cambridge University Press, **1994**, Chapter 8.
[17] D. Averin and K. Likharev in *Single-Charge Tunneling* (Eds.: H. Grabert and M.H. Bevoret), New York, Plenum Press, **1992**.
[18] M. Neitzert, Ph.D Thesis, University of Muenster, **1999**.
[19] J. Bardeen, *Phys. Rev. Lett.* **1960**, *6*, 57.
[20] J. Tersoff and D.R. Hammann, *Phys. Rev. Lett.* **1983**, *50*, 1988.
[21] J. Tersoff and D.R. Hammann, *Phys. Rev. B* **1985**, *31*, 805.
[22] M.P.J. van Staveren, H.B. Brom, L.J. de Jongh, Y. Ishii, *Phys.Rev.B* **1987**, *35*, 7749.
[23] Kappes, Schär, Röthlisberger, Yeretzian, Schumacher, *Chem.Phys.Lett.* **1988**, *143*, 251.
[24] E. Meyer and Heinzelmann in *Scanning Tunneling Microscopy II* (Eds.: H.-J. Güntherodt and Wiesendanger), Heidelberg, Springer, **1992**, Chapter 4.
[25] G. Meyer, N.M. Amer, *Appl. Phys. Lett.* **1988**, *53*, 1045.

[26] S. Alexander, L. Hellemans, O. Marti, J. Schneir, V. Elings, P.K. Hansma, M. Longmire, J. Gurley, *J. Appl. Phys.* **1989**, *65*, 164.
[27] R.M. Overney, Ph.D Thesis, Basel University, Basel, Switzerland, **1992**.
[28] Y. Martin, C.C Wiliams, H.K. Wickramasinghe, *J. Appl. Phys.* **1987**, *61*, 4723.
[29] B. Anczykowski, J.P. Cleveland, D. Krüger, V. Elings, H. Fuchs, *Appl. Phys. A*, **1998**, *66*, 885.
[30] . Y. Martin, H.K. Wickramasinghe, *Appl. Phys. Lett.* **1987**,*50*,1455.
[31] C. Schönenberger and S.F. Alvarado, *Z. Phys.* **1990**, *B 80*, 373.
[32] U.Ch. Fischer in *Scanning Probe Microscopy* (Eds.: R. Wiesendanger), Heidelberg, Springer, **1998**, chapter 7.
[33] J.O. Bovin, R. Wallenberg, D.J. Smith, *Nature*, **1985**, *317*, 47.
[34] W.D. Abraham, K. Sattler, E. Ganz, H.J. Mamin, R.E. Thomson, J. Clarke, *Appl. Phys. Lett.* **1986**, *49*, 853.
[35] A.M. Baró, A. Barolome, L. Vazquez, N. García, R. Reifenberger, E. Choi, R.P. Andres, *Appl.-Phys. Lett.* **1987**, *51*, 1594.
[36] T. Castro, Y.Z. Li, R. Reifenberger, E. Chio, S.B. Park, R.P. Andres, *J. Vac. Sci. Technol.* **1989**, *A7*, 2845.
[37] H. Hövel, Th. Becker, A. Bettac, B. Reihl, M. Tschudy, E.J. Williams, *IBM Research Report*, **1996**.
[38] D.M. Schaefer, A. Ramachandra, R.P. Andres, R. Reifenberger, *Z. Phys. D* **1993**, *26*, 249.
[39] S. Granjeaud, K. Yckache, M. Dayez, A. Humbert,C. Chapon, C.R. Henry, *Microsc. Microanal. Microstruct.* **1993**, *4*, 409.
[40] Y.Z. Li, R. Reifenberger, E. Choi, R.P. Andres, *Surface Science*, **1991**, *250*, 1.
[41] R. Wurster and B. Ocker, *Scanning*, **1993**, *15*, 130.
[42] D. Sarid, T. Henson, L.S. Bell, C.J. Sandroff, *J.Vac. Sci. Technol.* **1988**, *A6*, 424.
[43] T.W. Jing, N.P. Ong, C.J. Sandroff, *Appl. Phys. Lett.* **1988**, *52*, 104.
[44] E. Ganz, K. Sattler, J. Clarke, *Phys. Rev. Lett.* **1988**, 60, 1856.
[45] U. Müller, K. Sattler, J. Xhie, N. Venkateswaran, G. Raina, *J. Vac. Sci. Technol.* **1991**, *B9*, 829.
[46] M. Kuwahara, S. Ogawa, S. Ichikawa, *Surface Science*, **1995**, *344*, 1259.
[47] F.J.C.S. Aires, P. Sautet, G. Fuchs, J-L. Rousset, P. Melinon, *Microsc. Microanal. Microstruct.* **1993**, *4*, 441.
[48] A. Humbert, M. Dayez, S. Sangay, C. Chapon, C.R. Henry, *J. Vac. Sci. Technol.* **1990**, *A8*, 311.
[49] A. Humbert, M. Dayez, S. Granjeaud, P. Ricci, C. Chapon, C.R. Henry, *J. Vac. Sci. Technol.* **1991**, *B9*, 804.
[50] H.Y. Nie, T. Shimizu, H. Tokumoto, *J. Vac. Sci. Technol.* **1993**, *B12*, 1843.
[51] Y. Kuk, M.F. Jarrold, P.J. Silverman, J.E. Bower, W.L. Brown, *Phys. Rev. B*, **1989**, *39*, 11168.
[52] H. Kind, A.M. Bittner, O. Cavalleri, K. Kern, *J. Phys. Chem. B*, **1998**, *102*, 7582.
[53] Y. Xia and G.M. Whiteside, *Angew. Chem., Int. Ed.* **1998**, *37*, 550.
[54] J. Colchero, O. Marti, J. Mlynek, A. Humbert, C.R. Henry, C. Chapon, *J. Vac. Sci. Technol.* **1991**, *B9*, 794.
[55] D.M. Schaefer, A. Patil, R.P. Andres, R. Reifenberger, *Phys. Rev. B*, **1995**, *51*, 5322.
[56] N. Burnham and R. Colton, *J. Vac. Sci. Technol.* **1989**, *A7*, 2906.
[57] G. Schmid , *Cluster and Colloids: From Theroy to Applications*, Weinheim, VCH, **1994**.
[58] G. Schön, U. Simon, *Colloid Polym. Sci.* **1995**, *273*, 101.
[59] L.E.C. van de Leemput, J.W. Gerritsen, P.H.H. Rongen, R.T.M. Smokers, H.A. Wierenga, H. van Kempen, G. Schmid, *J. Vac. Sci. Technol.* **1991**, *B9*, 814.
[60] C. Becker, Th. Fries, K. Wandelt, U. Kreibig, G. Schmid, *J. Vac. Sci. Technol.* **1991**, *B12*, 810.
[61] G. Schmid, *Chem. Rev.* **1994**, *92*, 1709.
[62] M.T. Reetz, W. Helbig, S.A. Quaiser, U. Stimming, N. Breuer, R. Vogel, *Science*, **1995**, *267*, 367.
[63] J.R. Tucker, *J. Appl. Phys.* **1992**, *72*, 4399.
[64] R.P. Andres, J.D. Bielefeld, J.I. Henderson, D.B. Janes, V.R. Kolagunta, C.P. Kubiak, W. Mahoney, R.G. Osifchin, *Science*, **1996**, *273*, 1690.
[65] Z.L. Wang, *Adv. Mater.* **1998**, *10*, 13.
[66] A. Ulman, *An Introduction to Ultrathin Organic Films: From Langmuir-Blodgett to Self-Assembly*, New York, Academic Press, Inc. **1991**.
[67] G. Schmid, St. Peschel, Th. Sawitowski, *Z. f. anorg. Allg. Chem*, **1997**, *623*, 719.
[68] L.F. Chi, M. Hartig, T. Drechsler, Th. Schwaack, C. Seidel, H. Fuchs, G. Schmid, *Appl. Phys. A*, **1998**, *66*, 187.
[69] J.H. Fendler and F.C. Meldrum, *Adv. Mater.* **1995**, *7*, 607.
[70] L.F. Chi, S. Rakers, M. Hartig, H. Fuchs, G. Schmid, *Thin Solid Films*, **1998**, *327–329*, 520.
[71] J-P. Bourgoin, Ch. Kergueris, E. Lefevre, S. Palacin, *Thin Solid Films*, **1998**, *327–329*, 515.
[72] G. Schmid and L.F. Chi, *Adv. Mater.* **1998**, *10*, 515.

[73] N.D. Denkov, O.D. Velev, P.A. Kralchesky, I.B. Ivanov, N. Yoshimura, K. Nagayama, *Nature*, **1993**, *361*, 26.
[74] J.P. Spatz, S. Mößmer, M. Möller, *Chemistry*, **1996**, *2*, 1552.
[75] J.P. Spatz, S. Mößmer, M. Möller, T. Herzog, A. Plettl, P. Ziemann, *J. Luminescence*, **1998**, *76&77*, 168.
[76] J.C. Hulteen, R.P. Van Duyne, *J. Vac. Sci. Technol.* **1995**, *A12*, 1553.
[77] U.Ch. Fischer, H.P. Zingsheim, *J. Vac. Sci. Technol.* **1981**, *19*, 1881.
[78] J. Boneberg, F. Burmeister, C. Schäfle, P. Leiderer, D. Reim, A. Fery, S. Hermighaus, *Langmuir*, **1997**, *13*, 7080.
[79] F. Burmeister, C. Schäfle, Th. Mattes, M. Böhmisch, J. Boneberg, P. Leiderer, *Langmuir*, **1997**, *13*, 2983.
[80] L.F. Chi, T. Drechsler, St. Höppener, S. Rakers, Ch. Röthig, Th. Schwaack, F. Stanberg, H. Fuchs, *J. Surface Analysis*, **1998**, *4*, 168.
[81] M. Winzer, M. Kleiber, N. Dix, R. Wiesendanger, *Appl. Phys. A*, **1996**, *63*, 617.
[82] H. Gleiter, *Prog. Mater. Sci.* **1989**, *33*, 223.
[83] R. Houbertz, T. Feigenspan, F. Mielke, U. Memmert, U. Hartmann, U. Simon, G. Schön, G. Schmid, *Europhys. Lett.* **1994**, *28*, 641.
[84] G.H. Wang, H. Fuchs, R. Laschinski, R. Sander, R. Birringer, H. Gleiter, *Ultramicroscopy*, **1992**, *42–44*, 594.
[85] J.Y. Ying, G.H. Wang, H. Fuchs, R. Laschinski, H. Gleiter, *Material Letters*, **1992**, *15*, 180.
[86] J.Y. Ying, L.F. Chi, H. Fuchs, H. Gleiter, *Naonstructured Materials*, **1993**, *3*, 273.
[87] D.M. Eigler and E.K. Schweizer, *Nature*, **1990**, *344*, 524.
[88] J. Dagata, J. Schneir, H.H. Harary, C.J. Evans, M.T. Pastek, J. Bennett, *Appl. Phys. Lett.* **1990**, *56*, 2001.
[89] R.C. Barrett and C.F. Quate, *J. Appl. Phys.* **1991**, *70*, 2725.
[90] R. Gomer, *IBM J. Res. Dev.* **1986**, *30*, 428.
[91] J.I. Pascual, J. Méndez, J. Gómez-Herrero, A.M.Baró, N.Garacía, Vu Thien Binh, *Phys. Rev. Lett.* **1993**, *71*, 1852.
[92] N. Yokoi, S. Ueda, S. Namba, M. Takai, *Jpn. J. Appl. Phys.* **1993**, *32*, L129.
[93] U. Staufer, R. Wiesendanger, L. Eng, L. Rosenthaler, H.-R. Hidber, H.-J. Güntherodt, N. Garcia, *J. Vac. Sci. Technol.* **1988**, *A6*, 537.
[94] R.M. Silver, E.E. Ehrichs, A.L. de Lozanne, *Appl. Phys. Lett.* **1987**, *51*, 247.
[95] M.A. McCord, D.P. Kern, T.H.P. Chang, *J. Vac. Sci. Technol.* **1988**, *B6*, 1877.
[96] J. Schneir, R. Sonnenfeld, O. Marti, P.K. Hansma, J.E. Demuth, R.J. Hamers, *J. Appl. Phys.* **1988**, *63*, 717.
[97] Y.Z. Li, L. Vazquez, R. Piner, R.P. Andres, R. Reifenberger, *Appl. Phys. Lett.* **1989**, *54*, 1424.
[98] H.J. Mamin, P.H. Guethner, D. Rugar, *Phys. Rev. Lett.* **1990**, *65*, 2418.
[99] H.J. Mamin, S. Chiang, H. Birk, P.H. Guethner, D. Rugar, *J. Vac. Sci. Technol.* **1991**, *B9*, 1398.
[100] S. Hosaka, S. Hosoki, T. Hasegawa, H. Koyanagi, T. Shintani, M. Miyamoto, *J. Vac. Sci. Technol.* **1995**, *B13*, 2813.
[101] C.S. Chang, W.B. Su, T.T. Tsong, *Phys. Rev. Lett.* **1994**, *72*, 574.
[102] G.S. Hsiao, R.M. Penner, J. Kingsley, *Appl. Phys. Lett.* **1994**, *64*, 1350.
[103] D. Fujita, Q. Jiang, H. Nejoh, *J. Vac. Sci. Technol.* **1996**, *B14*, 3413.
[104] T.T. Tsong, *Phys. Rev. B*, **1991**, *44*, 13703.
[105] D. Fujita, H.-Y. Sheng, Z.-C. Dong, H. Nejoh, *Appl. Phys. A*, **1998**, *66*, 753.
[106] T.C. Chang, C.S. Chang, H.N. Lin, T.T Tsong, *Appl. Phys. Lett.* **1995**, *67*, 903.
[107] S. Hosaka, H. Koyanagi, A. Kikukawa, Y. Maruyama, *J. Vac. Sci. Technol.* **1994**, *B12*, 1872.
[108] J.Q. Song, Z.F. Liu, C.Z. Li, H.F. Chen, H.X. He, *Appl. Phys. A*, **1998**, *66*, 715.
[109] X.M. Hu, P. von Blanckenhagen, *Appl. Phys. A*, **1998**, *66*, 707.
[110] K. Mullen, E. Ben-Jacob, R.C. Jaklevic, Z. Schuss, *Phys. Rev. B*, **1988**, *37*, 98.
[111] T. Drechsler, Ph.D Thesis, University of Muenster, **1998**.
[112] H.R. Zeller and I. Giaever, *Phys. Rev.* **1969**, *181*, 789.
[113] P.J.M. van Bentum, R.T.M. Smokers, H. van Kempen, *Phys. Rev. Lett.* **1988**, *60*, 2543.
[114] H. van Kempen, P.J.M. van Bentum, R.T.M. Smokers, *Physica Scripta*, **1992**, *T 42*, 153.
[115] D.V. Averin and K.K. Likharev, *J. Low Temp. Phys.* **1985**, *59*, 347.
[116] K.K. Likharev and A.B. Zorin, *J. Low Temp. Phys.* **1986**, *62*, 345.
[117] C. Schönenberger, H. van Houten, H.C. Donkersloot, A.M.T. van der Putten, L.G.J. Fokkink, *Physica Scripta*, **1992**, *T 45*, 289.
[118] C. Schönenberger, H. van Houten, C.W.J. Beenakker, *Physica B*, **1993**, *189*, 218.
[119] C. Schönenberger, H. van Houten, J.M. Kerkhof, H.C. Donkersloot, *Appl. Surf. Sci.* **1993**, *67* 222.
[120] C. Schönenberger, H. van Houten, H.C. Donkersloot, *Europhys. Lett.* **1992**, *20*, 249.
[121] R. Wilkins, E. Ben-Jacob, R.C. Jaklevic, *Phys. Rev. Lett.* **1989**, *63*, 801.

[122] J.-C. Wan, K.A. McGreer, N. Anand, E. Nowak, A.M. Goldman, *Phys. Rev. B*, **1990**, *42*, 5604.
[123] K.A. McGreer, J.-C. Wan, N. Anand, A.M. Goldman, *Phys. Rev. B*, **1989**, *39*, 12260.
[124] P. Radojkovic, M. Schwartzkopff, M. Enachescu, E. Stefanov, E. Hartmann, F. Koch, *J. Vac. Sci. Technol.* **1996**, *B14*, 1229.
[125] J.G.A. Dubois, J.W. Gerritsen, S.E. Shafranjuk, E.J.G. Boon, G. Schmid, H. van Kempen, *Europhys. Lett.* **1996**, *33*, 279.
[126] M. Dorogi, J. Gomez, R. Osifchin, R.P. Andres, R. Reifenberger, *Phys. Rev. B*, **1995**, *52*, 9071.
[127] R.P. Andres, T. Bein, M. Dorogi, S. Feng, J.I. Henderson, C.P. Kubiak, W. Mahoney, R.G. Osifchin, R. Reifenberger, *Science*, **1996**, *272*, 1323.
[128] U.-W. Grummt, M. Geißler, T. Drechsler, H. Fuchs, R. Staub, *Angew. Chem.* **1998**, *110*, 3480.
[129] A. Bettac, L. Köller, V. Rank, K.H. Meiwes-Broer, *Surf.Sci.* **1998**, *402*, 475.
[130] P.N. First, J. A Stroscio, R.A. Dragoset, D.T. Pierce, R.J. Celotta, *Phys. Rev. Lett.* **1989**, *63*, 1416.
[131] A. Bifone, L. Casalis, R. Riva, *Phys. Rev. B*, **1995**, *51*, 11043.
[132] Y. Arakawa and H. Sakaki, *Appl. Phys. Lett.* **1982**, *20*, 939.
[133] U. Kreibig and M. Vollmer, *Optical Properties of Clusters*, Berlin, Springer, **1995**.
[134] J. Koglin, U.C. Fischer, H. Fuchs, *Phys Rev. B*, **1997**, *55*, 7977.
[135] T. Klar, M. Perner, S. Grosse, G. von Plessen, W. Spirkl, J. Feldmann, *Phys. Rev. Lett.* **1998**, *80*, 4249.
[136] H.F. Hess, E. Betzig, T.D. Harris, L.N. Pfeiffer, K.W. West, *Science*, **1994**, *264*, 1740.
[137] D. Gammon, E.S. Snow, B.V. Shanabrook, D.S. Katzer, D. Park, *Phys. Rev. Lett.* **1996**, *76*, 3005.
[138] U.C. Fischer, *J. Opt. Soc. Am*, **1986**, *B3*, 1239.
[139] J.M. Moison, F. Houzay, F. Barthe, L. Leprice, E. André, O. Vatel, *Appl. Phys. Lett.* **1994**, *64*, 196.
[140] K. Nishi, R. Mirin, D. Leonard, G. Medeiros-Ribeiro, P. Petroff, A.C. Gossard, *J. Appl. Phys.* **1996**, *80*, 3466.
[141] T. Saiki, K. Nishi, M. Ohtsu, *Jpn. J. Appl. Phys.* **1998**, *37*, 1638.
[142] Y. Yoda, S. Shinomori, K. Suzuki, Y. Arakawa, *Appl. Phys. Lett.* **1998**, *73*, 517.
[143] S. John, T. Van Noort, K. O. Van der Werf, B. G. De Grooth, N. F. Van Hulst, *Ultramicroscopy*, **1997**, *69*, 117.

6 Electrical and Electrochemical Analysis of Nanophase Materials

Zhong Shi and Meilin Liu

6.1 Introduction

Nanostructured materials exhibit a host of interesting new phenomena directly related to their reduced dimensionality. Not only the electronic, magnetic, and optical properties but also chemical, electrochemical, and catalytic properties of nanostructured materials are very different from those of the bulk form and depend sensitively on size, shape, and composition [1, 2]. The large surface-to-volume ratio and the variations in geometry and electronic structure have a dramatic effect on transport and catalytic properties. For example, the reactivity of small clusters has been found to vary by orders of magnitude when the cluster size is changed by only a few atoms [3]. Another example is hydrogen storage in metals. It is well known that most metals do not absorb hydrogen, and even among those that do, hydrogen is typically adsorbed dissociatively on surfaces with a hydrogen-to-metal ratio of 1. This limit can be significantly enhanced in small sizes. Cox et al. [4] showed that small positively charged clusters of Ni, Pd, and Pt generated in molecular beams and containing between two and 60 atoms can absorb up to eight hydrogen atoms per metal atom. The number of absorbed atoms decreases with increasing cluster size and approaches one for clusters having around 60 atoms. This shows that small clusters may be very useful in hydrogen storage devices. Yet another case of the effect of size on properties is observed in nanophase powders of lithium magnesium oxide, which display threshold catalytic activity at temperatures much below the bulk limits [5].

In a nanostructured electrode, a large portion of atoms is located at surfaces or interfaces, dramatically influencing the transport of ionic and electronic defects through solids. In addition, short-range rearrangement or redistribution of ionic and electronic defects may cause electrical or chemical polarization; this effect is more pronounced or even becomes dominant in nanophase dielectric materials. Each interface will polarize in its unique way when the system is subjected to an applied stimulus. Nanocomposites consisting of two or more phases, each having different electrical properties, may exhibit significant space charge or interfacial polarization. Chemical polarization results from changes in composition or stoichiometry of a material while electrical polarization is a consequence of acquiring electric dipoles due to redistribution of charges or of aligning existing dipoles (e.g., complex defects or defect associates) in the material. A nanoporous electrode has enormous surface or interface area for chemical or catalytic reactions. For example, 1 cm^2 of a 5 μm thick film, comprising primary of 20 nm particles of TiO_2, partially sintered together to form a microstructurally stable porous film, would have an actual surface area on the order of 750 cm^2. The high surface-to-volume ratio as well as the very small particle size provide nanostructured electrodes with unique electrocatalytic and photocatalytic properties [6].

In porous nanocrystalline electrodes, no significant potential drop seems to be present between the center and the surface of an individual semiconductor particle, due to the very small particle size in combination with a low donor density [7]. Nevertheless, it has been demonstrated that efficient photocurrent generation can still take place in this type of electrode, which is explained by the efficient capture of the photoholes by redox species in the solution [8]. In contrast to dense macroscopic semiconductor electrodes, most efficient photo-induced charge separation in porous nanocrystalline electrodes takes place close to the conductive support [8-11]. Södergren et al. [10] have derived a model in which transport of photo-generated electrons in the nanoporous film is driven by diffusion only. Despite the simplicity of this model, simulations for the spectral photocurrent response of nanoporous TiO_2 electrodes are in good agreement with experimental data [7]. The influence of the applied voltage has been introduced into this model by taking into account that the electron concentration in the conductive support is determined by the applied voltage [10, 11]. Porous nanocrystalline metal oxide electrodes have a great potential for many applications, such as dye-sensitised photoelectrochemical solar cells [7, 12, 13], electrochromic displays [14–16], fuel cells, chemical sensors, and lithium-ion batteries [17]. The electrochemical properties of nanostructured materials are critical to the performance of these devices.

In an electrochemical measurement, electrical potential or current can be readily controlled or measured. To date, various electrochemical methods have been successfully used to characterize nanophase materials. In this chapter, we will first introduce several methods for preparation of nanoelctrodes, then outline a number of electrochemical techniques commonly used to characterize the properties of nanophase materials, and fianlly present a few examples to illustrate how to characterize electrical and electrochemical properties of nanostructured materials using these techniques. For further details on electrochemical techniques and methods, readers are referred to a number of excellent review articles and monographs listed in the end of this chapter [18–24].

6.2 Preparation of nanostructured electrode

One of the key challenges in characterizing the electrical and electrochemical properties of a nanosturctured material is to make an electrical contact to the material to be studied. In this section, several techniques will be introduced for preparation of nanostructured electrodes or nanophase materials on a well-defined electrode. The advantages, together with limitations, of each technique will be discussed.

6.2.1 Powder microelectrode

For analytical and voltammetric applications, microelectrodes are getting increasingly popular [25, 26], largely because of the increasing availability of carbon or metal fibers and microlithographic techniques. Microelectrode techniques are very versatile in kinetic studies because of high rates of diffusion flux and less interference by resistive polarization and capacitive currents arising from the double-layer capacitance. To date, however, the materials used to fabricate microelectrodes are still limited to Pt, Ni, Au, and carbon [25, 26] because of the need to use ultrafine wires and the heat-sealing process usually employed in the preparation of microelectrodes. Recently, Cha et al. [27] have developed a powder microelectrode technique to study the kinetic and

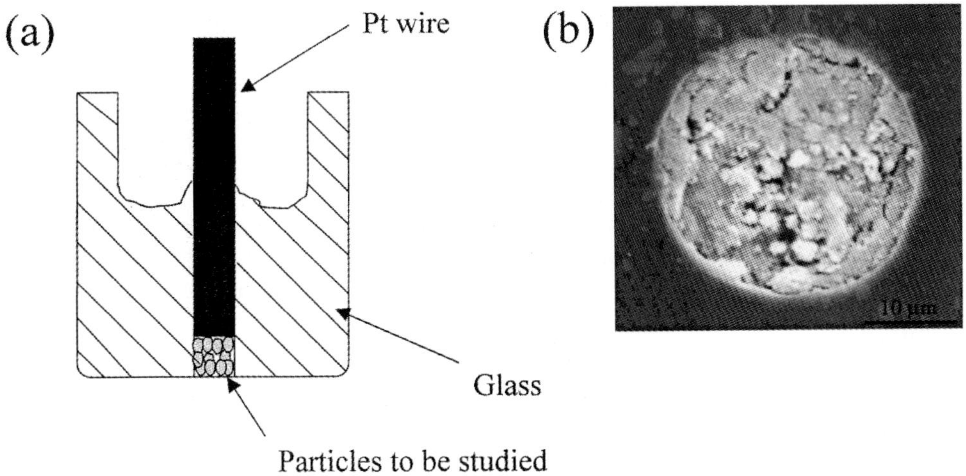

Figure 6-1. A schematic diagram of a powder microelectrode [27] and (b) an SEM micrograph of a powder microelectrode packed with graphite particles..

cyclic behavior of various powder materials. Schematically shown in Fig. 6-1a is a powder microelectrode, which can be prepared by etching the tip of a platinum microdisk electrode (diameter 10–25 μm) in aqua regia to form a micro cavity at the tip. The depth of the microcavity can be controlled by the etching conditions. Nanoparticles (or agglometrates of nanoparticles) can then be trapped into the microcavity of a microelectrode by grinding the etched tip on the surface of a flat plate (e.g., glass) covered with nanoparticles of interest. Microcavities can usually be packed easily and the powders within the microcavity can be directly, as shown in Fig. 6-1b studied using various electrochemical techniques.

The use of powder microelectrode offers the following advantages:

1) It is an easy and convenient approach to rapid screening of new electrode materials;
2) It provides valuable information directly from the sample particles to be studied, without the need of adding additives such as a binder or carbon black. While it has been mostly used in studying micro-size powders, it can be extended to the study of nano-size powder electrodes;
3) It offers an averaged behavior of a new material over a large number of particles, not the sole property of a single particle.

In addition to trapping of nanoparticles into a microcavaty, nanophase materials (or nanostructured electrodes) can also be prepared in a microcavity using electrodeposition, electrophoretic deposition, sol-gel, or self-assembly processes, which will be discussed in details below.

6.2.2 Electrodeposition or electrophoretic deposition

Recent advances have made electrodeposition an attractive approach for preparation of nanostructured materials [28, 29]. Electrodeposition has a number of advantages over other physical deposition techniques. The rate of deposition is fast; the process is not equipment-intensive; the low processing temperature (usually room temperature) minimizes interdiffusion or chemical reaction; the film thickness can be precisely controlled by monitoring the delivered charge; composition and defect chemistry can be controlled; films can be deposited onto complex shapes; nonequilibrium phases can be deposited; and the driving force can be precisely controlled by adjusting the applied potential. One of the major challenges is to ensure that not only are the electrodeposited materials small in dimension, but also their size distribution is relatively narrow. To date, a wide range of nanophase materials have been successfully deposited, including metals, alloys, semiconductors, polymers, and ceramics. Further, nanostructured metal-ceramic composites have been produced from suspensions of nano particles of alumina and metals (gold [30], copper [31], and Ni [32]) using electrophoretic deposition. In this case, it is necessary to form a stable and well-dispersed suspension of nanoparticles to be deposited.

The successful use of traditional direct-current electrochemical methods for producing nanostuctures is limited in two ways: (i) the substrate must be conductive and (ii) the controllable growth occurs only in the direction normal to the substrate. In order to make 0-, 2-, and 3-dimensional nanostructures the growth must be restricted in some way, for example, by applying a non-conductive mask. Another approach used to make electrochemical deposition spatially selective is to create regular (using lithography) or random arrays of metal nano-electrodes on conducting surfaces, and deposition will be favored on the conductive surface. Alternatively, one might deposit mass-selected clusters on a substrate. Here are two examples to show the advantages of an electrodeposition process.

1) Electrodeposition into nanopores
   Nanophase materials have been grown in the pores of nanoporous membranes such as anodized aluminum or track-etch polymers [33], which have cylindrical pores of uniform diameter. The pores are used as templates to prepare nanoparticles of desired material. When a polymer, metal, semiconductor, or carbon is synthesized electrochemically within one of these pores, a nanocylinder of the desired material is obtained. Depending on the material and the chemistry of the pore wall, this nanocylinder may be hollow (a tubule) or solid (a fibril). Martin and coworkers [34] have shown that metal nanotubes can also serve as ion-selective membranes. The nanotube diameter can be as small as 0.8 nm, and the length of the nanotube can span the complete thickness of the membrane. These membranes show selective transport analogous to ion-exchange polymers. The ion permselectivity is thought to occur because of excess charge density that is present on the inner walls of the nanotubes. Since the sign of the excess charge can be changed potentiostatically, a metal nanotube can be either cation- or anion-selective, depending on the applied potential [34].
2) Electrodeposition of nanocomposites
   Electrochemical deposition is a very attractive processing route for the synthesis of composite materials. The low processing temperatures allow codeposition of materials (ceramics, metals, polymers, semiconductors) that would not tolerate each other at high temperatures necessary for traditional processing, such as sintering,

vapor-phase processing, or solidification used to fabricate composites. A simple approach is to suspend particulate material in a plating electrolyte and codeposit this with a desired metallic matrix. This can be accomplished both by electroless deposition and by electroplating. Commercial applications of this approach include codeposition of alumina, silicon carbide, or diamond with a metal such as nickel [35]. The key challenge in this approch is to prevent agglomeration of the particles prior to codeposition.

Another electrochemical scheme for preparation of composites is electrochemical infiltration. Recently, Sheppard et al. has successfully used an electrochemical infiltration process to fill the pores (about 5 nm) of a silica xerogel film with nickel [36]. In this manner, room temperature processing was used to synthesize three-dimensionally interconnected nanoscale networks of metal and ceramic.

6.2.3 Formation of nanoparticles in polymers

Inorganic nanoparticles have often been prepared with an auxiliary medium, such as surfactant, the interface of monolayers and bilayer lipid membranes, polar headgroups of Langmuir-Blodgett films, clays and zeolites. One of the primary objectives of the various synthesis techniques is to control the particle size either by spatial restrictions, such as size of pores and entities in the media, or by reaction kinetic. Stabilizing nanoparticles is critical. The use of conducting polymers as matrices for nanoparticles not only enhances the stability, the interactions of nanoparticles with the macromolecules may also result in anomalous physical and chemical properties. A number of approaches used for preparing nanoparticles in a polymer matrix are briefly described below.

A popular method is a two-step, in-situ generation of nanoparticles in a polymer matrix. The first step is the incorporation of a metal ion in a polymer matrix by immersion of the polymer in an aqueous solution containing the metal ions. The second step is the formation of the particles in the polymer matrix by reacting the metal ions with a proper reducing agent. An example is the formation of CuS particles in poly(vinyl alcohol)-poly(acrylic acid) matrix [37]. The polymer mixture is first immersed into a $CuSO_4$ aqueous solution, where the acidic groups of poly(acrylic acid) serve as complexation sites for Cu^{2+} ions. Subsequently, the Cu^{2+} ions in the polymer matrix are reduced using Na_2S to form CuS particles (about 10 nm). Another similar approach is the preparation of nanoparticle particulate film. In this method, a polymer monolayer (instead of surfactant) is spread on an aqueous solution of metal salt, e.g., Cd, Zn, and Pb ions. Injection of a reactant gas (e.g., H_2S) into the gas-phase of the enclosed system [38] initiates particle growth. The polymer monolayer provides a matrix for the size-controlled growth of semiconductor particles that can be transferred, essentially intact, to a solid substrate. CdS particles were formed in the poly-(styrenephosphonate diethyl ester) (PSP) particulate monolayer using this approach.

Another widely used method is via polymerization of colloidal solutions containing metal ions and monomers. The particle size can be controlled by the reaction temperature and properties of the colloidal solution, thermal coagulation, and Ostwald ripening. As an example, polymerization of $Pb(MA)_2$ (lead methylacrylic acid) with styrene [39] has been used to form PbS nanoparticles in the polymer. Since there are two C=C bonds in each $Pb(MA)_2$ molecule, it is easier to copolymerize with styrene to form Pb-polymer microgel. This was subsequently treated with H_2S gas to obtain PbS nanoparticles in the polymer matrix.

Ion exchange is another useful method to prepare organic polymer and inorganic soild nanocomposite. Polymers such as Nafion or Surlyn have cation-exchange sites where metal ions, such as Cd^{2+} or Pb^{2+}, can be introduced into the polymer matrix. Treatment of such ion-exchanged films with chalcogenide sources results in precipitation of nanoparticles of compound semiconductors within the hydrophilic regions of the polymer. This method has been extended to synthesize layered semiconductor clusters such as CdS [40] as well as magnetic particles such as Fe_2O_3 [41].

A somewhat different approach has been explored by Schrock and coworkers [42] using direct incorporation of the metal ion Pb^{2+} into monomer units that eventually become part of a norbornene-derived copolymer following ring-opening metathesis polymerization (ROMP). Again, the polymer is designed to have carefully phase-separated and closely size controlled regions of high and low hydophobicity, and the semiconductor PbS is precipitated in the latter regions by exposure to H_2S from the gas phase.

Yet another approach is a hybrid of the polymer isolation and the surface-capped cluster or colloid approaches. In this method, a well-defined semiconductor cluster or colloid is prepared by the conventional capping techniques and dissolved in a solvent along with a soluble polymer. This mixed solution may then be deposited (e.g., by spin coating) onto a substrate and dried to produce a polymer film doped with the semiconductor clusters. This simple approach has provided some new examples of interesting photoconductive composites by using a photoconductive polymer such as polyvinylcarbazole (PVK) doped with clusters such as CdS [43].

6.2.4 Electrochemical self-assembly

Among various approaches used to prepare nanostructured electrodes or devices, such as crystal growth, ion implantation, and vapor phase molecular epitaxy, self-assembly is considered to be a very attractive method.

Molecular self-assembly uses the interactive forces of solid state lattice structures, chemical bonds, and van der Waals forces to form larger aggregates of atomic or molecular units with specific geometries, potentially leading to the design of a wide variety of nanostructures. This approach can be used to make nanostructures that are identical to one another (a truly monodisperse sample size).

Self-assembly of nanoparticles to the oppositely charged substrate surface is governed by a delicate balance of the adsorption and desorption equilibria. For example, the objective of immersion is to achieve efficient adsorption of one (and only one) monoparticulate layer of nanoparticles onto the oppositely charged substrate surface. It is equally important to prevent the desorption of the nanoparticles during the rinsing process. The optimization of the self-assembly in terms of maximizing the adsorption of nanoparticles from their dispersions and minimizing their desorption on rinsing requires the judicious selection of stabilizers and the careful control of the kinetics of the process. The self-assembly of CdS and PbS nanoparticles was found to be most efficient, for example, if the semiconductor particles were coated by a 1:3 mixture of thiolactic acid and ethyl mecaptane [44].

Let's take the self-assembly of alternating layers of polyelectrolytes and inorganic nanoparticles as an example to illustrate a layer-by-layer self-assembly process. First, a well cleaned substrate is primed by adsorbing a layer of surfactant or polyelectrolyte onto its surface. The primed substrate is then immersed into a dilute aqueous solution of a cationic polyelectrolyte, for a time optimized for adsorption of a monolayer, rinsed, and dried. The next step is the immersion of the polyelectrolyte monolayer covered substrate into a dilute dispersion of surfactant coated negatively charged semiconductor nanoparticles, also for a time optimized for adsorption of a monoparticulate layer, rinsed and dried. These operations complete the self-assembly of a polyelectrolyte monolayer – monoparticulate layer of semiconductor nanoparticle sandwich onto the primed substrate. Subsequent sandwich units are deposited analogously.

6.2.5 Mesoporous electrodes

Since the discovery of a new class of mesoporous aluminosilicates (M4S1) [45a], particularly MCM-41 having pores controllable from 2 to 10 nm and surface area greater than 700 cm^2/g, various mesoporous materials have been prepared using a liquid-crystal templating (LCT) mechanism. In particular, structures for chemically-selective catalysis or for fast charge and mass transport can be tailored by adjusting the electrostatic and steric properties of the surfactant molecules and the compensating ions as well as other processing parameters. This provides a unique way to preparation of novel mesostructures difficult to prepare otherwise. Recently, mesoporous SnO_2, TiO_2, and Sn-TiO_2 have been successfully prepared using LCT mechanisms for electrochemical applications [45b].

6.2.6 Composites electrodes consisting of nanoparticles

To evaluate nanophase materials as active electrode materials for lithium batteries, electrochemically functional composite electrodes consisting of the nanoparticles must be prepared. Lithium intercalation compounds, such as $LiNiO_2$, $LiCoO_2$, $LiMnO_2$, and various forms of carbon, have been extensively studied for lithium-ion batteries. Traditionally, nanoparticles of active electrode materials are mixed with an organic binder (e.g., PTFE) and a conductive additive (e.g., carbon black) to form a composite electrode, which is then subjected to various electrochemical testing to characterize the electrochemical behavior. Detailed procedures can be found elsewhere [46–49].

The difficulty associated with this traditional approach, however, is that it is not clear how the binder and the additive may influence the electrochemical properties of the electrochemical properties of the active nanoparticles in the composite. The performance of this composite electrode is influenced sensitively by many factors. For example, if the amount of carbon black is not sufficient or not well dispersed, the materials might be excellent. Accordingly, the study may not be conclusive. Further, the cycling of a redox composite electrode is very time-consuming because of low diffusion coefficients of Li^+ ions inside the solide phase. However, this method is still widely used in study of electrode materials for battery applications simply because there are no other alternatives.

6.3 Principles of electrochemical techniques

Both steady-state and transient techniques have been used to determine the electrical and electrochemical properties of nanophase materials,. In a steady-state experiment, a constant current or potential is applied while the steady-state response of the material is monitored. In a transient experiment, the system is perturbed from an equilibrium or a steady state and the relaxation of the response of the system, such as current, potential, charge, impedance, optical reflectance, or other measurable properties of the system, is monitored as a function of time. In many cases, the experimental conditions are created under which the rate of the system response is controlled only by the process to be studied. In a potential step method, such as chronoamperometry or chronocoulometry, the experimental conditions are set up so that the system response is controlled completely by diffusion of electroactive species in order to determine the diffusivity of the electroactive species. In this chapter, several techniques for determination of electrical and electrochemical properties will be discussed.

6.3.1 Impedance spectroscopy

Impedance spectroscopy (IS) has emerged as a powerful experimental tool in the study of electrical, electrochemical, and catalytic properties of materials, including transport of ionic and electronic defects, dielectric polarization, electrode kinetics, and other bulk or interfacial phenomena in various material systems [50].

6.3.1.1 Definition of impedance functions

In an impedance measurement, a linear system is perturbed by an applied small-amplitude alternating stimulus (e.g., a voltage) of angular frequency ω,

$$\tilde{V}(\omega) = V_0 \exp(j\omega t) \tag{6-1}$$

In responding to this periodic disturbance, the system will relax to a new pseudo-steady state. The response of the system (e.g., the current) will be at the same frequency of the perturbation, but may lead or leg behind the perturbation by a temporal phase angle θ,

$$\tilde{I}(\omega) = I_0 \exp[j(\omega t + \theta)] \tag{6-2}$$

The ratio of the alternating voltage to the alternating current is defined as the impedance transfer function (in ohms) of the linear system,

$$Z(\omega) = \frac{\tilde{V}(\omega)}{\tilde{I}(\omega)} = \mathrm{Re}\{Z(\omega)\} - j\,\mathrm{Im}\{Z(\omega)\} = Z'(\omega) - jZ''(\omega) \tag{6-3}$$

Impedance function is in general a complex function; the real part of impedance, $\mathrm{Re}\{Z\} = Z'$, characterizes the long-range transport (irreversible, energy dissipation processes) while the imaginary part of impedance, $\mathrm{Im}\{Z\} = Z''$, characterizes polarization or magnetization (reversible, energy storage processes) in materials.

In addition, several other transfer functions related to impedance are often used. For instance, the reciprocal of the impedance function is called admittance function, $Y(\omega) = [Z(\omega)]^{-1}$. The ratio of the alternating electric induction, $\tilde{D}(\omega)$, to the alternating electric field, $\tilde{E}(\omega)$, is defined as the dielectric permittivity transfer function of a linear system,

$$\varepsilon_r(\omega) = \frac{\tilde{D}(\omega)}{\varepsilon_o \tilde{E}(\omega)} = \mathrm{Re}\{\varepsilon_r(\omega)\} - j\,\mathrm{Im}\{\varepsilon_r(\omega)\} = \varepsilon_r'(\omega) - j\varepsilon_r''(\omega) \tag{6-4}$$

where ε_0 is the absolute permettivity (or the permittivity of free space) and ε_r is the relative permittivity of the material under study.

The real part of the relative permittivity, $\mathrm{Re}\{\varepsilon_r\}=\varepsilon_r'$, also called dielectric constant, characterizes the energy storage property of the material and depends in general on frequency. In fact, it is possible to determine the contributions of each polarization mechanism (electronic, ionic, dipole, and space charge polarization) from the frequency dependence of dielectric constant. In contrast, the imaginary part of the relative permitivity, $\mathrm{Im}\{\varepsilon_r\} = \varepsilon_r''$, characterizes the energy dissipation in the material due to relaxation of dipole and space charge polarization and to resonance of ionic and electronic polarization in the material. In fact, the AC conductivity of a dielectric material can be expressed as,

$$\tilde{\sigma}(\omega) = \omega\varepsilon''(\omega) \tag{6-5}$$

and the loss tangent is defined as

$$\tan\delta = \frac{\varepsilon''}{\varepsilon'} = \frac{\varepsilon_r''}{\varepsilon_r'} \tag{6-6}$$

where δ is the temporal phase angle between the charging current and the total current. Both $\tan\delta$ and ε_r'' reach a local maximum at each characteristic frequency for relaxation or resonance of dipoles in the material. Further, the reciprocal of the permittivity function is known as modulus function, $M_r(\omega) = [\varepsilon_r(\omega)]^{-1}$.

These functional correlations completely characterize the dynamic electrical characteristics of a linear system. The measurement and analysis of each of these transfer functions in a wide frequency range is called impedance, admittance, permittivity, and modulus spectroscopy, respectively. All of them are also generally referred to as immittance spectroscopy. These transfer functions are interrelated as follows,

$$Z(\omega) = \frac{1}{j\omega C_o \varepsilon_r(\omega)} = \frac{1}{Y(\omega)} = \frac{M_r(\omega)}{j\omega C_o} \tag{6-7}$$

where C_o, the capacitance of the empty cell used for transfer function measurement, is given by

$$C_o = \varepsilon_o \frac{A}{d} \tag{6-8}$$

for a cell with two parallel electrodes of area A separated by a distance d. Thus, once one transfer function is determined, the other transfer functions can be calculated. All of them are important because of their different dependence on and dispersion with frequency. The analysis of one particular transfer function may offer better resolution in determining certain material properties. In general, analysis of impedance functions is convenient for conductors (dominated by transport) while analysis of permittivity functions is convenient for dielectrics (dominated by polarization). However, we shall take the general term, impedance spectroscopy (IS), to stand for the measurement and analysis of some or all of the four transfer functions, i.e., immitance spectroscopy.

6.3.1.2 Basic assumptions

The fundamental assumption for impedance spectroscopy is that the response of the system is linear, time-invariant, and finite in the entire frequency domain. The assumption of linearity requires that the response of the system is independent of the amplitude of the applied alternating perturbation. Experimentally, the amplitude of the applied alternating perturbation should be sufficiently small to ensure the linearity of the system, particularly for systems involving interfacial reactions, which are highly non-linear in nature. The assumption of time-stability implies that the system is unchanged during impedance measurement. For dynamic systems or non-stationary systems, such as corrosion, impedance data should be acquired as fast as possible to minimize the effect of time-instability on the validity of impedance data. The response of the system is assumed to be finite and contains no singularities in the entire frequency domain, including the points where $\omega \rightarrow \infty$ and $\omega \rightarrow 0$.

While it is a basic assumption that the response is caused only by the alternating perturbation, the impedance of a system can be acquired not only in an equilibrium state ($\bar{V} = 0$ and $\bar{I} = 0$) but also in a steady state under the influence of a DC polarization (i.e., $\bar{V} \neq 0$ and $\bar{I} \neq 0$). That is, the system is perturbed by a small-amplitude alternating perturbation ($\tilde{V}$ or $\tilde{I}$) superimposed on a steady-state DC polarization ($\bar{V}$ or $\bar{I}$). In other words, the current and voltage can be described as

$$I = \bar{I} + \tilde{I} \tag{6-9a}$$

$$I = \bar{V} + \tilde{V} \tag{6-9b}$$

However, the impedance functions measured in a steady state depend, in general, on the amplitude of the DC polarization, $|\bar{V}|$ or $|\bar{I}|$ (or the conditions of the steady state) although they are independent of the amplitude of the alternating perturbation, $|\tilde{V}|$ or $|\tilde{I}|$. The effect of a DC polarization on transfer function is more pronounced for non-linear processes, such as electrochemical reactions at an interface. [51] In fact, the impedance arising from an electrochemical reaction obtained under different degree of DC polarization can be used for construction of Tafel plots, from which anodic and cathodic transfer coefficients can be determined. [52]

6.3.1.3 The validity of impedance data

The linearity, time-stability, and causality of a system can be verified by applying the Kramers-Kronig (K-K) transformations to the impedance data. The real and the imaginary part, or the phase and the amplitude, of a valid transfer function are inherently related by the K-K relations (Hilbert integral transform) [53–55],

$$Z''(\omega) = \frac{-2\omega}{\pi} \int_0^{\infty} \frac{Z'(x) - Z'(\omega)}{x^2 - \omega^2} dx \tag{6-10a}$$

$$Z'(\omega) - Z''(\infty) = \frac{2}{\pi}\int_0^\infty \frac{xZ''(x) - \omega Z''(\omega)}{x^2 - \omega^2}dx \tag{6-10b}$$

$$\theta(\omega) = \frac{2}{\pi}\int_0^\infty \frac{\ln|Z(x)|}{x^2 - \omega^2}dx \tag{6-10c}$$

A set of impedance data is invalid unless the calculated imaginary part of the impedance data from the real part of the impedance data using the K-K transformations matches the experimental data.

6.3.1.4 Equivalent circuit approximation

Under the conditions valid for impedance spectroscopy, the electrical behavior of an electrochemical system can be approximated by an equivalent circuit. The analogy between an equivalent circuit and an electrochemical system rests on the assumptions that (i) it is possible to model the electrochemical system with a series-parallel combination of linear passive elements, (ii) each circuit element is associated with an independent physicochemical process or property, and (iii) it is possible to determine the unique value of the elements from the impedance response.

Consider a simple electrochemical system, a pure ionic conductor (electrolyte) of thickness, d, sandwiched between two electronically conductive electrodes (of area A). Under the assumption that specific adsorption of reactants and products is absent and that there are no complications due to the ionic double layer and mass transfer, an equivalent circuit as shown in Fig. 6-2 can be used to approximate the electrical behavior of the simple electrochemical system, where R_b represents the resistance to the motion of ionic defects through the bulk phase of the electrolyte, C_{dl} the double-layer capacitance, R_{ct} the resistance to charge transfer at the electrolyte-electrode interfaces, and Z_W the impedance to mass transfer in the vicinity of the electrolyte-electrode interfaces. The impedance (in ohm-cm^2) of each element can be expressed as [56],

$$R_b = \frac{d}{\sigma_{i,el}} \tag{6-11a}$$

$$R_{ct} = \frac{RT}{F}\left(\frac{1}{\alpha_a + \alpha_c}\right)\frac{1}{J_o} \tag{6-11b}$$

$$Z_W = R_{mt}\frac{\tanh\sqrt{j\omega(L_d^2/D)}}{\sqrt{j\omega(L_d^2/D)}} \tag{6-11c}$$

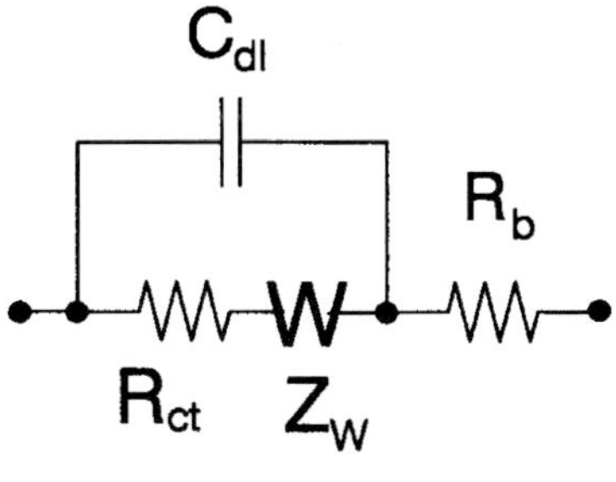

Figure 6-2. An equivalent circuit for a simple cell consisting of an electrolyte (a pure ionic conductor) and two identical electrodes.

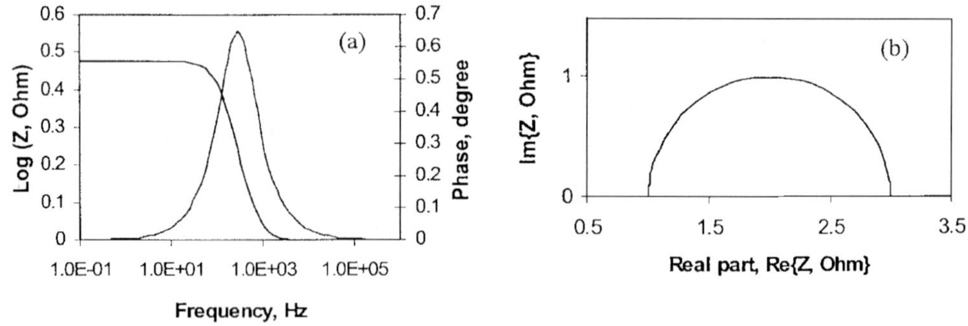

Figure 6-3. (a) Bode and (b) Nyquist presentation of an impedance response of the circuit shown in Fig. 6-2 when the rate of mass transfer is much faster than the rates of other processes occurring in the cell (so that the Z_W can be neglected).

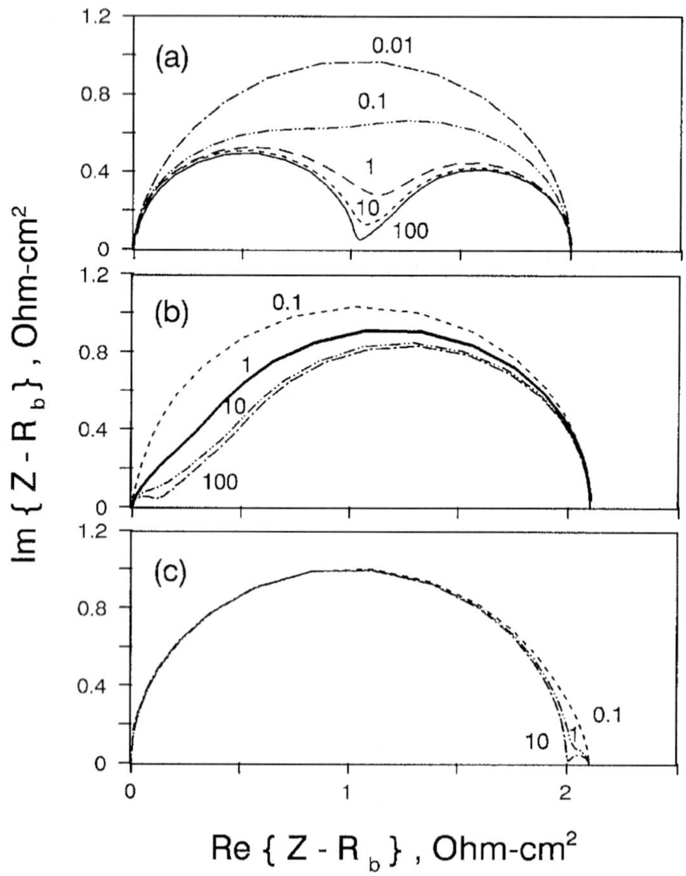

Figure 6-4. Predicted frequency dependence of interfacial impedance, $Z–R_b$, of the cell shown in Fig. 6-3. The number by each impedance spectrum represents the time constant (L_d^2/D) for mass transfer in seconds. The parameters used in the calculation where $R_{ct} = R_{mt} = 1.0\ \Omega\text{cm}^2$, $C_{dl} = 0.01\ \text{F/cm}^2$, and $L^2{}_d/D = 0.01, 0.1, 1, 10$, and 100s in (a); $R_{ct} = 0.1\ \Omega\text{cm}^2$, $R_{mt} = 1.0\ \Omega\text{cm}^2$, $C_{dl} = 0.1\ \text{F/cm}^2$, and $L^2{}_d/D = 0.1, 1, 10$, and 100s in (b); $R_{ct} = 1\ \Omega\text{cm}^2$, $R_{mt} = 0.1\ \Omega\text{cm}^2$, $C_{dl} = 0.01\ \text{F/cm}^2$, and $L^2{}_d/D = 0.1, 1, 10$, and 100s in (c).

The last expression represents the impedance arising from transport or diffusion of electroactive species through a Nernst diffusion layer of thickness L_d with a diffusion coefficient D (Nernst's hypothesis). The mass transfer time constant is L_d^2/D and the steady state mass-transfer resistance is $R_{mt} = \lim_{\omega \to 0} Z_W$.

In the simplest case, when the rate of mass transfer is much faster than the rate of other processes occurring in the system, Z_w may be neglected and the corresponding impedance function is shown in Fig. 6-3 in two different presentations, (a) Bode and (b) Nyqist plot. When the rate of mass transfer in the vicinity of the electrolyte-electrode interfaces affects the rate of other processes or the overall process, Warburg impedance Z_w can no longer be neglected in impedance analysis. Shown in Fig. 6-4 are impedance functions of the system with different assigned values for each circuit element illustrating the effect of Warburg impedance Z_W on the overall impedance response of the system, Z. When $R_{ct} = R_{mt}$ as shown in Fig. 6-4(a) or $R_{mt} >> R_{ct}$ as shown in Fig. 6-4(b), the overall impedance response of the system, $Z(\omega)$, is dramatically influenced by the mass transfer time constant (L_d^2/D) or the Z_W. When $R_{mt} << R_{ct}$ as shown in Fig. 6-4(c), however, the mass transfer time constant (L_d^2/D) or Z_W has little effect on the overall impedance response of the system, $Z(\omega)$. In addition, Fig. 6-4 also demonstrates that the Warburg impedance is dramatically influenced by diffusivity of electroactive species and the thickness of a diffusion boundary layer (L_d), implying that impedance spectroscopy can also be used to determine the diffusion coefficients of electroactive specices.

6.3.1.5 Applications

With modern electronics, impedance spectra can be readily obtained in the frequency range from 10^{-4} to 10^7 Hz. As a non-destructive method with high accuracy and reproducibility, impedance spectroscopy has been widely used as an *in-situ technique* to elucidate electrode kinetics and reaction mechanisms, to monitor electrical behavior of various materials, to separate the interfacial polarization from bulk response, to separate one process from another if their relaxation time constants are sufficiently different, and to separate ionic from electronic conduction in mixed ionic-electronic conductors (MIECs) [1, 57].

While its principle of operation is simple, the interpretation of an impedance spectrum can be extremely difficult, particularly when multiple relaxation processes are present and their time constants are not sufficiently different to permit deconvolution of the spectrum. In this case, the best approach is to simplify the system under study as much as possible through careful cell design, including the insertion of reference electrodes. For example, a four-probe impedance measurement may make it possible to acquire the impedance response of only one component or one interface of a test cell, not the overall response of the whole cell, significantly simplifying the analysis and interpretation of impedance data while improving the reliability of impedance analysis.

In an impedance measurement with three or more electrodes, the alternating perturbation is applied to the entire cell whereas the response of only a particular part of the cell, or a particular process occurring in the cell, is acquired.

6.3.2 Potential sweep method

Linear sweep voltammetry (LSV) or cyclic voltammetry (CV) is often the first method chosen by an electrochemist to study a new material for electrochemical and electrocatalytic applications. In LSV, the potential is ramped from an initial to a final value at a constant scan rate while the corresponding current is monitored. A linear sweep voltammogram is then just a plot of the measured current vs. the applied potential. In cyclic voltammetry, the potential is swept back and forth between two chosen limits while the current is recorded continuously. In either case, the potential E is usually changed linearly with time, i.e.,

$$E = E_i \pm vt \tag{6-12}$$

where E_i is the initial potential at $t = 0$, which is chosen in a range where no faradaic process takes place, and v is the scan rate, dE/dt.

Shown in Fig. 6-5 is a typical experimental arrangement for LSV and CV measurements using a powder microelectrode. The electrochemical cell consists usually of a vessel that can be sealed to prevent air entering the solution, with inlet and outlet ports to allow the saturating of the solution with an inert gas, N_2 or Ar. The removal

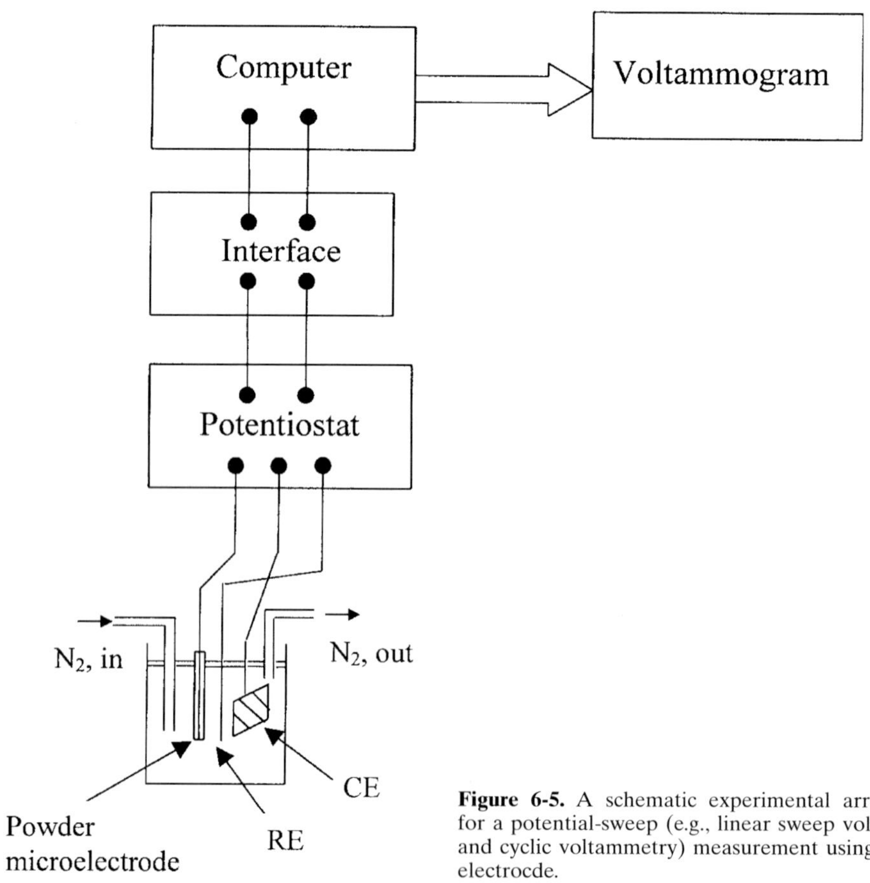

Figure 6-5. A schematic experimental arrangement for a potential-sweep (e.g., linear sweep voltammetry and cyclic voltammetry) measurement using a microelectrocde.

of O_2 is usually necessary to minimize or eliminate currents due to the reduction of O_2 which may interface with the measurement. A standard cell configuration consists of three electrodes immersed in an electrolyte: the working electrode (WE), counter electrode (CE), and reference electrode (RE). The potential at the WE with respect to the RE is controlled by a potentiostat, which is usually interfaced with a computer. In addition to linear sweep voltammetry, various waveforms may be superimposed on the potential at the WE; the current flowing between the WE and CE is measured.

When LSV or CV is used to study a system for the first time, it is usual to start by carrying out qualitative experiments in order to get a feel for the system, before proceeding to semi-quantitative and finally quantitative ones from which thermodynamic and kinetic parameters may be calculated. In a typical qualitative study it is usual to record voltammograms over a wide range of sweep rates and for various range of sweep potential. Commonly, there will be several peaks, and by observing how these appear and disappear as the potential limits and sweep rate are varied, and also by noting the differences between the first and subsequent cycles, it is possible to determine how the processes represented by the peak are related. At the same time, from the sweep rate dependence of the peak amplitudes the role of adsorption, diffusion, and coupled homogeneous chemical reactions may be identified. The difference between the first and the subsequent cyclic voltammograms frequently provides useful mechanistic information.

Consider a redox reaction

$$O + ne^- \rightleftharpoons R \tag{6-13}$$

occurring at the surface of a nano-structured electrode, a typical cyclic voltammogram of this reaction is shown in Fig. 6-6. The anodic current can be expressed as [18, 58]

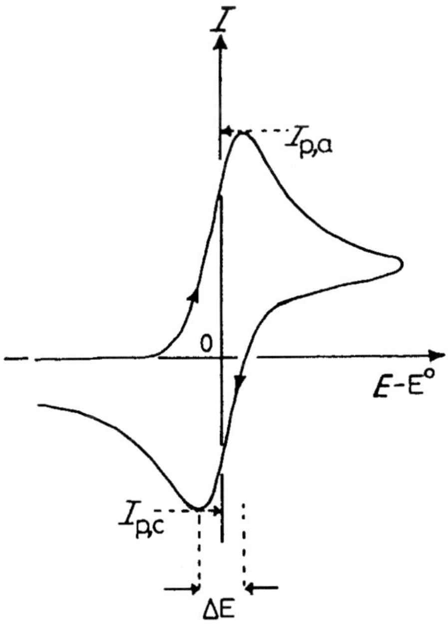

Figure 6-6. A schematic cyclic voltammogram expected form a reversible electrochemical reaction ($O + ne^- = R$) having a standard reduction potential E^0. E is the potential of the WE with respect to the RE and I is the current passing through the CE and the WE.

$$I = nFAc_R^{*}\left(\frac{\pi D_R F v}{RT}\right)^{1/2} \chi(\sigma t) \tag{6-14}$$

where A is the area of the electrode, c_R^{*} is the bulk concentration of R, n is the number of electrons transferred in the reaction, D_R is the diffusion coeffcient of R, v is the scan rate and $\chi(\sigma t)$ is a tabulated number that is a function of the electrode potential and contains the variation of the potential with time.

The relationship between the peak current and the concentration of the electroactive species for the anodic and cathodic peak current, $I_{p,a}$ and $I_{p,c}$, is

$$I_{p,a} = 0.4463nFA\, c_R^{*}\, (F/RT)^{1/2} v^{1/2} D_R^{1/2} \tag{6-15}$$

$$I_{p,c} = -0.4463nFA\, c_O^{*}\, (F/RT)^{1/2} v^{1/2} D_O^{1/2} \tag{6-16}$$

the dependence of the peak current on $v^{1/2}$ is indicative of diffusion control.

A cyclic voltammogram is characterized by (i) the amplitude of the separation of the potentials at which the anodic and cathodic peak currents occur, $\Delta E = E_{p,a} - E_{p,c}$, and (ii) the half wave potential, $E_{1/2}$, the potential mid-way between the peak potentials. For a reversible system, the value of ΔE is about $(0.057/n)$ V at 25 °C and independent of scan rate, though it is found in practice that ΔE increase slightly with v.

The half wave potential is related to E^{o} by

$$E^{o} = E_{1/2} + \left(\frac{RT}{nF}\right) \ln \left[\left(\frac{D_O}{D_R}\right)^{1/2}\right] \tag{6-17}$$

At the other kinetic extreme, a voltammogram such as the one shown in Fig. 6-7 represents a completely irreversible reaction. For example, $O + ne^{-} \rightarrow R$ where R cannot be re-oxidised to O (or anything else). In such a case, the (cathodic) peak current is given by

$$I_{p,c} = -2.99 \times 10^{5} \beta^{1/2} A\, c_O^{*}\, D^{1/2} v^{1/2} \tag{6-18}$$

and the potential at which this peak current occurs is

$$E_p = E^{o} - (RT/\beta F)\left[0.78 + \ln\left(\frac{D_O^{1/2}}{k^{0}}\right) + \ln\left(\frac{\beta F v}{RT}\right)^{1/2}\right] \tag{6-19}$$

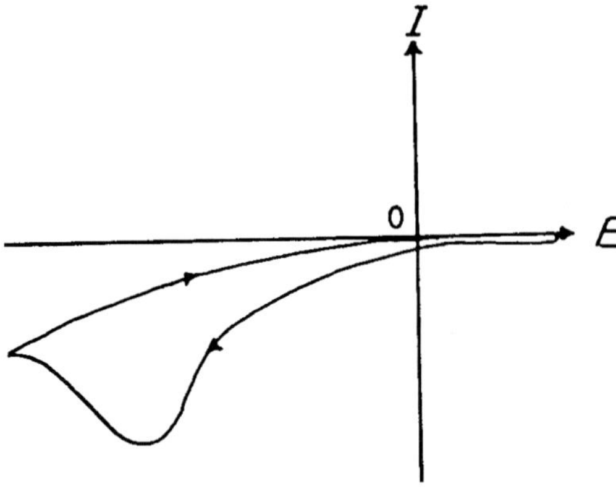

Figure 6-7. A schematic cyclic voltammogram expected from an irreversible process ($O + ne^{-} \rightarrow R$).

where β is an asymmetry parameter for a electrochemical process, and k^o is standard rate constant.

For such an irreversible reaction, it can be shown that

$$|E_p - E_{p/2}| = 1.857 \frac{RT}{\beta F} \tag{6-20}$$

and β can thus be obtained from the separation of peak potential and half-peak potential. Thus, CV can be used to indicate whether a reaction occurs, at what potential and may indicate, for reversible processes, the number of electrons taking part overall. In addition, for an irreversible reaction , the parameters n and β can be obtained.

In the case of the reversible system discussed above, the electron transfer rates at all potentials are significantly greater than the rate of mass transport, and therefore Nernstian equilibrium is always maintained at the electrode surface. In fact, it is quite common for most of real electrochemical reactions that they are reversible at low sweep rates to become irreversible at higher ones after having passed through a region known as quasi-reversible at intermediate values. At low potential sweep rates the rate of electron transfer is greater that that of mass transfer, and a reversible cyclic voltammogram is recorded. As the sweep rate is increased, however, the rate of mass transport increases and becomes comparable to the rate of electron transfer. The most noticeable effect of this is an increase in the peak separation. Diagnostic tests for a system in the reversible, irreversible, and quasi-reversible conditions are given in Table 6-1 to 6-3 [59].

Table 6-1. Diagnostic tests for cyclic voltammograms of reversible processes at 25 °C [59]

1. $\Delta E_P = E_P^A - E_P^C = \frac{59}{n}$ mV
2. $|E_P - E_{P/2}| = \frac{59}{n}$ mV
3. $|\frac{I_P^A}{I_P^C}| = 1$
4. $I_P \propto v^{1/2}$
5. E_P is independent of v
6. At potentials beyond E_P, $I^{-2} \propto t$

Table 6-2. Diagnostic tests for cyclic voltammograms of irreversible processes at 25 °C [59]

1. No reverse peak
2. $I_P^C \propto v^{1/2}$
3. E_P^C shifts $-30/\alpha_C n_\alpha$ mV for each decade increase in v
4. $|E_P - E_{P/2}| = \frac{48}{\alpha_c n_\alpha}$mV

Table 6-3. Diagnostic tests for cyclic voltammograms of quasi-reversible systems [59]

1. $|I_P|$ increases with $v^{1/2}$ but is not proportional to it
2. $|\frac{I_P^A}{I_P^C}| = 1$ provided $\alpha_C = \alpha_A = 0.5$
3. ΔE_P is greater than 59/n mV and increases with increasing v;
4. E_P^C shifts negatively with increasing v

6.3.3 Potential step method

In a potential step experiment, the potential of the working electrode, with respect to a reference electrode, is stepped from a rest potential (at which there is no significant faradaic processes take place) to a potential at which the rate of the overall reaction is controlled completely by the mass-transport of electroactive species while the current response or the cumulative charge is recorded as a function of time. Chronoamperometry and chronocoulometry are two examples of potential-step techniques. The potential may be either positive or negative with respect to the reference electrode in order to cause, respectively, an oxidation or reduction reaction to take place. For a planar electrode the relaxation of diffusion-limited current is described by the Cottrell equation [60]

$$I_d = -nFAc_O^*\sqrt{\frac{D_O}{\pi t}} \qquad (6\text{-}21)$$

where I_d is the diffusion-limited (cathodic) current for the reduction of O. Thus, a plot of I_d vs. $t^{-1/2}$ will yield a straight line and slope of the line can be used to determine the diffusion coefficient of the electroactive species, D_O. It is important to note that the data should be collected over a wide time range to ensure the reliability of data. Usually, relaxations over 1 ms to 10 s are commonly recorded for analysis. In some cases, such as the study of solid-state diffusion in a nanophase material, current relaxation over a longer time should be acquired.

A somewhat more elaborate variation of the chronoamperometric technique is the symmetrical double potential step experiment where the potential is returned to its initial value after the potential was stepped at a potential for a period of time, τ. An example of the application of the double potential step technique to a mass transport controlled electrode reaction is shown in Fig. 6-8. If both the oxidized and reduced

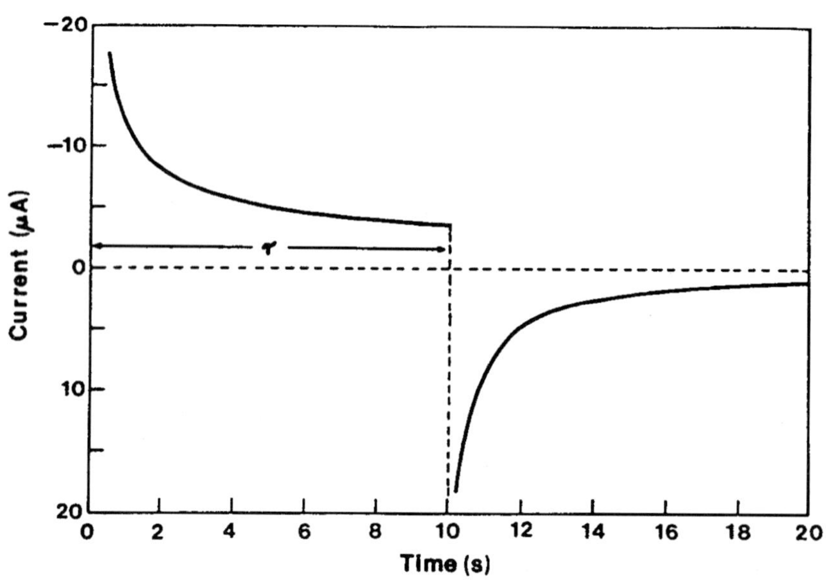

Figure 6-8. Double potential-step chronoamperometric response for an one-electron oxidation and reduction of $O + e^- = R$.

forms of the redox couple are stable and the potential to which the working electrode is returned on the reverse step (which is the initial potential in the symmetrical case) is sufficient to result in a mass transport limited oxidation or reduction of the species produced by the first potential step, then the current obtained on the reverse step, i_r, is given by [18]

$$i_r = nFAD_O^{1/2}\, c_O^{*}\, \pi^{-1/2}[1/(t-\tau)^{1/2} - 1/t^{1/2}] \tag{6-22}$$

Note that the current is negative if the second or reverse potential step results in an oxidation process and that D_O refers to the diffusion coefficient of the species initially present in solution (the oxidized species in this case). For an uncomplicated redox reaction, the ratio of the current that results from the forward step to the current produced by the reverse step is [18]

$$-i_r / i_f = [t_r / (t_r - \tau)] - [t_f / t_r] \tag{6-23}$$

where t_f and t_r are the periods of time over which the potential was stepped at the forward and reverse direction, respectively. A plot of $\frac{i_r}{i_f}$ versus $\frac{t_r}{\tau}$ constructed from the data in Fig. 6-8 is shown in Fig. 6-9. The beauty of the double potential step technique is similar to that of other reversal techniques like CV with current reversal; it can be used to probe the stability or reversibility of an electrogenerated redox couple. This is very important to electrode materials for rechargeable batteries. Variations of $\frac{i_r}{i_f}$ from theoretical values can often be used to obtain kinetic data as well as diagnostic information about a redox process. This technique is better suited than CV for studying quasi-reversible electron transfer reactions with coupled homogeneous chemical reactions because the magnitude of the applied potential steps can often be adjusted so as to render the electrode process mass transport limited.

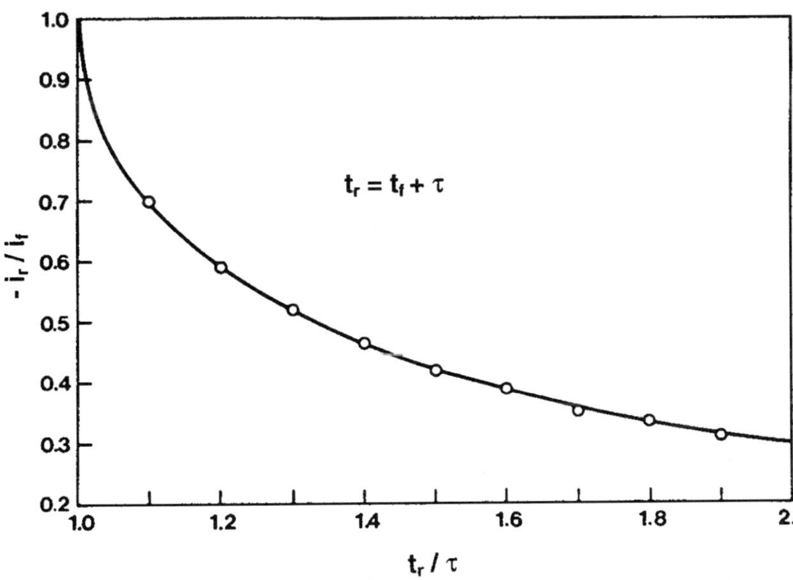

Figure 6-9. Plot of the chronoamperometric current ratio as a function of i_r / τ.

A variant of double potential step chronoamperometry that is often used to study adsorption is double potential step chronocoulometry [18]. Because cumulative charge (determined by the electronic integration of the current) is recorded as a function of time in chronoamperometry, and the influence of double layer charging process at the beginning of the potential step on the total charge will rapidly becomes negligible at a long time, the value of K can therefore be measured at quite long times. This permits rate constants perhaps an order of magnitude greater to be determined from charge, rather than current measurements (e.g., chromoamperometry) [61, 62]. Under semi-infinite conditions, the chronocoulometric response in the forward direction can be expressed as

$$Q_f(t \le \tau) = Q_{dl} + \frac{2nFAD_o^{1/2}C_o^* t^{1/2}}{\pi^{1/2}} \tag{6-24}$$

and the charge removed in the reverse direction can be expressed as

$$Q_r(t > \tau) = Q_{dl} + \frac{2nFAD_o^{1/2}C_o^*}{\pi^{1/2}}\left[\tau^{1/2} + \left(t - \tau^{1/2}\right) - t^{1/2}\right] \tag{6-25}$$

where Q_{dl} is the charge due to double layer charging and τ is the forward step duration time.

The cumulative charge due to electrolysis of the electroactive species at a diffusion-controlled rate in the forward and reverse direction can be expressed as

$$Q(t \le \tau) = \frac{2nFAD_o^{1/2}C_o^*}{\pi^{1/2}} t^{1/2} \tag{6-26}$$

and

$$Q(t > \tau) = \frac{2nFAD_o^{1/2}C_o^*}{\pi^{1/2}}\left[t^{1/2} - \left(t - \tau^{1/2}\right)\right] \tag{6-27}$$

A typical Q vs t transient of double potential-step chronocoulometry is shown in Fig. 6-10.

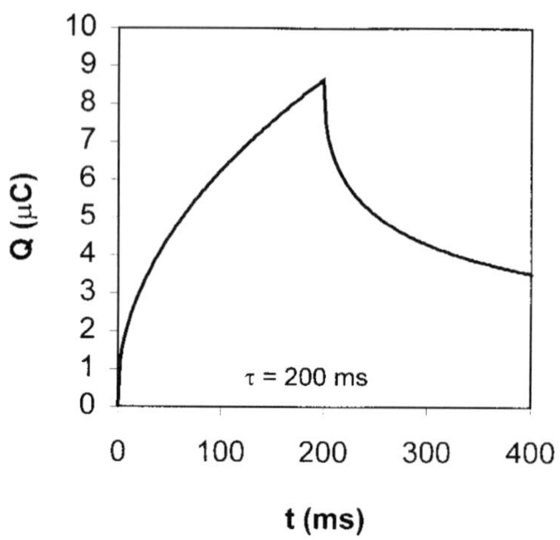

Figure 6-10. A typical response for a double-step chronocoulometric experiment.

6.3.4 Controlled-current techniques

Most studies of the insertion (extraction) of lithium ions into (from) an intercalation compound have been focused on the diffusion of lithium ions within the compound. Since the rate of solid-state diffusion is relatively slow, the overall charge or discharge rate of a Li-ion battery is often determined by the diffusion of Li^+ ions in the intercalation compounds. Thus, the chemical diffusion coefficient of Li^+ ion ($D_{Li}{}^+$) in an intercalation material (e.g., $Li_xMn_2O_4$ nanoparticles) is a critical parameter that determines the powder density of batteries based on them. The values of $D_{Li}{}^+$ can be determined using several methods, including galvanostatic intermittent titration technique (GITT) [63], current pulse relaxation (CPR) [64], chronoamperometry (CA), impedance spectroscopy [65], and electrochemical permeation method [66]. The reported values of $D_{Li}{}^+$ in $Li_xMn_2O_4$ are rather scattered, varying from 10^{-6} to 10^{-9} cm^2 s^{-1}, depending on the way in which MnO_2 was prepared and on the characterization technique employed. To obtain $D_{Li}{}^+$ using first three methods discussed above, one has to know the real surface area of the sample (A), and in some cases, the variation of the open-circuit potential with lithium concentration (dV_{oc}/dx) [67]. Unfortunately, precise determination of A and dV_{oc}/dx is difficult, particularly in the case of porous electrodes; this is why there is a large discrepancy in the reported diffusivity data.

6.3.4.1 Galvanostatic intermittent titration technique (GITT)

GITT combines transient (current relaxation) and equilibrium (coulometric titration) measurements, as schematically illustrated in Fig. 6-11. In a typical GITT experiment, galvanostatic currents are applied to a test cell for a time interval τ to produce a change in stoichiometry of the electrode material. After such a titration pulse, a new equilibrium voltage will be established. The charge and discharge currents are selected so that a change in lithium content from x = 0 to x = 1 (Δx = 1) for a insertion electrode material (e.g., $Li_xMn_2O_4$) would occur for a limited time (e.g., 40h). Applying a constant current to the cell composed of the intercalated electrode during a short time (e.g., 1800s) upon charging, the resulting cell potential transients are recorded. After interruption of the current pulse, the decay of the open-circuit potential is followed with time until the fluctuation of the open-circuit potential falls below 0.01V vs. Li/Li^+. This potential value is just recorded as an electrode potential. The application and interruption of the constant current continue until the lithium content x reaches 0.4, after which the measurement is performed in the reverse direction, i.e., discharging, until x vanishes. Similar to the charge curve, the resulting cell potential transients and electrode potentials are obtained.

In a GITT measurement, there are no limits on the initial and boundary conditions for Fick's second law and hence a linear relationship between the potential and the square root of time is expected in a short time range of the galvanostatic potential transient. The chemical diffusion coefficient of lithium ions in the $Li_xMn_2O_4$ electrode of planar symmetry can be calculated as a function of intercalated lithium content from GITT curve, as follows [63],

$$D_{Li^+} = \frac{4}{\pi}\left(\frac{V_m}{zFA}\right)^2 I_0 \left[\frac{(\frac{dE}{dx})}{(\frac{dE}{d\sqrt{t}})}\right]^2, \; t << \frac{l^2}{D_{Li^+}} \qquad (6\text{-}28)$$

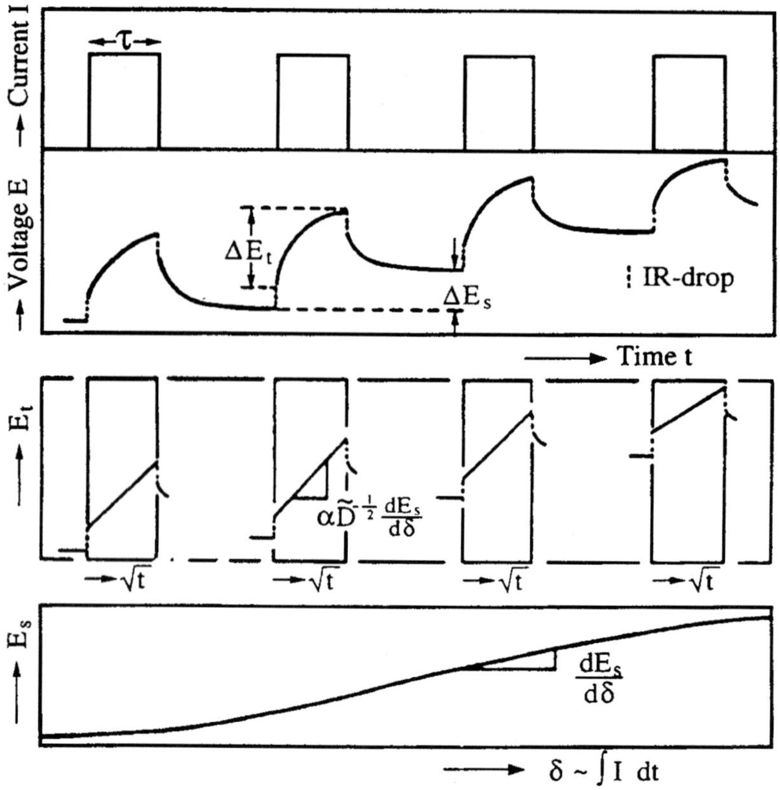

Figure 6-11. Principles of GITT for the evaluation of the thermodynamic and kinetic data of electrodes.

where V_m is the molar volume of the electrode material ($cm^3\ mol^{-1}$), z the valence of lithium ion ($z = 1$), F the Faraday constant, A the electrochemical active area of the electrode-electrolyte interface, I_0 the applied constant current, (dE/dx) the slope of the coulometric titration curve, and ($dE/d\sqrt{t}$) the slope of the E versus square root of time curve.

Another advantage of this technique is that it can determine the partial conductivities due to the motion of a minority ionic defect in a predominatly electronically conducting electrode. The partial ionic conductivity of a mixed ionic-electronic conductor can be calculated from the concentration and the diffusion coefficient of ionic defect together with the variations of the steady-state and transient voltage [68].

In addition, GITT provides the possibility to determine many other kinetic and thermodynamic parameters of the electrode as a function of stoichiometry [63], including partial ionic conductivity, electrical ionic mobility, Gibbs free energy of formation, et. al.

6.3.4.2 Current pulse relaxation (CPR) technique [64]

The chemical diffusion coefficients of lithium ions in $Li_xMn_2O_4$ cathodes can also be measured using current pulse relaxation techniques in a two-electrode electrochemical cell:

Li | 1M $LiClO_4$ in PC | $Li_xMn_2O_4$

The cell is first galvanostatically charged to 4.5 V vs. Li/Li^+ and then is discharged galvanostatically to a given lithium composition, x, at a constant current density (e.g., 0.1 mA cm^{-2}). At each composition, lithium is injected into the cathode by a constant current pulse (e.g., 0.5 mA cm^{-2}) for a short period of time (e.g., 10s). The diffusion coefficient of lithium ion can be calculated from the decay rate of the transient voltage using the following equation,

$$D_{Li}^{+} = \left[\frac{V_M (dE/dx)}{nFA} \frac{i\tau}{\Delta E / \Delta(t^{-1/2})} \right]^2 \tag{6-29}$$

where dE/dx is the slope of the equilibrium potential-composition curve, i the intensity of current pulse, τ pulse duration, and $\Delta E/\Delta(t^{-1/2})$ the slope of the plot of the relaxation potential versus $t^{-1/2}$ after the current pulse is interrupted. The quasi open-circuit voltage (OCV) curves of the cells can be measured using a galvanostatic intermittent charge-discharge method [69].

6.3.4.3 Electrochemical permeation method

Another method to study solid-diffusion is electrochemical permeation method, which was originally developed for the determination of the diffusion coefficients of hydrogen in steel [70, 71] and gas molecules in Nafion [72, 73]. Consider a MnO_2 nanostructured electrode with a thickness of L as shown in Fig. 6-12a. First, both surfaces are kept at the same potential so that Li^+ concentration is uniform throughout the electrode ($c = c_O$, $0 < x < L$). At $t = 0$, the potential of one surface ($x = 0$) is stepped at a lower potential to increase Li^+ concentration at $x = 0$ ($c_{x=0} = c_L$ at $t \geq 0$). Lithium ions diffuse through the electrode toward the other surface ($x = L$) due to a concentration gradient induced by the potential step. After an elapse of time, Li^+ ions permeate to the other surface ($x = L$), where they are removed into the solution because Li^+ concentration at $x = L$ is electrochemically kept constant ($c_{x=L} = c_O$). The flux due to permeation of Li^+ ions can be monitored as a change in oxidation current at $x = L$. The current (I) will initially increases with time, and then reaches a steady-state value (I_∞). When the diffusion obeys Fick's law, the current transient is predicted by a "build-up" curve shown in Fig. 6-12b.

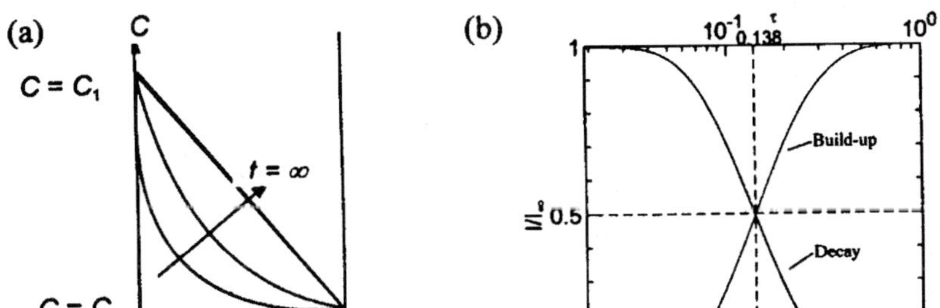

Figure 6-12. (a) Principle and (b) theoretical curves for current transient of an electrochemical permeation measurement [66].

After the steady state is attained, a potential step in the opposite direction gives a current transient shown by a "decay" curve in Fig. 6-12b. The D_{Li^+} values can be determined from the current transients using the following relation [66]

$$D_{Li^+} = \frac{0.138L^2}{t_{1/2}} \tag{6-30}$$

where $t_{1/2}$ is the time when the current reaches a half of the steady-state value ($I/I_\infty = 0.5$). It should be noted that both the real electrode area (A) and the variation of the open-circuit potential with lithium concentration (dV_{oc}/dx) are not needed for the calculation of D_{Li^+}, an advantage over other techniques.

6.3.5 Electrochemical quartz crystal microbalance

The quartz crystal microbalance (QCM) or nanobalance (QCNB) is a piezoelectric device capable of extremely sensitive mass measurements. The quartz crystal typically oscillates in a mechanically resonant shear mode by application of an alternating, high frequency electric field using electrodes which are usually deposited on both sides of the quartz disk. The mass sensitivity arises from a dependence of the oscillation frequency on the total mass of the (usually disk-shaped) crystal, its electrodes, and any materials attached to the electrode surface.

The in-situ QCM was first applied to electrochemical problems by Nomura and coworkers [74, 75] to determine Cu(II) and Ag(I) formed from electrodepositon. The applications of QCM to electrochemical systems will be distinguished from nonelectrochemical applications by referring to the former as EQCM (electrochemical QCM) experiments. EQCM has evolved as a powerful technique capable of detecting very small mass changes at an electrode surface where electrochemical reactions are taking place. This relatively simple technique only requires, in addition to conventional electrochemical equipment, an inexpensive radio-frequency oscillator, a frequency counter, and commercially available AT-cut quartz crystals. EQCM has been successfully used in the investigation of phenomena such as underpotential deposition, electrolyte adsorption, mass changes accompanying ion and solvent movement in redox polymer films, and electrochemically driven self-assembly. Recently, EQCM has been applied to the study of nanostructured materials [76]. In this section, we will provide the readers with a brief introduction to EQCM and with ways of obtaining useful information from in-situ EQCM measurements. Further details about EQCM are referred to several review articles [77–79].

6.3.5.1 Principle of operation

Due to the converse piezoelectric effect, the application of an electric field across a piezoelectric materials induces a mechanical strain [80]. The vibrational motion of the quartz crystal results in a transverse acoustic wave that porpagates back and forth across the thickness of the crystal between the crystal faces. Accordingly, a standing-wave condition can be established in the quartz resonator when the acoustic wavelength is equal to twice the combined thickness of the crystal and electrodes. The frequency f_0 of the acoustic wave fundamental mode is given by

$$f_0 = \frac{v_{tr}}{2d} \tag{6-31a}$$

where v_{tr} is the transverse velocity of sound in AT-cut quartz (3.34×10^4 m s^{-1}) and d is the resonator thickness.

Though mass change at electrode surfaces are related to the change in the resonant frequency of the EQCM in an electrochemical process, it is not easy to get the quantitative correlation between frequency changes and mass changes. If one assumes that acoustic velocity and density of the foreign layer are identical to those of quartz, a change in thickness of the foreign layer is tantamount to a change in the thickness of the quartz crystal. Under these conditions, a fractional changes in mass results in a fractional change in the resonant frequency; appropriate substitutions yields the well-known Sauerbrey equation:

$$\Delta f = \frac{2f_0^2 \Delta m}{A(\mu_Q \rho_Q)^{1/2}} \tag{6-31b}$$

where Δf is the measured frequency change, f_0 the frequency of the quartz resonator prior to a mass change, Δm the mass change, A the piezoelectrically active area, ρ_Q the density of quartz, and μ_Q the shear modulus of quartz. This equation is the primary basis of EQCM measurement wherein mass changes occurring at the electrode interface are evaluated directly from the frequency changes of the quartz resonator. It is generally considered adequate as long as the thickness of the film added to the QCM is less than 2% of the quartz crystal thickness. Typical operating frequencies of the EQCM vary from 5 to 10 MHz, with the mass detection limits up to 1 ng cm^{-2}.

6.3.5.2 *Electromechanical model of the EQCM*

In general, the mechanical vibrations in a piezoelectric crystal can be described in terms of electrical equivalent [81]. This also serves to enhance understanding of EQCM, particularly the conditions under which the Sauerbrey equation is valid. The equivalent circuit for a quartz resonator has an inductor (L_1), a capacitor (C_1), and a resistor (R_1) connected in series (Fig. 6-13). The relationship between this circuit and the quartz crystal is especially useful because the LCR branch is identical to a tank circuit, in which oscillations can be sustained by cycling of current between the capacitor and the inductor. The equivalent electrical parameters can be expressed in terms of crystal properties as follows, along with some typical experimental values for each parameter,

$$C_0 = \frac{\varepsilon_Q \varepsilon_0 A}{d} \approx 10^{-12}\ \text{F} \tag{6-32a}$$

$$C_1 = \frac{8Ah^2}{\pi^2 dc} \approx 10^{-14}\ \text{F} \tag{6-32b}$$

$$R_1 = \frac{d^3 D_Q}{8Ah^2} \approx 100\ \Omega \tag{6-32c}$$

$$L_1 = \frac{d^3 \rho_Q}{8Ah^2} \approx 0.075\ \text{H} \tag{6-32d}$$

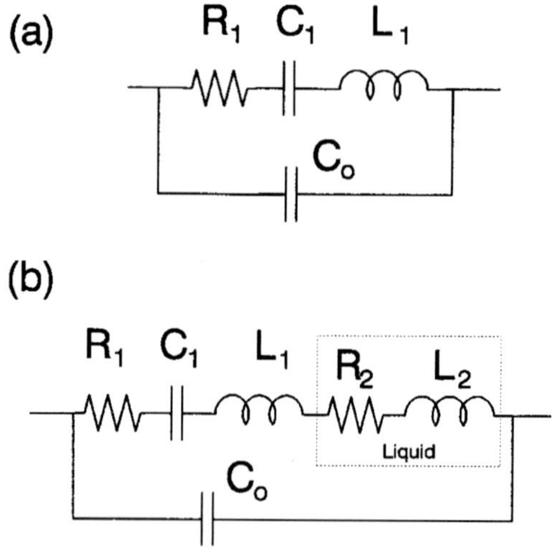

Figure 6-13. (a) Butterworth-van Dyke equivalent electrical circuit used to describe the mechanical properties of a quartz resonator. The components L_1, C_1, and R_1 in the circuit represent, respectively, the inertial mass, compliance, and energy dissipation in the crystal, and C_0 represents the static capacitance of the quartz crystal. (b) An equivalent circuit used to describe the mechanical properties of a quartz resonator immersed in a liquid. The inductance L_2 and resistance R_2 represent the mass and viscosity components of the liquid.

where ε_Q is the dielectric constant of quartz, D_Q a dissipation coefficient corresponding to the energy losses during oscillation, h the piezoelectric stress constant, and c the elastic constant.

6.3.5.3 Experimental apparatus and operation

Typically, a quartz crystal (diameters of 0.5 and 1.0 in.) with appropriately sized excitation electrode is mounted to the bottom of a glass cylinder, that assumes the role of the working electrode compartment of an otherwise conventional electrochemical cell. The two excitation electrodes are electrically connected to an oscillator circuit that contains a broadband RF amplifier, so that the electrode facing solution is at hard ground. The circuit is designed so that the crystal is in a feedback loop, therefore driving the crystal at a frequency at which the maximum current can be sustained in zero-phase angle condition. Several oscillator designs are available, and a key requirement of the circuit is that it provides sufficient gain to allow for oscillation of the crystal in a viscous medium. The output of the oscillator is then connected to a conventional frequency meter for measurement. A critical feature of the EQCM is the potentiostat, which can be either a Wenking potentiostat or a more conventional potentiostat. The difference between these two potentiostats is in the working electrode: the Wenking potentiostat functions with the working electrode at hard ground, whereas current commercial potentiostat functional with the working electrode at virtual ground. Finally, a computer is used to collect frequency and electrochemical data simultaneously, as well as control the waveform applied to the working electrode. This arrangement allows simultaneous measurement of the electrochemical charge, current, voltage, and EQCM frequency. Frequency counters are capable of sampling the frequency output of the oscillator at 100-ms intervals. This capability enables analysis of the kinetics of a wide range of electrochemical processes, including electrodeposition and dissolution, nucleation and growth, and ion-solvent insertion in redox polymer films.

6.3.5.4 Data interpretation

Interpretation of EQCM data is rather straightforward. Since the electrochemical charge represents the total number of electrons transferred in a given electrochemical process, it corresponds to mass changes occurring at the electrode surface. Accordingly, under ideal conditions, the frequency change measured with the EQCM will be proportional to the electrochemical charge and will be related to the apparent molar mass by

$$\Delta f = MW \cdot \frac{C_f Q}{nF} \tag{6-33}$$

where MW is the apparent molar mass (g mol^{-1}), Q the electrochemical charge, and C_f (Hz g^{-1}) the sensitivity constant derived from the Sauerbrey relationship. Thus, a plot of Δf vs. Q is particularly useful in the determination of MW/n, the molar mass per electron transferred.

An alternative approach to data analysis involves the relationship between the electrochemical current and the first derivative of the frequency change with respect to time, as given in Eq. (6-34)

$$i = \frac{\mathrm{d}(\Delta f)}{\mathrm{d}E} \cdot \frac{n v F}{MW \cdot C_f} \tag{6-34}$$

where v is the potential scan rate in units of V s^{-1}. This format is particularly useful for cyclic voltammetry experiments, as $\mathrm{d}(\Delta f)/\mathrm{d}t$ should appear similar in form to the voltammograms if the electrochemical events are accompanied by corresponding mass changes.

6.4 Application to nanostructured electrodes

In this section, electrode materials for lithium ion batteries e.g., lithium transition metal oxides and carbon, will be used as examples to illustrate how to characterize the electrical and electrochemical properties of nanophase materials using the techniques discussed earlier. Further, the relationship between these fundamental properties (such as chemical diffusion coefficient of Li^+ ion, potential window, and reversibility) and the performance of battery electrodes (such as energy density, rate of charge/discharge, cell voltage, and rechargeability) will be elaborated.

6.4.1 Characterizing the reversibility of battery electrode materials

Once a nanostructured electrode is prepared, the first and the most important criterion in determining its viability as an electrode material for a secondary lithium battery is its reversibility, which is often studied using cyclic voltammetry and chronopotentiometry [82]. Cyclic voltammetry is a powerful tool for probing kinetic and thermodynamic properties of a new electrode material. The peak shape and the number of observable redox processes within a given potential range often provide additional criteria for the selection of viable electrode material. Electrode 'cycling' or multiple double-step chronopotentiometry is then applied to promising compounds in order to obtain information regarding the available capacity and possible cycle life of an electrode. In general, voltammetric techniques are valuable in initial screening or in discriminating 'good' from 'bad' battery electrode materials. However, the behavior of a composite or thin-film electrode depends not only on the active materials but also on the electrode porosity and additives (such as binders and conductivity enhancers). Required voltammetric scan rates are typically in the μV/s range in order to compensate for slow mass-transfer of electroactive ions in solid electrode materials. When a powder microelectrode is used, cyclic voltammograms are typically characterized by symmetric, well-defined peaks resembling those observed at thin homogeneous films. Depending on the particle size and transport properties of the material with respect to incorporated ions, voltammetric scan rates in the mV/s range may be used. This method should therefore be advantageous when studying poorly conductive battery electrode materials such as manganese oxides. However, effects due to a distribution of particle size are expected to result in broadened voltammetric signals.

In a typical cyclic voltammogram, the current is proportional to the area of working electrode. Assuming that the nanoparticles to be studied are spheres, the current observed on a nanoparticle is proportional to the square of the particle diameter. Typically the currents observed on a microelectrode are about 3 to 5 orders of magnitude smaller than those observed in cyclic voltammetry on ordinary electrodes. This reduction in currrent greatly reduces the distortion of voltammograms due to iR drop. Because the particles trapped in a microcavity are very small, they can be readily oxidized or reduced entirely when the potential is cycled at a very slow rate (e.g., 1 to 10 mV/s). Thus, the multiscan cyclic voltammetry is just like an accelerated charge-discharge cycling test. The reversibility and the cycle stability of nanoparticles can be examined by continuous scanning reversed at different potentials.

From the viewpoint of kinetics it is interesting to examine the half-peak width of CV and the relationship between the peak currents and the sweep rates. Typically, the peak current is proportional to the square root of the scan rate, indicating that reversible Li^+ intercalation/deintercalation process is a diffusion-controlled process. Often, it is the transport of lithium ions in the solid state that determines the current density or rate capability.

EQCM can also be used to study the reversibility of nanostructured materials. Since EQCM can directly monitor the mass change during an intercalation/de-intercalation process of Li^+ ion within a nanophase material. The mass change in a nanostructured electrode material can be measured sensitively by EQCM technique, and the reversibility of the electrode material can be readily evaluated by the change in the mass after each cycle. Since the change in mass is determined from the change in resonance frequency which may also be influenced by many other factors (such as the viscosity of solvent), only those nanostructured materials which can be deposited on the substrate of quartz can be studied using this method.

The simplest yet important and widely-used technique in the study of nanomaterials as battery electrode is galvanostatic charge-discharge cycling. In fact, each battery electrode material must be tested using this method to determine the viability as an electrode material. VO_2 nanocrystalline [48], spinel $Li_xMn_2O_4$ nanotubules [45], and $Li_xMn_2O_4$/ PEO nano-composites [49] have been studied as positive electrode materials for rechargeable lithium batteries. However, the voltages of these nanophase materials in lithium batteries range from 2 to 4 V vs. Li^+/Li. Limit of discharge voltage is even at 1V for VO_2 whereas the same materials with larger particle size exhibit voltages ranging from 3 to 4.3 V vs. Li^+/Li.

6.4.2 Characterizing the transport properties

Another critical property, which determines the rate capability or power density of nanostructrued electrode materials in lithium ion batteries, is the diffusion coefficient of the Li^+ ions in the solid framework of the host structure.

The determination of ion diffusion coefficients in solid materials has always been a challenging task. In particular, experimental techniques that have been traditionally developed for diffusion processes occurring in liquid ionic media may not be applicable to study of nanophase solids.

Various experimental methods have been used to determine the chemical diffusion coefficient of Li^+ ions in intercalation electrodes. Most of the methods used in the early studies utilized modifications of the Cottrell transient techniques developed on the basis of the solution to the Fick's second equation for an instantaneous planar source of diffusing species in a semi-infinite geometry [18]. Basically, the techniques consists of applying short (a few seconds) galvanostatic (or potentiostatic) pulses to promote excess concentration of the diffusion species at the electrode surface and then following the consequent voltage (or current) relaxation as a function of time after switching off the current [83, 84]. Alternatively, the technique involves the application of a galvanostatic pulse of very short duration (< 1s) to the cell and the analysis of the resulting overvoltage-time transient during the pulse. Here the overvoltage is a function of the concentration of the diffusing lithium ions at the electrode-electrolyte interface.

For example, GITT was used for the determination of the chemical diffusion coefficient of the Li^+ ions in Li_xC_6 carbon electrodes [85]. The technique consisted of applying to (the electrode) voltage steps and monitoring the charge increment q(t) versus time t, thus allowing the evaluation of the value of $D_{Li}{}^+$ for the whole composition range of the Li_xC_6 electrode.

The chemical diffusion coefficient of Li^+ in $Li_{1-x}NiO_2$ was determined using impedance spectroscopy [65]. This alternating current technique is particularly suitable for kinetic studies in thin-film insertion electrodes [86] where the semi-infinite diffusion boundary conditions, necessary for the direct current transient techniques, can no longer be assumed. The interpretation of the impedance results requires the construction of an equivalent circuit that is representative of the electrochemical system under study. The most commonly used circuit to interpret the impedance response of an electrode-electrolyte interface is one shown in Fig. 6-4.

6.5 Summary

Nanophase materials are not only facilitating from a scientific point of view but also have significant technological implications. For example, nanostructured batteries, fuel cells, and chemical sensors may dramatically improve energy efficiency, environmental quality, and a new generation of electrical vehicles. One of the key challenges in characterizing electrical and electrochemical properties of nanophase materials is to establish an electrical contact to the materials to be studied. A number of preparative approaches have been described for preparation of nanostructured electrodes. Among them, preparation of nanophase electrodes on a microelectrode or powder microelectrode makes it possible to directly investigate the properties of the nanophase materials, excluding the effect of binders and other additives which would be otherwise added to form an electrochemically functional electrode. Electrodeposition or electrophoretic deposition is gaining popularity in fabrication of nanostructure electrodes because of its low deposition temperature, easy control, and fast deposition rate. Formation of nanoparticles in a polymer matrix provides a convenient way of preserving the microstructural stability of nanoparticles. Electrochemical self-asssembly has also been successfully used for fabrication of nanoelectrodes and devices. Further, composite electrodes, consisting of nanoparticles, binders, and conductive additives, are also widely used in screening and evaluation of nanoparticles for battery applications. The use of a composite electrode is effective for study of nanophase materials to which an electrical contact can not be readily established otherwise. Once a nanoelectrode is prepared, several electrochemical techniques can be used to study its kinetic and thermodynamic properties, including impedance spectroscopy, potential sweep methods, potential step methods, current-controlled techniques, and electrochemical quartz crystal microbalance. While each technique has its own unique advantages over other techniques, it is often found that the combination of several techniques is the best approach to improve accuracy and reliability. For instance, the combination of impedance spectroscopy, current-controlled techniques, and potential-step techniques is uniquely suited for investigations into solid-state diffusion of ionic defects in intercalation compounds. Similarly, the combination of impedance spectroscopy, open cell voltage measurements, and steady-state permeation measurements is a powerful approach to separating ionic from electronic transport in mixed ionic-electronic conductors.

References

[1] W.P. Kirk and M.A. Reed, *Nanostructures and Mesoscopic Systems*, New York, Academic Press, **1992**.
[2] A.D. Berry, R.J. Tonucci, and P.P. Nguyen, *MRS Symp. Proc.* **1996**, *431*, 387.
[3] R.L. Whetten, D.M. Cox, D.J. Trevor, and A. Kaldor, *Phys.Rev.Lett.* **1985**, *54*, 1494.
[4] D.M. Cox, P. Fayet, R. Brickman, M.Y. Hahn, and A. Kaldor, *Catal. Lett.* **1990**, *4*, 271.
[5] K. Bowen, *Z.Phys. D* **1992**, *26*, 46.
[6] G.K. Boschloo, A. Goossens, and J. Schoonman, *J.Electroanal.Chem.* **1997**, *428*, 25.
[7] B. O'Regan, J. Moser, M. Anderson and M. Grätzel, *J.Phys.Chem.* **1990**, *94*, 8720.
[8] A. Hagfeldt, U. Björkstén and S.-E. Lindquist, *Sol. Energy. Mater. Sol. Cells* **1992**, *27*, 293.
[9] U. Björkstén, J. Moser and M. Grätzel, *Chem. Mater.* **1994**, *6*, 858.
[10] S. Sõdergren, A. Hagfeldt, J. Olssen and S.-E. Lindquist, *J.Phys.Chem.* **1994**, *98*, 5552.
[11] A. Hagfeldt, S.-E. Lindquist and M. Grätzel, *Sol. Energy. Mater. Sol. Cells* **1994**, *32*, 245.
[12] B. O'Regan and M. Grätzel, *Nature* **1991**, *353*, 737. B.O'Regan and M.Grätzel, *Nature*, **1991**, *353*, 737.
[13] M. Nazeeruddin, A. Kay, I. Rodicio, R. Humphry-Baker, E. Müller, P.Liska, N.Vlachopoulos and M. Grätzel, *J.Am.Chem.Soc.* **1993**, *115*, 6382.
[14] X. Marguerettaz, R. O'Neill and D. Fitzmaurice, *J.Am.Chem.Soc.* **1994**, *116*, 2629.
[15] A. Hagfeldt, N. Valchopoulos, and M. Grätzel, *J.Electrochem.Soc.* **1994**, *141*, L83.
[16] D. Liu and P.V. Kamat, *J.Electrochem.Soc.* **1995**, *142*, 835.
[17] S.Y. Huang, L. Kavan and M. Grätzel, *J.Electrochem.Soc.* **1995**, *142*, L142.
[18] A. J. Bard, L. R Faulkner, *Electrochemical methods : fundamentals and applications*, New York: Wiley, **1980**.
[19] P. T. Kissinger and W. R. Heineman (editors), *Laboratory techniques in electroanalytical chemistry*, New York: Dekker, **1984**.

[20] R. Varma and J.R. Selman (edited), *Techniques for characterization of electrodes and electrochemical processes*, New York: Wiley, **1991**.
[21] H. D. Abruna , *Electrochemical interfaces : modern techniques for in-situ interface characterization*, New York: VCH Pub., 1991.
[22] J. R. Scully, D. C. Silverman, and M. W. Kendig (editors). *Electrochemical impedance: analysis and interpretation*, Philadelphia: ASTM, **1993**.
[23] Petr Vanysek, *Modern techniques in electroanalysis*, New York: John Wiley & Sons, **1996**.
[24] A.J. Bard, Ed., *Electroanalytical Chemistry: a series of advances*, New York, Marcel Dekker, from **1966**. (20 volumes have published).
[25] M. Fleischmann, S. Pons, D.R. Rolison and P.P. Schmidt (Eds.), *Ultramicroelectrodes*, Datatech Systems, **1987**.
[26] M. Montenegro, M. Queiros and J. Daschback (Eds.), *Microelectrodes, Theory and Applications*, NATO ASI Series E (Applied Sciences), Dordrecht, Kluwer, **1991**.
[27] C.S. Cha, C.M. Li, H.X. Yang and P.F. Liu, *J.Electroanaly. Chem.* **1994**, *368*, 47.
[28] P.C. Searson, *Solar Energy Materials and Solar Cells* **1992**, *27*, 377.
[29] P.C. Searson and T.P. Moffat, *Critical Reviews in Surface Chemistry* **1994**, *3*, 171.
[30] C. Buelens, J.P. Celis and J.R. Roos, *Trans.Inst.Met.Fin.* **1985**, *63*, 6.
[31] C.C. Lee and C.C. Wan, *J. Electrochem. Soc.* **1988**, *135*, 1930.
[32] P.R. Oberle, M.R. Scanlon. R.C. Cammarata, and P.C.Searson, *Appl. Phys. Lett.* **1995**, *66*, 19.
[33] C.R. Martin, *Science* **1994**, *266*, 1961.
[34] M. Nishizawa, V.P. Menon, and C.R. Martin, *Science* **1995**, *268*, 700.
[35] M. Nastasi, D.M. Parkin, and H. Geleiter (eds.), *Mechanical Properties and Deformation Behavior of Materials Having Ultra Fine Microstructures*, Boston, Kluwer, **1993**.
[36] T.J. Lee, K.G. Sheppard, A. Ganburg, and L. Klein, in *Electrochemical Microfabrication II* (Eds.: M. Datta, K. Sheppard, and J. Dukovic,), Proceedings Volume 94-2, The Electrochemical Society, Pennington, NJ, **1995**.
[37] D.Y. Godovski, A.V. Vokkov, I.V. Karachevtser, M.A. Moskvino, A.L. Volynskii and N.F. Bakeev, *Polymer Science USSR* **1993**, *A35*, 1308.
[38] Y. Yuan, I. Cabasso, and J.H. Fendler, *Chem. Mater.* **1990**, *2*, 226.
[39] M. Gao, Y. Yang, B. Yang, F. Bian, and J. Sher, *J.Chem.Soc. Chem.Commun.* **1994**, 2779.
[40] L. Spanhel, E. Arpac, H. Schmidt, *J Non Cryst Solids* **1992**, *147-48*, 657.
[41] R. F. Ziolo, E. P. Giannelis, B. A. Weinstein, M. P. O'Horo, B. N. Ganguly, V. Mehrotra, M. W. Russell, D. R. Huffman, *Science* **1992**, *257*, 219.
[42] V. Sankaran, C.C. Cummines, R.R. Schrock, R.E. Cohen, R.J. Silbey, *J. Am. Chem. Soc.* **1990**, *112*, 6858.
[43] Y. Wang, N. Herron, J. Caspar, *Physics and Chemistry of Nanometer Scale Materials Mater Sci Eng B Solid State Adv Technol* Lausanne, Elsevier Sequoia SA **1993**, *B19*,61
[44] N.A. Kotov, I. Dékány, J.H. Fendler, *J.Phys.Chem.* **1995**, *99*, 13065.
[45] (a) C. T. Kresge, M. E. Leonowicz, W. J. Roth, J. C. Vartuli and J. S. Beck, *Nature* **1992**, *359*, 710; J. S. Beck et al., *J. Am. Chem. Soc.*, **1992**, 114, 10834; (b) F. Chen and M. Liu, *Chem. Comm.* **1999**.
[46] C. Natarajan, K. Setoguchi, G. Nogami, *Electrochim Acta* **1998**, *43*, 3371.
[47] I. Exnar, L. Kavan, S.Y. Huang, M.Gratzel, *J Power Sources* **1997**, *68*, 720.
[48] C. Tsang and A. Manthiram, *J.Electrochem. Soc.* **1997**, *144*, 520.
[49] K.A. Striebel, S.-J. Wen, D.I. Ghantous, and E.J. Cairns, *J.Electrochem.Soc.* **1997**, *144*, 1680.
[50] M. Liu, *Recent Research Developments in Electrochemistry*, Transworld Research Network, **1999**.
[51] M. Liu and Z. L Wu, *Solid State Ionics* **1998**, *107*, 105.
[52] H. Hu and M. Liu, *Solid State Ionic* **1998**, *109*, 259.
[53] M. Urquidi-Macdonald, S. Real, and D. D. Macdonald, *J. Electrochem. Soc.* **1988**, 133, 2018.
[54] Kendig and Mansfeld, *Corrosion 1983*, 39, 466.
[55] J. R. Macdonald and M. K. Brachman, *Rev. Mod. Phys.* **1956**, 28, 393.
[56] M. Liu, *J. Electrochem. Soc.* **1998**, 145, 142.
[57] M. Liu and H. Hu, *J. Electrochem. Soc.* **1996**, 143, L109.
[58] P.A. Christensen and A. Hamnett, *Techniques and mechanisms in electrochemistry*, Blackie Academic & Professional, **1994**, p.172.
[59] R. Greef, R. Peat, L.M. Peter, D. Pletcher, and J. Robinson, *Instrumental methods in electrochemistry, Ellis Horwood Limited,* **1985**.
[60] F.G. Cottrell, *Z. Physik. Chem.* **1902**, *42*, 385.
[61] J.H. Christie, G. Lauer, and R.A. Osteryoung, *J. Electroanal.Chem.* **1964**, *7*, 60.
[62] J.H. Christie, G. Lauer, R.A. Osteryoung, and F.C. Anson, *Anal. Chem.* **1963**, *35*, 1979.
[63] W. Weppner and R.A. Huggins, *J. Electrochem. Soc.* **1977**, *124*, 1569.
[64] S. Basu and W.L. Worrell, in *Fast Ion Transport in Solids* (Eds.: P. Vashishta, J.N. Mundy and G.K.Shenoy), Elsevier North Holland, **1979**, p.149.

[65] P.G. Bruce, L. Lisowaka-Oleksiak, M.Y. Saidi, and C.A. Vincent, *Solid State Ionics* **1992**, *57*, 353
[66] M.Ineba, S.Nohmi, A. Funabiki, T. Abe, Z. Ogumi, *Materials for Electrochemical Energy Storage and Conversion II – Batteries, Mater. Res. Soc. Symp. Proc.*, MRS Warrendale PA USA, v496, **1998**, p.493–498.
[67] T. Yamamoto, S. Kikkawa, M. Koizumi, *Solid State Ionics* **1985**, *17*, 63.
[68] P.G. Bruce, *Solid State Electrochemistry*, Cambridge University Press, **1995.**
[69] L. Guohua, H. Ikuta, T. Uchida and M. Wakihara, *J. Electrochem. Soc.* **1996**, *143*, 178.
[70] A. Turnbull, M. Saenz de Santa Maria, and N.D. Thomas, *Corros. Sci.* **1989**, *28*, 89.
[71] J. McBreen, L. Namis, and W. Beck, *J. Electrochem. Soc.* **1996**, *113*, 1218.
[72] Z. Ogumi, T. Kuroe, and Z. Takehara, *J. Electrochem. Soc.* **1985**, *132*, 2601.
[73] Z. Ogumi, Z. Takehara, and S. Yoshizawa, *J. Electrochem. Soc.* **1984**, *131*, 769.
[74] T. Nomura, and O. Hattori, *Anal. Chim. Acta* **1980**, *115*, 323.
[75] T. Nomura and M. Iijima, *Anal. Chim. Acta* **1981**, *131*, 97.
[76] M. Hepel, *J.Electrochem.Soc.* **1998**, 145, 124.
[77] M.D. Ward and D.A. Buttry, *Science* **1990**, *249*, 1000.
[78] D.A. Buttry, in *Electroanalytical Chemistry* (Ed.: A. J. Bard), Vol. 17, New York, Marcel Dekker, **1991**, p.2.
[79] D.A. Buttry and M.D. Ward, *Chem. Rev.* **1992**, *92*, 1355.
[80] P. Curie and J. Curie, *C. R .Acad. Sci.* **1980**, *91*, 294.
[81] V.G. Bottom, *Introduction to Quartz Crystal Unit Design*, New York, Van Nostrand Reinhold, **1982**.
[82] C. Julien and G.-A. Nazri (Eds.) *Solid State Batteries: Materials Design and Optimization*, Dordrecht, Kluwer, **1994**.
[83] F. Bonino, M. Lazzzri, C.A. Vincent, and A.R. Wandless, *Solid State Ionics* **1980**, *1*, 311.
[84] F. Bonino and B. Scrosati, In *Solid State Batteries* (Eds: C.A.C.Sequeira and A.Hooper), NATO ASI Series, Dordrecht, Martin Nijhoff, **1985**, p.15.
[85] Y.-M. Choi, S.-I. Pyun, J.-S. Bae and S.-I. Moon, *J.Power Sources* **1995**, *56*, 25.
[86] C. Ho, D. Raistrick, and R.A. Huggins, *J.Electrochem.Soc.* **1980**, *127*, 343.

7 Optical Spectroscopy of Nanophase Material

C. Burda, T. Green, C. Landes, S. Link, R. Little, J. Petroski, M. A. El-Sayed

7.1 Introduction

The electronic properties of a material change drastically as the density of states is reduced as a consequence of reducing the size and the dimensionality [1–6]. The energy eigenstates are now determined by the system's boundaries and therefore surface effects become very important [1–4, 7]. A transition from the bulk band structure to individual localized energy levels occurs in clusters of subnanometer to nanometer size and the detection of quantum size effects has been of great interest to scientists in the search for novel materials with new properties [5, 8–10]. Possible future applications of nanoparticles include the areas of data communication and high density optical data storage [4, 7, 11], solar energy conversion [12], and the use of nanoparticles as catalysts because of their high surface to volume ratios [4].

Closely related to size induced changes in the electronic structure are the optical properties of nanoparticles [3, 13–18]. Optical spectroscopic methods probe the energy differences between electronic states as well as the lifetimes of excited states and their respective energy relaxation channels using time-resolved techniques [3, 14, 18]. The quantum size effect on the optical absorption spectra is best known for semiconductor nanoparticles. The decrease in particle size shifts the absorption edge from the infrared to the visible region of the electromagnetic spectrum as the band gap energy of the semiconductor increases [3, 14–18]. In a molecular type of description this is equivalently to an energy decrease of the highest occupied molecular orbital (HOMO) and an energy increase of the lowest unoccupied molecular orbital (LUMO) [14–16] due to the spatial confinement of the charge carrier wavefunctions. By changing the size of semiconductor nanoparticles one can therefore tune the color of their colloidal solutions as well as their oxidation reduction properties [17].

Generally, semiconductor nanoparticles are luminescent [19–30]. Depending on the surface properties some luminescence bands are found to be redshifted from the absorption onset [19, 20, 25–30]. One sharp peak with only a small Stokes shift corresponds to the band gap or near band gap emission resulting form the recombination of the electron-hole pair. A much broader band at longer wavelength is observed for particles with many surface defects and originates from the trapped charge carrier recombination [19, 20, 25–30]. The surface consisting of many defects resulting from sites of uncompensated charges lead to quenching of the band gap or near band gap emission and lead to strong deep trap long-wavelength emission. However, it has been shown that the surface can be passivated by an overcoat layer consisting of a semiconductor material [24] of a larger band gap or adsorbent molecules [25]. Luminescence quantum yields of close to unity at room temperature have been achieved in this manner [24].

The origin of the photoluminescence of semiconductor nanoparticles with energy that does not correspond to the band gap energy and lifetimes much longer than picoseconds, has been the subject of great deal of discussion in the literature [19–30]. The long luminescent lifetimes are in sharp contrast with those of charge carrier recombination processes occuring (short picosecond time scale) [31]. It has therefore been suggested that shallow and deep (surface) traps are responsible for the long lifetimes as trapping competes with the radiative recombination of the electron and hole [19, 20, 25–30]. Emission resulting from the recombination of trapped electrons and holes is much slower and accounts for the long luminescent lifetime component. On the other hand, it has been proposed [21–23] that the long lived emission results from an optically dark triplet state. Band gap recombination is then a spin-forbidden transition and would explain the observed long luminescence lifetimes [21–23].

The ultrafast dynamics of the electron-hole pair can be separated by femtosecond pump-probe spectroscopy using the method of competitive quenching [32–35]. By adding molecules to the surface of the semiconductor nanoparticles they can act as electron donors [34] or acceptors [32, 33] after photoexcitation. Removing an excited electron from the nanoparticle by electron transfer to the absorbed molecule before electron trapping can occur, isolates the hole on the nanoparticle and its relaxation can be probed by the recovery of the transient bleach of the band gap absorption.

Spatial separation of the electron and hole can also be achieved in quantum-dot quantum-well (QDQW) heterostructures such as CdS-HgS-CdS [36, 37] consisting of a wide band gap semiconductor core and clad with a narrow shell of a material of small band gap (the well). After photoexcitation with 400 nm femtosecond laser pulses, nonradiative relaxation results in the spatial separation of the electron and hole pair. The electron is found [37] to relax rapidly into the HgS well while the hole remains localized in the CdS shell for a much longer time. The slow relaxation of the hole is attributed [37] to the difference in its effective mass in the HgS well as compared to that in the CdS shell. This introduces an interfacial barrier. The energy of this type of charge-separated (optically dark) state was calculated theoretically [38] and strong experimental evidence for its presence has been observed by femtosecond pump-probe spectroscopy [37].

Metallic nanoparticles have fascinated scientists because of their colorful colloidal solutions long before semiconductors and their applications became an integral part of modern technology. Gold nanoparticles were used as a pigment of ruby-colored stained glass dating back to the 17th century [39]. Faraday [40] recognized that the red color is due to metallic gold in colloidal form and Mie [41] was the first to explain this phenomenon theoretically in 1908 by solving Maxwell's equation for the absorption and scattering of electromagnetic radiation by spherical particles. His theory has found wide applicability since then because it allows calculating particle extinction spectra as long as the material dielectric function is known [42–46].

The physical origin of the light absorption by metallic nanoparticles in a certain size range is the coherent oscillation of the valence band electrons induced by an interaction with the electromagnetic field [13]. These resonances are known as surface plasmons and are indeed a small particle effect as they are absent in the individual atoms as well as in the bulk [13, 42–44]. However, the size dependence of the surface plasmon absorption is not as easily explained as in the case of semiconductor nanoparticles [13], where a shift in the HOMO and LUMO results in a larger band gap and a blueshift of the absorption onset. Studies of the electron phonon relaxation time following the different plasmon excitations are carried out for gold nanorods and nano-dots. It will further be demonstrated how the surface plasmon absorption in colloidal gold nanostructures can be used as a sensitive monitoring tool to probe the stability of capping miscelles.

This chapter reviews the optical spectroscopy of some colloidal metal and semiconductor nanoparticle solutions. In Section 7.3, the results of the optical properties and electron-phonon relaxation processes in gold nanoparticles are discussed. In addition the changes in the optical properties of the platinum nanoparticles during its growth is discussed. In Section 7.4, the electron and hole dynamics in semiconductor quantum dots and quantum-dot-quantum-wells are discussed. The dynamics of surface trapping and well trapping are detailed.

7.2 Experimental

The size and shape distributions of the nanoparticles formed in solution at different times of growth or irradiation were determined from the TEM images of the evaporated solution on carbon coated copper grids at the Georgia Tech Microscopic Facility. A Hitachi HF-2000 field emission TEM operating at 200 kV was used. Normally, 300 or more particles are counted to determine the size distribution of each sample.

The femtosecond dynamics were determined with an amplified Ti-Sapphire laser system (Clark MXR, CPA 1000) which was pumped by a diode-pumped, freqeuncy-doubled Nd:Vanadate laser (Coherent Verdi). This produced laser pulses of 100 fs duration (FWHM) and an energy of 1 mJ at 790 nm. The repetition rate was 1 kHz. A small part (4%) of the fundamental was used to focus in a 2 mm sapphire plate to generate a white light continuum which was used between 430–780 nm. The excitation beam was modulated by an optical chopper with a frequency of 500 Hz. The probe light was split into a reference and a signal beam. The samples were irradiated in cylindrical cuvettes of 2 mm optical path length, placed in a spinning sample holder. After passing the monochromator (Acton Research) both beams were detected by two photodiodes. The kinetic traces were obtained using a sample-and-hold unit and a lock-in-amplifier (Stanford Research Systems). The typical measured optical density (OD) changes were in the range of 50 mOD. For spectral measurements a CCD camera (Princeton Instruments) attached to a spectrograph (Acton Research) was used. The group velocity dispersion of the white light continuum was compensated.

Nanosecond experiments were carried out with an optical parametric oscillator (Spectra Physics, MOPO-730) which was pumped by a Nd:YAG laser (Spectra Physics, GCR-250). The output pulses had a pulse duration of about 7 ns, a repetition rate of 10 Hz, and a wavelength range from 225 nm to 1.8 μm. The pulse energy was in the mJ range.

Steady state absorption measurements were carried out on a Beckman DU 650 spectrometer and steady state photoluminescence were determined on a PTI Quantamaster fluorometer.

7.3 Metal nanostructures

7.3.1 Size and shape dependence of the plasmon absorption of gold nanoparticles

Colloidal solutions of spherical gold nanoparticles exhibit a deep red color due to the well known surface plasmon absorption and have therefore been of scientific interest since the turn of this century [1, 2, 13, 39–46]. Nevertheless, questions like definite quantum size effects due to increased energy level spacing or the transitions from isolated atoms to clusters and finally to bulk matter and the related electronic, optical, and thermodynamic properties are still of great concern to many chemists, physicist, and also materials scientists [5, 13, 47]. In addition, the ability to control the shape of metallic nanoparticles [48–51] and the wealth of new instrumentation available today to investigate surface properties as well as time-resolved events in the femtosecond to picosecond time domains have sparked a renewed interest in metal nanoparticles. This section describes the size and shape dependence of the optical properties of gold nanoparticles in their ground state (steady-state spectroscopy). The focus is mainly on colloidal gold nanoparticles in aqueous solution with mean particle sizes ranging between 10 and 100 nm. Section 7.3.2 then deals with time-resolved measurements of the electron dynamics.

The gold nanospheres presented here were prepared by reduction of gold ions in aqueous solution with sodium citrate under reflux [52]. This procedure developed by Turkevich [52] yields fairly monodisperse solutions of gold nanoparticles with an average diameter around 10–20 nm. A typical TEM image of spherical gold nanoparticles synthesized in this way is shown in Fig. 7-1 (left). The average particle diameter was determined to 15 nm for this particular sample. Larger gold nanoparticles can easily be obtained by reducing the gold ions with hydroxylamine hydrochloride in the pres-

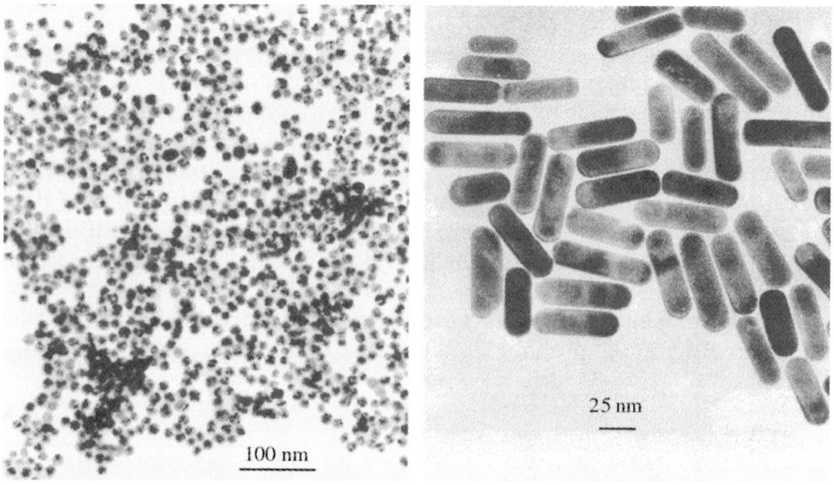

Figure 7-1. TEM images of gold nanospheres (left) and nanorods. The nanospheres were prepared by reduction of gold ions with sodium citrate in aqueous solution. The mean diameter of the gold nanospheres is 15 nm. The gold nanorods were synthesized by an electrochemical method with the aid of organic surfactant molecules forming a protective micelle around the rods. The mean length and width of the gold nanorods are 60 and 18 nm, respectively (average aspect ratio of 3.3).

ence of previously prepared nanoparticles [52]. As this reducing agent is not able to initiate new nucleation centers, all the already existing particles grow uniformly in size resulting in no change in the size distribution. This is just one example of how to obtain colloidal gold nanoparticles and many other preparation methods have been developed over the years [13, 53]. Among the most interesting is the synthesis of molecular type nanocrystals passivated by an overlayer of organic thiol molecules [54, 55]. Furthermore, gold nanorods have been prepared electrochemically with the aid of shape inducing organic surfactant molecules, which form a protecting micelle around the gold nanorods [49]. A TEM picture of gold nanorods with a mean length of 60 nm and a mean width of about 18 nm is also given in Fig. 7-1 (right). The ratio of the length divided by the width is the aspect ratio, R, of the gold nanorods, which is an important quantity when describing the optical properties of these nanorods and is for the sample in Fig. 7-1 equal to 3.3. Gold nanorods can also be obtained by electrodeposition of gold into the pores of an aluminum oxide membrane [50, 51].

The deep red color of solutions containing spherical gold nanoparticles mentioned above originates from the surface plasmon absorption of these small gold particles [13, 41–46]. This surface plasmon resonance is caused by the coherent oscillation of the (free) conduction electrons induced by light. The surface of the nanoparticle plays an important role because, although all electrons are oscillating with respect to the positive ion core, the main effect producing the restoring force is the surface polarization. Mie [41] already described this phenomenon theoretically in 1908 when he applied Maxwell's equations to spherical particles with a bulk dielectric function $\varepsilon(\omega)$ ($\varepsilon(\omega) = \varepsilon_1(\omega) + i\varepsilon_2(\omega)$), where ε_1 and ε_2 are the real and imaginary part of the complex dielectric function, surrounded by a medium with a dielectric constant ε_m (assumed to be frequency independent) and interacting with an electromagnetic field. The total extinction coefficient σ_{ext} for N particles of Volume V is composed of a series of absorption and scattering modes. In the limit of $d << \lambda$ where d is the particle diameter and λ the wavelength of the light only the dipole absorption contributes significantly and Mie's theory reduces to the following well-known form (dipole approximation, quasi-static limit) [13, 42–44]:

$$\sigma_{ext} = \frac{18 \cdot \pi \cdot N \cdot V \cdot \varepsilon_m^{3/2}}{\lambda} \cdot \frac{\varepsilon_2}{[\varepsilon_1 + 2 \cdot \varepsilon_m]^2 + \varepsilon_2^2} \tag{7-1}$$

For larger nanoparticles (gold $d > 25$ nm), the dipole term contributes to the extinction and higher order oscillations are excited [13]. While Eq. (7-1) is independent of particle size the next terms within Mie's theory depend explicitly on the particle diameter rendering a size dependent absorption spectrum (extrinsic size effect) [13]. The plasmon band shifts to longer wavelengths while its width increases. On the other hand, the plasmon band width also increases with decreasing size for nanoparticles smaller than about 25 nm in the case of gold. Obviously, Mie's theory as presented in Eq. (7-1) cannot account for such a size dependence. Therefore, it is assumed that the bulk dielectric function itself becomes size dependent $\varepsilon(\omega, d)$ (intrinsic size effect) [13].

There exist many theories on how a size dependent dielectric function is introduced [56–65]. As it is impossible to account for all of them only two examples will briefly be given here: The first approach to this problem was suggested by Kreibig [56, 57] who argued that the dielectric function becomes size dependent due to an enhanced electron–surface scattering in particles smaller than the mean free path of the conduction electrons. Kreibig's model predicts a 1/d dependence of the plasmon band width for small nanoparticles in agreement with experiments. The second theory explains the size dependence of the plasmon band width in the quasi-static limit by considering the chemical nature of the nanoparticle environment (chemical interface damping CID [61]). Charge transfer processes involving energy levels of the metal-adsorbate complex lead to energy and momentum dissipation of the coherent electron oscillations. The energetic positions of these levels depend on the particle size as well as on the specific molecules by which the nanoparticles are surrounded [61].

Optical absorption spectra of colloidal gold nanospheres of different sizes produced as described above are shown in Fig. 7-2 [66]. The surface plasmon absorption around 520 nm is clearly visible. The inset illustrates how the plasmon band width varies with nanoparticle diameter over a size range covering both intrinsic ($d < 25$ nm) and extrinsic ($d > 25$ nm) size effects. The predicted increase in the plasmon band width for particles smaller as well as larger than about 25 nm is therefore in excellent agreement with the experimental results in Fig. 7-2. Furthermore, the plasmon band width can be related to the dephasing of the coherent electron oscillation if the resonance is assumed to be homogeneously broadened. A dephasing time T_2 for the loss of coherence on the order of 4 fs is obtained in this manner [66].

Much more drastic than the effect of particle size on the optical absorption of gold nanoparticles is the effect of particle shape. In the case of rod-shaped nanoparticles the surface plasmon absorption splits into a transverse and longitudinal mode corresponding to the coherent electron oscillation perpendicular and along the major axis of the rod, respectively [13, 42–44]. The optical absorption spectrum of a collection of randomly orientated gold nanorods with aspect ratio R can be computed by the following equation (dipole approximation) [42].

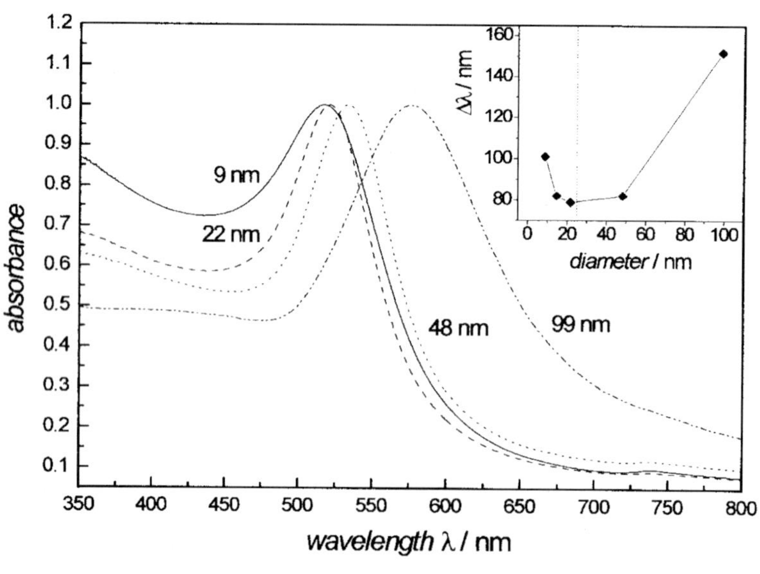

Figure 7-2. Size dependence of the optical absorption spectra of colloidal gold nanospheres [66]. As illustrated in the inset, the plasmon band width increases for nanoparticle sizes below about 25 nm because of a size dependent metal dielectric function [intrinsic size effect]. The width also increases again for particles larger than 25 nm due to the contribution from quadrupole (and octople etc.) extinction [extrinsic size effect]. Furthermore, for larger particles the surface plasmon maximum shifts to longer wavelength with increasing particle size because of the excitation of higher order absorption and scattering modes peaking at lower energies.

$$\sigma_{ext} = \frac{2 \cdot \pi \cdot N \cdot V \cdot \varepsilon_m^{3/2}}{3 \cdot \lambda} \cdot \sum_j \frac{(1/P_j^2) \cdot \varepsilon_2}{(\varepsilon_1 + \frac{1-P_j}{P_j} \cdot \varepsilon_m)^2 + \varepsilon_2^2} \tag{7-2}$$

P_j are the depolarization factors for the three axes A, B, C of the nanorod with A > B = C.

$$P_A = \frac{1-e^2}{e^2} \cdot \left[\frac{1}{2 \cdot e} \cdot \ln\left(\frac{1+e}{1-e}\right) - 1\right] \tag{7-3}$$

$$P_B = P_C = \frac{1-P_A}{2} \tag{7-4}$$

$$e = \sqrt{1 - \left(\frac{B}{A}\right)^2} = \sqrt{1 - \frac{1}{R^2}} \tag{7-5}$$

Optical absorption spectra of two gold nanorod solutions are shown in Fig. 7-3a. The transverse surface plasmon absorption spectrally coincides with the absorption maximum of nanospheres while the longitudinal resonance is shifted to longer wavelengths. The position of the maximum of the longitudinal surface plasmon absorption is extremely sensitive to the nanorod aspect ratio *R*. For the two samples in Fig. 7-3a a difference in aspect ratio of 0.6 results in a wavelength shift of about 80 nm from around 660 nm to 740 nm. Experimentally a linear dependence of the absorption maximum of the longitudinal resonance on the nanorod aspect ratio is found [49, 50, 67] as demonstrated in Fig. 7-3b for a series of prepared samples. Also included in Fig. 7-3b is the dependence of the maximum of the transverse surface plasmon absorption, which is independent of thc aspect ratio for these samples.

The sensitivity of the longitudinal surface plasmon resonance on the particle shape has proven to be very useful in studying the thermal [67] and photothermal [68] stability of these gold nanorods in solution. It was found that the capping micelles surrounding the nanorods selectively dissolve in the aqueous medium as the solution temperature increases, with the longest micelles being the least stable dissolving at the lowest temperature. This results in selective destruction of the nanorods having the largest aspect ratio and causes the longitudinal plasmon absorption to shift to higher energies [67]. On the other hand, by photothermal heating the gold nanorods directly with a laser of moderate energy and having a frequency in resonance with the nanorod absorption, the nanorods undergo a shape transformation into nanospheres of comparable volume thus resulting in the complete disappearance of the longitudinal surface plasmon absorption. The photoisomerization of gold nanorods is explained by melting of the nanorods after laser excitation [67]. This is easily possible as the extinction cross sections of gold nanoparticles are on the order of about half of the particle size (orders of magnitude larger than that for the best organic dyes) [69] and because of the well-known fact that the melting temperatures of nanoparticles are much lower than the bulk values [70–77].

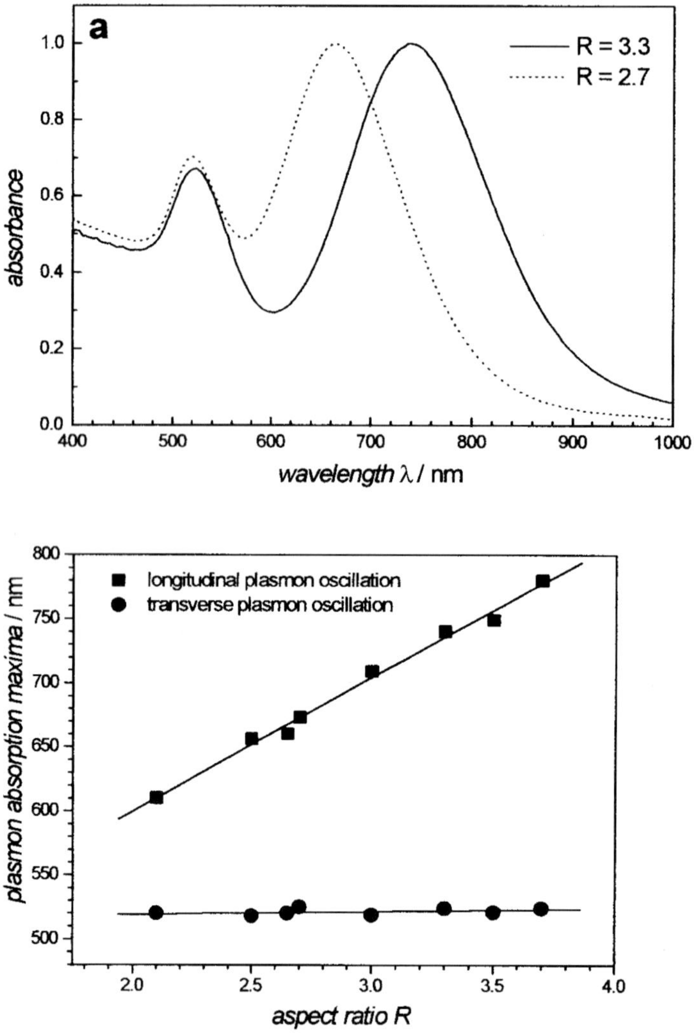

Figure 7-3. (a) Optical absorption spectra of two gold nanorod samples with average aspect ratios of 2.7 and 3.3. The surface plasmon absorption is split into a transverse and longitudinal mode absorbing around 520 nm and at longer wavelength, respectively. (b) While the maximum of the transverse surface plasmon oscillation (circles) is only weakly dependent on the nanorod aspect ratio R the maximum of the longitudinal absorption band (squares) is found to increase linearly with increasing aspect ratio.

7.3.2 Electron dynamics in gold nanoparticles

In recent years time-resolved femtosecond studies on semiconductor and metallic nanoparticles has found great interest as the dynamics of the excited charge carriers can be followed directly by pump-probe spectroscopy. The lifetimes of the photoexcited nanoparticle system are of fundamental interest in designing materials for possible future applications in optoelectronic devices such as optical switches or solar cells [4, 7, 11, 12]. In metallic nanoparticles, where a considerable energy gap between the highest occupied orbital (HOMO) and the lowest unoccupied orbital (LUMO) exceeding the thermal energy at room temperature only opens up for particle sizes below ~ 2 nm [47], the optical response is mainly related to the temperature of the excited electrons [78–86]. A change of the electronic temperature as determined by the Fermi electron distribution directly results in changes of the optical constants of the material as expressed by its complex dielectric function [78–81]. For the noble metals copper, silver, and gold their intense surface plasmon absorptions in the visible, which are usually described by Mie theory [13, 41–44] using the complex dielectric function of the metal (see Section 7.3.1), have been found to be a very sensitive tool to monitor the time evolution of the hot electron gas excited by an ultrashort laser pulse [78–86, 87–95].

Figure 7-4 shows the transient absorption spectra of 15 nm gold nanospheres (a) and gold nanorods having an average aspect ratio of 3.8 (b) recorded at different delay times after excitation with 400 nm femtosecond laser pulses. The ground state absorption spectra are also given in the figures scaled to arbitrary units for comparison. The plasmon absorption band(s) (longitudinal mode at 520 nm and transverse mode at 750 nm for the nanorods) show a bleach (negative absorption) centered at the wavelength of the ground state plasmon maximum with positive absorptions at higher and lower energies (partly hidden for the nanorods due to the limited spectral window of the CCD camera). This shape of the transient absorption spectra is explained by a broadening of the plasmon band at higher electronic temperatures with a simultaneous decrease in absorption intensity [78–84]. The recorded signal is then the difference spectrum between a broader and less intense plasmon band after laser excitation (heating) and the ground state plasmon oscillation. The transient response decays as the hot electrons thermally equilibrate with the nanoparticle lattice by electron-phonon collisions [78–84]. The energy deposited by the pump laser pulse is finally released to the surrounding medium by phonon-phonon interactions with the solvent molecules leading to a complete recovery of the plasmon band bleach.

By monitoring the bleach at its maximum, where the transient signal is strongest and therefore most sensitive, as a function of delay time between excitation and probe pulse it is possible to determine the electron-phonon and phonon-phonon relaxation times. This is shown in Fig. 7 5 for 15 nm nanospheres after excitation at 400 nm using different laser pump powers between 50 and 160 nJ. The measured decay curves are fitted with a biexponential function giving increasing electron-phonon relaxation times of 1.5, 2.0, 3.3, and 3.6 ps with increasing excitation powers of 50, 80, 100, 160 nJ, respectively. The offset is modeled by a lifetime of 100 ps for all four traces corresponding to the phonon-phonon relaxation time. A plot of the electron-phonon relaxation times against the laser pump power gives a limiting lifetime of 690 $\pm$ 100 fs for zero pump power corresponding to an electron-phonon coupling constant of 2.9 $\pm$ 0.5 * 10^{16} $Wm^{-3}K^{-1}$ [81, 82], which is similar to the value for bulk gold [96–102]. The increase in the measured bleach recovery times has been explained by the temperature dependence of the electron heat capacity [79] and is also observed in thin metal films. For more detailed information the reader is referred to references [78–82, 96–106]. An important experimental fact to point out here for the following results is that electron-phonon relaxation times measured for different particle sizes and shapes can only be compared with each other if the same initial change in electronic temperature is induced by the exciting laser pulse. This means that comparable laser powers need to be used for different samples having about equal optical density [80].

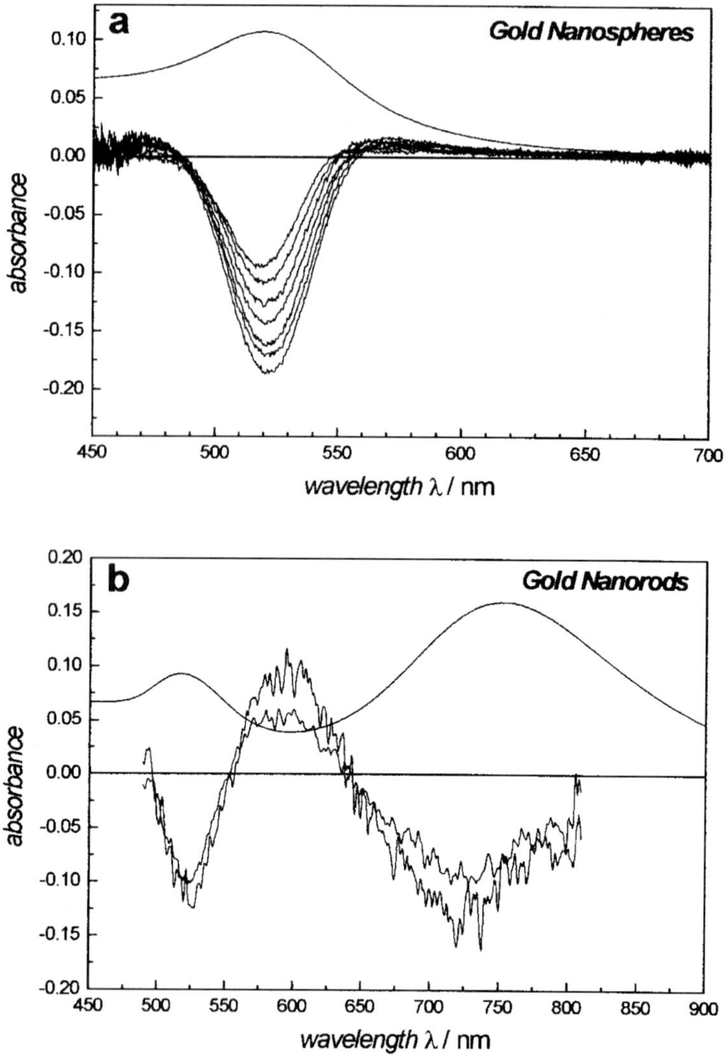

Figure 7-4. Femtosecond transient absorption spectra of gold nanospheres (a) [80] (average particle diameter of 15 nm) and gold nanorods (b) [109] (average aspect ratio of 3.8) recorded at different delay times between the excitation pulse centered at 400 nm and a white light continuum probe pulse. The plasmon absorption of the gold nanoparticles dampens due to the excited electron gas, which results in a transient bleach of the plasmon band(s) accompanied by absorption at both shorter and longer wavelengths than the respective plasmon resonance. The bleach features recover as the heated electron gas thermally equilibrates with the lattice by electron-phonon interactions followed by the phonon-phonon coupling with the surrounding solvent (water in this case for both the nanospheres and nanorods). The ground state absorption spectra scaled to arbitrary units for comparison and measured by steady-state optical absorption spectroscopy are also included in (a) and (b) (upper part of the figures).

Electron-phonon relaxation times ranging between 1 and 4 ps have been reported by several authors for spherical gold nanoparticles embedded in different media [78–84, 88–92] and are also obtained for silver [87, 94] and copper particles [86, 87, 93]. Using femtosecond laser pulses it is, however, also possible to follow the influence of

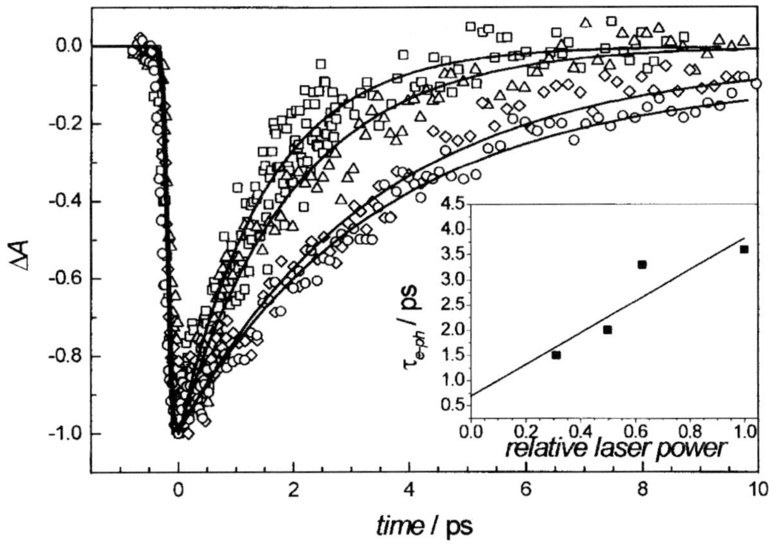

Figure 7-5. Excitation power dependence of the electron-phonon relaxation time measured for 15 nm gold nanospheres after excitation at 400 nm with 100 fs laser pulses. The probe wavelength is the bleach maximum at 520 nm where the transient absorption signal is most sensitive. The excitation power was varied between 50 and 160 nJ with an estimated beam diameter of about 125 μm at the sample. The measured electron-phonon relaxation times increase with increasing laser pump power from 1.5 to 3.6 ps as determined by biexponential fits of the data points. The long component of 100 ps accounting for the offset at longer delay times corresponds to the phonon-phonon relaxation time. The inset shows a plot of the obtained lifetimes against the relative laser power, which yields a limiting electron-phonon relaxation time of about 690 $\pm$ 100 fs for zero pump power corresponding to an electron-phonon coupling constant of 2.9 $\pm$ 0.5 * 10^{16} $Wm^{-3}K^{-1}$.

electron-electron collisions on the thermalization of the initial non-Fermi electron distribution created by the pump pulse to a Fermi distribution with a defined electronic temperature [80, 85, 94]. Evidence for an electron thermalization time longer than the pulse duration (100 fs for the experiments presented here) is shown in Fig. 7-6 for 15 nm gold nanospheres and excitation at 630 nm. A clear deviation from a simple monoexponetial decay behavior is observed for short time delays (< 2 ps) as illustrated by the dotted line. Better agreement is obtained (solid line) when using a model developed by Sun et al. [105, 106], with which the early electron dynamics in thin gold films (thickness of the order of the nanoparticle's diameter) can be explained. This approach yields an electron thermalization time of 500 fs and an electron-phonon relaxation time of 750 fs with about equal amplitudes for the kinetic trace shown in Fig. 7-6. This is again in close agreement with results obtained on thin gold films (bulk gold) [101, 102, 105, 106]. Furthermore, the influence of a finite electron thermalization to a Fermi distribution is most pronounced at very low excitation powers and when pumping away from the threshold for interband transitions (~ 2.4 eV in gold coinciding with the plasmon resonance at 520 nm) [80].

While in the ground state the shape-dependence of the plasmon band width (related to the phase coherence) is thought to be caused by an increased electron-surface scattering due to the limitation of the electron mean free path [13, 66]. It is of great interest to investigate if electron-surface scattering is also dominant for the energy relaxation of the hot electron gas. With an electron mean free path of about 50 nm [107, 108], the plasmon bleach recovery is therefore measured for several sizes

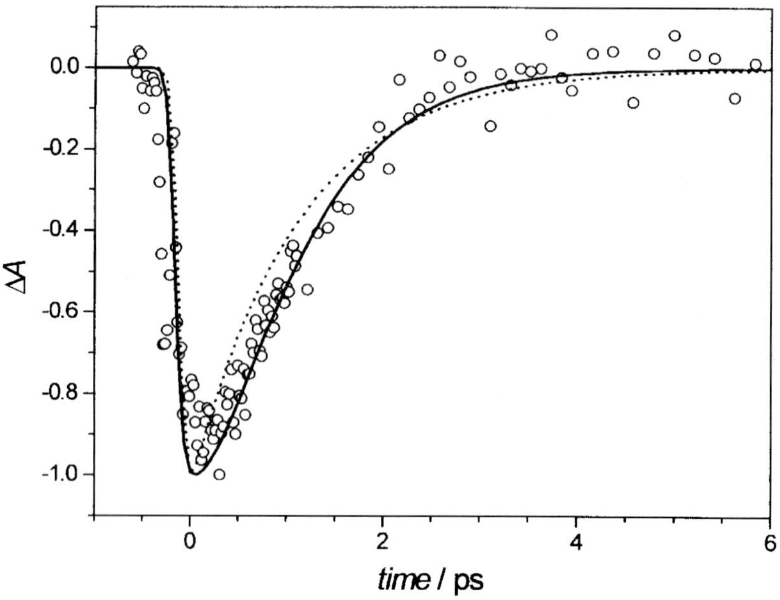

Figure 7-6. The effect of electron-electron thermalization on the bleach recovery measured for 15 nm gold nanospheres and monitored at 520 nm after excitation at 630 nm. The observed transient signal decays much slower within the first 2 ps than expected from a purely monoexponential decay due to electron-phonon relaxation alone (dotted line). However, by taking electron-electron interactions into account the solid line is obtained, which yields an electron-electron thermalization time of 500 fs and an electron-phonon relaxation time of 750 fs with an amplitude ratio of about 1. This effect is most pronounced if exciting away from the interband transitions in gold (> ~ 520 nm) and when using very low excitation powers.

ranging between ~ 10 to 50 nm [80] expecting a decrease in the measured lifetime for smaller nanoparticles due to an increased electron-surface scattering (if those collisions are inelastic). Figure 7-7 shows the results of the femtosecond studies on 22 and 48 nm gold nanospheres (a) with the respective TEM images given in (b) and (c). The bleach recovery was followed at the bleach maximum and the excitation wavelength was 630 nm. The measured lifetimes of 400 fs and 1.6 ps for the electron-electron and electron-phonon interactions in 22 nm gold particles compare well with the values of 450 fs and 1.7 ps obtained for the larger 48 nm gold particles. The small difference is within the experimental error and similar lifetimes are also obtained for 9 nm nanospheres [80]. A size dependence of the electron dynamics is therefore not detectable in the size range of greater than 10 nm [80].

The effect of particle shape on the electron-phonon relaxation time [109] is displayed in Fig. 7-8 where the bleach recovery of the transverse and longitudinal modes of the surface plasmon oscillation of gold nanorods having an average aspect ratio of 3.8 is compared with each other and with that of 15 nm spherical gold nanoparticles. Under the same excitation conditions (same pump power and sample extinction at 400 nm) electron-phonon relaxation times of 2.9, 3.1, and 3.1 ps are measured when monitored at the transverse mode, the longitudinal mode, and the surface plasmon absorption of the spheres, respectively [109]. This shows that shape also has no effect on the cooling of the hot electrons excited by a femtosecond laser pulse. The measured lifetimes are furthermore independent of the mode of the surface plasmon oscillation (transverse vs. longitudinal) in the gold nanorods.

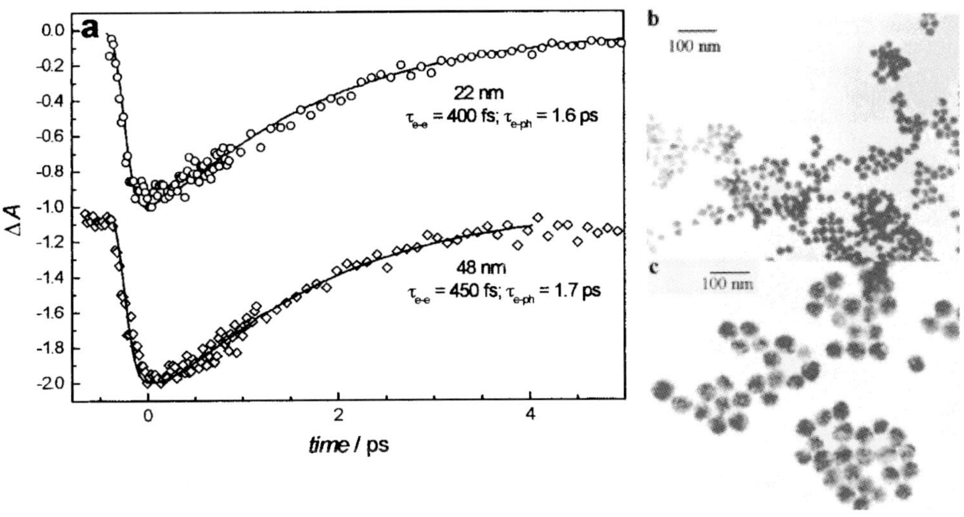

Figure 7-7. Size dependence of the electron dynamics in gold nanospheres [80]: The transient bleach decay (a) is followed at the bleach maximum after excitation with 630 nm laser pulses for the 22 and 48 nm gold nanospheres pictured in the TEM images (b) and (c), respectively. The measured electron-electron and electron-phonon relaxation times of 400 fs and 1.6 ps for the 22 nm particles and 450 fs and 1.7 ps for the 48 nm are independent of particle size within the accuracy of the experiment. An enhanced electron-surface scattering is thought to be responsible for the faster dephasing (T_2) of the coherent plasmon oscillation in metal nanoparticles smaller than the mean free path of the conduction electrons (~ 50 nm in gold). However, from these results it follows that the energy relaxation (T_1) of the hot electrons is not dominated by (inelastic) electron-surface collisions.

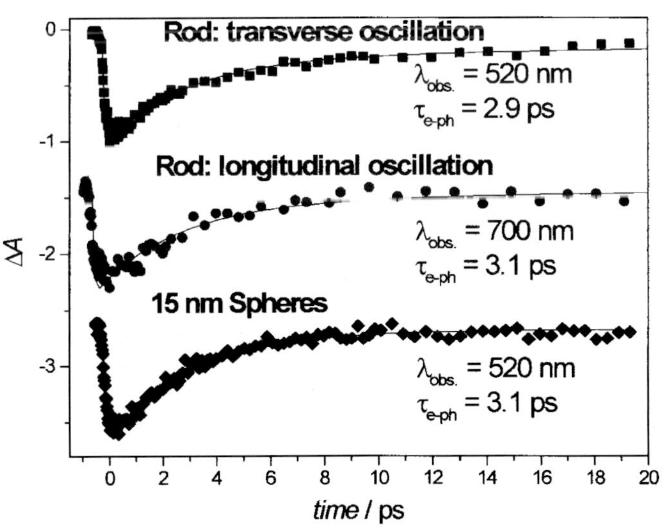

Figure 7-8. Shape dependence of the electron dynamics in gold nanospheres and nanorods: The bleach recoveries of the transverse and longitudinal plasmon oscillations are followed for the same gold nanorod solution (average aspect ratio of 3.8) at 520 and 700 nm and are compared to the relaxation dynamics in 15 nm gold nanospheres under the same experimental conditions. Very similar electron-phonon relaxation times are obtained, which leads to the conclusion that the electron-phonon interactions in gold nanoparticles are independent of the particle shape and the specific plasmon mode (transverse or longitudinal).

In conclusion, the electron-phonon relaxation in gold nanoparticles of the investigated size range is independent of particle size and particle shape. In addition, the transient behavior is very similar to the results found for the electron-electron and electron-phonon interactions in bulk gold as measured in thin films [96–106], which indicates that the bulk electronic band structure is already fully developed in these relatively large particles and that possible specific surface states are of no major importance for the energy relaxation (T_1). A cancellation of two competing effects (decreasing density of energy states and increasing electron-phonon coupling with decreasing nanoparticle size) cannot, of course, be ruled out. This is at least in sharp contrast to the ground state surface plasmon absorption itself, which strongly depends on particle size [13, 66]. The plasmon band width, which is directly related to the dephasing time (T_2) of the coherent electron oscillation, increases for decreasing sizes below 20 nm due to enhanced electron-surface scattering and increases for larger particles due to the contribution of higher order oscillatory modes (see Section 7.3.1).

7.3.3 The optical properties of platinum nanoparticles during the growth process

Recently, synthetic control of nanoparticle shapes in a colloidal platinum solution was achieved by varying the initial ratio of the platinum salt to that of the polyacrylate capping material [48, 110]. The growth of the nanoparticles proceeds *via* reduction of the platinum salt (K_2PtCl_4) by hydrogen gas over approximately 12 hours [111, 112]. Using a 1:1 molar ratio, the dominant shape in the solution is cubic, consisting of six {100} faces. The average size was found to be approximately 11 nm. Increasing the polyacrylate concentration five-fold results in the dominant shape in the solution being tetrahedral which is made up of four {111} faces with an average size of 7 nm. Figure 7-9 presents a high resolution transmission electron microscopic (HRTEM) image of these shapes. The shapes are well defined, although some atomic level steps as well as rounding of some of the edges is evident [113]. Also present in the solutions are truncated octahedra which consist of six {100} and eight {111} faces.

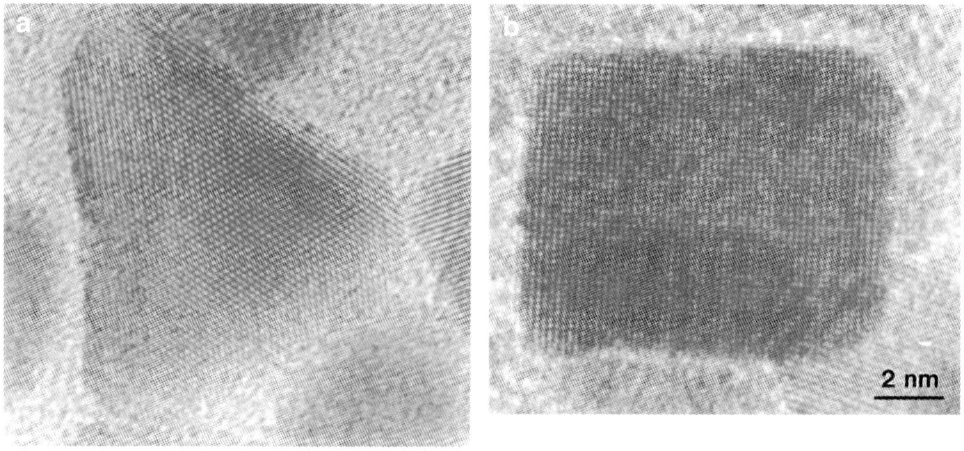

Figure 7-9. High resolution TEM images of a a) tetrahedral nanoparticle oriented along [110] showing the {111} faces and b) a cubic platinum nanoparticle oriented along [001] showing the {100} faces. The atomic roughness of the faces is apparent in both of these nanoparticles.

The shape formation and growth mechanism of these platinum nanoparticles has been found to depend on the capping material (due to its buffering nature) as well as the pH of the solution [114]. Using TEM, the shape distribution of platinum nanoparticles at different stages of their growth as a function of time was determined for the case of the 1:1 ratio, 1:5 ratio and nanoparticles made without the addition of a capping material. These distributions are plotted in Fig. 7-10. It was found that the smallest nanoparticles formed during the early stages of growth or at high polymer concentration displayed distributions with a dominance of tetrahedral shapes. These tetrahedral nanoparticles are transformed into truncated octahedra and eventually into cubic shapes as the growth continues or at low polymer to Pt complex concentration ratio. The mechanism proposed is one in which the initially rapid reduction of Pt^{2+} produces an initial growth that gives very small nanoparticles having the most stable {111} faces present in tetrahedra and truncated octahedra. The competition between polymer capping and H_2 reduction of the Pt^{2+} complex occurring on the most catalytically active {111} surface [115] determines the fate of these tetrahedral nanoparticles. If the capping material remains bonded to the surface, tetrahedral nanoparticles of small size result. The capping material can be removed by neutralization which occurs from the lowered pH of the solution which frees the platinum surface for further reaction. The rapid reduction of the Pt^{2+} on the uncapped {111} surface leads to its disappearance and the formation of a {100} face due to the deposition of Pt atoms. This can result in truncated octahedral nanoparticles formed. The truncated octahedral nanoparticles continue to grow until transformed into cubic nanoparticles. When the platinum supply is depleted before the cubic growth is complete, this results in the rounding of the shapes as seen in Fig. 7-9.

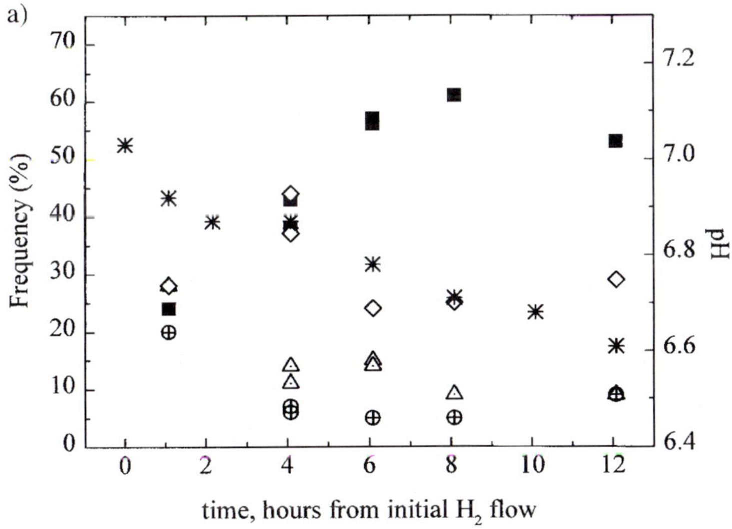

Figure 7-10. (a) Time dependence of the shape distribution of the different Pt nanoparticles collected from TEM images for a 1:1 Pt^{2+} to polyacrylate ratio, (b) for a 1:5 ratio, and (c) without the addition of the polyacrylate capping material. The changes in the percentages of cubic (■), tetrahedral (Δ), truncated octahedral (◇) and unidentified (⊕) nanoparticles are shown, as well as the change in the pH (*) over the same time period. This figure shows that at low polymer concentration (a), cubes are formed at the expense of the tetrahedra as the pH decreases with time. While at high polymer concentration

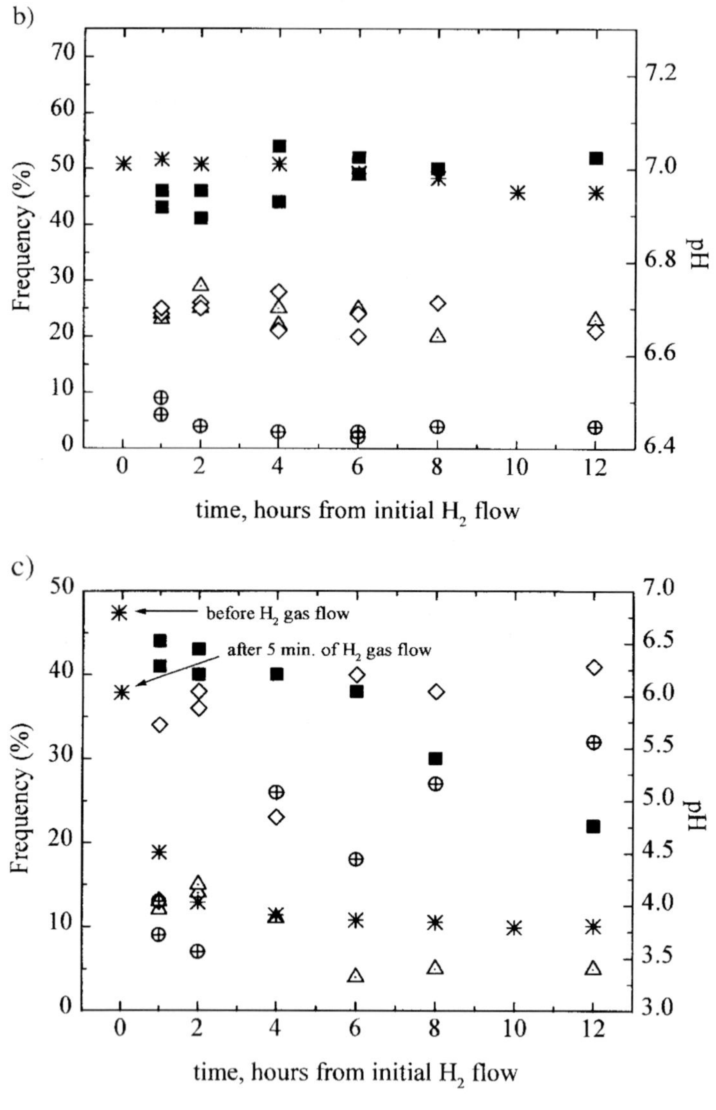

Figure 7-10. (b), the distribution as well as the pH remain independent of time. Without the addition of polymer (c), the pH changes to 6.03 after only 5 minutes of H_2 gas flow which shows the rapid initial appearance of the H^+ signifying a corresponding initial rapid formation of Pt atoms leading to the nucleation process. This figure also shows the instability of the shapes of these nanoparticles when it is uncapped as the percentage of the unidentified shapes (⊕) increases with time. (The unidentified nanoparticles refer to those that are oriented irregularly on the carbon film support so that their shapes cannot be directly identified in the TEM images.) Taken from Petroski, et al, *J. Phys. Chem. B*, 1998, 102, 3316.

Along with monitoring the pH and the size and shape changes during the growth period, the absorption spectra were also taken at these same time intervals. The optical properties of platinum metals have not undergone the same sort of intense study as the free electron metals such as gold due to the lack of an absorption band in the visible region.

Colloidal dispersions of nanometals exhibit absorption bands or broad regions of absorption in the ultraviolet-visible range due to the excitation of plasma resonances or interband transitions. Certain metals such as gold, silver, or copper have distinct absorption bands in the visible region due to the surface plasma resonances leading to brightly colored solutions. Other metals such as the platinum metals exhibit only broad absorption continua which extend throughout the visible-near ultraviolet range, causing these colloidal solutions to be brown to black.

Mie theory can be used to calculate the absorption spectra of fairly dilute dispersions of spherical particles of colloidal dimensions from the wavelength dependence of the optical constants (the refractive index n or the optical-frequency relative permittivity ε) of the particles relative to the surrounding medium [41]. In a study by Creighton and Eadon [45], Mie calculations were performed for various elements, including the platinum metals. Their calculations predicted a plasmon band to be in the ultraviolet region, specifically, at 215 nm for spherical particles of 10 nm in diameter. These authors also considered the effect of shape of the particles by using differing aspect ratios of prolate spheroid. The effect of the departure from spherical shape is to split the dipole resonance into two absorption bands, in which the induced dipole oscillates respectively along and transverse to the spheroidal axis. The second absorption band is predicted to occur at longer wavelengths in the visible region. Further, the Pt band decreases in intensity as the size of the particle (or increasing aspect ratio) increases. Experimental results of the colloids can vary due to broadening of the spectra because of polydispersity, partial aggregation, or departures from spherical particle shape. The predicted plasmon band for platinum nanoparticles has been observed experimentally in both aqueous [112] and organic media [116].

In the experimental absorption spectra taken during the 12 hour growth period (see Fig. 7-11) of uncapped colloidal particles, a maximum peak is observed at ~215 nm. There are two interesting characteristics of the plasmon band. The first is an addi-

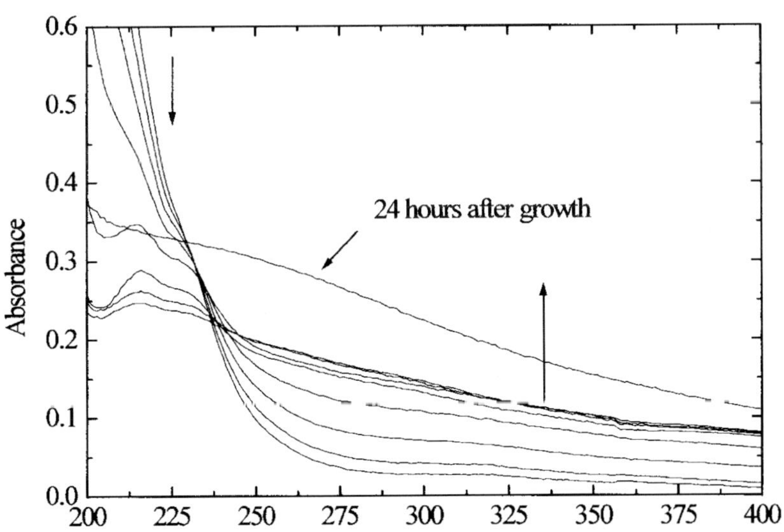

Figure 7-11. Absorption spectra for the uncapped platinum nanoparticles during the 12 hour growth period. A maximum peak is observed at 215 nm and a second peak is appears at 228 nm at approximately 4 hours into the growth process before disappearing after growth is complete.

tional peak at ~228 nm which has not been previously reported. The second is the observation that the 215 nm peak reaches a maximum at approximately four hours into the growth process. After this time, the bands start to decrease in intensity until it disappears altogether after about 24 hours. This peak also starts to shift slightly to longer wavelength (~ 217 nm) after the maximum intensity has been reached. The observed red-shift of the 215 nm band may be explained by the increasing size of the particle since the shift occurs after the maximum intensity has been achieved. It is known [8] that the maximum wavelength blue-shifts with decreasing particle diameter. The ratio of the two bands does not change during the growth period. The new peak at 228 nm can have several explanations. Obviously, simple Mie theory for spherical particles cannot be used in its simplified form to explain shaped particles. The Mie theory may be over-simplified in using only the dipole term in the Mie series and not correcting for the quadrupole and higher-order terms in the Mie summation, which may be significant in the case of changing shape and size.

The effect of the initial platinum salt concentration on the plasmon band was observed at three different starting concentrations and the results of the maximum band at 4 hours into the growth process is shown in Fig. 7-12. Three initial concentrations of Pt salt are used: 4×10^{-5}, 8×10^{-5}, and 16×10^{-5} M. The spectra have been normalized for clarity. The first observation which can be made from this comparison is that the 215 nm peak red shifts to 217 nm with increasing concentration, which is the same shift observed in the later stages of growth (see above).

Studying the absorption spectra for the growth of the platinum nanoparticles in the presence of the polyacrylate capping material does lead to one complication in that the Pt salt and the polyacrylate both absorb in the ultraviolet region. In the uncapped case stated above, the Pt salt concentration decreases during growth as it is reduced which causes this peak to diminish after two or three hours allowing the plasmon band(s) to be clearly observed. In the case of the capped particles, the polyacrylate

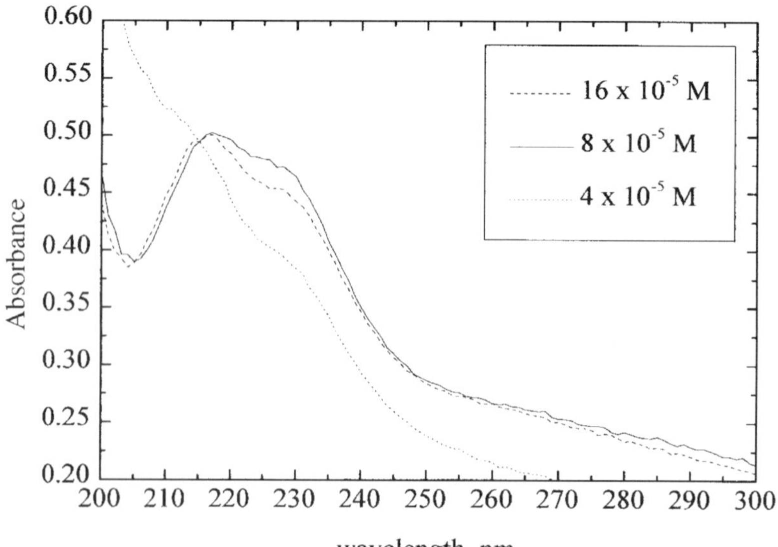

Figure 7-12. Comparison of the absorption spectra for uncapped platinum nanoparticles at varying initial concentration of platinum salt showing the change in the 230 nm peak. a) 4×10^{-5} M, b) 8×10^{-5} M (same as Fig. 7-11), and c) 16×10^{-5} M. The spectra have been normalized for clarity.

concentration remains relatively stable over time, so the growing plasmon band region is always obscured. Therefore, the plasmon band was revealed by subtracting out the polyacrylate absorption from the spectra during the growth process.

Figure 7-13 shows the UV-VIS absorption spectra for a 1:1 ratio platinum:polyacrylate sample, but a spectrum of the polyacrylate was taken before adding the Pt salt solution and that spectrum was subtracted from the growth spectra. Many of the spectral features in Fig. 7-13 are similar to the uncapped cases in Fig. 7-11 and 7-12. Again, the maximum of the band at ~215 nm appears at approximately four hours into growth and then begins to disappear towards the end of growth. The second peak at ~228 nm is not as prominent in the 1:1 ratio spectra as in the uncapped case. This could be a consequence of the polyacrylate subtraction or it could be attributed to surface enhancement due to the bonding of the polyacrylate with the surface of the platinum nanoparticles.

Figure 7-14 shows the UV-VIS absorption spectra for the growth process for the 1:5 ratio platinum sample where a spectrum of the polyacrylate was taken before adding the Pt salt solution and using that spectrum for the subtraction. Comparisons can be made to the other spectra, though this case did not lead to a very good subtraction due to the high concentration of the capping material present. The maximum of the band at ~215 nm again appears at approximately 4 hours into growth and then begins to disappear towards the end of growth. The second peak at 228 nm is barely noticeable, resembling more of a shoulder than a peak. This could be a consequence of the polyacrylate subtraction or it could be attributed to surface enhancement due to the bonding of the polyacrylate with the surface of the platinum nanoparticles.

Surface enhancement of nanoparticles has a definite effect on the surface plasmon band. The dielectric constant of the polymer is much smaller than that of the water and therefore a decrease in the dielectric difference on the surface of the particles

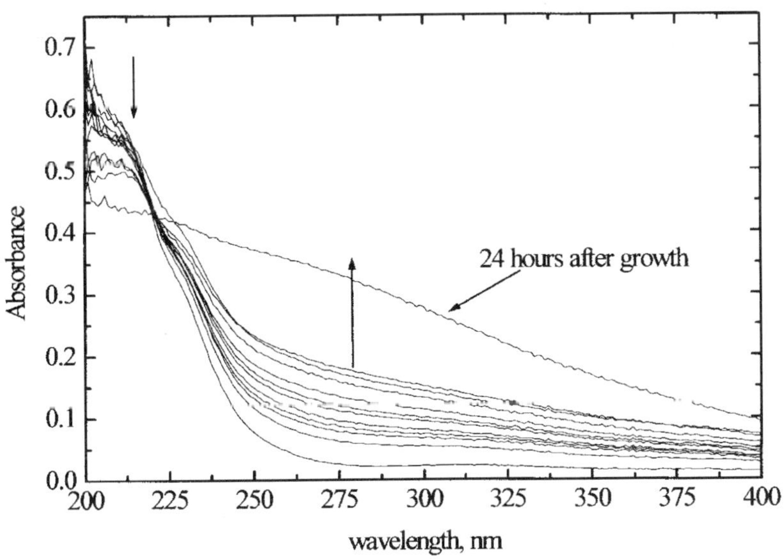

Figure 7-13. Absorption spectra of a 1:1 ratio platinum solution with subtraction of a polyacrylate spectrum taken before starting the growth process. A maximum peak is observed at 215 nm and a second peak is appears at 228 nm at approximately 4 hours into the growth process before disappearing after growth is complete.

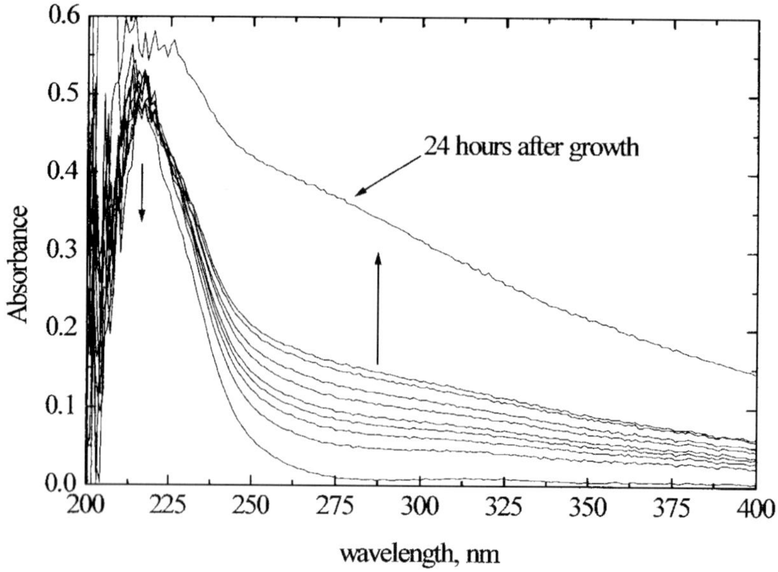

Figure 7-14. Absorption spectra of a 1:5 ratio platinum solution with subtraction of a polyacrylate spectrum taken before starting the growth process. A maximum peak is observed at 215 nm and a second peak appears at 228 nm at approximately 4 hours into the growth process before disappearing after growth is complete.

may cause the absorption intensity to decrease. This is observed by the effect of the surface capping material on these bands since the increasing concentration of the capping material decreases the intensity. Since the size and shape of the particles change during the growth process, but the ratio of the bands does not, the answer would seem to be linked to the capping material, which is relatively constant in the solution.

Figure 7-15 is a comparison of the maximum absorption spectra from Fig. 7-12, 7-13, and 7-14. The spectra have not been normalized. In this figure, a slight shift of the 215 nm band to 217 nm is evident in going from capped to uncapped particles. Also observed is the successive decrease in the intensity of the second band at 228 nm with increasing amounts of capping material. It should be noted that the overall absorbance attributed to the platinum plasmon band remains relatively constant in these three cases for the maximum intensity band.

The disappearance of the plasmon band after growth of the nanoparticles is completed may be attributed to the aggregation occurring in solution, which is observed by the chain-like formations in the TEM images beginning at the fourth hour of growth. It should be noted that this is also around the point in which the pH has stopped decreasing and remains relatively constant in the case of the uncapped particles (see Fig. 7-10). Though the individual sizes of the nanoparticles are still within the Mie theory range, together they may exceed this size used in the calculation which may cause the decrease and eventual disappearance. The chains of particles formed may be an extreme large prolate spheroid which was predicted [45] to greatly decrease the intensity of the absorption spectrum. As was stated earlier, there are two factors that can determine the plasmon band and its position: the surface plasmon and the interband transitions. As opposed to the free-electron metals like gold which primarily owe their visible range peak to the surface plasmon and their ultraviolet range

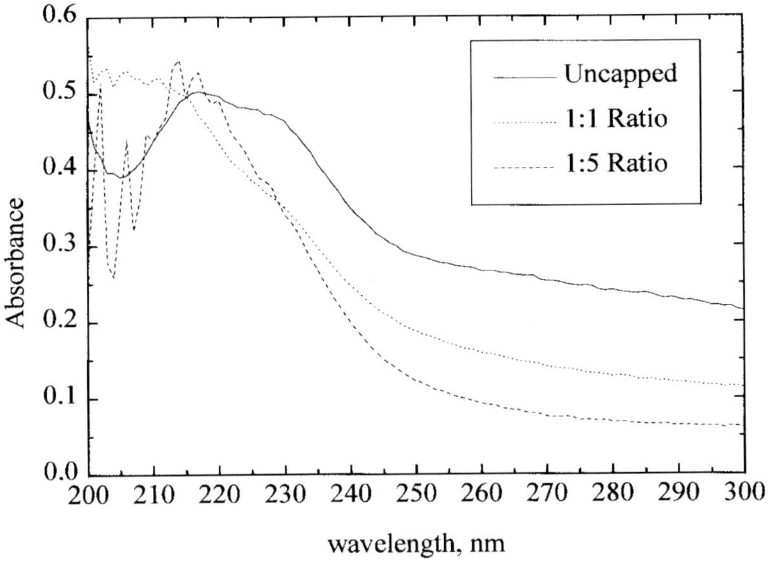

Figure 7-15. Comparison of the maximum peaks for uncapped, 1:1 ratio and 1:5 ratio. The shift in the 215 nm peak as well as the change in intensity of the 230 nm peak and of the spectra overall is evident. The spectra have not been normalized.

peak to the interband transitions, the absorbance of these less free-electron metals are a mixture of these two. It is possible that the peak at 228 nm has more of a surface plasmon character and the predicted and observed plasmon band at 215 nm has more interband transition character. This is in accordance with the fact that this band does not significantly change with time or varying concentrations. However, the band at 228 nm does seem to depend greatly on the way the solution is made.

Calculations are underway to understand the transient nature of the plasmon band for these platinum nanoparticles. Simple Mie theory is not enough to explain the phenomena outlined in this work. A Mie theory model corrected for both shape-dependence and the changing concentration of different shapes as a function of time may provide a more accurate model of the observed optical spectra.

First, it is necessary to address the shape-dependence of the absorption spectrum since it is possible that the different crystal faces present in the samples may exhibit different optical characteristics. Fuchs presented an expansion of spherical Mie theory to describe ionic cubes, which may be appropriate for describing the absorption of cubic platinum nanocrystals [117]. Therefore, optimizing Fuchs' expression as a function of crystal shape, in particular as a function of the changing concentration of tetrahedral shapes, may explain the 228 nm peak.

A second means of expanding the model is to address higher order multipole interactions within the crystals. The simplified model only addresses dipole interactions. Particles of lower symmetry (relative to spheres) might be expected to have more significant contributions from higher order multipole interactions [118]. For example, Fuchs presented 6 different multipole resonances contributing to the absorption spectra of ionic cubic crystals [117]. Hummel *et al* reported a theoretical treatment of Mie spectra for spherical aluminum particles of varying size, and included higher order multipole considerations [118]. Since there seems to be some relation between particle shape and higher order multipole contributions, a better model may need to include both considerations.

7.4 Semiconductor nanostructures

Quantum confinement of excitons in semiconductors occurs as the particle size becomes smaller than the exciton Bohr radius. For such small sizes, the surface effects and the interaction with the surrounding medium become important. These interesting size effects occur on the nanometer scale, allowing tunable optical properties of the nanostructures. Devices and applications make use of such properties such as enhanced and fast optical nonlinearity [119], high luminescence efficiency [119] and single electron transfer [120]. The difficulty with the realization of many such applications has involved the stability of the delocalized state. The increased surface effects for small sizes contribute to greater sensitivity to surface defects such as vacancies and dangling bonds. Such surface defects allow the relaxation via exciton localization or trapping. The possibility for both delocalized and localized states has caused confusion over excitonic effects and surface trap effects [121]. For instance, photoluminescence may originate from both exciton and trap states. The following chapter is aimed to describe the optical properties and dynamics in some semiconductor nanostructures. The size dependent absorption and emission properties of colloidal II-VI semiconductors were extensively studied. Aspects of exciton dynamics as a function of the surface properties will be discussed and experiments, providing optical information about semiconductor nanostructures will be summarized.

7.4.1 CdS quantum dots and interfacial charge transfer dynamics

Semiconductor quantum dot particles consist of a stabilized core with hundreds to thousands of atoms arranged in a crystalline structure similar to their bulk material. Particle stabilization is achieved by static repulsions [122], ionic or covalent capping agents [25], micelles, or zeolite cages [123, 124]. Quantum dots may exist as films, powders, or in solutions and may provide materials with absorption in the IR, visible, or UV and can be controlled by changing the particle size, not its chemical composition or structure. It is these tunable properties, intermediate of the bulk materials and individual molecules, which are responsible for the extensive research in quantum dot systems.

The surfaces of quantum dots play a significant role with respect to physical properties due to their diameters falling within the nanometer size range. In such a size regime, the atoms located on the surface may constitute up to 40% of the total number of atoms comprising the particle. Therefore, the dynamics and the optical properties of quantum dots are very sensitive to surface derivations which may also enhance their utility in certain applications.

In potential applications for semiconductor quantum dots, such as in microelectronics, solar cells, or as photocatalysts [125], a critical feature of generation and/or separation of charges is required. The generation of charge carriers may be induced upon photoexcitation whereby an electron in a high excited state is formed along with a respective hole. Their separation, including transfer to acceptors (or donors), is achieved by the competition with the charge carrier recombination through trapped states. Thus the understanding of these processes and their characterization is critical with respect to their potential applications.

The methods typically used in semiconductor quantum dot synthesis result in structures with defects in the core and on the surface of the particles. Although thermodynamically controlled techniques and size selective precipitation methods have dramatically improved the quality of the particle crystallinity and integrity of the surface [126, 127], defects remain a common characteristic. These imperfections are responsible for energetically trapped states within the band gap transition. Depending upon the kinetics of the charge separation through electron donors (or acceptors), these states may compete with electron transfer processes. The use of optical spectroscopy on a system of quantum dots with a relatively large percentage of defects in the presence of adsorbed electron acceptors may characterize electron and hole dynamic processes and help understand the potentials or limitations in future QD applications.

CdS QDs made according to the procedure outlined by Henglein et al. [25] have an advantage for this system due to the presence of core and surface defects which provide trapped states. In addition, methyl viologen (MV^{2+}), which is known to act as a good electron acceptor, may be added to remove an electron from the system thereby isolating the trapped hole. By investigating the dynamics of a QD CdS versus a QD CdS–MV^{2+}, the path of electron or hole trapping processes may be determined by effectively removing the electron from the QD. The rates for charge generation and separation, electron transfer, electron trapping, and hole trapping may thus be monitored through pump probe transient absorption experiments which follow the generation and recovery of the lowest energy excitonic transition using the band gap absorption.

Femtosecond transient absorption spectroscopy may determine the rates of carrier trapping and electron transfer processes in a CdS QD solution in the presence and absence of (MV^{2+}) electron acceptors (Fig. 7-16) [32]. In order to determine the electron and hole trapping rates, 100 fs laser pulses at 400 nm were used to pump a CdS solution and promote an electron from the valence band (VB) to the conduction band (CB) generating an electron-hole pair. Within 300 fs after photo-excitation, the carriers were found to occupy the lowest energy excitonic transition (band edge), which resulted in a bleach (optical hole) at 480 nm. In order for this bleach to recover, *both* the electron and hole must be removed from the conduction and valence bands or LUMO and HOMO, respectively.

The presence of defects at the nanocrystal surface and internal lattice structure cause trapped states which are at energies within the band gap. The electron and hole pair may become localized into these low energy states effectively recovering the band edge transition. The formation of the transient absorption bleach observed at 480 nm thus measures the generation of electron and hole carrier occupation of the band edge state, and the recovery rate of this bleach measures the rate of the disappearance of the *slowest* trapping process. In CdS particles (without MV^{2+}), the recovery of the bleach occurs in 30 ps. When MV^{2+} molecules were added to the solution, the recovery of the bleach was accelerated to 7.5 ps, and in addition a rise in absorption centered at 650 nm was observed for the appearance of $MV^{+\bullet}$ radical cation.

MV^{2+} is known to be an efficient emission quencher for CdS nanocrystals [27, 128, 129]. Upon electron transfer from CdS to MV^{2+}, the radical cation $MV^{+\bullet}$ is formed which exhibits a broad absorption centered near 650 nm. Within 300 fs after photo-excitation of CdS-MV^{2+}, a broad rise in absorption was observed and remained for a

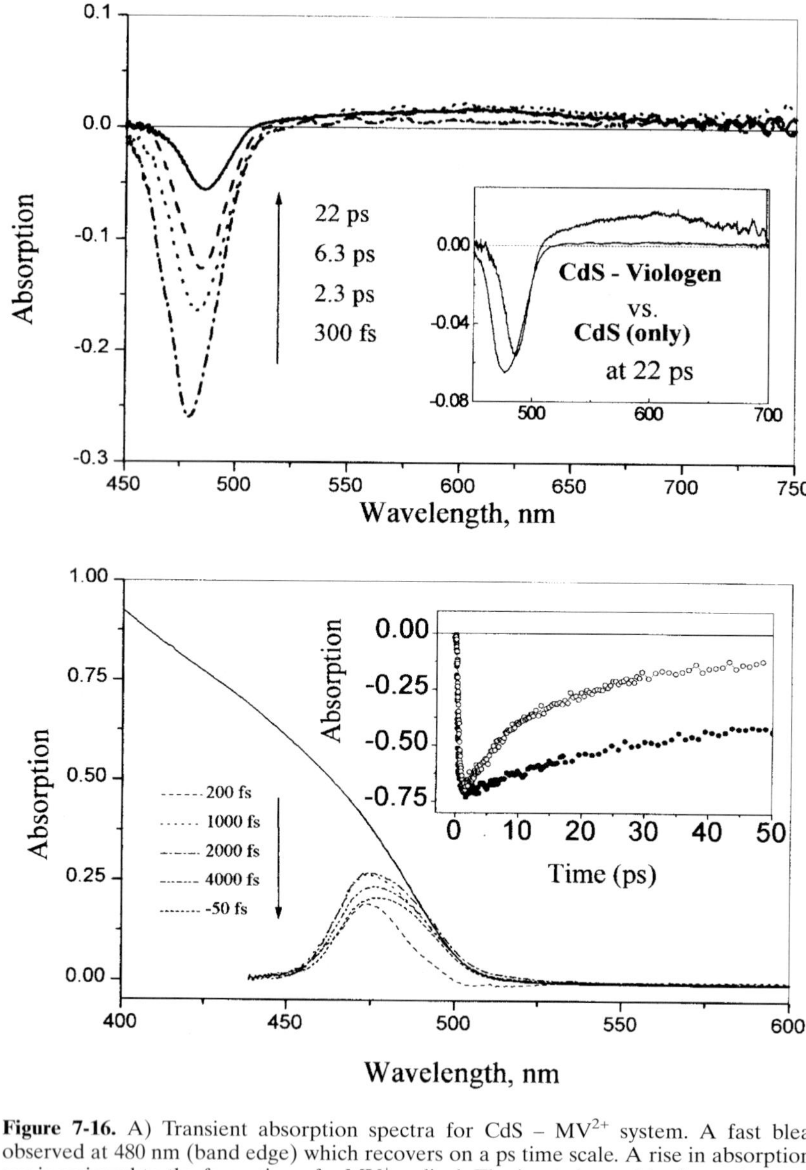

Figure 7-16. A) Transient absorption spectra for CdS – MV^{2+} system. A fast bleach formation is observed at 480 nm (band edge) which recovers on a ps time scale. A rise in absorption centered at 650 nm is assigned to the formation of a MV^{+} radical. The inset shows this absorption at 22 ps overlapped with the transient spectrum for CdS QD at 22 ps magnified 10X to illustrate that the absorption feature is not from a solvated electron. B) Steady state absorption (solid line) overlapped with transient absorption spectra (inverted for a CdS QD solution) illustrating that the bleach formation is from the band edge transition. The formation of the bleach occurs within the pump pulse (short dash at –50 fs corresponds to center of pump pulse) indicating fast formation of band edge transition followed by ps decay. The inset shows the kinetics for the 480 nm bleach formation and decay for CdS QD solutions and CdS-MV^{2+} solutions.

time period longer than our experimental window (100 ps). These results are consistent with a rapid electron transfer from CdS to MV^{2+}, where the electron remains with the quencher for a relatively long time period (nano-milliseconds).

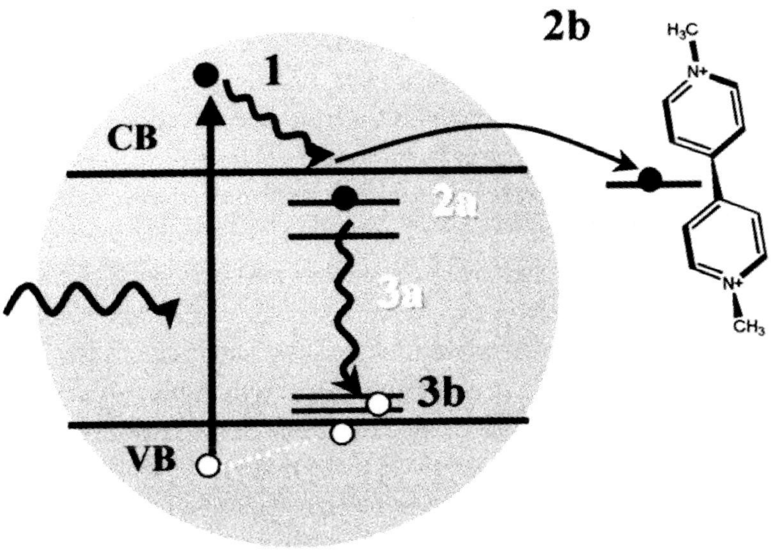

Figure 7-17. Schematic illustration of the electron and hole trapping processes (path a) and electron transfer to MV^{2+} (path b) when added to solution. 1) Upon photoexcitation with 400 nm pump pulses, an electron is excited to a high electronic state followed by rapid relaxation to the band edge transition. This is observed as a bleach formation at 480 nm within 200 fs. 2a) In the absence of MV^{2+}, the electron and hole are trapped. The electron trapping process is slowest and thus the rate determining step in the bleach recovery (40 ps). 3a) The electron ultimately recombines with the hole. 2b) Upon the addition of MV^{2+}, electron transfer occurs within 300 fs where the electron remains with the MV^{+} radical effectively removing the carrier from the CdS QD. 3b) The recovery of the bleach is thus determined by the hole trapping dynamics which occur in less than 10 ps.

The formation of the bleach at 480 nm and the absorption at 650 nm along with the decay of both features provide a model for the carrier trapping vs. electron transfer in the CdS-MV^{2+} system (Fig. 7-17). The electron trapping in the bare CdS nanocrystals occurs in 30 ps. The addition of MV^{2+} to CdS causes a rapid (300 fs) electron transfer where the electron remains with the MV^{2+} molecule and is effectively removed from the nanocrystal. The difference observed in the trapping processes for the electron and the hole may be explained by a higher density of states for the hole due to a higher effective mass. A second system using CdSe with napthoquinone (NQ) quenchers was studied to compare its dynamics with CdS-MV^{2+}.

7.4.2 Colloidal CdSe quantum dots

7.4.2.1 Colloidal CdSe quantum dots and interfacial electron transfer observed by optical spectroscopy

Colloidal CdSe quantum dots (QDs) in the size range of 1 to 100 nm diameter have been actively studied in recent years to understand the dependence of their electronic properties on size [1–3, 7, 8]. Such QDs are large enough to build up the bulk crystal structure but, on the other hand, they are too small to form continuous Bloch bands of electronic states. When the QD diameter is comparable to or smaller than the diameter of the bulk exciton (5.9 nm for CdSe) large changes in the electronic structure occur [3]. The three-dimensional confinement splits the continuous band into a series of discrete quantum states. The dependence of the electronic structure on the size of colloidal CdSe QDs was intensively studied by Bawendi et al. [19–23, 130] As a first consequence, the lowest optical absorptions are shifted to higher energies. Secondly, the excited electron dynamics of such QDs can change significantly [19, 22, 131, 132]. For such QDs electron trapping by surface traps becomes very important in determining the electron-hole dynamics and recombination, and thus the emission properties of these particles. For this reason, the controlled preparation and surface modification of semiconductor QD systems is a field of considerable interest.

The preparation method, developed by Murray et al. in 1993 [126], gives the most homogeneous CdSe QD sample. In our case we prepared 4 nm diameter QD with a standard deviation of 10% in size. The average shape is very close to spherical, although shapes with prolate deviations are observed. Figure 7-18 shows a typical transmission electron microscope picture of the sample on a carbon coated copper grid.

The absorption (dashed) and photoluminescence (solid) spectra in Fig. 7-19 confirm that a sample with relatively narrow size distribution was obtained. The absorption shows a relatively sharp onset at 580 nm. The photoluminescence has a narrow peak at 570 nm and a broad shoulder between 620 and 780 nm. The narrow emission band at 570 nm originates from charge carrier recombination from shallow trap states, often referred to as near band gap emission. The broad emission shoulder at the longer wavelength represents the radiative recombination of deep trapped charge carriers.

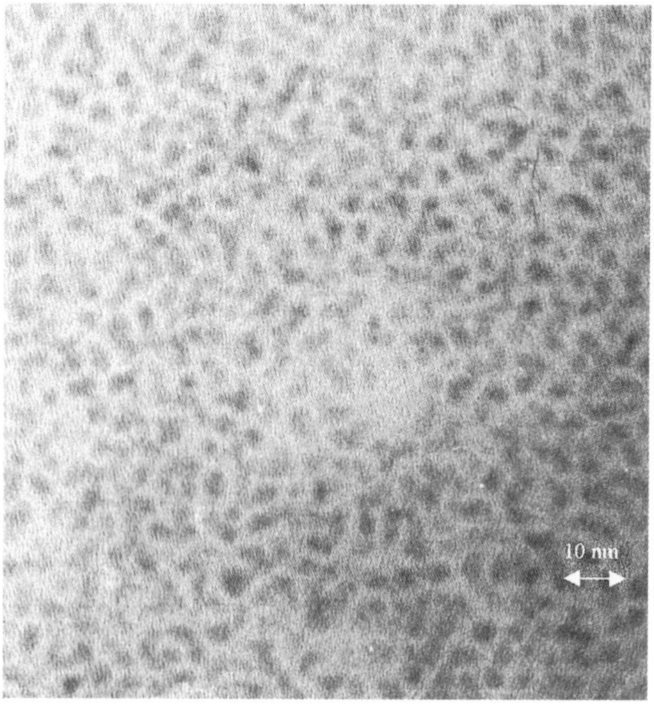

Figure 7-18. TEM image of the CdSe quantum dots (QDs) sample with an average diameter of 4.0 ± 0.3 nm. Slight prolate deviations are visible for individual QDs.

In the following we present some spectroscopic studies on the CdSe QDs with the electron acceptor naphthoquinone (NQ) and the electron donor thiophenol (TP) on the surface. The spectral dynamics of CdSe-NQ and CdSe-TP is compared.

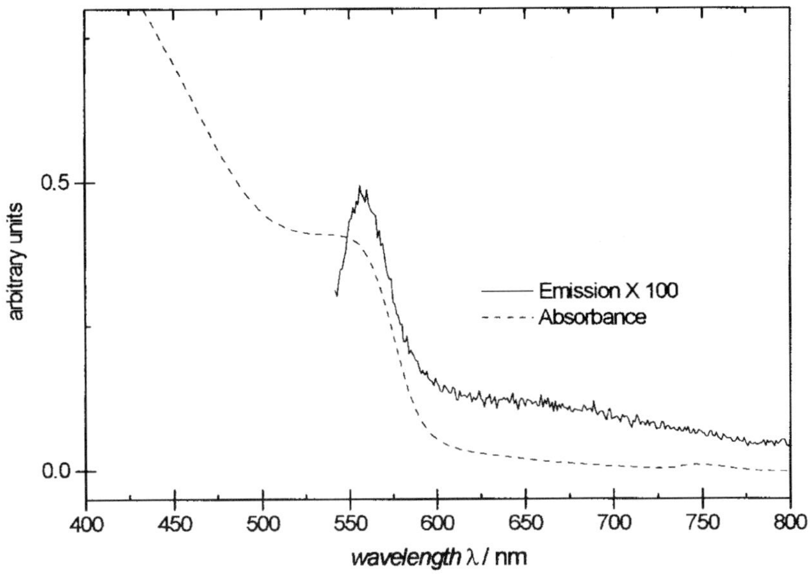

Figure 7-19. Absorption (dashed) and emission spectrum (solid) of the sample shown in the TEM above. Toluene was used as solvent and the temperature was 298 K.

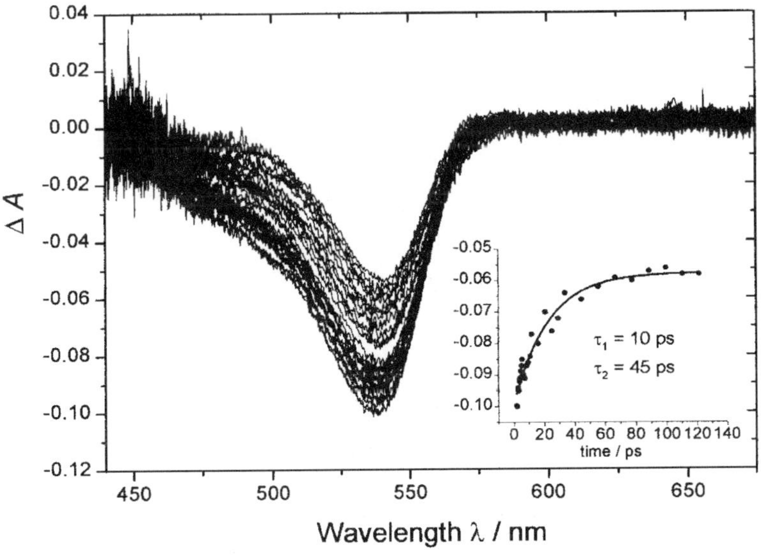

Figure 7-20. The time dependence of the bleach spectra of the CdSe QD in colloidal solution with thiophenol adsorbed on its surface. The inset shows the decay of the observed bleach at its maximum 550 nm (taken from ref. [34]).

By addition of the thiophenol (10 μl per ml QD solution), the steady-state emission of CdSe QD was completely quenched [34], since the electron donor thiophenol (TP) led to neutralization of the hole in the valence band of the excited QD [34]. With femtosecond transient spectroscopy, we monitored the bleach recovery of the CdSe QDs in the presence of TP. In Fig. 7-20, the transient pump-probe spectra of the CdSe QD-TP system are shown. The bleach recovery kinetics were not accelerated. The measured bleach recovery time, observed at 550 nm, became slightly longer than in the unperturbed QD (τ_1 = 10 ps, τ_2 > 45 ps versus τ_1 = 2.5 ps, τ_2 > 40 ps). The results suggest that in CdSe QDs, as in CdS QDs, electron trapping is the rate determining process (τ_{trap} = 40 ps) of the bleach recovery.

The addition of 1,2-naphthoquinone (NQ) to the CdSe QD suspension led to efficient quenching of the steady-state near band gap emission. With femtosecond transient spectroscopy, we monitored the bleach recovery of the CdSe QDs in the presence of NQ. The resulting transient femtosecond spectra of CdSe QDs in the presence of NQ (Fig. 7-21) showed the formation of an absorption between 600 and 680 nm. This was assigned to the previously reported absorption of the radical anion of NQ. We determined that the rate of formation for this radical anion absorption (200 fs, observed at 650 nm) had the same rate constant as the formation of the bleach (200 fs, observed at 550 nm). In addition, the decay times for both the absorption of the NQ anion and the bleach recovery were the same (2.8 ps). The bleach recovery time was reduced from the approximately 100 ns in the bare CdSe QD (without quinones) to less than 3 ps in the presence of NQ. This was attributed to the electron shuttling effect of the surface quinones, which first accept the electron and subsequently shuttle it back to the hole in the QD valence band. The simultaneous observation of the QD bleach and the NQ^- anion absorption is presented in Fig. 7-21.

It is interesting to compare how the surface molecules can effect the relaxation dynamics. In Fig. 7-22 the temporal evolution of the bleach maximum is presented for CdSe/NQ (solid squares), CdSe/TP (diamonds) and pure CdSe QDs (open circles). The electron acceptor NQ increases the bleach revovery rate and TP decreases it compared to the pure CdSe.

The large exchange interaction in CdSe causes a more complicated valence band structure and, as a result, the possibility of "dark excitons" (since the transition from the ground state is spin-forbidden). Bawendi and co-workers [21, 22] have provided evidence for the dark nature of the energetically lowest exciton state of CdSe QDs. These spin-forbidden states in 4 nm CdSe might decelerate the relaxation to deeply localized traps.

As mentioned above, the observed emission quenching of CdSe QDs by TP suggests that the TP neutralizes the holes in the valence band of the photoexcited CdSe, thus inhibiting electron-hole recombination. This charge transfer across the CdSe interface and subsequent neutralization of the valence band hole is also responsible for the slowing of the bleach recovery relative to the bare QD. The TP preferably binds to electron trap sites and thus eliminates at least partly the fast trapping processes for the electron. As the electron relaxation dynamics is rate limiting for the bleach recovery, it is clear that TP slows down the bleach recovery dynamics. The Stokes-shifted photoluminescence from this 4 nm CdSe QD sample supports the suggestion of a relaxation via surface trap states. The multi-exponential bleach decay traces of the pure QD indicate competitive kinetics between internal state relaxation and surface trapping.

NQ is a classical electron acceptor. By-passing of the internal dark states of CdSe accelerates the electron relaxation by shuttling the electron from the conduction band across the interface (< 200 fs) to NQ and from there back into the valence band (< 3 ps). The electron shuttling effect of the NQ is revealed by the faster bleach recovery dynamics in comparison to the CdSe and CdSe-TP systems. Since shuttling of the electron accelerates the bleach recovery, it can be concluded that the electron dynamics is rate limiting in the relaxation processes of CdSe, similar as in CdS [32]. An interesting aspect is that the charge separation across the CdSe interface by NQ reduces the overlap of the charge carrier wavefunctions, allowing in this case faster relaxation since the

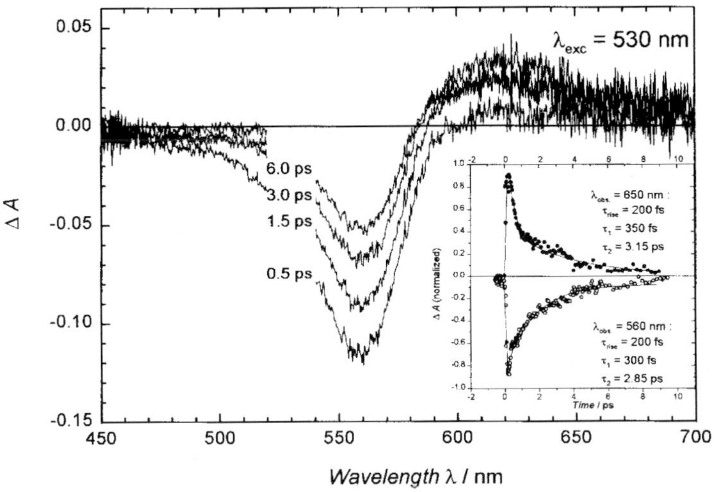

Figure 7-21. The simultaneous observation of the CdSe bleach (560 nm) and the $NQ^{\cdot-}$ radical anion absorption (~630 nm). The similar decay times give evidence for the electron shuttling effect.

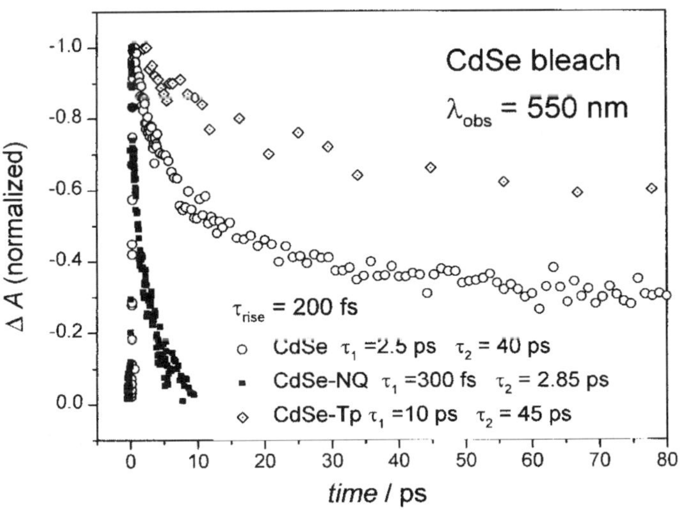

Figure 7-22. The effect of the adsorption of the electron acceptor naphthoquinone (dark squares) and the electron donor thiophenol (diamonds) on the transient bleach of the CdSe QD band gap absorption (circles) at 550 nm [33].

driving force favors the charge recombination and exchange interaction is minimized. On the other hand, the neutralization of the hole by the TP on CdSe QDs introduces an unfavorable localization of the hole wavefunction, which slows the relaxation dynamics even more. The TP eliminates electron traps on the CdSe QD surface.

In summary, the electron transfer *via* the CdSe QD interface is demonstrated in both directions, from and to the surface molecule. The comparison of the effects of the addition of the electron donor (TP) with the effect of the addition of the electron acceptor (NQ) to the QD on the electron-hole dynamics leads to the following conclusion: The added organic components on QD surfaces can act either as an efficient electron shuttle and accelerate the charge carrier relaxation (e.g. NQ on CdSe), as an electron robber and change the relaxation pathway of the photoexcited electron (MV^{2+} on CdS), or as an electron donor and hole trap to slow down the relaxation process (TP on CdSe). Figure 7-23 summarizes the proposed electron transfer mechanisms for the different composite systems.

In (a), the relaxation of the excited electron (step 1) and its combination with the hole (step 2) in the valence band (which leads to bleach recovery) occurs *via* the NQ and by-passes surface trapping [9, 10] and/or changes in the state multiplicity [11, 12]. As a result an acceleration of the bleach recovery is observed.

The electron-hole dynamics in the CdS-MV^{2+} system [2] is summarized in (b). In this case, the electron is rapidly transferred from the conduction band of the QD to the electron acceptor (MV^{2+}) (step 1). In an aerated solution at room temperature, the MV^{+} does not shuttle the electron back to the hole in the valence band. The hole is thus trapped in 7.5 ps by the surface states (step 2). This led to the conclusion [2] that the observed bleach recovery time of 30 ps in the bare QD must be rate limiting by the electron trapping and not by the hole trapping.

In (c), the hole in the valence band of the excited CdSe particles is first neutralized by the electron transfer from the electron donor (step 1). The removal of the excited electron (and thus the bleach recovery) takes place by surface trapping (step 2). The fact that the bleach recovery is not greatly affected by the addition of TP suggests that surface trapping in the CdSe QD is faster than the back electron transfer from the QD conduction band to the TP. It is then concluded, that the bleach recovery in the CdSe QD, like that in the CdS QD, is rate limiting by the electron and not by the hole trapping.

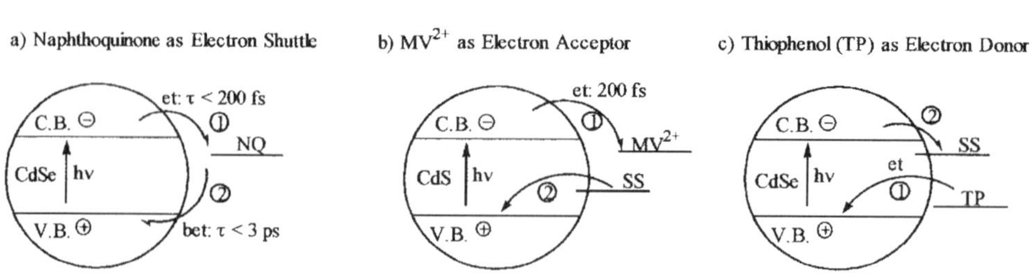

Figure 7-23. The electron-hole dynamics in CdS (b) and CdSe NP in presence of electron acceptor (a) or electron donor (c). For (b) and (c) it is proposed that the electron-hole dynamics is determined by trapping into surface states (SS) [33].

7.4.2.2 High pump power transient absorption spectroscopy on CdSe QDs: the effect of multiple excitons in nanoparticles

In the pump-probe experiments discussed so far, we observe the two main exciton bleaching (negative absorption) bands since low laser pump power was applied. As the pump laser intensity is increased, we also observe in addition a new high pump power induced transient absorption on the low energy side of the two exciton bleach bands in our femtosecond pump-probe experiment [133]. Figure 7-24 presents the transient absorption spectra at different laser pump powers. The power of the pump pulse was adjusted to 2 (top), 6 (middle), and 16 μJ (bottom) while the beam characteristics and the solution remained unchanged from those used in the 2 μJ experiment. The transient spectra at higher pump powers still show the bleaching of the transition at 560 nm. In addition, at higher pump powers superimposed transient absorptions are observed on both sides of the bleach.

This new absorption can be discussed in terms of three possible causes: The absorption from a metastable state, the effect of the electric field of the many electron and hole pairs formed in each particle at high pump intensities, or the formation of bound electron-hole dimers (biexcitons) with greatly increased binding energy as compared to that found in the bulk.

1. The metastable state could be the dark state proposed by Bawendi et al. [26, 27] or a surface trap state populated via the exciton level. In this case, the red-shifted absorption would be expected to have a rise time, which is equal to the observed decay time of the bleaching spectrum. This is not observed. The transient appears on an ultrashort time range (< 200 fs) while the bleaching decays on the picosecond time scale.
2. At the excitation levels we are using in the high power experiments, more than 50 electron-hole pairs can be formed in the QD. At this high density of electrons and holes, it is possible that an induced Stark shift or broadening of the absorption could take place. This seems to be a good possibility, especially in the light of the recent Stark field studies published by Colvin and Alivisatos [134] in which a modulated external field of up to 64 kV/cm was used. From these studies, the dipole moment of the lowest energy exciton is found to be 32 D. A multiple excitation of a QD could lead to an even higher internal electric field leading to large internal Stark fields. The transient absorption maxima should be more red shifted when laser pulses of 16 μJ are used than when only 6 μJ pulses are applied. This is not observed. Furthermore, the transient absorption maximum should shift with increasing delay time between the pump pulse and the monitoring light pulse (i.e., as the number of excitons decays with time).

   It could be that the shift results from the interaction between the free exciton dipole [134] (32 D) and an internal field of a net surface dipole resulting from rapidly trapped surface excitons. Trapping might not completely randomize the initial non-isotropic excitation process. Since the particle is almost spherical, a good number of the dipoles would cancel one another due to the fact that while the initial dipole axes are parallel to the laser pump field, the signs of the induced dipoles could be positive or negative.

   In this model we might use the experimentally observed shift to calculate the size of the net surface dipole as follows. We use the electrostatic dipole approximation and the observed 120 meV shift ΔE to write down:

$$\Delta E = \frac{2\mu_s \mu_{Exc.}}{4\pi\varepsilon r^3} \qquad (7\text{-}6)$$

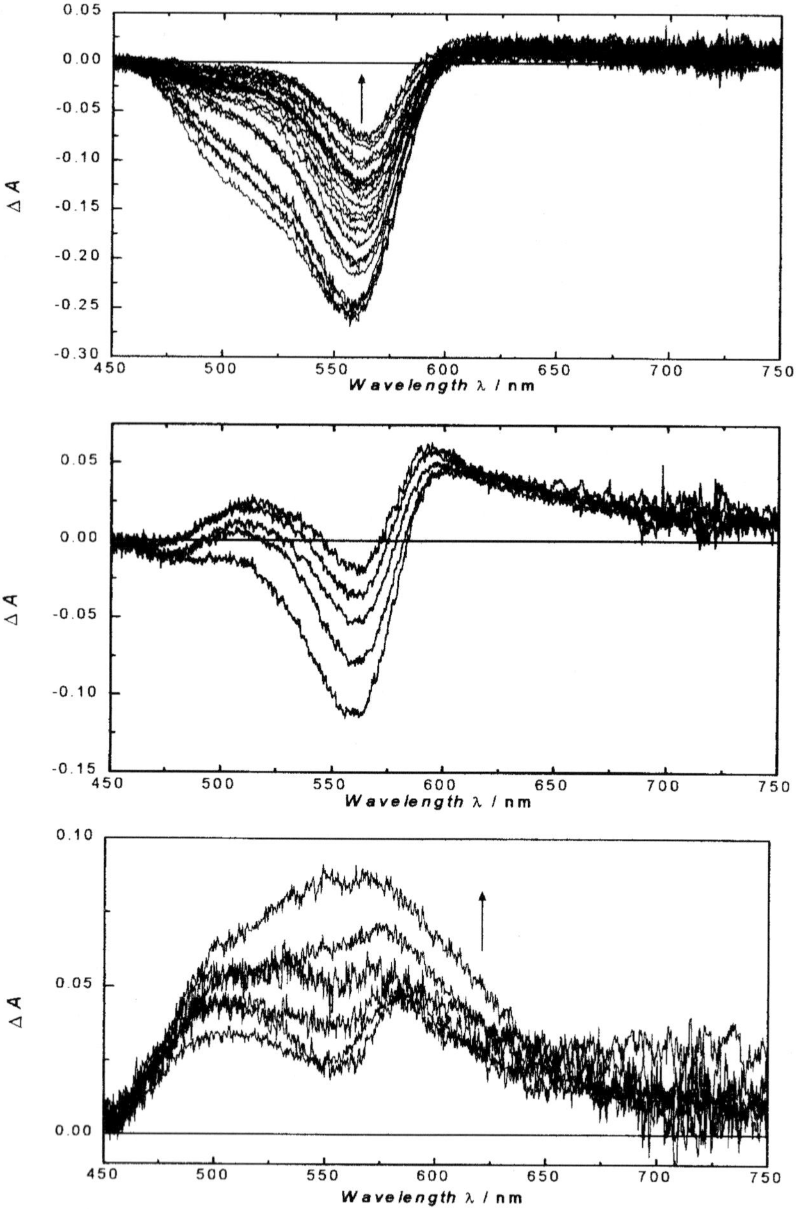

Figure 7-24. Transient absorption spectra of CdSe NPs with an average diameter of 4 nm, pumped with 400 nm fs-laser pulses with a laser power of 2 (top), 6 (middle), and 16 μJ (bottom) per pulse. **Top:** The delay times of the spectra are 200 fs up to 110 ps. The measured absorption changes can be explained by the state filling effect and surface trapping of the exciton. **Middle:** The positive induced absorption feature at longer wavelengths is attributed to the influence of a high density of excited charge carriers to the probe transition. It leads to new high pump power induced absorptions. **Bottom:** The bleaching is at this laser power covered by a broad absorption and is only visible by the growth of the broad absorption at 560 nm, where initially an absorption minimum was observed. The change in the spectral shape of the CdSe NPs from a bleaching (top) to an absorption (bottom) is a laser power dependent effect and a higher charge carrier density is responsible for this transient behavior [133].

where μ_S is the net surface dipole moment and $\mu_{Exc.}$ is that of the exciton (32 D) [134]. If the distance between the two dipoles was set at $r = 2$ nm as an average value in our 4 nm QD and ε is $10 \times \varepsilon_0$ (ten times the vacuum permittivity), one can calculate a maximum value of the average surface dipole moment $\mu_S = 24$ D.

The above model can also explain the weak sensitivity of the observed shift to the excitation power. As we increase the power, we merely excite more particles as the trapping sites in each particle saturate rapidly. Thus the intensity of the new absorptions increase but the interaction between the free and the net trap dipole remains constant. Furthermore, the lifetimes of trapped excitons are very long, thus explaining the independence of the shift on the delay time.

3. The last assignment of the transient absorption to biexcitons seems to be an interesting interpretation of the observed spectral changes. This might be supported by the fact that such absorptions have been observed for the bulk [135], in QDs of 11 nm diameter in glass [136], and has been theoretically predicted by Park et al. [137]. Theory by Hu et al. predicts that biexcitons are significantly stabilized in QDs compared to the bulk semiconductor due to the quantum confinement and its resulting enhancement of exchange and coulomb interaction [138, 139].

Hu et al. also observed experimentally photoinduced transient absorption at the low and the high energy sides of the bleach maximum [138]. They also simulated their experimental results and assigned [139] the induced transient absorptions to transitions to the biexciton ground and excited states.

The biexciton binding energy is calculated from the equation $E_b = 2E_{exc} - E_{biexc}$. It is clear that the calculated value depends on which excitons form the observed biexciton with absorption at 590 nm. If the energy of the maximum of the strongest bleach band is used for E_{exc}, a binding energy is calculated to be 120 ± 20 meV. This is at least 20 times larger than that observed in the bulk (1.2–4.1 meV) [135, 140] and four times larger than that observed for the 11 nm particles in a glass (32 meV) [136]. This might be a manifestation of the more severe quantum confinement of the biexciton in the 4 nm CdSe QDs. According to the theory by Hu et al. [138, 139], the binding energy of the confined biexciton should increase with decreasing QD radius.

If the observed absorption assigned to the biexciton is due to the interaction of the two lowest energy "dark" excitons [21, 22] then the binding energy is reduced by 30–40 meV to 70 meV. However, the dynamics of the formation of the transient absorption at 590 nm argues against this assignment. The absorption and thus the formation of the biexciton is observed within our pulse width (100 fs). The relaxation to the spin forbidden dark level takes place in a time > 400 fs [33], thus cannot explain a much more rapid formation of the biexciton. If the observed transient absorption is due to a "hetero-biexciton" resulting from the interaction between the bright exciton and a net dipole of the rapidly surface trapped electrons and holes then the 120 meV is simply the Stark red shift of the free exciton absorption as discussed above.

7.4.3 Core-shell heteronanostructures

In this section, the emission from the intrinsic state of CdS is distinguished from the emission from surface localized states. The distinction is based upon the elimination of surface localized states by surface passivation [141, 142] with inorganic materials such as $Cd(OH)_2$, ZnS, CdO and HgS to form core-shell structures. The passivation fills defects and dangling bonds. The effect of the relative core-shell band gap is determined by using shell materials of larger band gap ($Cd(OH)_2$ and ZnS) and shell materials of smaller band gap (CdO and HgS) than the CdS core. Surface coverage by organic materials of much larger band gaps results in the resistance to oxidation and/or reduction by the nanoparticle charge carriers. The passivation by such organics has been demonstrated by the enhanced nanoparticle luminescence. However, the organic capping produces Stokes shifted luminescence of long lifetimes, typical of trapped states. The inorganic materials have recently demonstrated smaller Stokes-shifted emission [143, 144, 149] with shorter lifetimes [146]. The inorganic materials are therefore more effective than the organic capping materials at sustaining the intrinsic state.

For better analysis, the size distribution and surface structure are critical. The synthesis has therefore focused on narrowing the size distribution and controlling the surface structure. Reverse micelle allows the achievement of both of these goals. A water in oil microemulsion is employed to prepare CdS nanoparticles and ZnS, $CdOH_2$, and CdO capped CdS nanoparticles. The capping of CdS by a HgS spherical shell in an aqueous colloid is also synthesized for comparison. For stability, the CdS-HgS core shell structure is encapsulated by an outer CdS cladding. The resulting structure is the CdS-HgS-CdS core-shell-cladding system. The relative band gaps of the core-shell-cladding materials results in an electronic quantum well within the CdS QD. This structure was first prepared by Mews et al. [150] and is known as a quantum dot quantum well (QDQW). The band gap offsets in such a structure causes the relaxation of the exciton into the well material. Unlike the wider band gap shell materials such as $Cd(OH)_2$ and ZnS (which confine the exciton in the core), the smaller band gap shell materials like CdO and HgS confine the exciton within the shell causing novel properties. The influence of these different surface shells on the absorption and emission of intrinsic quantum-confined states is described.

7.4.3.1 CdS nanoparticles capped with $Cd(OH)_2$

The formation of the wider band gap $Cd(OH)_2$ shell around the smaller band gap CdS core leads to the following properties: 1) a red shift in the absorption edge, (Fig. 7-25) 2) increased exciton emission, (Fig. 7-26) 3) slight fluctuations in surface state sensitivity to surface charge and 4) blue shift in surface-state energy with increasing capping thickness. These changes result from the additional $Cd(OH)_2$ shell. As the $Cd(OH)_2$ shell forms around the CdS nanoparticles, the surface S^{2-} vacancies are partially filled by $(OH)^-$ forcing the electron back into core intrinsic states and eliminating many surface localized states. The elimination of surface dangling bonds causes an increased exciton emission. This increased contribution to the exciton emission is shown in Fig. 7-27, in which the ratio of the intensities of the band gap excitonic emission to the deep trap emission is plotted against the pH of the solution.

7.4.3.2 CdS nanoparticles capped with ZnS

ZnS surface passivation provides more understanding of the mechanism by which the wider band gap inorganic shells increase the exciton emission in the CdS nanoparticles. Unlike $Cd(OH)_2$, ZnS passivates both Cd^{2+} and S^{2-} deficient sites. Figure 7-28 presents the absorption spectra of the CdS-ZnS (core-shell) heteronanostructures of different sizes. In agreement with quantum confinement, the absorption band red shifts with increasing size. The ZnS shell enhances the excitonic emission just as the $Cd(OH)_2$ shell does. Figure 7-29 presents the excitation and emission spectra for a CdS-ZnS nanoparticle. Both exciton and trap emissions are observed. However, as a

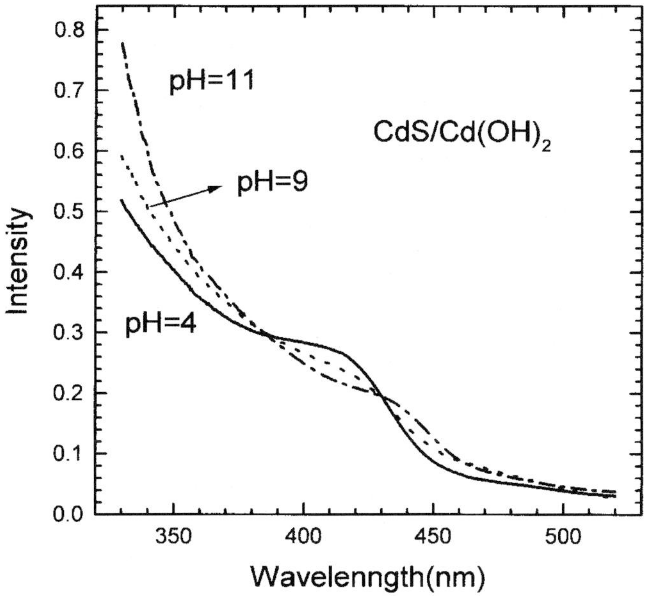

Figure 7-25. The absorbance spectrum of 3.8nm CdS nanoparticles for different $Cd(OH)_2^+$ shell thicknesses, as taken from reference 153.

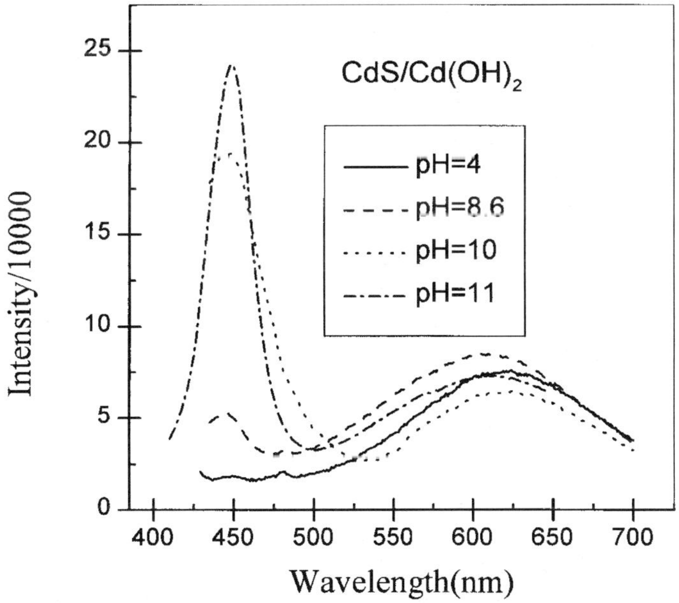

Figure 7-26. The emission spectrum of CdS-$Cd(OH)_2$ core-shell nanoparticles as taken from reference 153.

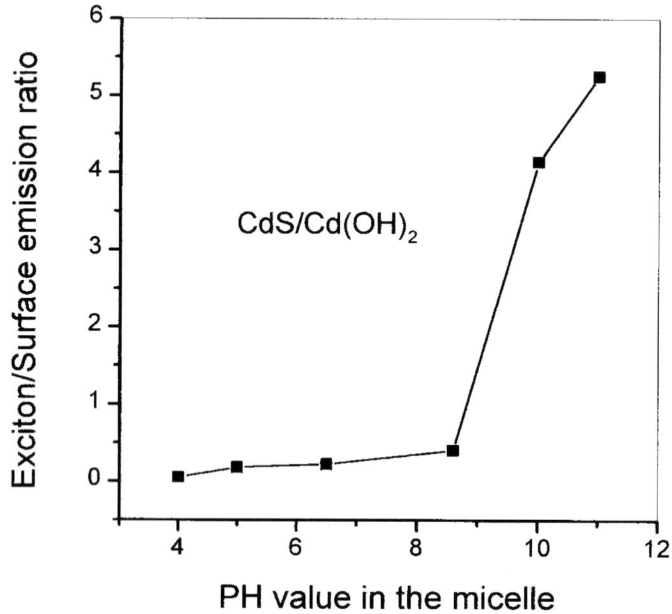

Figure 7-27. The emission intensity vs. pH for CdS nanoparticles as taken from reference 153.

result of the ZnS shell, the exciton emission is greatly enhanced. The surface trapped emission can almost be neglected for the heavily capped CdS. As with the $Cd(OH)_2$, the increased band edge (exciton) emission results from passivation of Cd^{2+} deficient sites. With ZnS, the S^{2-} deficient sites are also filled by Zn^{2+}. Unlike $Cd(OH)_2$, several layers of ZnS are necessary for effective passivation and enhanced carrier confinement. This necessity of multilayers for core confinement was also observed by Alivisatos [147]. Alivisatos observed that 1–3 layers of CdS was not enough for passivating CdSe nanoparticles. After capping CdSe with 3 layers of CdS, an additional layer of organic surfactants was needed to produce band edge emission. Obviously, the band gap difference between CdSe and CdS is not large enough to prevent tunneling of the exciton. These observations support the importance of the band gap differences of core and shell materials in order to effectively passivate the core.

Although the CdS excitonic emission is enhanced by both $Cd(OH)_2$ and ZnS passivations, the final enhancement (i.e. the observed enhancement after adding excess capping shells of both materials) is greater for ZnS heavily capped particles (almost twice the emission quantum yield of $Cd(OH)_2$ capped CdS). The lower passivation efficiency of $Cd(OH)_2$ results from its inability to fill both Cd^{2+} and S^{2-} vacancies whereas ZnS accommodates both vacancies. Two other difficulties associated with $Cd(OH)_2$ passivation are the steric limitations and the charge imbalance due to the different sizes and charges of $(OH)^-$ and S^{2-}. On the other hand, one layer of $Cd(OH)_2$ capping is more effective than a monolayer of ZnS at passivating the surface states, due to the tunneling across the ZnS monolayer. The larger polarizability and the smaller band gap of ZnS require thicker ZnS shells to effect a barrier from the surroundings, but a sufficiently thick ZnS barrier is more effective than a $Cd(OH)_2$ barrier.

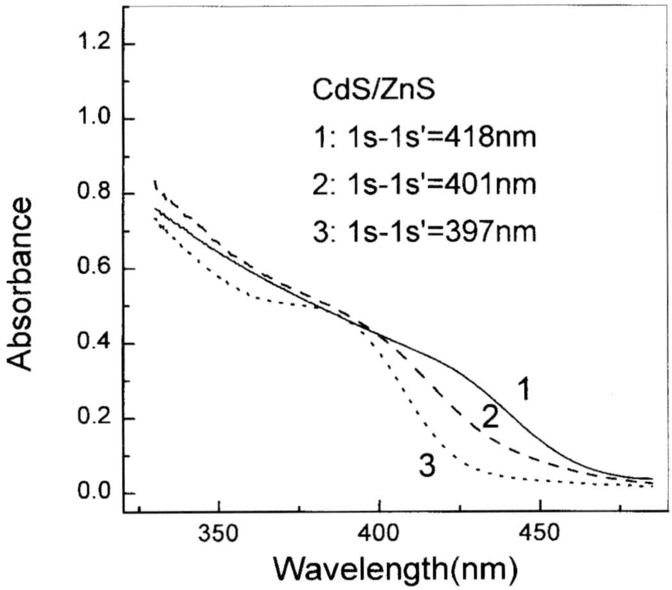

Figure 7-28. The absorbance spectra of CdS-ZnS core-shell for different CdS core sizes as taken from reference 153.

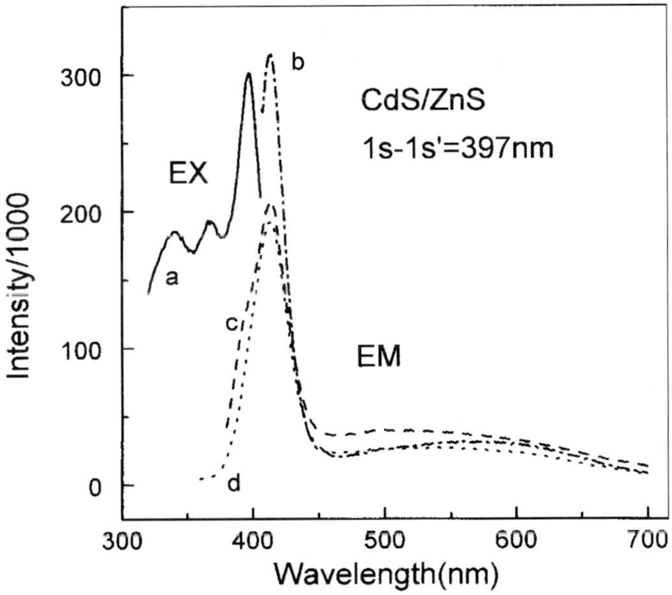

Figure 7-29. The excitation and emission spectra of CdS-ZnS, as taken from reference 153.

The improved passivation by ZnS allows better observation of the excitonic phenomena. In fact, the thicker ZnS shells allow the observation of structure in the excitation spectra (Fig. 7-29). The structure of the excitation spectra reflects the transitions to upper excited exciton states for the CdS-ZnS structure. The phonon bottleneck effect [148] and decoupling from the surface states caused by the thicker ZnS shell (5 layers) cause the stability of these exciton states. There are three peaks in the emission excitation spectrum. The first peak corresponds to the 1s–1s transition [149]. The other two peaks at higher energies are the 1p–1p and 1d–1d. The energies of these excited states are listed in Table 1 along with calculated results [149] based on the effective mass approximation. The 1s–1s transition has been easily observed in uncapped CdS nanoparticles. The new observation for these ZnS capped CdS particles is the distinct resolution of the 1p–1p and 1d–1d states.

Table 1. Quantized levels from allowed transitions assignment for exciton emission (theoretical value [149] in parenthesis).

Transitions	Sample 1 (3.3nm)	Sample 2 (3.4nm)	Sample 3 (3.7nm)
1s–1s′	3.122 (3.12)	3.091 (3.09)	2.966 (2.97)
1p–1p′	3.387 (3.68)	3.332 (3.54)	3.13 (3.37)
1d–1d′	3.667 (4.39)	3.635 (4.29)	3.396 (3.89)

7.4.3.3 CdS capped with HgS with an outer CdS cladding

HgS is a material with a smaller band gap than the CdS core. The absorbance changes upon formation of CdS-HgS-CdS from a CdS core are revealed in Fig. 7-30. As observed, the CdS quantum dots have an absorbance onset at 500 nm. The absorbance shoulder occurs at 465 nm. The formation of the HgS quantum shell about the CdS core produces a red shift of absorbance. The onset of the absorbance red shifts to 600 nm. Formation of the outer cladding to yield the CdS-HgS-CdS heteronanostructure causes more red-shifting with a new absorbance shoulder at 630 nm. The absorbance onset shifts to 700 nm. These observations are consistent with the previous results of Mews and coworkers [150].

The emission spectra of passivated CdS and CdS-HgS-CdS samples are compared. Emission from CdS QDs is presented in Fig. 7-31. As discussed in the previous section, two emission bands are observed for CdS. The exciton band occurs at 495 nm and is relatively narrow. The other emission is broad and has a large Stokes shift to 600 nm. The lower energy emission at 600 nm originates from surface traps which localize the exciton. The emission from the CdS-HgS-CdS quantum dot quantum well is given in Fig. 7-31 for excitation at 465 nm. The 495 nm band of bare CdS is eliminated in the QDQW system. Most importantly, a new band occurs at 700 nm. This new band is characteristic of the QDQW. Mews and coworkers [152] demonstrated that the emission originates from the HgS quantum well by ns transient hole burning and fluorescence line narrowing. The dynamics of this new band are investigated by fs time resolved absorbance spectroscopy [36, 37].

Time resolved absorbance spectroscopy was employed to get a better understanding of the modification of the electronic structure resulting from the incorporation of the HgS spherical shell within the CdS nanoparticle. Using hole burning techniques with a femtosecond laser, we attempted to burn an optical hole in CdS QDs and compare its temporal behavior with that in the CdS/HgS/CdS QDQW. In Fig. 7-32, the ps time dependence of the optical hole (bleach band) for both the CdS QD (top) and CdS-HgS-CdS QDQW (bottom) is presented. The excitation in both spectra is 400 nm. No optical hole is observed at the excitation in either particle at zero delay time. The optical holes are observed at lower energies. While the band shape of the optical

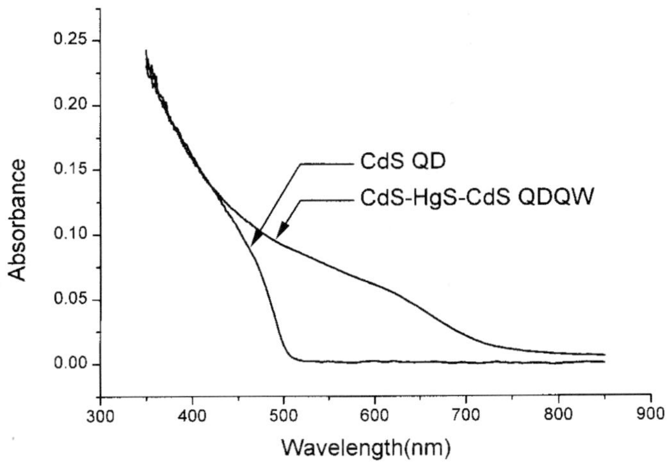

Figure 7-30. The absorbance spectra of Cd, and CdS-HgS-CdS heteronanostructures as taken from reference 36.

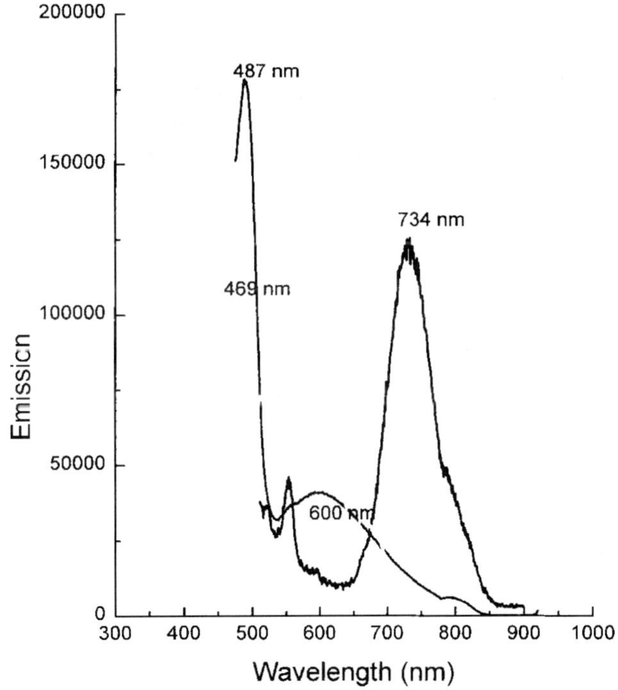

Figure 7-31. The emission spectra of CdS and CdS-HgS-CdS as taken from reference 36.

hole for the QDQW shows spectral diffusion as its shape and peak shift from high to low energies with time, the optical hole for the CdS QDs does not show such a large time dependent spectral shift. The optical hole in the CdS QDs appears in the lowest energy absorption region immediately after excitation and undergoes a very slight red shift within 2.5ps [32]. We could not resolve the spectral diffusion process from the energy of the pump laser (400 nm) and the band gap bleach due to very rapid processes involved in the QD. The optical hole in the QDQW develops instantly over a broad range. However, the broad optical hole in the CdS-HgS-CdS QDQW exhibits spectral diffusion at a much slower rate. Initially, this broad optical hole and its dynamics in the QDQW were attributed to the presence of different interfacial traps [36].

For more understanding of the detailed dynamics of the spectral diffusion in the QDQW, a more detailed examination of this process was carried out, using faster measurements. As observed in Fig. 7-33, the 50 fs delay hole burning spectrum (shown by the broken line) gave far better resolution than the steady state absorbance. The second derivative of the 50 fs spectrum (as given in the inset of Fig. 7-33) provides even more information. This derivative spectrum clearly shows two maxima, one at 525 nm (2.36 eV) and another at 625 nm (1.91 eV). The second derivative also showed a minimum at 600 nm. Further examination reveals that the 525 nm maximum corresponds to a deconvoluted shoulder in the 50 fs bleach spectrum. The 625 nm maximum could result from the overlapping absorbance of the first two allowed exciton transitions, predicted by Bryant et al. [38] at 1.89 eV ($1P_{3/2}$ – 1P) and 1.93 eV ($1P_{1/2}$ – 1P). This agreement is supported by the observation that the maximum in the bleach spectrum shifts from 625 nm (1.98 eV) at 50 fs delay to 650 nm (1.91 eV) at 2 ps delay. The observed minimum (600 nm) in the 50 fs bleach coincides with an optically inactive exciton state ($2S_{3/2}$ – 1S) near 600 nm.

In order to probe the dynamics of the different excitonic states in more detail, kinetic studies were performed by observing the formation and decay times of the optical hole (bleach) at different energies, while pumping is carried out at 400 nm.

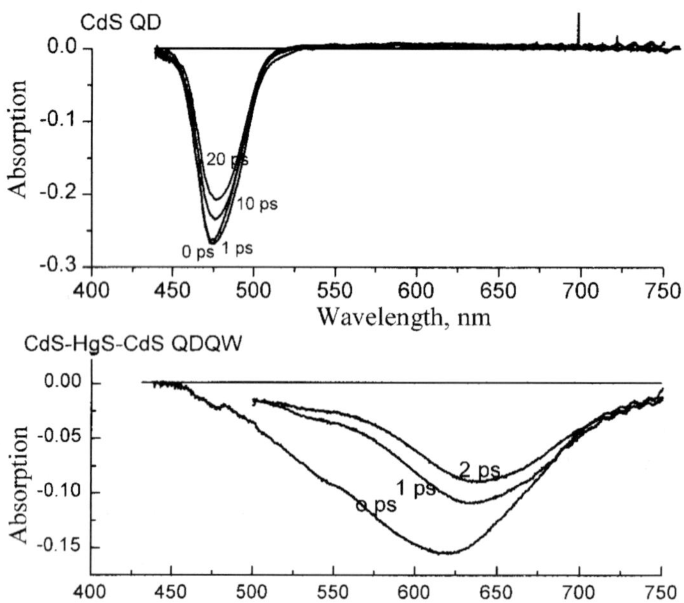

Figure 7-32. Ccomparison picosecond bleach spectra of CdS and CdS-HgS-CdS as taken from reference 37.

The results are shown in Fig. 7-34. It is clear that both the decay times and the formation times (inset) show different behaviors for the bleach dynamics examined in the 2.1–2.6 eV (the high energy optically allowed) region and in 1.8–2.1 eV (the low energy allowed) region. This suggests that the relaxation processes of the electron and hole (giving rise to the spectral diffusion) seem to have a state at an energy of 2.08 eV, which acts as a "bottleneck" in the excitation relaxation process from high to lower energy. The high energy bleach forms and decays much faster than the low energy

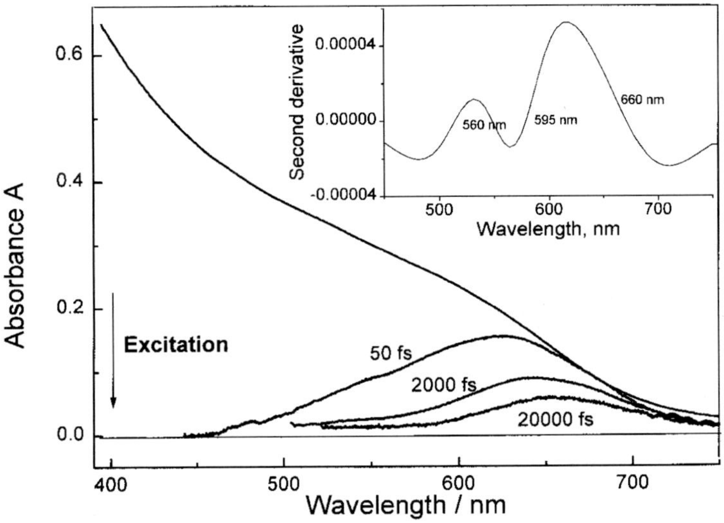

Figure 7-33. The femtosecond bleach spectra of CdS-HgS-CdS with second derivative extrema as taken from reference 37.

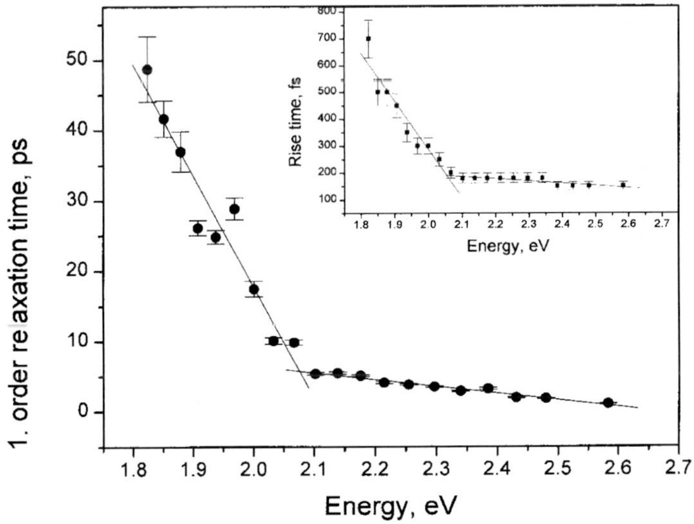

Figure 7-34. Rise and decay kinetics in the vicinity of the predicted dark state as taken from reference 37.

one. The energy of this state falls within the two different bands in the derivative spectrum of the broad bleach spectrum shown in the inset of Fig. 7-33. Furthermore, the decay times of the high-energy region do not correspond to the formation times of the low energy region. The high energy region (>2.1eV) decays with a rate distinctly faster than the low energy region. There is a clear change in dynamics near 2.1 eV. This is the minimum of the inset spectrum. One can determine that the energy of the dark state(s) is 2.10 ± 0.05 eV, which agrees very well with the predicted energy of the dark state [38]. This calculation shows that this state (the charge-separated state) has its electron already in the HgS well while its hole is in the CdS clad. The crossing of the hole from the CdS clad to the HgS well is a slow process due to the large difference in its effective mass of the two materials.

Acknowledgement

The continued support of this work by the Office of Naval Research (ONR grant No. N00014-95-1-0306) and National Science Foundation (CAG 9479397 and DMR-9632823) is greatly appreciated. T.G. and S.L. thank the MDI for partial support from the ONR Molecular Design Institute at Georgia Tech. C. B. thanks the Swiss National Science Foundation for partial financial support. We thank Z. L. Wang for his help on the electron microscopy in this work.

References

[1] A. Henglein, *J. Phys. Chem.* **1993**, *97*, 8457.
[2] A. Henglein, *Chem. Rev.* **1989**, *89*, 1861.
[3] A. P. Alivisatos, *J. Phys. Chem.* **1996**, *100*, 13226.
[4] G. Schmid, Clusters & Colloids: From Theory to Application, Weinheim, VCH, **1994**.
[5] J. A. A. J. Perenboom, P. Wyder, P. Meier, *Phys. Rep.* **1981**, *78*, 173.
[6] A. E. Hughes, S. C. Jain, *Adv. Phys.* **1979**, *28*, 717.
[7] P. V. Kamat, D. Meisel, Studies in Surface Science and Catalysis, Vol. 103, Semiconductor Nanoclusters – Physical, Chemical, and Catalytic Aspects, Amsterdam, Elsevier, **1997**.
[8] L. E. Brus, *J. Chem. Phys.* **1983**, *79*, 5566.
[9] L. E. Brus, *J. Chem. Phys.* **1984**, *80*, 4403.
[10] L. E. Brus, *Appl. Phys. A* **1991**, *53*, 465.
[11] A. S. Edelstein, R. C. Cammarata, Nanoparticles: Synthesis, Properties and Applications, Bristol, Institute of Physics Publishing, **1996**.
[12] M. Graetzel in Electrochemistry in Colloids and Dispersions (Ed.: R. A. Mackay, J. Texter), Weinheim, VCH, **1992**.
[13] U. Kreibig, M. Vollmer, Optical Properties of Metal Clusters, Berlin, Springer, **1995**.
[14] Y. Wang in Advances in Photochemistry, Volume 19 (Ed.: D. C. Neckers, D. H. Volman, G. Buenau), New York, John Wiley, **1995**.
[15] Y. Wang, N. Herron, *J. Phys. Chem.* **1991**, *95*, 525.
[16] J. R Heath, J. J. Shiang, *Chem. Soc. Rev.* **1998**, *27*, 65.
[17] H. Weller, *Angew. Chem. Int. Ed. Engl.* **1993**, *32*, 41.
[18] H. Weller, A. Eychmueller in Advances in Photochemistry, Volume 20 (Ed.: D. C. Neckers, D. H. Volman, G. Buenau), New York, John Wiley, **1995**.
[19] M. G. Bawendi, W. L. Wilson, L. Rothberg, P. J. Carroll, T. M. Jedju, M. L. Steigerwald, L. E. Brus, *Phys. Rev. Lett.* **1990**, *65*, 1623.
[20] M. G. Bawendi, P. J. Carroll, W. L. Wilson, L. B. Brus, *J. Chem. Phys.* **1992**, *96*, 946.
[21] M. Nirmal, D. J. Norris, M. Kuno, M. G. Bawendi, A. L. Efros, M. Rosen, *Phys. Rev. Lett.* **1995**, *75*, 3728.
[22] A. L. Efros, M. Rosen, M. Kuno, M. Nirmal, D. J. Norris, M. G. Bawendi, *Phys. Rev. B* **1996**, *54*, 4843.
[23] M. Kuno, J. K. Lee, B. O. Dabousi, F. V. Mikulec, M. G. Bawendi, *J. Chem. Phys.* **1997**, *106*, 9869.

[24] B. O. Babbouni, J. Rodriguez-Viejo, F. V. Mikulec, J. R. Heine, H. Mattoussi, R. Ober, K. F. Jensen, M. G. Bawendi, *J. Phys. Chem. B* **1997**, *101*, 9463.
[25] L. Spanhel, M. Haase, H. Weller, A. Henglein, *J. Am. Chem. Soc.* **1987**, *109*, 5649.
[26] A. Eychmueller, A. Haesselbarth, L. Katsikas, H. Weller, *Ber. Bunsenges. Phys. Chem.* **1991**, *95*, 79.
[27] A. Haesselbarth, A. Eychmueller, H. Weller, *Chem. Phys. Lett.* **1993**, *203*, 271.
[28] W. Hoheisel, V. L. Colvin, C. S. Johnson, A. P. Alivisatos, *J. Chem. Phys.* **1994**, *101*, 8455.
[29] A. V. Barzykin, M. A. Fox, *Israel J. Chem.* **1993**, *33*, 21.
[30] M. O'Neil, J. Marohn, G. McLendon, *J. Phys. Chem.* **1990**, *94*, 4356.
[31] C. F. Klingshirn, Semiconductor Optics, Berlin, Springer, **1997**.
[32] S. L. Logunov, T. C. Greem, S. Marguet, M. A. El-Sayed, *J. Phys. Chem. A* **1998**, *102*, 5652.
[33] C. Burda, T. C. Green, S. Link, M. A. El-Sayed, *J. Phys. Chem. B* **1999**, *103*, 1783.
[34] C. Burda, T. C. Green, S. Link, M. A. El-Sayed, Microcrystalline and Nano-crystalline Semiconductors, MRS Proceedings **1998**.
[35] C. Burda, R. B. Little, S. Link, T. C. Green, M. A. El-Sayed, Society for Optical & Quantum Electronics Proceedings **1998**.
[36] V. Kamalov, R. B. Little, S. L. Logunov, M. A. El-Sayed, *J. Phys. Chem.* **1996**, *100*, 6381.
[37] R. B. Little, C. Burda, S. Link, S. L. Logunov, M. A. El-Sayed, *J. Phys. Chem. A* **1998**, *102*, 6581.
[38] W. Jaskolski, G. Bryant, *Phys. Rev. B*, **1998**, *57*, 4237.
[39] M. Kerker, *J. Colloid Interface Sci.* **1985**, *105*, 297.
[40] M. Faraday, *Philos. Trans.* **1857**, *147*, 145.
[41] G. Mie, *Ann. Physik* **1908**, *25*, 377.
[42] G. C. Papavassiliou, *Prog. Solid State Chem.* **1980**, *12*, 185.
[43] M. Kerker, The Scattering of Light and Other Electromagnetic Radiation, New York, Academic Press, **1969**.
[44] C. F. Bohren, D. R. Huffman, Absorption and Scattering of Light by Small Particles, New York, Wiley, **1983**.
[45] J. A. Creighton, D. G. Eadon, *J. Chem. Soc. Faraday Trans.* **1991**, *87*, 3881.
[47] T. G. Schaaff, M. N. Shafigullin, J. T. Khoury, I. Vezmer, R. L. Whetten, W. G. Cullen, P. N. First, C. Gutierrez-Wing, J. Ascensio, M. J. Jose-Yacaman, *J. Phys. Chem. B* **1997**, *101*, 7885.
[48] T. S. Ahmadi, Z. L. Wang, T. C. Green, A. Henglein, M. A. El-Sayed, *Science* **1996**, *272*, 1924.
[49] Y. Yu, S. Chang, C. Lee, C. R. C. Wang, *J. Phys. Chem. B* **1997**, *101*, 34, 6661.
[50] B. M. I. v. d. Zande, M. R. Bohmer, L. G. J. Fokkink, C. Schonenberger, *J. Phys. Chem. B* **1997**, *101*, 852.
[51] C. R. Martin, *Chem. Mater.* **1996**, *8*, 1739
[52] J. Turkevich, P. C. Stevenson, J. Hillier, Discussions of the Faraday Society No. *11*, **1951**, 55.
[53] D. A. Handley, Colloidal Gold: principles, Methods, and Applications Vol. 1, New York, Academic Press, **1989**.
[54] M. M. Alverez, J. T. Khoury, T. G. Schaaff, M. N. Shafigullin, I. Vezmer, R. L. Whetten, *Chem. Phys. Lett.* **1997**, *266*,91.
[55] M. Brust, D. Bethell, D. J. Schiffrin, C. J. Kiely, Adv. Mater. **1995**, *7*, 795.
[56] U. Kreibig, C. v. Fragstein, Z. Phys. **1969**, *224*, 307.
[57] U. Kreibig, *Z. Phys.* **1970**, *234*, 307.
[58] U. Kreibig, U. Genzel, *Surf. Sci.* **1985**, *156*, 678.
[59] M. M. Alvarez, J. T. Khoury, T. G. Schaaff, M. N. Shafigullin, I. Vezmer, R. L. Whetten, *J. Phys. Chem. B* **1997**, *101*, 3706.
[60] A. Kawabata, R. Kubo, *J. Phys. Soc. Japan* **1966**, *21*, 1765.
[61] N. J. Persson, Surf. Sci. **1993**, *281*, 153.
[62] C. Yannouleas, R. A. Broglia, *Ann. Phys.* **1992**, *217*, 105.
[63] M. Cini, *J. Opt. Soc. Am.* **1981**, *71*, 386.
[64] L. Genzel, T. P. Martin, U. Kreibig, *Z. Phys. B* **1975**, *21*, 339.
[65] W. A. Kraus, G. C. Schatz, *J. Chem. Phys.* **1983**, *79*, 6130.
[66] S. Link, M. A. El-Sayed, *J. Phys. Chem B.*, **1999**, *103*, 4212.
[67] M. B. Mohamed, S. Link, M. A. El-Sayed, *J. Phys. Chem. B* **1998**, *102*, 9370.
[68] S. Link, C. Burda, M. B. Mohamed, B. Nikoobakht, M. A. El-Sayed, *J. Phys. Chem A* **1999**, *103*, 1165.
[69] S. Link, Z. L. Wang, M. A. El-Sayed, *J. Phys. Chem B.*, **1999**, *103*, 3529.
[70] M. Schmidt, R. Kusche, B. v. Issendorff, H. Haberland, *Nature* **1998**, *393*, 238.
[71] L. J. Lewis, P. Jensen, J.-L. Barrat, *Phys. Rev. B* **1997**, *56*, 2248.
[72] F. Ercolesi, W. Andreoni, E. Tosattie, *Phys. Rev. Lett.* **1991**, *66*, 911.
[73] F. G. Shi, *J. Mater. Res.* **1994**, *9*, 1307.
[74] S. Iijima, T. Ichihashi, *Phys. Rev. Lett.* **1996**, *56*, 616.

[75] D. J. Smith, A. K. Petford–Long, L. R. Wallenberg, J.-O. Bovin, *Science* **1996**, *233*, 872.
[76] P. Buffat, J.-P. Borel, *Phys. Rev A* **1976**, *13*, 2287.
[77] Z. L. Wang, J. M. Petroski, T. C. Green, M. A. El-Sayed, *J. Phys. Chem. B* **1998**, *102*, 6145.
[78] T. S. Ahmadi, S. L. Logunov, M. A. El-Sayed, *J. Phys. Chem.* **1996**, *100*, 8053.
[79] T. S. Ahmadi, S. L. Logunov, M. A. El-Sayed, J. T. Khoury, R. L. Whetten, *J. Phys. Chem. B* **1997**, *101*, 3713.
[80] S. Link, C. Burda, Z. L. Wang, M. A. El-Sayed, *J. Chem. Phys.*, **1999**, *111*, 1255.
[81] J. K. Hodak, I. Martini, G. V. Hartland, *J. Phys. Chem. B* **1998**, *102*, 6958.
[82] J. K. Hodak, I. Martini, G. V. Hartland, *Chem. Phys. Lett.* **1998**, *284*, 135.
[83] M. Perner, P. Bost, G. v. Plessen, J. Feldmann, U. Becker, M. Mennig, H. Schmidt, *Phys. Rev. Lett.* **1997**, *78*, 2192.
[84] M. Perner, T. Klar, S. Grosse, U. Lemmer, G. v. Plessen, W. Spirkl, J. Feldmann, *J. Luminesc.* **1998**, *76* & *77*, 181.
[85] J.-Y. Bigot, J.-C. Merle, O. Cregut, A. Daunois, *Phys. Rev. Lett.* **1995**, *75*, 4702.
[86] T. V. Shahbazyan, I. E. Perakis, J.-Y. Bigot, *Phys. Rev. Lett.* **1998**, *81*, 3120.
[87] T. W. Roberti, B. A. Smith, J. Z. Zhang, *J. Chem. Phys.* **1995**, *102*, 3860.
[88] B. A. Smith, D. M. Waters, A. E. Faulhaber, M. A. Kreger, T. W. Roberti, J. Z. Zhang, *J. Sol-Gel Sci Technol.* **1997**, *9*, 125.
[89] A. E. Faulhaber, B. A. Smith, J. K. Andersen, J. Z. Zhang, *Mol. Cryst. Liq. Cryst.* **1996**, *283*, 25.
[90] B. A. Smith, J. Z. Zhang, U. Giebel, G. Schmid, *Chem. Phys. Lett.* **1997**, *270*, 139.
[91] M. J. Feldstein, C. D. Keating, Y.-H. Liau, M. J. Natan, N. F. Scherer, *J. Am. Chem. Soc.* **1997**, *119*, 6638.
[92] H. Inouye, K. Tanaka, I. Tanahashi, K. Hirao, *Phys. Rev. B* **1998**, *57*, 11334.
[93] T. Tokizaki, A. Nakamura, S. Kaneko, K. Uchida, S. Omi, H. Tanji, Y. Asahara, *Appl. Phys. Lett.* **1994**, *65*, 941.
[94] N. Del Fatti, C. Flytzanis, F. Vallee, *Appl. Phys. B* **1999**, *68*, 433
[95] R. D. Averitt, S. L. Westcott, N. J. Halas, *Phys. Rev. B* **1998**, *58*, 10203.
[96] R. H. M. Groeneveld, R. Sprik, A. Lagendijk, *Phys. Rev. B* **1992**, *45*, 5079.
[97] R. H. M. Groeneveld, R. Sprik, A. Lagendijk, *Phys. Rev. B* **1995**, *51*, 11433.
[98] H. E. Elsayed-Ali, T. Juhasz, G. O. Smith, W. E. Bron, *Phys. Rev. B* **1991**, *43*, 19914.
[99] T. Juhasz, H. E. Elsayed-Ali, H. Hu, W. E. Bron, *Phys. Rev. B* **1992**, *45*, 13819.
[100] T. Juhasz, H. E. Elsayed-Ali, G. O. Smith, C. Suarez, W. E. Bron, *Phys. Rev. B* **1993**, *48*, 15488.
[101] W. S. Fann, R. Storz, H. W. K. Tom, J. Boker, *Phys. Rev. B* **1992**, *46*, 13592.
[102] W. S. Fann, R. Storz, H. W. K. Tom, J. Boker, *Phys. Rev. Lett.* **1992**, *68*, 2834.
[103] R. W. Schoenlein, W. Z. Lin, J. G. Fujimoto, G. L. Eesley, *Phys. Rev. Lett.* **1987**, *58*, 1680.
[104] S. D. Brorson, J. G. Fujimoto, E. P. Ippen, Phys. Rev. Lett. **1987**, *58*, 1962.
[105] C.-K. Sun, F. Vallee, L. H. Acioli, E. P. Ippen, J. G. Fujimoto, *Phys. Rev. B* **1993**, *48*, 12365.
[106] C.-K. Sun, F. Vallee, L. H. Acioli, E. P. Ippen, J. G. Fujimoto, *Phys. Rev. B* **1994**, *50*, 15337.
[107] N. W. Ashcroft, N. D. Mermin, Solid State Physics, Philadelphia, Saunders College, **1976**.
[108] C. Kittel, Introduction to Solid State Physics, New York, Wiley, 1996.
[109] S. Link, C. Burda, M. B. Mohamed, B. Nikoobakht, M. A. El-Sayed, *Phys. Rev. B*, submitted.
[110] T. Ahmadi, Z. L. Wang, A. Henglein and M. A. El-Sayed, Chemistry of Materials, **1996**, *8*, 1161.
[111] L. D. Rampino and F. F. Nord, *J. Am. Chem. Soc.*, **1942**, *63*, 2745.
[112] A. Henglein, B. G. Ershov and M. Malow, *J. Phys. Chem.*, **1995**, *99*, 14129.
[113] Z. L. Wang, T. Ahmadi and M. A. El-Sayed, *Surface Sci.*, **1997**, *380*, 302.
[114] J. M. Petroski, Z. L. Wang, T. C. Green and M. A. El-Sayed, *J. Phys. Chem. B*, **1998**, *102*, 3316.
[115] L. M. Falicov and G. A. Somorjai, *Proc. Natl. Acad. Sci. USA*, **1985**, *82*, 2207.
[116] J. F. Rivadulla, M. C. Veraga, M. C. Blanco, M. A. Lopez-Quintela, and J. Rivas, *J. Phys. Chem. B.* **1997**, *101*, 8997.
[117] R. Fuchs, *Phys. Rev. B.* **1975**, *11*, 1732.
[118] R. E. Hummel, R. Wißmann, eds; U. Kreibig, M. Vollmer, rev. Handbook of Optical Properties, vol. II. 1997. CRC Press – Boca Raton. 145–190.
[119] S. Schmitt-Rink, D. A. B. Miller, D. S. Chemla, *Phys. Rev. B* **1987**, *35*, 8113.
[120] H. Grobert and M. H. Devoret in Single Charge Tunneling (Ed.:Plenum), New York, **1992**.
[121] A. P. Alivisatos, *Science* **1996**, *271*,933.
[122] B. Bischoff, and M. Anderson, *Chem. Mater.*, **1995**, *7*, 1772–1778.
[123] M. Steigerwald, and L. Brus, *Acc. Chem. Res.*, **1990**, *23*, 183.
[124] N. Herron, Y. Wang, and T. Bein, *J. Am. Chem. Soc.*, **1989**, *11*, 350.
[125] G. Schon, U. Simon, *Colloid Polymer Science*, **1995**, *273*, 101–117.
[126] C. B. Murray, D. J. Norris, M. G. Bawendi, *J. Am. Chem. Soc.* **1993**, *115*, 8706.
[127] T. Vossmeyer, L. Katsikas, M. Giersig, I. G. Popovic, K. Diesner, A. Chemseddine, A. Eychmueller, H. Weller, *J. Phys. Chem.*, **1994**, *98*, 7665.

[128] D. Dounghong, J. Ramsden, M. Gratzel, *J. Am. Chem. Soc.*, **1982**, *104*, 2977.
[129] Y. Nosaka, H. Miyama, M. Terauchi, T. Kobayashi, *J. Phys. Chem.*, **1988**, *92*, 255.
[130] D. J. Norris, M. G. Bawendi, *J. Chem. Phys.* **1995**, *103*, 5260.
[131] A. P. Alivisatos, A. L. Harris, N. J. Levinos, M. L. Steigerwald, L. E. Brus, *J. Chem. Phys.* **1988**, *89*, 4001.
[132] S. Hunsche, T. Dekorsy, V. Klimov, H. Kurz. *Appl. Phys. B* **1996**, *62*, 3.
[133] C. Burda, S. Link, T. Green, M. A. El-Sayed, *J. Phys. Chem.* submitted.
[134] V. L. Colvin, A. P. Alivisatos, *J. Chem. Phys.* **1992**, *97*, 730.
[135] K. H. Pantke, J. Erland, J. M. Hvam, *J. Cryst. Growth* **1992**, *117*, 763.
[136] H. Klimov, S. Hunsche, H. Kurz, *Phys. Rev. B* **1994**, *50*, 8110.
[137] S. H. Park, R. A. Morgan, Y. Z. Hu, M. Lindberg, S. W. Koch, N. Peyghambarian, *J. Opt. Soc. Am.* **1990**, *7*, 2097.
[138] Y. Z. Hu, M. Lindberg, S. W. Koch, *Phys. Rev. B* **1990**, *42*, 1713.
[139] Y. Z. Hu, S. W. Koch, M. Lindberg, N. Peyghambarian, E. L. Pollock,F. Abraham, *Phys. Rev. Lett.* **1990**, *64*, 1805.
[140] Landolt-Boernstein; Physics of II–VI and I–VII Compounds, Vol. 17b, O. Madelung, Ed., Heidelberg, Springer, **1982**.
[141] Y. Kanemitsu, S. Okamoto, *Phys. Rev.* **1997**, *B 55*, R 7375.
[142] R. N. Bhargava, D. Gallagher, X. Hong, A. Nurmikko, *Phys. Rev. Lett.* **1994**, *72*, 416.
[143] M. A. Hines, P. J. Guyot-Sionnest, *J. Phys. Chem.* **1996**, *100*, 468.
[144] A. R. Kortan, R. Hull, R. Opila, M. G. Bawendi, M. L. Steigerwald, P. J. Carroll, L. E. Brus, *J. Am. Chem. Soc.*, **1990**, *112*, 1327.
[145] C. F. Hoener, K. A. Allen, A. J. Bard, A. Campion, M. A. Fox, T. E. Mallouk, S. E. Webber, J. M. Whites, *J Phys. Chem.*, **1992**, *96*, 3812.
[146] A. Eychmuller, A. Hasselberth, L Katsikas, H. J. Weller, *J. Luminescence* **1991**, *48–49*, 745.
[147] X. Peng, M. C. Delchlamp, A. V. Kadavanich, A. P. Alivisatos, *J. Am. Chem. Soc.*, in press.
[148] H. Benisty in Phonons in Semiconductor Nanostructure (Ed.: J. P. Leburton, J. Pascual, C. S. Torres) NATO Series; Dordrecht, Kluwer, **1992**.
[149] A. D. Yoffe, *Adv. Phys.* **1993**, *42*, 173.
[150] A. Mews, A. Eychmuller, M. Giersig, D. Schoos, H. J. Weller, *J. Phys. Chem.* **1994**, *98*, 934.
[151] V. Klimov, S. Hunche, H. Kurz, *Phys. Status Solidus B* **1995**, *188*, 259.
[152] A. Mews, A. V. Kadavanich, U. Banin and A. P. Alivisatos, *Phys. Rev. B*, **1995**, *53*, R13242.
[153] B. Zou, R. B. Little, J. P. Wang and M. A. El-Sayed, *Intl. J. Quantum Chem.*, submitted.

8 Nuclear Magnetic Resonance – Characterization of Self-Assembled Nanostructural Materials

Li-Qiong Wang, Gregory J. Exarhos, and Jun Liu

Abstract

Self-assembly has been widely used for the preparation of novel nanostructural materials. To both accelerate the dynamics of this processing route and develop new nanostructures, it is critical to understand the attendant interfacial interactions which occur in solution between the different precursor components, and how such molecular level interactions affect nanostructural ordering. Other chapters within this book contain extensive discussions of experimental techniques for characterizing nanoscale materials, such as small angle x-ray or neutron scattering, high-resolution electron microscopy, and surface force microscopy. This chapter will focus on nuclear magnetic resonance (NMR) methods which are sensitive to local chemical environments and provide complementary information on the molecular scale, in contrast to other analytical techniques. The first section provides a brief introduction of fundamental NMR principles and their applications. It is followed by examples to illustrate how NMR can be used to derive information related to long-range ordering on the nanometer scale, the molecular conformation on a sub-nanometer scale, and their correlation to the interfacial binding.

8.1 Introduction

Self-assembly processes have been used widely to synthesize novel nanostructural materials [1]. Through the use of these techniques, atoms, molecules, and particles arrange themselves into highly ordered structures which exhibit structure-selective properties (e.g. molecular transport, molecular sieving, etc.). The fundamental forces which drive the self-assembly process can be both intra- and inter- molecular. For example, bifunctional molecules containing a hydrophilic head group and a hydrophobic tail group can be induced to form self-assembled monolayers (SAM) on a substrate [2, 3]. Because of the capability to control the molecular arrangement on the surface and interfacial properties, SAM are widely used to engineer materials properties, such as wetting, adhesion, and friction. The monolayers are also used to mediate the molecular recognition processes and to direct oriented crystal growth [4]. On crystalline surfaces that will not chemically react with the SAM (such as alkyl thiols on gold), the main driving force is the intermolecular van der Waals interaction, although the substrate symmetry has an effect on the overlayer structure. These kinds of SAM have been used as model systems to understand surface properties and conformations of more complex molecules which may influence ordering phenomena on oxide surfaces, such as silica [5, 6]. In addition to the van der Waals forces between the hydrocarbon chains, the intermolecular binding, and the interfacial chemistry between the SAM molecules and the substrate play a critical role in these systems. The SAM formation may involve cross-linking the adjacent molecules, as well as chemically binding the molecules to the substrate.

A recent example of self-assembled materials is that of ordered mesoporous ceramics which have first been reported by Mobil Oil Research [7]. The ordered mesoporous materials were synthesized using self-assembled surfactant micellar structures as templates. These materials have a very high surface area (> 1000 m^2/g), ordered pore structure (mostly hexagonal packed cylindrical pore channels), and extremely narrow pore size distribution. The pore diameter can be adjusted from 2 to 15 nanometers (nm) (30 nm if polymeric templates are used [8, 9]). The preparation methods involve mixing ceramic precursors (such as sodium aluminate, silicate precursors, and silica) in a surfactant (cetyltrimethylammonium chloride, CTAC, or cetyltrimethylammonium bromide, CTAB) solution and mixing the agents at temperatures below 150°C. In principle, surfactant molecules form ordered micellar phases. The most common phase consists of rod-like micelles packed in hexagonal arrays. The ceramic precursors bind to the head groups of surfactant molecules, and finally co-condense, forming a continuous ceramic phase. Subsequently, the surfactant molecules can be removed by thermal or chemical treatment. Because of the unique properties, the new mesoporous materials will have many potential applications [10] in the chemical industry (catalysis) [11], semi-conductor industry (porous films) [12] and industries involved with environmental remediation (adsorption, separation and ion exchange) [13].

SAM have been extensively studied by diffraction techniques and by surface force microscopy (see Ref. [3] for a review). Other techniques, such as X-ray photoelectron spectroscopy (XPS), X-ray absorption fine structure spectroscopy (EXAFS), and infrared spectroscopy (IR), also have been used. X-ray diffraction (XRD) provides information on molecular spacing, orientation, molecular tilting, and degree of ordering [3]. Atomic force microscopy (AFM) allows direct imaging of domains and aggregates [14]. Only under special conditions can resolution to the molecular level be achieved. Studies of surface composition and surface chemical reactivity can use XPS and IR [15]. EXAFS provides structural information concerning the local environment, such as bond length, bond angle, and spatial geometry [16]. The most widely used methods for characterizing ordered mesoporous type materials include transmission electron microscopy (TEM) and small angle X-ray diffraction or (XRD). The structure and symmetry for long range ordered materials can be identified easily from the image at different orientations and the typical diffraction patterns expected of a known structure [17, 18].

Several techniques widely used for studying inorganic materials, such as electron microscopy, have some limitations because the organic components in many materials have inherently low stability and low contrast under electron-beam irradiation in vacuum. Techniques to overcome some of the problems include fixation, staining, and thin sectioning [19]. The specimen preparation is usually tedious and can introduce artifacts. Cryo-electron microscopy [20, 21] has become more widely used in the last decade because of the availability of better microscopes (microscopes that can perform low dose imaging) and well-designed sample transport mechanisms. The specimen can be rapidly frozen to a very low temperature, so that residual water is prevented from crystallizing which can cause a destructive effect on the fine structures of organic materials. The frozen specimen is transported to the microscope and imaged with low beam intensity, usually using phase contrast to form the image. Continued progress in the imaging technique and computer-aided image reconstruction and analysis can further improve the resolution associated with cryo-electron microscopy techniques.

Although organic materials are difficult to analyze by microscopy alone, they are amiable to a wide range of spectroscopy techniques, including NMR. This chapter shows how NMR can be used to study the nature of the cogent interfacial interactions when combined with other techniques, how these interactions affect the self-assembly of the organic components, and how the subsequent molecular conformation directs structural evolution on a sub-nanometer scale. This fundamental knowledge not only allows development of better materials, but also helps in understanding their properties.

As compared with other techniques, NMR is sensitive to the local environments that are complementary to the longer-range structure afforded by imaging and diffraction techniques. The ability to resolve structure over short distances is especially important in the study of polymers, amorphous materials, biological compounds, and other systems characterized by the absence of long-range order. Unlike most other techniques, NMR can provide information about the dynamics of the system in addition to structural information. The relaxation, molecular conformation, and mobility of functional molecules in self-assembly can be obtained from a variety of dynamic measurements. NMR is non-destructive and requires little sample preparation. Moreover, NMR is unparalleled by other techniques in terms of its ability to selectively probe a specific interaction through selection of an appropriate excitation scheme. For example, the nuclear dipole-dipole interaction can be studied selectively by using modern pulse sequences that excite the dipole-dipole interaction, yielding highly specific information for a target nuclide

Due to the low detection sensitivity for some species, the need for a comparatively large sample size can be a possible disadvantage. However, large samples are often a necessary requirement to ensure representative results for these heterogeneous materials.

8.2 Basic principles of solid state NMR

A cadre of authoritative book chapters and review articles are available on solid state NMR theory and practice [22]. In brief, nuclear magnetic resonance arises from the adsorption and emission of radio-frequency energy by nuclear spins as they oscillate and reorient under internal and external magnetic fields. The local environments around the observed nuclei and the mobility of the nearby atoms or molecules greatly influence the oscillation frequency and reorientation times of the observed nuclei. Due to its extreme sensitivity to the local environment and the mobility of the molecules in the sample, NMR spectroscopy allows researchers to obtain detailed structural and dynamic information for many different types of materials. Based on the fact that NMR signal intensity is proportional to the number of contributing nuclei, the NMR method also can be used for quantitative compositional analysis of unknown materials.

Following are the three basic interactions in solid-state NMR and a brief discussion of how they yield specific information about local symmetry and bonding. The detailed theory and measurements which underlie these interactions are not described in this chapter.

8.2.1 Chemical shift interaction

The chemical shift interaction arises from the magnetic electron-nucleus interaction (electrons shielding nuclei from the applied magnetic field), which alters the local field experienced by the nucleus and, therefore, affects its oscillation frequency. The change in the oscillation frequency is measured as the chemical shift of a specific nucleus. The magnitude of the chemical shift is proportional to the applied magnetic field and the shielding constant determined by the electron distribution near the specific nucleus. For example, the methyl proton has a different chemical shift from the methylene proton, due to a difference in shielding constants. Because the electron distribution near the nuclei in general is not symmetric, the chemical shift has directional properties. The chemical shifts seen in NMR spectra depend largely on the microscopic orientation of the atomic or molecular species with respect to the applied magnetic field. For molecules randomly oriented in a rigid lattice, the distribution of orientations produces a range of chemical shifts, hence a broad spectral line. By contrast, the rapid and isotropic molecular motion in liquids averages the anisotropic interactions and thus narrows the spectral lines. To reduce the line broadening in the solid sample due to the anisotropic effect, and to increase the sensitivity, magic angle spinning (MAS) is usually used. In MAS, the sample is rotated about an axis inclined at 54.7° with respect to the applied magnetic field. (The magic angle of 54.7° is derived by removing the angular term in the chemical shift interaction.) This technique has been widely used for spin-1/2 nuclei, such as ^{13}C, ^{15}N, ^{19}F, ^{29}Si, and ^{31}P [23].

Under the MAS condition where the linewidth due to the chemical shift anisotropy is larger than the spinning frequency of the samples, a set of sharp spectral lines are produced that are spaced at the spinning frequency and centered on the isotropic (averaged) chemical shift. From the MAS NMR spectra, the isotropic chemical shift corresponds to the values that would be obtained in isotropic solution. These chemical shifts are fingerprints of specific functional groups or species and can be used to identify the functional groups or species in solid materials. The analysis of the spinning side-band patterns can provide the three-dimensional chemical shift tensors that can be used to obtain structural information, including the coordination and symmetry of the molecules.

8.2.2 Dipole-dipole interaction

The dipole-dipole interaction is due to the interaction of magnetic moments between two nuclei, and is dependent upon the magnitude of the magnetic moment of the nearby nuclei, the distance between two nuclei, and the directional term, including the angle between the internuclear vector and the applied magnetic field. MAS NMR can reduce the spectral line broadening, due to the anisotropic dipole-dipole interaction through rapid spinning at the magic angle to remove the similar angular term, as in the chemical shift interaction. Because the dipole-dipole interaction is also distance dependent, the structural parameters, including the internuclear distances, can be extracted from the dipolar-based experiments. For certain solids containing ^{1}H or ^{19}F, the spectral linewidth is significantly broadened, due to the large magnetic moments. In order to remove the large dipole interaction from the nearby protons, proton dipolar decoupling is used. This decoupling is accomplished by applying a perturbing field at the proton resonance frequency that results in a reduced linewidth.

Although dipole-dipole interactions can significantly broaden spectral lines, the dipole-dipole interaction is the basis for the very commonly used cross-polarization (CP) technique [24]. Due to the difference in nuclear spin temperatures, CP is widely used to enhance the sensitivity of the observed nuclei by transferring magnetization from an abundant-spin system (usually protons) to the spin system of the observed nucleus. Cross-polarization occurs if the nuclei can interact with each other via moderately strong dipole-dipole couplings. The stronger the dipole coupling, the faster the cross-polarization process will occur. CP-MAS and proton decoupling are often used together to obtain high-resolution solid-state NMR spectra.

Many methods are available to measure the magnetic dipole-dipole interactions. Detailed information on the theory and measurement of these dipolar couplings can be found in the literature [25–34]. For example, spin-echo double resonance (SEDOR), and rotational-echo double resonance (REDOR) experiments have been used to measure the heteronuclear coupling. Information on internuclear distances can be extracted from the heteronuclear dipolar couplings where magnitudes are proportional to the inverse third power of the internuclear distance and, therefore, are extremely sensitive to the separation of the coupled spins. As compared with the conventional CP-MAS NMR, REDOR provides the information not only on the individual nuclei, but also the connectivity to other nearby nuclei. Therefore, a combination of conventional MAS NMR with dipolar-based techniques is required to determine structures of unknown solid materials.

8.2.3 Electric quadrupole interaction

Electric quadrupole interactions are due to the interactions of non-spherical nuclear charge distributions with electrostatic field gradients generated by asymmetric electron distributions in molecules or at lattice sites (the chemical-bonding environment). This interaction affects only nuclei with $I > 1/2$ in non-cubic environments. For quadrupolar nuclei with spin > 1/2, such as ^{11}B, ^{17}O, ^{23}Na, ^{27}Al, and ^{69}Ga, an additional source of spectral broadening arises from the anisotropic quadrupolar effect.

The detailed information on the theory and measurement of quadrapole interactions has been discussed extensively in the literature [35–42]. Double-rotation (DOR) and dynamic-angle spinning (DAS) NMR techniques, where the sample rotates about two axes, can be used to reduce the spectral linewidths by eliminating the quadrupolar effects in addition to the chemical shift anisotropy and dipole-dipole coupling effects. Currently, the multiple-quantum correlation NMR technique is used to obtain high-resolution solid-state NMR spectra for quadrapole nuclei under the conventional magic-angle spinning conditions.

Instead of removing the quadrupolar interaction, a static wide-line NMR experiment for quadrupolar nuclei (using a large spectral window) can be used to obtain the quadrupole coupling constant. At the same time, simulations of the wide-line spectra provide molecular motion and dynamic information. For example, a ^2H wide-line NMR spectrum of a static solid gives a pake pattern (due to the quadrupolar interaction) that can be simulated to give the information about the molecular motion. The distinctive quadrupole splitting patterns for ^2H in different liquid crystal environments can be used to construct the complex phase diagrams because the splitting is controlled by the symmetry of the specific phase.

8.3 Application of NMR in characterization of self-assembled materials

8.3.1 Phase identification

As discussed in other chapters, information on long-range ordering usually is derived primarily from TEM, XRD, or optical microscopy. The crystalline structures and phase diagrams of surfactant micellar structures, used in the synthesis of mesoporous materials and nanoscale composites, have been extensively studied [18]. However,

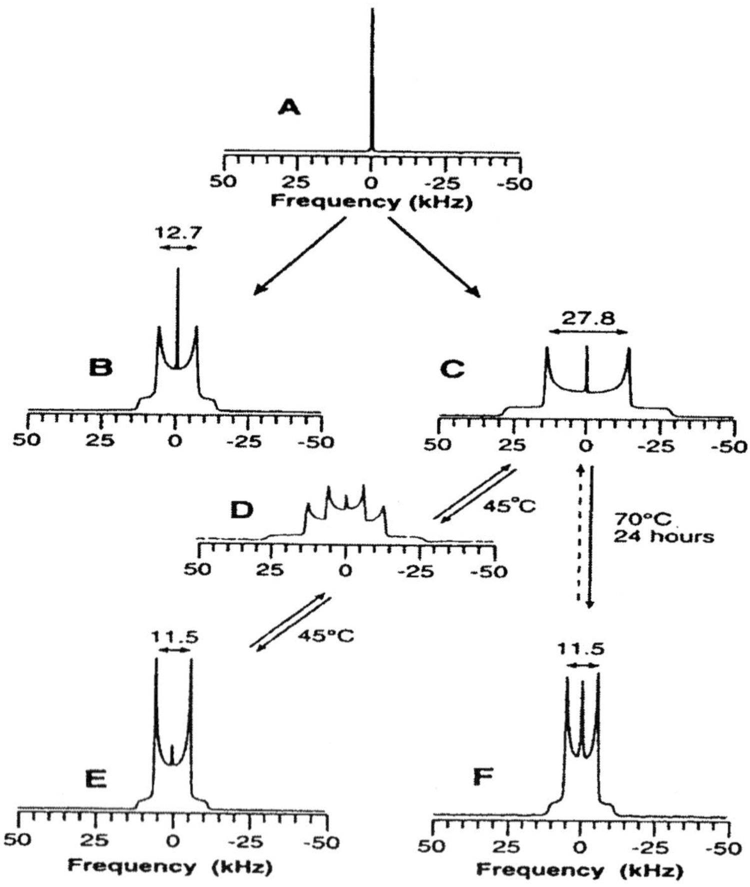

Figure 8-1. ^{2}H NMR spectra of α-deuterated CTAB in different surfactant concentrations with and without adding silicate solution. (A) an isotropic CTAB aqueous solution at 25 °C, representative of either cylindrical micelles (12.8% CTAB) or spherical micelles (6.8% CTAB). (B) A hexagonal silicatropic liquid crystal at 25 °C formed by the addition of the double four-ring silicate oligomer solution and trimethylbenzene to an isotropic 12.8% CTAB solution (A). (C) A lamellar SLC at 25 °C formed by the addition of the D4R silicate solution and TMB to an isotropic 6.8% CTAB solution (A). (D) An intermediate SLC mesophase formed from (C) after heating for 10 min at 45 °C; a reversible first-order phase transformation occurs, as evidenced by superimposed ^{2}H powder patterns from coexisting lamellar and hexagonal phases. (E) A hexagonal SLC phase formed from (C) after equilibration for approximately 8 hours at 45 °C. (F) A hexagonal SLC is formed irreversibly from the lamellar sample (C) after heating for 24 hours at 70 °C [taken from Ref. 45].

it is desirable to study the materials in their natural states, so that true structural information can be obtained in real time. Unlike microscopy techniques, NMR does not require special sample preparation and handling, and, therefore, can be used to follow the phase transformations in surfactant-silicate systems. In ^{2}H and ^{15}N NMR [43, 44], the distinctive quadrupole splitting patterns seen for ^{2}H in different liquid crystal environments are controlled by the symmetry of the liquid crystal phase and can be utilized to determine the phase diagrams. Anisotropic molecular motions lead to partial time-averages of ^{2}H quadrupole interactions, making the ^{2}H NMR quadrupole splitting patterns sensitive to the long-range organization of surfactant molecules.

A beautiful example was provided in the formation of mesophase surfactant-silicate composites [45]. Figure 8-1 shows typical ^{2}H NMR spectra of α-deuterated CTAB in different surfactant concentrations with and without adding silicate solution. The ^{2}H NMR spectra for an isotropic CTAB aqueous solution at 25/C, representative of either cylindrical micelles (12.8 % CTAB) or spherical micelles (6.8 % CTAB), give a single narrow peak independent of the shape of the micelles. The isotropic ^{2}H NMR signal results from the isotropically mobile spherical or cylindrical micelles in

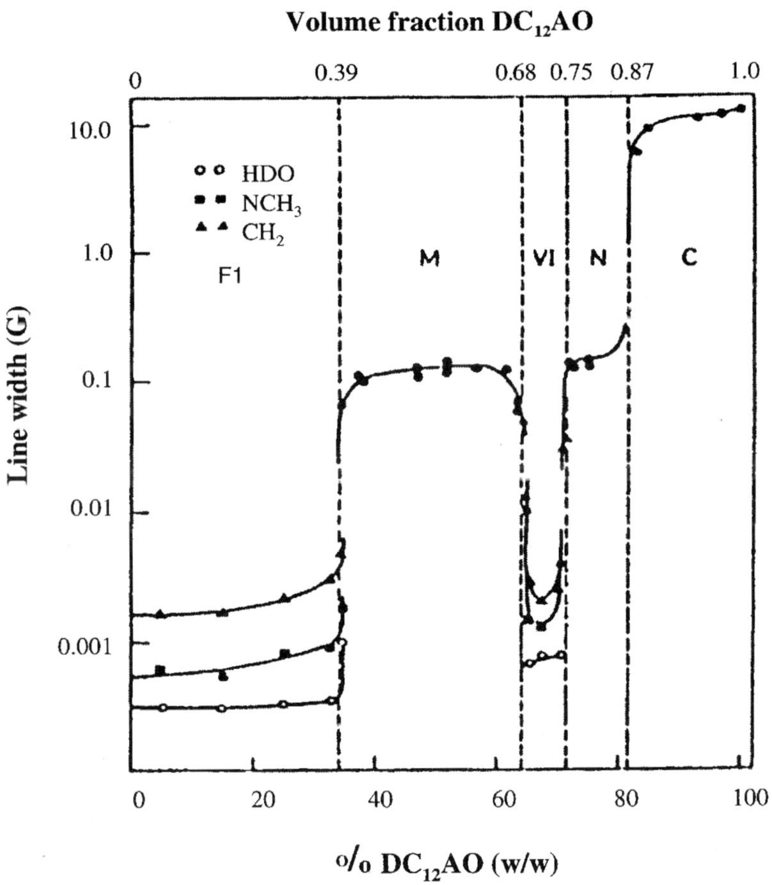

Figure 8-2. ^{2}H NMR linewidths in dimethyldodecylamine oxide-D_2O vs. surfactant concentration showing phase transitions [taken from Ref. 46]. Phase: F1 = micellar, M = hexagonal, VI = cubic, N = lamellar, C = crystalline.

dynamic equilibrium with monomeric surfactant molecules in solution. After these spherical or cylindrical micelles aggregate to form liquid crystal phases, the ^2H spectra show the representative quadrupole splitting for the hexagonal and lamellar liquid crystals. The splitting for the lamellar phase is about twice as large as that for the hexagonal phase, due to the larger order parameter associated with the lamellar liquid crystal mesophase. The addition of silicate solution to the isotropic surfactant solution results in the long-range order in typical mesophase syntheses. The meso liquid crystal phase shows a slightly larger quadrupole splitting than the liquid crystal phase without silicate, due to the larger ordering from the interaction between the silicate and surfactant species.

The advantage of NMR is clearly seen in that all the different phases can be captured as is, including the intermediate phase. In fact, the chemical shift and the linewidth can be quantified and used to map out the phase boundaries [46]. Figure 8-2 shows the linewidth of ^2H as a function of surfactant concentration. As the surfactant concentrations increase, the aggregate structures change from micellar to hexagonal, to cubic, to lameller, and finally to solid crystalline phases. The NMR spectrum not only shows different linewidth in different regions, but also shows sharp transitions across phase boundaries.

The ^{14}N NMR spectra also can be used to study surfactant mesophase formation. The distinctive quadrupole splitting enables the identification of different phases formed during the synthesis [44]. The advantage of using ^{14}N NMR over the ^2H NMR is that samples do not need to be deuterated. However, the NMR resonance of ^{14}N is much broader, and the results are less reliable.

8.3.2 Chain conformation

NMR is very sensitive to the local chemical environment and the molecular conformation of long-chain molecules. The relative populations of *trans* and *gauche* conformations influence the ^{13}C chemical shift of the interior methylene carbons of the alkyl chains. Detailed information on local molecular conformation affects a range of surface and interfacial properties, such as wetting, mobility, and binding, and is difficult to obtain from imaging and scattering techniques.

The carbon atoms of n-alkanes gave a resonance at 30 ppm in solution where equilibrium populations of *trans* and *gauche* conformations exist, but in the crystalline state a down-field shift of about 3–4 ppm is observed for an *all-trans* conformation [47]. Recently, well-oriented monolayers were reported to display an intense peak at 33 ppm. This peak is characteristic of the ^{13}C chemical shift for the interior methylene carbons of the alkane chains in *all-trans* conformation [6]. Thus, the chemical-shift position of the resonance associated with the interior methylene can be used to measure the degree of the order for the long-chain molecules.

Many NMR experiments involve MAS and high-power proton decoupling to remove the line-broadening due to the chemical shift anisotropy and ^1H-^{13}C heteronuclear dipolar coupling in order to obtain high resolution ^{13}C solid-state NMR spectra [48]. Figure 8-3 shows an example of high resolution solid-state ^{13}C NMR spectra for long chain surfactant molecules [6]. In this experiment, zirconium octadecylphosphonate (ODPA) is placed on nonporous silica (Cab-O-Sil) as self-assembled monolayers (Fig. 8-3A). The large resonance at 34 ppm in the ^{13}C NMR spectrum is associated with the interior methylene of ODPA in *all-trans* conformation, although only a small, more mobile component is present for the disordered chains at 30 ppm.

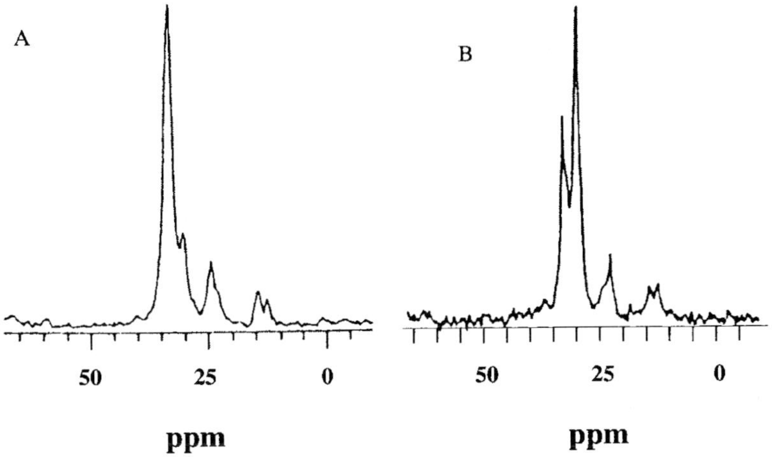

Figure 8-3. ^{13}C CPMAS NMR spectra [taken from Ref. 6] of (A) zirconium octadecylphosphonate (ODPA) deposited on fumed silica, (B) octadecyltrichlorosilane (OTS) deposited on silica at room temperature. Both ODPA and OTS are chemically bonded to the silica surfaces. All-trans conformations are resolved in both systems.

Therefore, the conclusion is that a well-ordered metal phosphate monolayer is formed by the self-assembly technique. As a comparison, the ^{13}C spectrum of alkyl silanes on silica support is also shown in Fig. 8-3B. The resonance peak for the backbone methylene is located at 30 ppm; therefore, the monolayer is much more disordered than ODPA.

The examples in Fig. 8-3 involve chemical binding of the monolayers with the substrate. Ordered SAM can also be formed without chemical binding at the interface. One such widely studied system is composed of alkane thiol monolayers on a gold substrate [49]. For NMR studies, colloidal gold particles, rather than a flat substrate, are used. A combination of ^{13}C chemical shift, relaxation, and line shape data shows that the alkane thiol surfactant on the gold in the case of long chain (n = 16, 18) is in semicrystalline state. At room temperature, domains of *all-trans* chains coexist with a smaller population of more mobile chains containing *gauche* conformers. The high curvature of the surface of the very small gold particles (diameter about 2 nm) prevents efficient packing of neighboring chains, as in the case of alkane thiol monolayers on planar gold. Instead, the spacing between the colloids, as measured by TEM and XRD, suggests that intercalation of the chains of neighboring gold particles allows crystallization into an extended *all-trans* conformation. The tail group of the chain, however, exhibits completely conformational disordered at room temperature as the short chain (n = 8) alkane thiol on the gold.

The molecular ordering is dependent on the temperature. In-situ heating is an ideal NMR experiment to study the order-disorder transition in self-assembled monolayers [50]. Figure 8-4 gives the variable temperature ^{13}C solid-state MAS NMR spectra for octadecyl amine (ODA) on clay mineral [51], and alkane thiols on gold [50]. At a low temperature, one large peak at 33 ppm is observed, corresponding to the interior methylene in *all-trans* conformation. When the temperature is increased beyond 50 °C, the maximum peak intensity begins to shift to 31 ppm, indicating the chains start to become disordered.

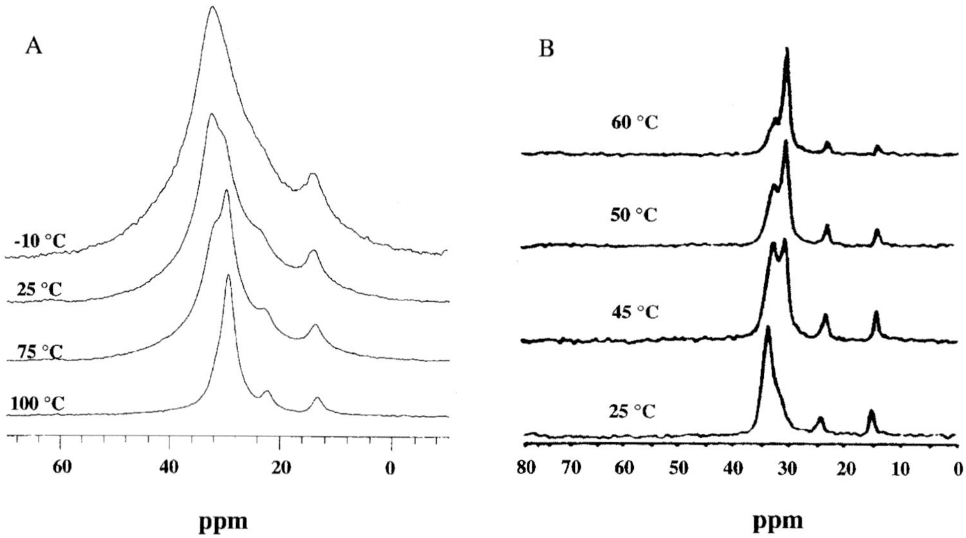

Figure 8-4. Variable temperature ^{13}C solid-state MAS NMR spectra for octadecyl amine on clay mineral (Fig. 8-4A), and alkane thiols on gold (Fig. 8-4B) [taken from Ref. 50, 51]. At a low temperature, one large peak at 33 ppm observed corresponding to the interior methylene in *all-trans* conformation. When the temperature is increased beyond 50 °C, the maximum peak intensity begins to shift to 31 ppm, indicating the chains start to become disordered. ^{13}C CPMAS

The molecular conformation is also related to the geometric environments. Figure 8-5 shows the ^{13}C spectra of the surfactant molecules [51] in: (A) lamellar and hexagonal biphase mesoporous silicate, and (B) lamellar clay minerals. The TEM micrographs of the corresponding materials are also shown. Two different conformations are resolved in NMR. The resonances at about 33 ppm and 30 ppm are assigned to the internal methylenes of surfactants with *all-trans* conformations, and to those with a significant number of *gauche* conformations, respectively. The geometric confinement in lamellar materials including both the mesoporous silicates and clays can force the molecules to pack more efficiently to form the *all-trans* conformation, while the molecules in cylindrical pore channels exhibit more freedom and less efficient packing. This conclusion can be supported by the line-shape analysis and contact-time measurements that show surfactant molecules in the layered structure with *all-trans* conformation are less mobile than those in the hexagonal structure with mixed trans-gauche conformations. This study also shows a possible direct correlation between the chemical shift of NMR resonance and the microstructure of the corresponding phase, therefore making it possible to quantitatively identify the amount of surfactant in each given phase.

8.3.3 Interfacial binding

Some of the most valuable information NMR can provide is the nature of interfacial binding, which usually dictates the self-assembly processes and the properties of the materials. For example, cationic surfactant containing the ammonium group, both CTAC [$(CH)_3NCH_2CH_2(CH_2)_{11}CH_2CH_2CH_3$] and amphoteric surfactant containing

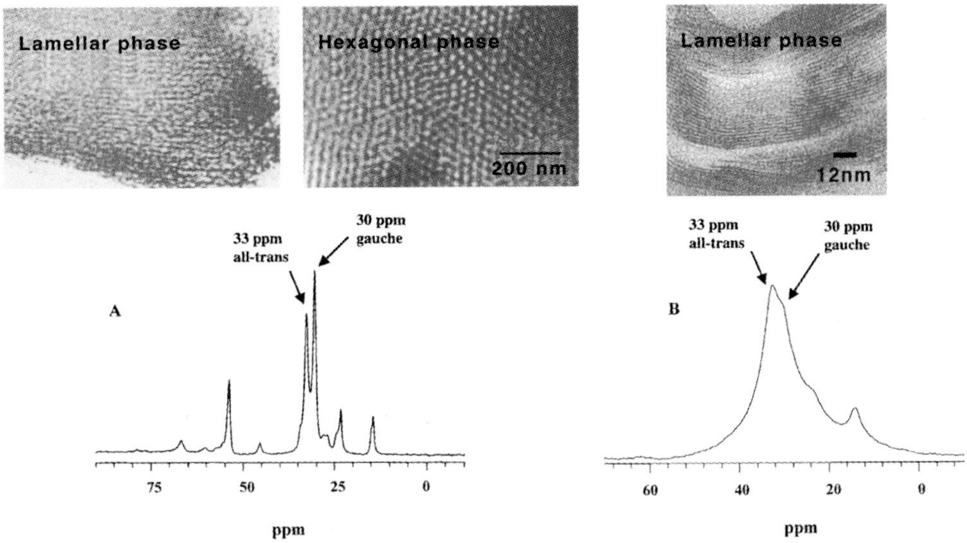

Figure 8-5. ^{13}C NMR spectra of surfactant molecules in (A) lamellar and hexagonal biphase mesoporous materials, and (B) lamellar clay minerals, along with TEM micrographs [51]. Two different conformations are resolved in NMR. The resonances at about 33 ppm and 30 ppm are assigned to the internal methylenes of surfactants with *all-trans* conformations, and to those with a significant number of *gauche* conformations, respectively. Geometric confinement in lamellar materials (with small layer spacings) can force the materials to pack efficiently to form *all-trans* conformation.

the carboxylic and ammonium group CTPB [$HOOCCH_2N(CH_3)_3CH_2(CH_2)_{11}CH_3$], have been used to template the growth of mesoporous zirconia. As shown in Fig. 8-6, ^{13}C single-pulse (SP) NMR clearly suggests that with CTAC and silica the binding group is the ammonium group, but with CTAPB and zirconia, the binding group is the carboxylic group [52]. For CTAC and silica, the largest line broadening and loss of mobility associated with C1, C2, and C3 indicate the silicate is directly bonded to the ammonium group. On the other hand, the ammonium group in the amphoteric surfactant is not involved in bonding. The largest line shift and broadening is associated with the carboxylic group, indicating the carboxylic portion is the bonding group.

More information can be obtained by carefully comparing the SP ^{13}C spectra of the pure surfactant, the intermediate disordered phase that is formed in the early stage of the reaction, and the final ordered phase (Fig. 8-6A) [53]. Compared to the pure surfactant molecules, the bonding of the surfactant and the silicate, in the ordered and disordered phases, caused considerable peak broadening, especially for the C1, C2, and C3 groups associated with the methyl and methylene group in the head-group area. The electrostatic binding of the silicate and the surfactant also caused a ~ 1 ppm downfield shift for the NMR resonance peak associated with the methyl group next to the head group and substantial broadening for the peak corresponding to the methylene group adjacent to the head group. An additional peak at 57.4 ppm appeared in the ordered phase, suggesting that methyl groups lost their symmetry due to geometrical constraints in the head-group area in that phase. The relative peak intensities observed in the ^{13}C MAS NMR spectra of surfactant-silicate depend on whether CP or SP excitation is used, an observation that is interpreted in terms of the mobility of the surfactant on the silica surfaces. The ^{13}C MAS SP spectra without ^1H decoupling show

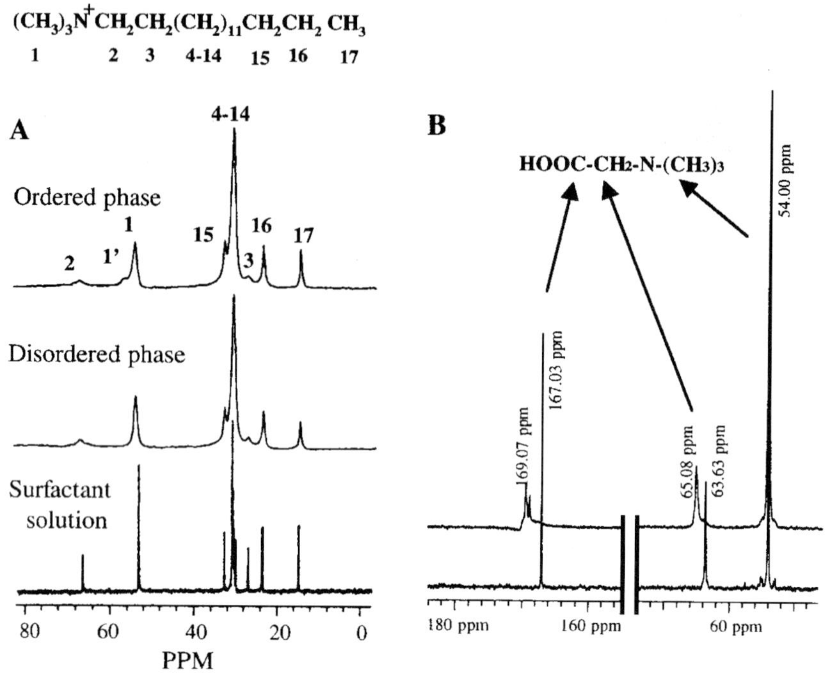

Figure 8-6. NMR spectra of surfactant molecules [taken from Ref. 52, 53]. (A) Single pulse MAS spectrum of the pure surfactant solution, disordered aggregates, and hexagonal phase. The bonding of the inorganic species causes the line shift and broadening of carbon groups near the head area. The methyl group (indicated by the peak split for C1) also lose the stereochemical symmetry because of the local ordering at the interfaces. (B) In the amphoteric surfactant, the ammonium group affected by bonding. The largest line shift and broadening are associated with the carboxylic group. To simplify the results for the amphoteric surfactant, the co-surfactant, which has the same structure as the head group of the surfactant, was studied instead. This procedure eliminated all contributions from the tail groups while providing the same information on the chemical bonding of the head group.

the methylene group next to the head group for the ordered mesoporous materials exhibits a marked lack of motion, as compared to other segments of the surfactant molecules. It also shows the methyl group at the tail of surfactant is very mobile.

This conclusion is further supported by experiments using a combination of NMR line-shape and relaxation-time analyses, T^{CH} (cross-polarization time constant) and $T_{1\rho}{}^{H}$ (the proton spin-lattice relaxation time constant in the rotating frame) with variable-temperature NMR. The mobility of the functional group in surfactant can be measured. Figure 8-7 clearly shows the rate of polarization varies among all individual carbon groups of the functional groups and side groups in the ordered phase at a given temperature. In these experiments, less mobile carbon groups, or carbon groups in a more anisotropic environment, exhibit a faster CP rate. Therefore the motion of molecular segments near the silicate surfaces is restricted; the motion of molecular segments away from the silicate surface is not. The cross-polarization time constant (T_{CH}) and the proton spin-lattice relaxation time constant in the rotating frame ($T_{1\rho}{}^{H}$) were derived from the variable-contact-time measurements, giving rise to more quantitative descriptions of cross polarization and relaxation behaviors for the surfactant in mesoporous materials. Because cross-polarization is most efficient for the static ^{13}C-^{1}H dipolar interactions, the less mobile carbon groups exhibit the faster cross-

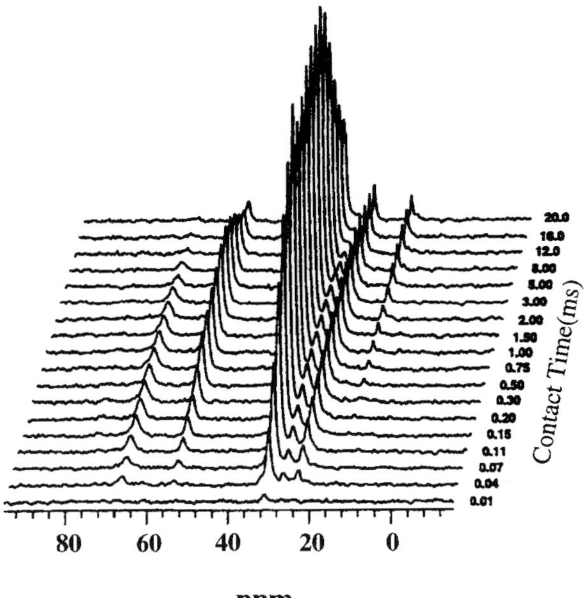

Figure 8-7. Contact time array experiment of the surfactant molecules in the hexagonal phase. The mobility of all the functional groups and segments can be derived. The mobility of the carbon groups near the head area is affected, but the groups away from the head-group area remain disordered and mobile [taken from Ref. 53].

polarization rate or the shorter T_{CH}. The information on the mobility for each segment of the surfactant derived from variable contact-time measurements is in agreement with the line-shape analysis obtained from ^{13}C MAS spectra with and without ^{1}H decoupling.

The NMR experiments provide a physical picture for the formation of mesoporous silica under basic conditions: the silicate species were first electrostatically bonded to the surfactant. Initially, not only was the overall structure disordered, but also the local environment of the interfacial region between the surfactant and the silicate remained disordered. Further condensation of silicate ions caused the global structural ordering and local molecular ordering of the surfactant at the interface, but away from the interface, the surfactant tails remained disordered and mobile.

8.4 Materials design, characterization, and properties

The materials design and properties critically depend on characterization and understanding the self-assembly on the atomic and the molecular level, as well as the nanometer scale, as discussed in previous sections. This relationship can be demonstrated by the development of some novel materials – functionalized monolayers on mesoporous supports [54]. These hybrid mesoporous materials have great potential for industrial and environmental applications [13, 54, 55]. In these materials, functional molecules are attached to the mesoporous support in a way similar to the preparation of SAM on flat substrates, except that short chain molecules are used. The driving forces for the monolayer formation are the covalent binding to the substrate and the cross-linking between adjacent molecules. This approach provides a unique opportunity to rationally engineer the surface properties. The hybrid mesoporous materials demonstrate exceptional selectivity and capacity for adsorbing heavy metal ions from contaminated waste streams. Materials capable of recognizing other species and molecules are also under development. Many other applications are being evaluated.

The materials design has several requirements: close packing of monolayers on the walls of the pore channels, stable binding to the substrate and between the molecules, accessibility of the surface groups, and proper surface properties (wetting, for example). High population density and close packing of the monolayer are desired for high loading density of the target molecules. Stable interfacial binding is needed for the durability of the materials. Because the monolayer chemistry is performed within pore channels of a few nanometers, care must be taken so that bulk hydrolysis and polymerization do not block the pore channels and bury the functional groups.

NMR proves to be an ideal technique to study the interfacial reaction and the molecular organization. First, ^{29}Si NMR can be used to establish how the molecules are bound to the substrate and linked to one another (Fig. 8-8). It is important to recognize that relative peak intensities in ^{29}Si CP-MAS are not strictly quantifiable, due to differences in relaxation behavior. Therefore, the Bloch decay pulse sequence (single pulse excitation) with long recycle times was used to obtain data and quantify the molecular composition of these materials. Several different bindings can be identified. The large peak at –111 ppm is from the silica support. For low functionalized monolayers coverage, three additional peaks from –50 to –80 ppm are identified, corre-

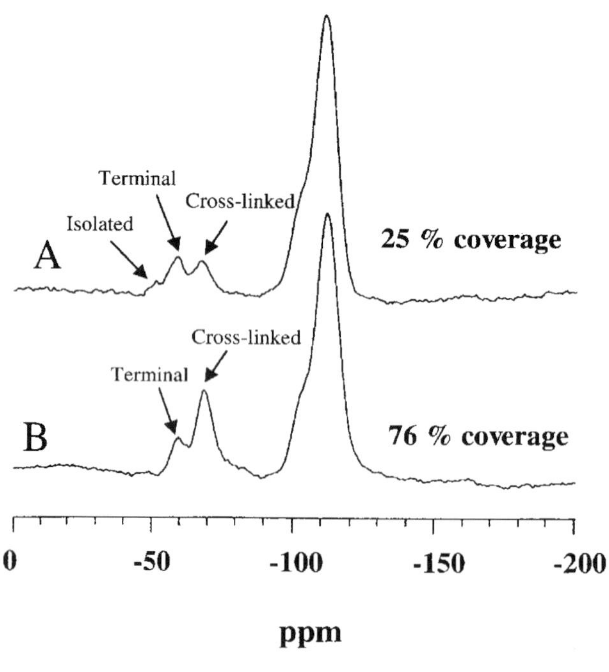

Figure 8-8. ^{29}Si NMR spectra of organic monolayers on mesoporous silica. (A) At 25% coverage, (B) at 76% coverage [taken from Ref. 54].

sponding to three different environments for the siloxane groups in the functionalized monolayers [54]: (i) isolated groups that are not bound to any neighboring siloxanes, (ii) terminal groups that are bound only to one neighboring siloxane, (iii) cross-linked groups that are bound to two neighboring siloxanes. Among the three, the most dominant peak comes from terminal group (ii). The dominance of the terminal group suggests the molecules are not closely bound to one another, and they are rather isolated. For high functionalized monolayer coverage, the molecules are closer to one another, and the most predominant peak corresponds to the cross-linked siloxane group (iii). The isolated siloxane group (i) is absent. This configuration signals a transition from dispersed islands to close packing.

The ^{13}C NMR further provides information on the behavior of the monolayers on the surface. Single-pulse (SP) ^{13}C NMR spectra along with the peak assignments [Si-CH_2(3)-CH_2(2)-CH_2(1)-SH] for 25%, 76%, and mercury-laden 76% functionalized monolayers samples, respectively, are shown in Fig. 8-9. For 25% functionalized monolayers coverage on mesoporous silicates, two resonance peaks were observed, corresponding to the methylene carbon group C3, directly bonded to the Si atom and to the other two methylene carbons (C2 and C1). An additional peak was observed for 76% functionalized monolayers coverage. This peak was assigned to the methylene carbon (C1) next to the -SH group, based on the chemical shifts reported for $CH_3(CH_2)_7SH$ [49]. The difference in NMR spectra taken for the organic monolayers at different coverage was attributed to the different molecular conformations. At low surface coverage, the carbon chains can adapt to a wide range of conformations; therefore, the peaks for C2 and C1 cannot be distinguished because of conformational heterogeneity. At higher population densities, all of the carbon chains are near one

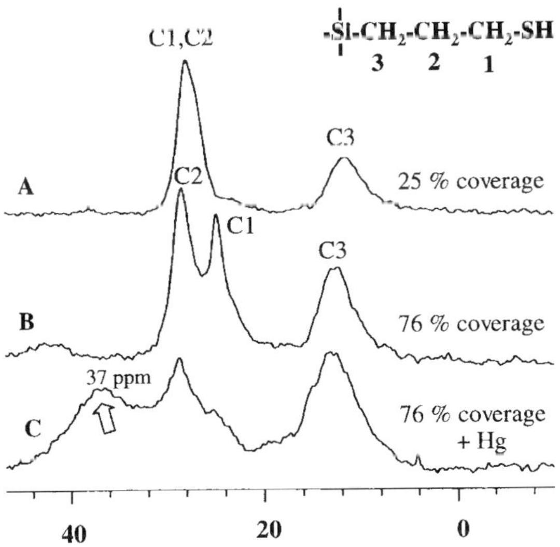

Figure 8-9. ^{13}C NMR spectra of organic monolayers on mesoporous silica. (A) At 25% coverage, C1 and C2 cannot be separated because of conformational heterogeneity. (B) At 76% coverage, C1 and C2 are clearly resolved. (C) At 76 % coverage containing mercury, a new peak at 37 ppm is assigned for C1 because of the mercury thioalkoxide [taken from Ref. 54].

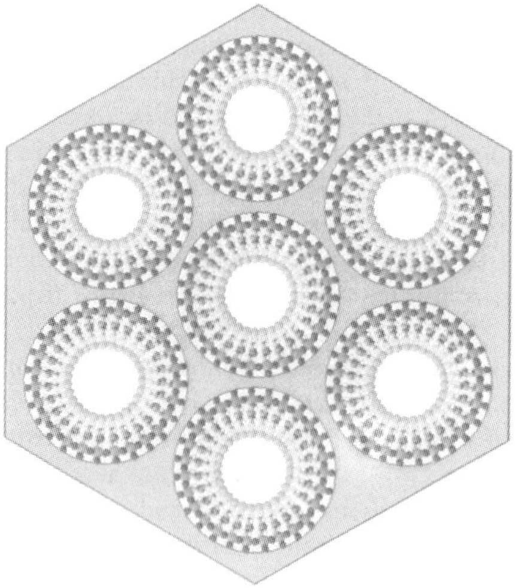

Figure 8-10. A schematic model is given for the functional monolayers on mesoporous supports based on the NMR results, along with results from EXAFS and molecular modeling [Adapted from Ref. 54].

another and have a more upright orientation with respect to the silica surface. The molecules have a higher degree of ordering that narrows the linewidths in the ^{13}C spectrum and allows better resolution of the peaks for all three carbons.

The ^{13}C spectrum for the 76% functionalized monolayer coverage with mercury (Fig. 8-9C) shows the three resonances, corresponding to the C1, C2, and C3 methylene carbons observed in Fig. 8-9B, are still discernible but become much broader. A new broad peak appears at 37 ppm, and the peak at 24.7 ppm decreases significantly. This result suggests strong chemical bonding between the mercury and thiol group, which causes the shift of the peak corresponding to C1 attached to the thiol group. The next C2 group is also affected, but to a lesser degree. The peak at 24.7 ppm indicates the thiol groups are not yet saturated with mercury.

The NMR results, along with results from EXAFS and molecular modeling, provide a solid model (Fig. 8-10) for the functional monolayers on mesoporous supports, based on which the properties of the materials can be predicted. According to this model, the maximum mercury loading is about 700 mg/g of absorbing material, which is exactly what has been observed thus far. The model assumes that all the surface groups are accessible, and, therefore, the surface properties (wetting) should be directly related to the population density of the functional groups. Experimentally a one-to-one correlation between wetting angle of water and the packing density of functional groups has been obtained. The model also allows direct calculation of the final pore size based on the initial pore size and the molecular length of molecules in the monolayers. The calculated result agrees well with results obtained from a nitrogen adsorption technique [56].

8.5 Conclusion

This chapter has focused on the characterization of self-assembled materials using nuclear magnetic resonance (NMR) techniques. The three basic interactions, including chemical shift, dipole-dipole, and electric quadrupole in solid-state NMR under an applied magnetic field, are briefly described. Examples illustrate how an understanding of these interactions can be used to deduce highly specific information about local symmetry and bonding.

High-resolution NMR is sensitive to the short-range structure that is complementary to the longer-range structure provided by imaging and diffraction techniques. The ability to resolve structure over the short distances is especially important for the study of polymers, amorphous materials, biological compounds, and other systems characterized by the absence of long-range order. NMR can provide not only structural information but also the information on the mobility and conformation of functional molecules for the system of interest. The examples in this chapter were chosen to illustrate how NMR techniques can be used to determine phase diagrams which are influenced by long-range ordering on the nanometer scale, the molecular conformation on a sub-nanometer scale, and their combined effects on the interfacial binding. The fundamental knowledge not only allows development of ordered nano-materials, but also helps in understanding structure-property relationships in these ordered mesostructures.

Acknowledgment

This work has been supported by the Division of Materials Sciences Office of Basic Energy Sciences U.S. Department of Energy (USDOE). Pacific Northwest National Laboratory (PNNL) is a multiprogram national laboratory operated for the USDOE by Battelle Memorial Institute (BMI) under Contract DE-AC06-76RLO 1830

References

[1] J. Liu, A. Y. Kim, L. Q. Wang, B. J. Palmer, Y. L. Chen, P. Bruinsma, B. C. Bunker, G. J. Exarhos, G. L. Graff, P. C. Rieke, G. E. Fryxell, J. W. Virden, B. J. Tarasevich, L. A. Chick, *Advances in Colloidal and Interface Science*, **1996**, *69*, 131.

[2] G. M. Whitesides, *Scientific American*, **1995**, *273*, 146.

[3] A. Ulman, *Chem. Rev.* **1996**, *96*, 1533.

[4] B.C. Bunker, P.C. Rieke, B. J. Tarasevich, A. A. Campbell, G. E. Fryxell, G. L. Graff, L. Song, J. Liu, J. W. Virden, and G. L. McVay, *Science* **1994**, *264*, 48.

[5] G. E. Fryxell et al. *Langmuir*, **1996**, *12*, 5064; A. R. Bishop and R. G. Nuzzo. *Cur. Opin. Col. Inter. Sci.* **1996**, *1*, 127.

[6] W. Gao and L. Reven. *Langmuir*, **1995**, *11*, 1860.

[7] J. S. Beck, J. C. Vartuli, W. J. Roth, M. E. Leonowicz, C. T. Kresge, K. D. Schmitt, C. T-W. Chu, D. H. Olson, E. W. Sheppard, S. B. McCullen, J. B. Higgins, and J. L. Schlenker, *J. Am. Chem. Soc.* **1992**, *114*, 10834; C. T. Kresge, M. E. Leonowicz, W. J. Roth, J. C. Vartuli, and J. S. Beck, *Nature*, **1992**, *359*, 710.

[8] M. Temlin, A. Franck, A. D. Chesne, H. Leist, Y. Zhang, R. Ulrich, V. Schadler, U. Wiesner, *Science*, **1997**, *278*, 1795.

[9] D. Zhao, J. Feng, Q. Huo, N. Melosh, G. H. Fredrickson, B. F. Chmelka, G. D. Stucky, *Science*, **1998**, *279*, 548.

[10] J. S. Beck and J. C. Vartuli. *Cur. Opin. Sol. St. Mater. Sci.* **1996**, *1*, 76; J. Liu, A. Y. Kim, L. Q. Wang, B. J. Palmer, Y. L. Chen, P. Bruinsma, B. C. Bunker, G. J. Exarhos, G. L. Graff, P. C. Rieke, G. E. Fryxell, J. W. Virden, B. J. Tarasevich, and L. A. Chick. *Adv. Colloid. Interface. Sci.* **1996**, *69*, 131; N. K. Raman, M. T. Anderson, and C. J. Brinker, *Chem. Mater.* **1996**, *8*, 1682.

[11] A. Sayari, *Chem. Mat.* **1996**, *8*, 1840.

[12] P. J. Bruinsma, N. J. Hess, J. R. Bontha, J. Liu, S. Baskaran, *MRS Symp. Proceedings*, **1997**, *443*, 105.

[13] W. W. Gibbs, *Scientific American*, **1997**, 47; W.C. Holton, *Environmental Health Perspective*, **1998**, *106*, A74.

[14] S. Manne and H. E. Gaub, *Science*, **1995**, *270*, 148.
[15] D. Briggs, Handbook of X-Ray and Ultraviolet Photoelectron Spectroscopy, London, Heyden & Son Ltd, **1977**.
[16] E. A. Stern and S. M. Heald, *Rev. Sci. Instrum*, **1979**, *50*, 1579, 1979; D. E. Sayers and B. A. Bunker, In X-Ray Absorption, Principles, Applications, Techniques of EXAFS, SEXAFS and Xanes (Ed.: D. C. Koningsberger and R. Prins), New York, Wiley, **1988**, p. 211.
[17] O. Terasaki, J. M. Thomas, G. R. Millward and D. Watanable, *Chem. Mater.***1989**, *1*, 158.
[18] B. Cabane, In Surfactant Solutions: New Methods of Investigation (Ed.: R. Zana), New York, Marcel Dekker, Inc, **1987**, p. 57.
[19] P. Goodhew, Specimen Preparation in Materials Science, Amsterdam, North-Holland Publishing Company, **1976**.
[20] Y. Talmon, *Colloid. Surf. Sci.* **1986**, 19.
[21] P. K. Vinson, J. R. Bellare, H. T. Davis, W. G. Miller and L. E. Scriven, *J. Colloid Interface Sci.* **1991**, *142*, 74.
[22] U. Haeberlen, High Resolution NMR in Solids, New York, Acad. Press, **1976**; C. P. Slichter, Principles of Magnetic Resonance, New York, Springer-Verlag, **1978**; A. Abragam, Principles of Nuclear Magnetism, Oxford, Clarendon Press, **1983**; B.C. Gerstein and C.R. Dybowski, Transient Techniques in NMR of Solids, New York, Academic Press Inc. **1985**; M. Mehring, Principles of High Resolution NMR in Solids, Berlin, Heidelberg, New York ,Springer-Verlag, **1983**.
[23] C. P. Slichter, Principles of Magnetic Resonance, New York, Springer-Verlag, **1978**.
[24] A. Pines, M. G. Gibby, and J. S. Waugh, *J. Chem. Phys.* **1973**, *59*, 569.
[25] D. P. Weitekamp, A. Bielecki, D. Zax, K. Zilm, and A. Pines, *Phys. Rev. Lett.* **1983**, *50*, 1807.
[26] C.S. Yannoni and R. D. Kendrick, *J. Chem. Phys.* **1981**, *74*, 747.
[27] J. Baum and A. Pines, *J. Am. Chem. Soc.* **1986**, *108*, 7447.
[28] M. Engelsberg and R. E. Norberg, *Phys. Rev.* **1972**, *B 5*, 3395.
[29] J. A. Reimer and T. M. Duncan, *Phys. Rev. B* **1983**, *27*, 4895.
[30] N. Boden, M. Gibb, Y. K. Levine, and M. Mortimer, *J. Magn. Reson.* **1974**, *16*, 471.
[31] D. E. Kaplan and E. L. Hahn, *J. Phys. Radium*, **1958**, *19*, 821.
[32] P. K. Wang, C. P. Slichter, and J. H. Sinfelt, *Phys. Rev. Lett.* **1984**, *53*, 82.
[33] S. E. Shore, J. P. Ansermet, C. P. Slichter and J. H. Sinfelt, *Phys. Rev. Lett.* **1987**, *58*, 953.
[34] T. Gullion and J. Schaefer, *J. Magn. Reson.* **1989**, *81*, 196.
[35] A. Samoson, E. Lippmaa, A. Pines, *Mol. Phys.* **1988**, *65*, 1013.
[36] Y. Wu, B. Q. Sun, A. Pins, A. Samoson, E. Lippmaa, *J. Magn. Reson.* **1990**, *89*, 297.
[37] A. Llor and J. Virlet, *Chem. Phys. Lett.* 1988, *152*, 248.
[38] K. T. Mueller, B. Q. Sun, G. C. Chingas, J. W. Zwanziger, T. Terao, A. Pines, *J. Magn. Reson.* **1990**, *86*, 470.
[39] J.-P. Amoureux, C. Fernandez, and L. Frydman, *Chem. Phys. Lett.* **1996**, *259*, 347.
[40] A. J. Vega, *J. Magn. Reson.* **1992**, *96*, 50.
[41] A. J. Vega, *J. Solid State NMR.* **1992**, *1*, 17.
[42] C. A. Fyfe, K. C. Wong-Moon, Y. Huang, H. Grondey, and K. T. Mueller, *J. Phys. Chem.* **1995**, *99*, 8707.
[43] G.J.T. Tiddy, *Phys. Lett.* **1980**, *57*, 1; E.S. Blackmore and G.J.T. Tiddy, *Liquid Crystals* **1990**, *8*, 131; A. Belmajdoub, J.C. Boubel, D. Canet, *J. Phys. Chem.* **1989**, *93*, 4844; U. Henriksson, E.S. Blackmore, G.J.T. Tiddy, O. Soderman, *J. Phys. Chem.* **1992**, *96*, 3894.
[44] A. Steel, S.W. Carr, M.W. Anderson, *J. Chem. Soc. Chem. Commun.* **1994**, 1571.
[45] A. Firouzi, D. Kumar, L.M. Bull, T. Besier, P. Sieger, Q. Huo, S.A. Walker, J.A. Zasadzinski, C. Glinka, J. Nicol, D. Margolese, G.D. Stucky, B.F. Chmelka, *Science* **1995**, *267*, 1138.
[46] C. L. Khtrapal, A. C. Kunwar, A. S. Tracey and P. Diehl, in Nulcaer Magnetic Resonance Studies in Lyophobic Liquid Crystals, Berlin-Heidelberg, Springer-Verlag, **1975**.
[47] W.L. Earl, D.L. VanderHart, *Macromolecules* **1979**, *12*, 762; A.E. Tonelli, F.C. Schilling, *Acc. Chem. Res.* **1981**, 223.
[48] R.C. Zeigler, G.E. Maciel, *J. Phys. Chem.* **1991**, *95*, 7345
[49] A. Badia, W. Gao, S. Singh, L. Demers, L. Cuccia, and L. Reven, *Langmuir* **1996**, *12*, 1262.
[50] W. Gao, L. Dickinson, C. Grozinger, F. G. Morin, L. Reven, *Langmuir* **1997**, *13*, 115.
[51] L.-Q. Wang, J. Liu, G. J. Exarhos, in preparation.
[52] A. Kim, P. Bruinsma, Y. Chen, L.-Q. Wang, and J. Liu, *Chem. Commun.* **1997**, 161.
[53] L.-Q. Wang, J. Liu, G. J. Exarhos, and B. C. Bunker. *Langmuir* **1996**, *12*, 2663.
[54] X. Feng, G. E. Fryxell, L.-Q. Wang, A. Y. Kim, J. Liu, and K. M. Kemner, *Science* **1997**, *276*, 923; J. Liu, X. Feng, G. E. Fryxell, L-Q. Wang, A. Y. Kim, M. Gong, *Advanced Materials* **1998**, *10*, 161.
[55] L. Mercier and T. J. Pinnavaia, *Adv. Mater.* **1997**, *9*, 500.
[56] working in progress.

Part II Nanocrystal Systems

9 Photoluminescence from Single Semiconductor Nanostructures

Stephen Empedocles, Robert Neuhauser, Kentaro Shimizu and Moungi Bawendi

Abstract

We review some recent results in the spectroscopy of single CdSe nanocrystal quantum dots. By eliminating the effects of inhomogeneous broadening and ensemble averaging, single nanocrystal spectroscopy has revealed many new and previously unexpected physical phenomena. Among those discussed in this chapter are ultra-narrow emission lineshapes (~600 × narrower than ensemble spectra), a highly polarizable emitting state in the presence of strong local electric fields, linebroadening as a result of environmental fluctuations and shifting of the emission spectra over a wide range of energies (from less than 300 μeV to 80 meV). In addition, polarization spectroscopy of single nanocrystals has revealed the presence of a theoretically predicted 2 dimensional transition dipole moment oriented in the x-y plane of the nanocrystals. As a result, it is, in principle, possible to use polarization spectroscopy to determine the 3 dimensional orientation of individual nanocrystals. These and other studies of single quantum dots have provided us with significant insight into the detailed physics and dynamics of this unique and fascinating physical system.

9.1 Introduction

Low dimensional materials such as quantum wells have had a profound impact in the field of semiconductor physics, yielding numerous fundamental observations as well as important optoelectronic applications. Quantum dots, the zero-dimensional analog of quantum wells, represent the ultimate in semiconductor based quantum confined systems. These structures, often referred to as “artificial atoms”, are predicted to have discrete, atomic-like energy levels, and a sprectrum of ultra-narrow transitions that is tunable with the size of the quantum dot [1–5].

Of particular interest are nanocrystal quantum dots synthesized as colloids. This interest has been fueled by the high degree of reproducibility and control that is currently available in the fabrication and manipulation of these quantum confined structures. CdSe nanocrystals, which are often studied as a prototypical colloidal quantum dot system, can be synthesized in macroscopic quantities with sizes that are tunable during synthesis (15–100 Å) in a narrow size distribution (< 5% rms) [6]. In this size range, the nanocrystals are smaller than the diameter of the bulk Bohr exciton (112 Å for CdSe). As a result, the electronic structure is dominated by quantum confinement effects in all three dimensions [7, 8], making these nanocrystals truly zero dimensional structures.

Once fabricated, nanocrystal quantum dots can be dissolved in a number of polar and non-polar solvents, they can be embedded into a variety of polymers and even incorporated into thin films of bulk semiconductors [9, 10]. They can also be manipulated into close packed glassy thin films [11], ordered three dimensional superlattices (colloidal crystals) [12] or linked to form quantum dot molecules [13]. Such a high degree of flexibility and control makes nanocrystal quantum dots ideally suited for the study of zerodimensional physics.

Size dependent optical properties with band edge absorption and emission wavelengths that are tunable across the visible range (~400–700 nm) makes CdSe nanocrystals of particular interest for the study of fundamental physics as well as potential optoelectronic device application. Unfortunately, the characteristics that make quan-

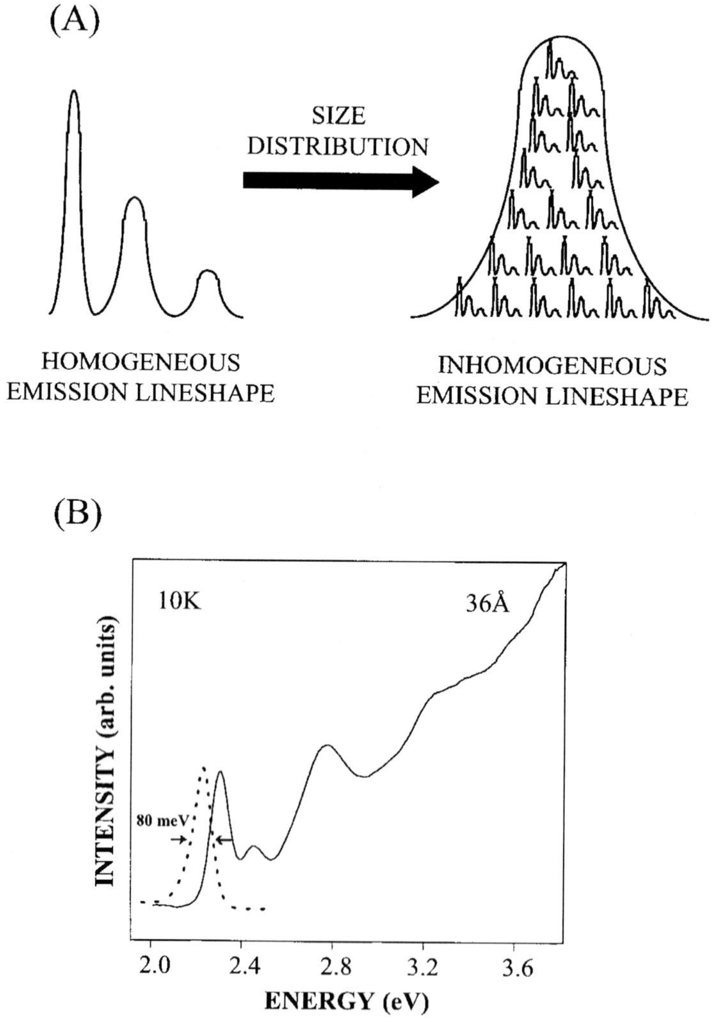

Figure 9-1. Inhomogeneous broadening. (a) Measured ensemble emission and absorption spectra a convolution of the single nanocrystal lineshape with the distribution of sizes within the ensemble sample. (b) Inhomogeneously broadened excitation and emission spectra (solid and dotted lines respectively) for a 36 Å ensemble nanocrystal sample.

tum dots interesting also make them inherently difficult to study. Variations in size and shape within ensemble samples can result in extensive inhomogeneous spectral broadening (Fig. 9-1a). The result is a loss of spectral information in ensemble samples. For instance, theory predicts that quantum dots should have a spectrum of discrete, atomic-like energy states [7, 8]. However, while the discrete nature of the excited states in CdSe nanocrystals has been verified [3, 4, 14], transition linewidths appear significantly broader than expected (Fig. 9-1b). This is true even when size selective optical techniques are used to extract homogeneous linewidths [14–19]. In addition, incoherent dynamic and orientational effects may exist which can be completely hidden when averaged over an ensemble.

In order to eliminate the effects of inhomogeneous broadening and ensemble averaging, it has recently become possible to image and take emission spectra from single quantum dots. Similar to single molecule spectroscopy [20, 21], single quantum dot spectroscopy has been extremely successful in extracting new microscopic information from these ensemble systems. Single quantum dot spectroscopy has been used to study CdSe nanocrystals [22–27] as well as quantum dots fabricated using more traditional epitaxial techniques [30–33]. In this chapter, we discuss some recent results obtained through the spectroscopy of single CdSe nanocrystal quantum dots at cryogenic temperatures.

9.2 Sample Preparation

Two types of nanocrystals were studied. The first were synthesized by the method of Murray et al. [6], as single domain wurtzite crystallites which are slightly prolate along their unique crystal axis (aspect ratio 1.1–1.2). The second type, referred to as "overcoated" nanocrystals were prepared in the same manner, with the addition of a final layer of ZnS [34, 35]. The addition of a ZnS capping layer has been found to have many effects on the physical characteristics of these nanocrystals, the most apparent of which is an increase in the fluorescence quantum yield, reported as high as 50% at room temperature [34, 35]. At the same time, ZnS overcoating has been shown to have only a minor effect on the electronic structure of CdSe nanocrystals. Sizes quoted in this chapter refer to the average diameter of the nanocrystals (or CdSe cores for overcoated nanocrystals) in each ensemble distribution.

Single nanocrystal samples were prepared by spin coating an extremely dilute solution of nanocrystals in poly(methyl methacrylate)/toluene (0.5% PMMA by weight) onto a crystalline quartz substrate. The concentration of nanocrystals in the initial solution ($\sim 10^{-9}$ mol/L) was chosen to produce an areal density of less than 1 nanocrystal/μm^2 in order to allow spatial resolution of individual nanocrystals using far-field optics. The sample substrate was then mounted on the cold finger of a liquid helium cryostat. All data presented in this chapter was taken at 10 K.

9.3 Single Nanocrystal Imaging

Single nanocrystal images were obtained using a far-field, epifluorescence imaging microscope (Fig. 9-2). A typical image of nanocrystals can be seen in Fig. 9-3a. Each bright spot corresponds to the fluorescence from a single nanocrystal. On the timescale of a single image (< 0.5 second), the fluorescence from individual nanocrystals is

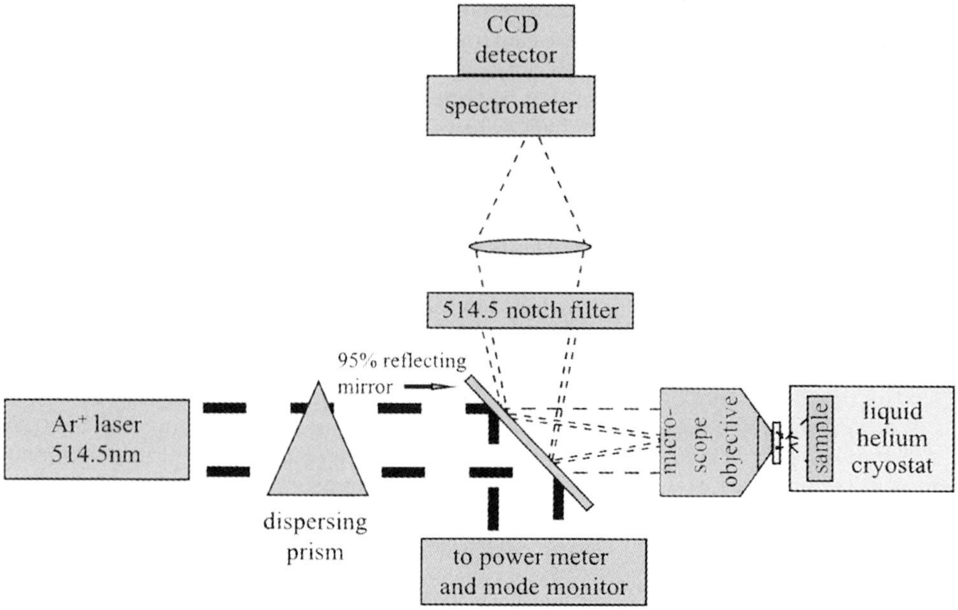

Figure 9-2. Far-field epifluorescence microscope. Excitation light (514 nm unless otherwise stated) was transmitted through a 95% reflecting mirror (at an angle of 45°) and focused by a long working distance microscope objective (NA = 0.7) to a ~30 μm spot on the sample surface. The fluorescent image was collected by the same objective lens, reflected off the 95% reflecting mirror and passed through a wavelength specific filter to remove any excitation light while allowing all of the fluorescence to pass. The image was then focused onto the entrance slit of a spectrometer and detected with a liquid nitrogen cooled charge coupled device (CCD) camera. Both images and spectra were collected on the same detector by switching between the diffraction grating and a mirror at 0 degrees for spectra and images respectively.

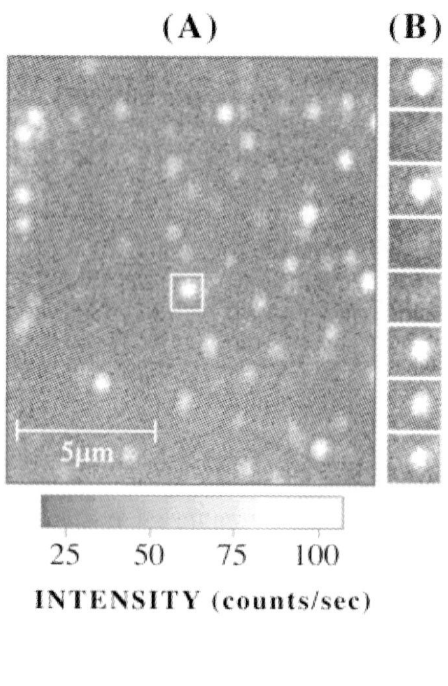

Figure 9-3. Single nanocrystal image. (a) Image of single 45 Å nanocrystals taken with a 0.5 second integration time and an excitation intensity of 60 W/cm^2. (b) Consecutive 0.5 second images of the nanocrystal indicated by the white dotted box in (a). The images show binary fluorescence blinking over time.

occasionally seen to flicker on and off (fluorescence intermittency) (Fig. 9-3b). The pattern of on/off behavior is strongly dependent on the particular nanocrystal as well as the excitation intensity and is thought to result from ionization of the nanocrystal [23]. At 10 K, a ZnS overcoating of 10–14 Å, which acts as a barrier to ionization, can decrease intermittency to the point where blinking occurs on the timescale of many seconds to minutes. In this case, images of single nanocrystals appear relatively stable over time. This binary fluorescent blinking, rather than a stepwise or continuous dimming of the emission from a single point, is strong evidence that the spots seen in Fig. 9-3 originate from single nanocrystals.

9.4 Polarization Spectroscopy

Polarization spectroscopy has long been used to gain insight into the orientation of excitation and emission transition dipole moments in ensemble molecular and solid state systems. However, in many ensemble systems, there is no order in the orientation of individual chromophores. As a result, polarization information is lost. In single chromophore spectroscopy, however, individual transition dipoles can be measured one at a time. As a result, information can be obtained regarding the orientation of individual chromophores even within an amorphous matrix [36, 37]. In single nanocrystals, this information can be extremely valuable since many theoretical predictions have been made regarding the effects of various perturbations, such as electric [38] and magnetic fields [5, 39], depending on the relative orientation of the nanocrystal.

Emission polarization can be studied in single nanocrystals by measuring changes in the detected emission intensity as a linear polarizer (analyzer) is rotated in the detection pathway. Polarization data was taken with an analyzer in front of the CCD detector that was rotated in 15 degree increments between consecutive images. Polarization data was also taken in spectral mode, however, in this case, a depolarizer was added after the analyzer so that no polarization selectivity from the diffraction grating would contribute to the observed results.

In order to verify that the experimental apparatus had no intrinsic polarization selectivity which might affect the single nanocrystal data, two non-polarized samples were studied: Amorphous ensemble nanocrystal films and 1 μm latex spheres filled with an ensemble of randomly oriented dye molecules (Nile Red). Both of these samples should have no intrinsic polarization dependence due to the randomly oriented nature of the individual chromophores within each ensemble system. In order to prevent selectively exciting one orientation over another, thus artificially producing a polarization dependence, the *calibration measurements only* were made with circularly polarized excitation. In both samples, no emission polarization dependence was found, confirming that there is no polarization selectivity in the experimental apparatus.

The inset of Fig. 9-4 shows images of two single nanocrystals, which are excited with linear polarized light, as a function of analyzer angle. The intensity of the two nanocrystals can be seen oscillating out of phase with each other as the analyzer is rotated. Also plotted in Fig. 9-4 are the normalized emission intensities as a function of angle for these two nanocrystals. The intensities can be clearly seen oscillating with a high degree of polarization $[(I_{max}-I_{min})/I_{max} = 81\%$ and $93\%]$ as a sine-squared function of the analyzer angle with a 180 degree period. Rotation of the excitation polarization by 90 degrees had no observable effect on the degree of emission polarization or phase. The high degree of polarization and lack of a polarization memory

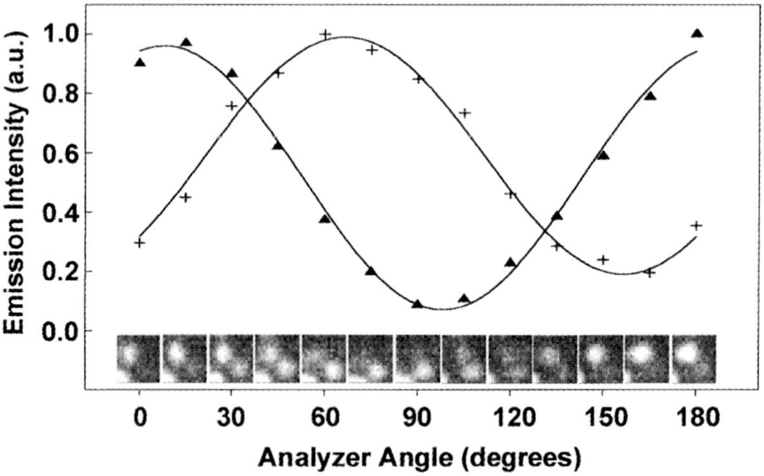

Figure 9-4. Single nanocrystal polarization spectroscopy. Normalized emission intensity of two single 52 Å overcoated nanocrystals as a function of analyzer angle. The solid lines correspond to the fit to a sine-squared function of analyzer angle with a 180 degree period. Data in Fig. 9-4 is the average of 2 complete 180 degree rotations. Inset) Images of the two nanocrystals in Fig. 9-4 as a function of analyzer angle (angle can be read directly off the x-axis below each image). Data was taken with a 5 second integration time, 60 W/cm^2 excitation intensity and 0 degree excitation polarization. Data in Fig. 9-4 are from ref. [27].

suggests the existence of a highly oriented electronic transition dipole. What is surprising is that the degree of polarization is strongly dependent on the particular nanocrystal. Figure 9-5 plots the intensity of three singly nanocrystals as a function of analyzer angle. As can be seen, even within the same sample, different nanocrystals reveal different degrees of polarization.

The data in Fig. 9-5 is quite different from what is typically observed in other single chromophore systems. In the single molecules studied to date, a unidirectional transition dipole exists, creating a single "bright axis" along which the emitted light is coupled. The strength of emission as a function of polarization is proportional to $|\boldsymbol{\mu} \cdot \boldsymbol{E}|^2$, where $\boldsymbol{\mu}$ is the transition dipole vector and $\boldsymbol{E}$ is the polarization of the emitted light. The intensity of the detected signal is then proportional to $\cos^2(\theta)\cos^2(\phi)$ where θ is the angle between the emission polarization and the projection of $\boldsymbol{\mu}$ onto the sample plane (the plane normal to the detection axisw), and ϕ is the tilt angle between $\boldsymbol{\mu}$ and the sample plane (out-of-plane angle). When a polarizer is rotated in the emission pathway, the detected intensity oscillates between $I_{\min} = 0$ and $I_{\max} = |\mu|^2 \cdot \cos^2(\phi)$*. The result is 100% polarized emission regardless of the orientation of the molecule. Less than 100% polarization has been observed in single pentacene molecules in p-terphenyl, however, in this case the loss of polarization was the result of birefringence of the crystalline p-terphenyl matrix turning the linear polarized light into elliptical light [40, 41].

* This result is only strictly true for a radiating dipole in free space, observed through a lens with an infinitely small collection angle. In reality, the large numerical aperture of the microscope objective used (NA = 0.7) allows some detection of light emitted parallel to the collection axis, which means that, in principle, some "bright axis" transitions can be found which do not show 100% polarization. In practice, however, these dipoles are rarely seen since they also emit the least amount of light in the direction of the detector. As a result, virtually all single chromophores with a "bright axis" show 100% polarization. Later in this chapter, we will explicitly address the effects of collection angle and the radiation pattern of a dipole on a dielectric surface.

The differences observed in Fig. 9-5 are not an artifact of the experimental or sample preparation procedure such as those described in [40, 41]. To verify this, single DiI molecules, which have previously been shown to have a "bright axis" [36, 37] were studied. In our experimental apparatus, the average degree of polarization for the 10 single DiI molecules studied was found to be ~100%, consistent with a "bright axis" transition (Fig. 9-6b inset). In addition, results nearly identical to those seen in Fig. 9-5 have also been obtained for nanocrystals that were deposited onto the substrate surface in hexane with no surrounding matrix.

The varying degrees of polarization shown in Fig. 9-5 are actually the result of the nature of the emission transition dipole. Due to the prolate shape and unidirectional wurtzite crystal structure, theoretical calculations have predicted two potential transition dipole orientations for the band edge states of CdSe nanocrystals [42]. The first is similar to what was described above for single molecules, with a unidirectional "bright axis" oriented parallel to the wurtzite c-axis. Transitions involving these states should have a polarization dependence similar to what is seen for single DiI molecules, with a high degree of polarization observed for every nanocrystal. The other potential orientation is a doubly degenerate 2 dimensional transition dipole oriented isotropically in the x-y plane of the nanocrystal. In this case, what remains is a unidirectional "dark axis" along which no light is emitted. This "dark axis" is oriented parallel to the c-axis.

Transitions involving states with a "dark axis" should show very different polarization dependence than "bright axis" states. The detected emission intensity for a "dark axis" transition is proportional to $(1-\cos^2(\theta)\cos^2(\phi))$ where θ and ϕ are defined as above relative to the "dark axis" orientation. In this case, as the emission analyzer is

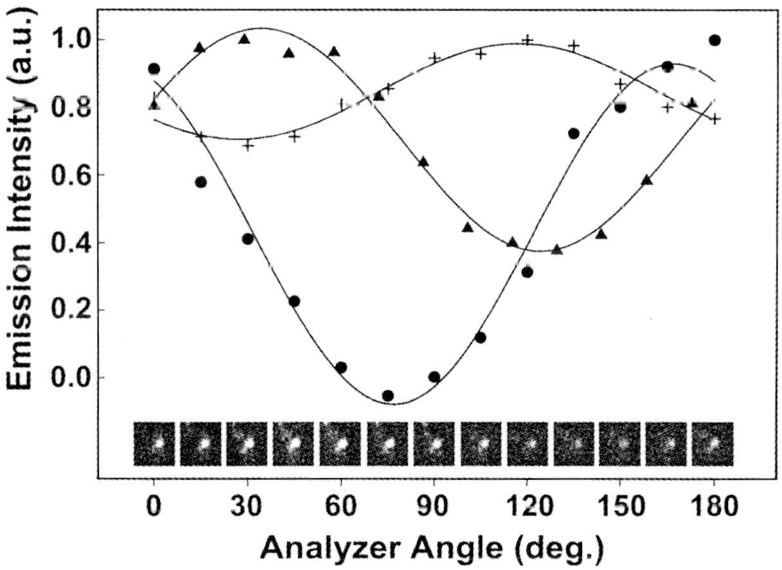

Figure 9-5. Single nanocrystal polarization spectroscopy. Normalized emission intensity of three single 52 Å overcoated nanocrystals as a function of analyzer angle. The solid lines correspond to the fit to a sine-squared function of analyzer angle with a 180 degree period. Data in Fig. 9-5 is the average of 6 complete 180 degree rotations. Inset) Image of one nanocrystal from Fig. 9-5 (triangles) as a function of analyzerangle (angle can be read directly off the x-axis below each image). All data was taken with a 5 second integration time, 60 W/cm^2 excitation intensity and 0 degree excitation polarization. Data in Fig. 9-5 are from ref. [27].

rotated, the strength of the transition oscillates between $I_{max}=|\mu|^2$ and $I_{min} = |\mu|^2 \cdot \cos^2(\phi)$. As a result, the degree of polarization will vary from 0 to 100% depending on the out-of-plane angle (ϕ). The data in Fig. 9-5 suggests that emission in CdSe nanocrystals results from a "dark axis" state. This is not surprising since the two lowest optically active states in CdSe nanocrystals are predicted to have "dark axis" transition dipoles. Since the lowest excited state is formally spin forbidden, emission is likely to proceed through some form of mixing or phonon mediated transition involving one of these near-lying optically active states.

An interesting characteristic of "dark axis" transitions is that, since the absolute value of μ can be measured directly ($I_{max}=|\mu|^2$), the out of plane angle (ϕ) can be estimated based on the degree of polarization observed. As a result, it is, in principle, possible to use polarization spectroscopy to directly measure the 3 dimensional orientation of individual nanocrystals from the phase (in-plane angle) and degree of polarization (out-of-plane angle) of each. Note, however, that there is no way to distinguish between angles into (ϕ) and out of ($-\phi$) the sample plane.

Figure 9-6 displays data from 176 single nanocrystals showing the phase and degree of polarization for each. As can be seen, a broad distribution of parameters is observed which covers for entire range of phase angles and degrees of polarization. It should be noted that while there appears to be a relatively isotropic distribution of phase angles, the distribution of polarizations is somewhat center weighted. There are two additional effects which contribute to the predicted polarization dependence of both "bright" and "dark" axis transitions and therefore affect the shape of the

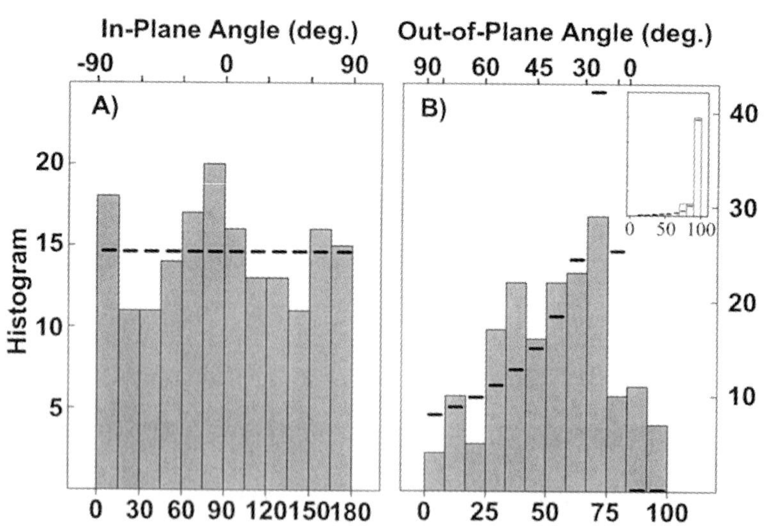

Figure 9-6. Polarization statistics. (A) Histogram of polarization phase angles for 176 single 52 Å overcoated nanocrystals. Top axis represents the corresponding in-plane orientation (θ) of the nanocrystal c-axis. (B) Histogram of degrees of polarization for the same 176 nanocrystals. Top axis represents the corresponding out of plane angle (ϕ) of the c-axis relative to the sample plane. Note that the top axis is non-linear and does not directly correspond to the histogram values below. Solid lines in (A) and (B) correspond to the calculated probability histogram for an isotropic distribution of "dark axis" dipole orientations, including the effects of surface and finite collection angle. Inset) Histogram of the degrees of polarization of 10 single DiI molecules. Solid lines represent the calculated distribution of "bright axis" orientations. Data in Fig. 9-6 are from ref. [27].

expected probability distribution in Fig. 9-6b. First, we have only considered the radiation pattern from an isolated dipole in free space. However, the pattern of radiation from a dipole in a thin film on a dielectric interface has been found, both theoretically [43, 44, 45] and experimentally [46] to be significantly different than that of a free dipole. Second, we have so far assumed an infinitely small collection angle in our detection optics. This is not a good assumption, since the large numerical aperture (NA = 0.7) of the microscope objective used allows the detection of photons emitted with an angle of as much as ~45 degrees from normal. The result is that it is possible to detect some photons which are emitted with a polarization perpendicular to the sample plane. In the simple discussion of the polarization dependence of "bright" and "dark" axis states, this additional contribution was neglected. The sum of these two effects slightly alters the expected polarization dependence of both "bright" and "dark" axis transitions (although the general conclusions remain the same). The calculated probability histogram (solid lines) in Fig. 9-6 and 9-6(inset) include the effect of the surface and collection angle by integrating the theoretical radiation patterns of a dipole in a thin film over the full collection angle of the microscope objective. The fits are consistent with an isotropic distribution of "dark" and "bright" axis orientations respectively.

From the data in Fig. 9-6a and b, the in-plane and out-of-plane orientation of each nanocrystal c-axis can be calculated. The top axes have been added to indicate the corresponding angles and include the contribution of the surface and finite collection angle described above. A knowledge of the orientation of single nanocrystals relative to the experimental frame of reference can be an extremely valuable tool in interpreting single nanocrystal experiments. Man theoretical predictions which can be easily studied on the single nanocrystal level rely on a precise knowledge of the orientation of the c-axis relative to applied perturbations such as electric [38] and magnetic fields [5, 39]. In addition, changes in the orientation of the transition dipole over time could indicate either rotation of the nanocrystal within the sample or significant changes in the electronic structure.

9.5 Single Nanocrystal Spectroscopy

In order to take spectra from single nanocrystals, the entrance slit of the spectrometer is used to spatially isolate individual nanocrystals along a vertical region of the image (Fig. 9-7a and b). As can be seen, it is usually possible to align several nanocrystals at different vertical positions within the narrowed slit. The mirror in the spectrometer is then replaced with a diffraction grating and light from each vertical position is dispersed onto the CCD (Fig. 9-7c). What results is an image in which single nanocrystals are spectrally dispersed along the x-axis and are vertically separated along the y-axis corresponding to their spatial position within the entrance slit. In this way, it is possible to simultaneously obtain spectra from several different nanocrystals. In Fig. 9-7c, 4 single nanocrystal spectra are collected, however, careful manipulation of the sample concentration can allow the simultaneous detection of up to 100 spectra.

Figures 9-8a and b show a comparison between a representative sample of single nanocrystal spectra and the corresponding ensemble spectra from those samples. Note that Fig. 9-8a and b show data from 2 different size samples. Due to quantum confinement, there is a shift of the emission energy of both the single nanocrystal spectra as well as the ensemble spectra when the average size is changed. As expected, the ensemble emission spectrum can be reproduced by a convolution of the average

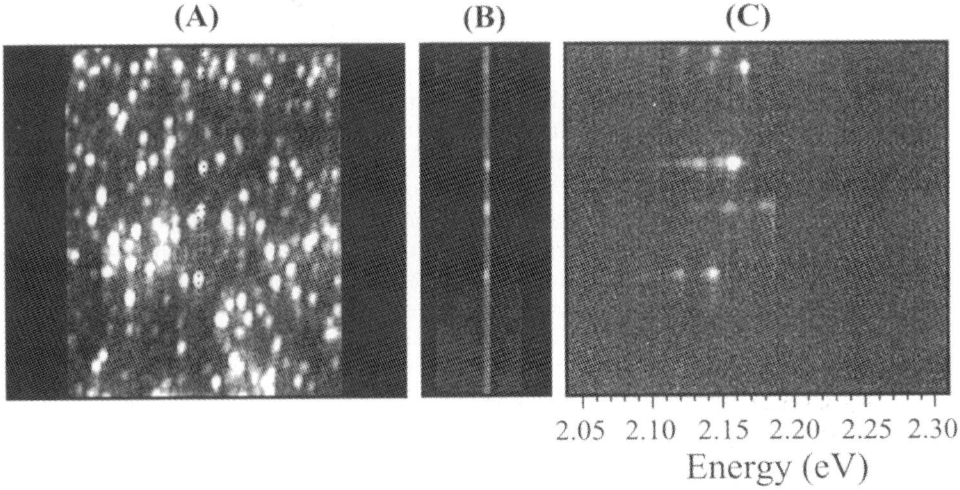

Figure 9-7. Single nanocrystal imaging and spectroscopy. (a) Image of single 45 Å nanocrystals at 10K with the entrance slit on the detector/spectrometer system open to 7 mm. White dotted lines represent the center of the entrance slit. (b) Image of the same region as in (a) with the entrance slit closed to 0.125 mm. Several bright spots can still be seen between the narrowed slits. (c) Spectrally dispersed image of the entrance slit in (b). This image consists of many single nanocrystal spectra plotted along the x-axis, each separated along the y-axis according to their vertical position between the entrance slit. For each single nanocrystal between the narrowed slit, there is a corresponding single nanocrystal spectrum. Note that the center nanocrystal spectrum is very dim, suggesting that this nanocrystal stopped emitting sometime during the 30 second integration time.

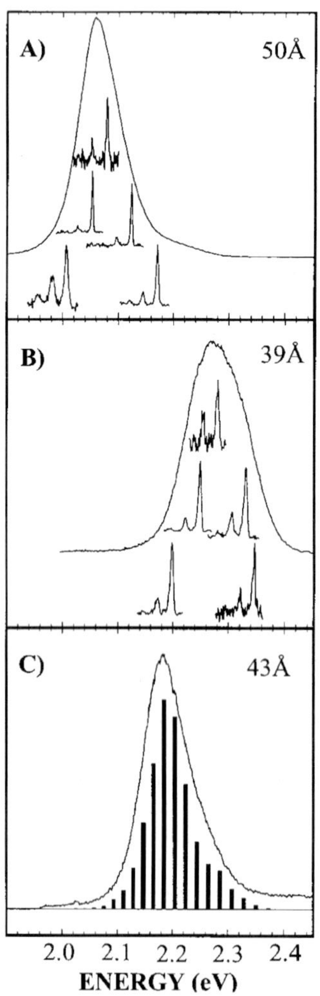

Figure 9-8. Ensemble vs single nanocrystal spectra for three different sample sizes. a) and b) Ensemble spectra with a representative set of single nanocrystal spectra obtained from these samples. c) Ensemble spectrum with a histogram of integrated intensities for 513 single nanocrystal spectra obtained from that sample. The histogram includes the scaled contribution of zero, one and two phonon lines from each nanocrystal. All single nanocrystal spectra were taken with a 60 second integration time and 100 W/cm^2 excitation intensity. Data are from ref. [22].

single nanocrystal lineshape with the distribution of zero-phonon energies measured within the sample (Fig. 9-8c). It is the highly parallel detection scheme described above that allows such comparisons to be made. The close agreement between the ensemble average distribution and the average of the distribution of single dot spectra is an important result. It demonstrates that the nanocrystals observed in these experiments are not somehow anomalous, but are representative of the ensemble distribution.

All single nanocrystal spectra show qualitatively similar spectral characteristics including narrow peaks and a longitudinal-optical (LO) phonon progression with peak spacing comparable to the bulk LO phonon frequency. Differences in phonon coupling are observed between different nanocrystals (see Fig. 9-8a and b) with an average value that is equal to the phonon coupling measured in ensemble samples [22]. The range of phonon couplings measured in these experiments is something that can only be seen on the single nanocrystal level and represents a clear example of ensemble averaging, in which information about the individual members of a distribution is lost when averaged over the ensemble.

9.6 Spectral Diffusion

Examination of many single nanocrystal spectra reveals a wide range of linewidths and non-Lorentzian lineshapes. In addition, the lineshape of individual nanocrystals is found to be extremely sensitive to experimental parameters such as excitation intensity [Fig. 9-9a-d]. This is true even though excitation occurs far from the emitting state (> 200 meV) so that traditional power or saturation broadening should be insignificant. There is also a strong lineshape dependence on integration time [Fig. 9-9e]. This is similar to what is seen in single molecule spectroscopy where linewidths are found to be dependent on acquisition time [47 51]. In single molecule spectroscopy, lineshapes are thought to arise from rapid shifting of the emission energy (spectral diffusion) on a timescale that is fast relative to the acquisition time of the experiment. These spectral shifts result from changes in the local environment which interact with single molecules through short range strain fields [49]. In these cases, the lineshape of single molecule spectra are determined by the dynamics of the local environment and not the intrinsic physics of the molecule.

Figure 9-10 demonstrates that single nanocrystal lineshapes are also dominated by spectral diffusion. Figure 9-10a shows a high resolution spectrum of the zero phonon line of a single nanocrystal with a 10 second integration time. Spectral features include a linewidth of ~2 meV as well as a second peak, shifted by ~3 meV. While it may be tempting to interpret the lineshape of this spectrum in terms of intrinsic nanocrystal physics, Fig. 9-10b indicates that this is not accurate. Figure 9-10b shows several spectra of the same nanocrystal with a 0.1 second integration time. At such short integration times, a single resolution limited peak is seen shifting in energy across the full spectral range occupied by the 10 second spectrum. A histogram of peak positions from 150 consecutive 0.1 second spectra demonstrates that the entire lineshape of the 10 second spectrum is a consequence of spectral diffusion [figure 9-10c].

The observation of spectral diffusion in these nanocrystals is actually quite surprising. While single molecules are exquisitely sensitive to changes in their surrounding environment [21, 20], this was not expected to be true for nanocrystals. The energies of the delocalized exciton in CdSe nanocrystals do not appear to be sensitive to the

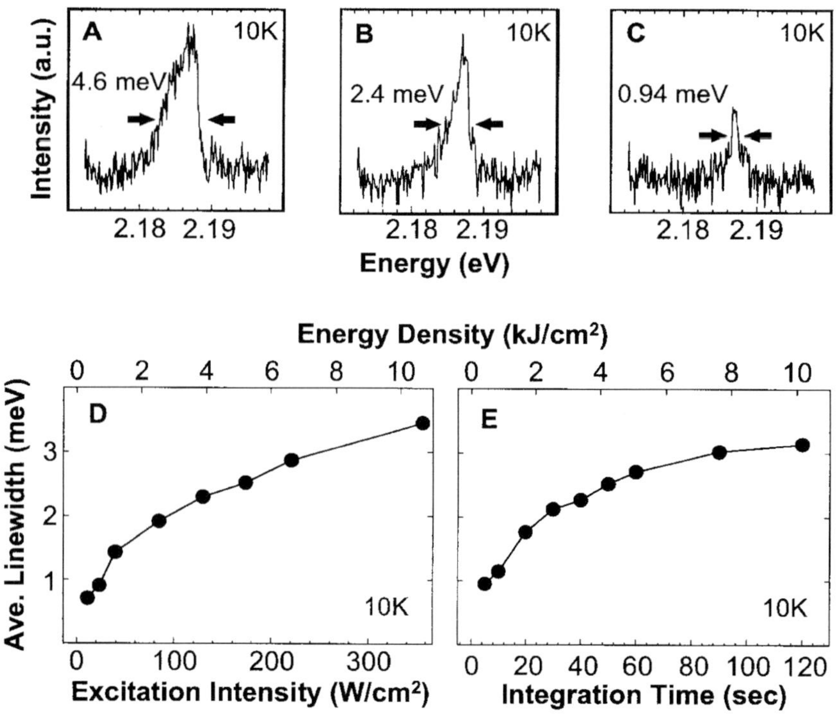

Figure 9-9. Effect of experimental conditions on single nanocrystal lineshapes. (a)–(c) Three spectra of the same single 45 Å nanocrystal with a 1 minute integration time and 314, 150 and 65 W/cm^2 excitation intensity respectively. (d) Average single nanocrystal linewidth as a function of excitation intensity for a sample of 34 overcoated nanocrystals (56.5 Å) with a integration time of 30 seconds. (e) Average single nanocrystal linewidth as a function of integration time for 40 nanocrystals from the same sample as (d), with 85 W/cm^2 excitation intensity. The top axes of (d) and (e) indicate the excitation energy density (time x intensity) at each point in order to allow a direct comparison the effects of time and intensity on the average single dot linewidth. Data in (a)–(c) are from ref. [22]. Data in (d) and (e) are from ref. [26].

chemical nature of the nanocrystal surface [52] and should therefore be relatively unaffected by small changes in the surrounding environment. As such, the perturbing force required to produce the observed spectral shifts must be quite large.

There are two main implications of the data in Fig. 9-10. First, it is clear that the lineshape of a single nanocrystal contains information about changes in the surrounding environment and not the intrinsic physics of the nanocrystal. Second, changes in the lineshape of a single nanocrystal spectrum resulting from different experimental conditions (Fig. 9-9) are likely to be the result of changes in spectral diffusion. In the case of broadening as a function of excitation intensity, this can be observed directly (Fig. 9-11). The magnitude and frequency of these small spectral shifts are found to increase with increasing excitation intensity. In addition, the linewidth of a single nanocrystal spectrum is actually dependent on the wavelength of excitation, with excitation closer to the band edge resulting in narrower emission spectra [26]. This suggests that the contribution of excitation intensity is in the form of energy released as the exciton relaxes to its emitting state. This is further supported by the close overlap of the time and intensity curves in Fig. 9-9d and e when plotted against excitation energy density (top axis). A direct comparison reveals that changing the excitation

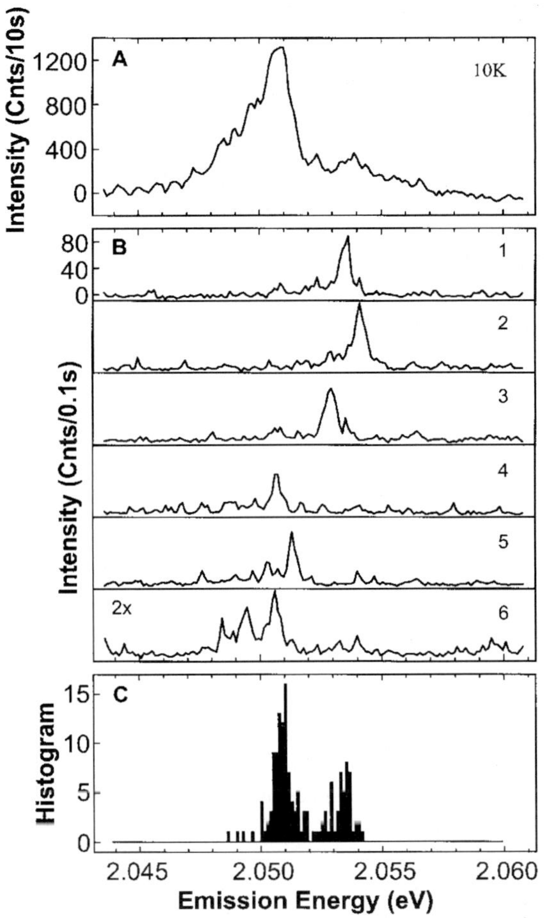

Figure 9-10. Contribution of spectral diffusion to single nanocrystal lineshapes. (a) Spectrum of a single 56.5 Å overcoated nanocrystal with a 10 second integration time at 10K. (b) A representative sample of 6 spectra from the same nanocrystal as in (a), with a 0.1 second integration time. Of 150 spectra taken in rapid succession (total ~15 sec), frames 1–6 indicate spectrum number 1, 16, 35, 59, 84 and 150 respectively. (c) Histogram of peak positions from each of the 150 spectra described in (b). Excitation intensity for all spectra in figure 10 was 200 W/cm^2. All data in Fig. 9-10 are from ref. [26].

intensity has the same effect on the average single nanocrystal linewidth as changing the integration time by the same relative amount. This overlap suggests that the observed broadening is related to the number of excitations (and therefore the amount of excess excitation energy released) during the acquisition of a single spectrum.

The effect of excess excitation energy is not the result of heating. Heating is expected to be insignificant in these experiments due to rapid dissipation of heat into the surrounding matrix ($\sim 10^5$ excitations/second with $\Delta T \approx 6$K/excitation which dissipates in $< 10^{-12}$ seconds). Instead, phonons emitted as the exciton relaxes to its lowest excited state may couple directly to the local environmental changes responsible for spectral diffusion. While we have not yet discussed the nature of these environmental changes, evidence which will be presented later in this chapter suggests that spectral diffusion is the result of changing local electric fields around individual nanocrystals.

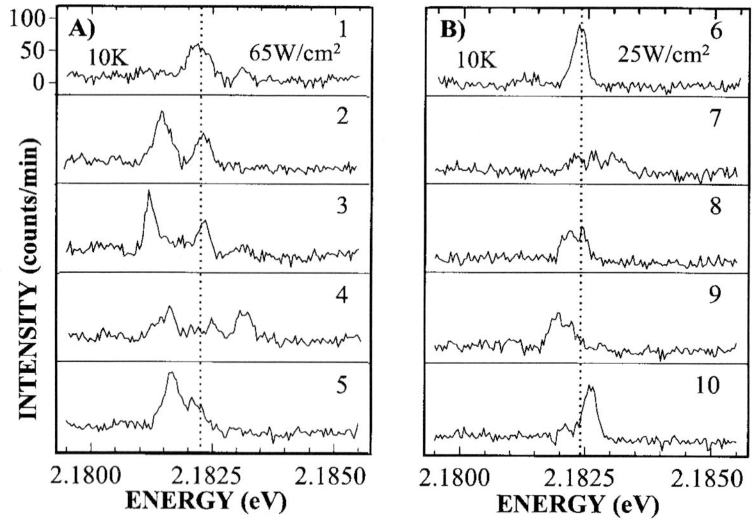

Figure 9-11. Effect of excitation intensity on small spectral shifts. (a) Five consecutive 60 second spectra of a single 43 Å overcoated nanocrystal at 65 W/cm^2 excitation intensity. (b) The next five consecutive spectra of the same nanocrystal at 25 W/cm^2, showing a decrease in the number and magnitude of the small spectral shifts observed. Data from Fig. 9-11 are from ref. [22].

These fields are thought to result from charges trapped on or near the surface of the nanocrystal. The role of the released excess excitation energy may be to facilitate movement of these charges between local trap sites, resulting in fluctuations in the local electric field which produce the observed spectral diffusion shifts.

By decreasing the excitation intensity, it is possible to reduce spectral diffusion in some nanocrystals to the point where shifts that are observable above our resolution limit occur on a timescale of many seconds to minutes. At these intensities, resolution limited linewidths as narrow as 120 μeV have been observed at 10K (Fig. 9-12.). These peaks are ~600 × narrower than the full ensemble fluorescence. This finding is consistent with theoretical predictions and reinforces the description of these nanocrystals as “artificial atoms”.

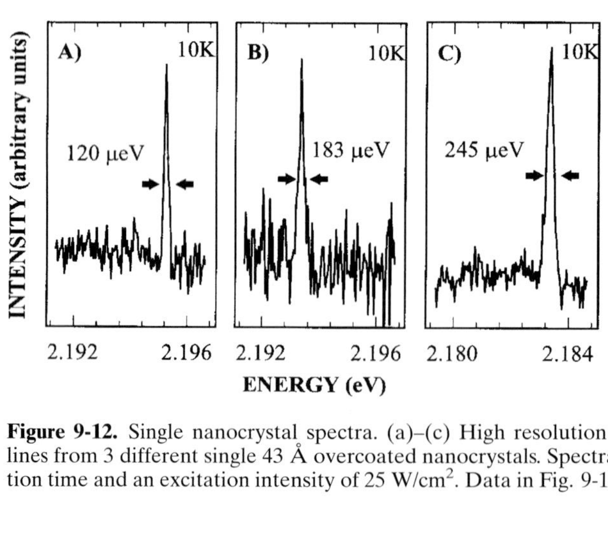

Figure 9-12. Single nanocrystal spectra. (a)–(c) High resolution spectra of ultra-narrow zero phonon lines from 3 different single 43 Å overcoated nanocrystals. Spectra were taken with a 60 second integration time and an excitation intensity of 25 W/cm^2. Data in Fig. 9-12 are from ref. [22].

9.7 Large Spectral Diffusion Shifts

While the small spectral shifts described above occur on a very fast timescale, a second type of spectral shifting is also observed. These shifts result in much larger changes in the emission energy and occur on a timescale from seconds to many minutes, even under high excitation intensities. Figure 9-13 shows consecutive low resolution spectra of a single nanocrystal which reversibly shifts in energy more than 80 meV over 16 minutes. The large decrease in emission intensity as the spectrum shifts to the red may be the result of the absorbing states shifting out of resonance with the exciting laser**.

The most interesting feature of Fig. 9-13 is that as the spectrum shifts to lower energies, there is a significant increase in coupling to LO phonons which is measured as the ratio of the integrated intensity of the zero and one phonon peaks. Phonon coupling is a very sensitive measure of the overlap between the electron and hole wavefunctions within the nanocrystal core (Fig. 9-14). A separation between the electron and hole results in a temporary electric field within the nanocrystal. This field tends to distort the ionic CdSe crystal lattice by pulling the positively charged Cd^{2+} ions toward the electron and the negatively charged Se^{2-} ions toward the hole (Fig. 9-14b). When the exciton recombines, the temporary field is removed and the distorted crystal lattice relaxes to its equilibrium state resulting in vibrations (Fig. 9-14c). The result

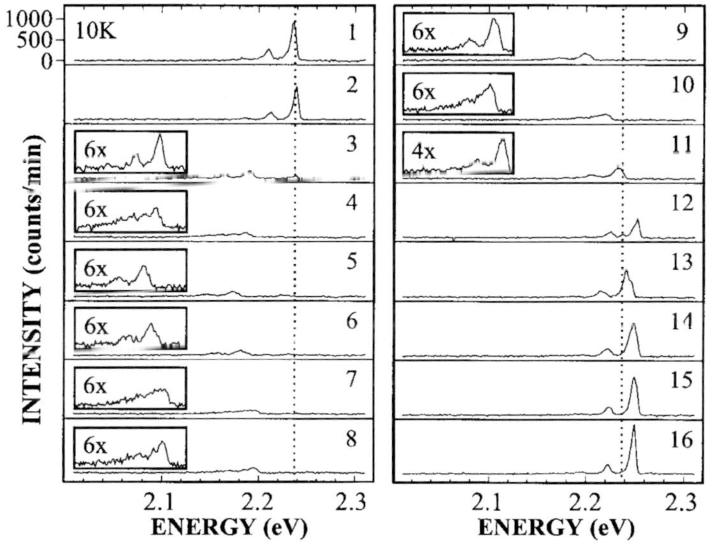

Figure 9-13. Spectral diffusion. Sixteen consecutive 60 second spectra of the same single 39 Å nanocrystal with an excitation intensity of 2.5 kW/cm^2. Insets show magnification of the y-axis by the indicated amounts. Data for Fig. 9-13 are from ref. [22].

** In these experiments the nanocrystals were excited far above the band edge, in a region with a high density of states and should therefore be relatively insensitive to small spectral diffusion shifts. However, even in ensemble excitation spectra, shifts as large as those observed in Fig. 9-13 could be enough to move the inhomogeneously broadened absorbing state out of resonance with the exciting laser.

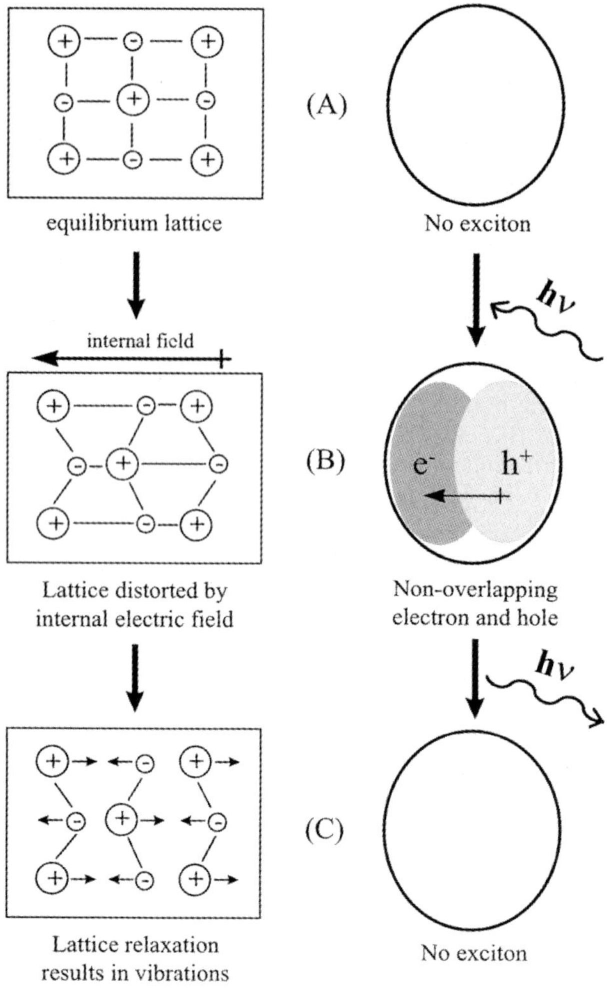

Figure 9-14. Frolich coupling to optical phonons. (a) In the absence of an exciton the crystal lattice sits at an equilibrium position. (b) An exciton with a separation between the electron and hole wavefunction creates a temporary internal electric field which distorts the ionic crystal lattice. (c) When the exciton recombines, the internal field disappears and the atoms relax to their equilibrium positions, creating lattice vibrations.

is enhanced coupling of the emitted photon to optical phonons. This interaction is called Frolich coupling and the degree of coupling is related to the amount of separation between the electron and hole wavefunctions.

The changes in phonon coupling observed in Fig. 9-13 suggests that these large spectral diffusion shifts are accompanied by a change in the overlap of the electron and hole wavefunctions. One possible source of such a change is a local electric field which could pull the electron and hole wavefunctions toward opposite sides of the nanocrystal. In fact, the spectral characteristics seen in Fig. 9-13 are very similar to what is observed in low temperature ensemble Stark experiments [53]. In the presence of an applied electric field, ensemble emission spectra are seen to shift to lower energies with a corresponding increase in phonon coupling. The observed similarities suggest that perhaps changing local electric fields around individual nanocrystals may play a role in spectral diffusion.

9.8 Stark Spectroscopy

To further investigate the role of electric fields in spectral diffusion, it is possible to make direct measurements of the quantum confined Stark effect in single nanocrystals [25]. In addition to potentially contributing to the understanding of spectral diffusion, the quantum confined Stark effect can also be used to probe the nature of the excited states in these nanocrystals. For instance, delocalized exciton states within the nanocrystal core should be highly polarizable, while localized surface trap states should have a strong dipole character. These two types of states should be easily distinguished by the application of an electric field. In the past, however, inhomogeneous broadening and ensemble averaging have greatly complicated the interpretation of ensemble Stark measurements. For example, while the presence of an excited state dipole has been suggested in ensemble Stark absorption studies [54, 55], nearly identical Stark data have also been interpreted without the need for a polar state [56, 57]. By eliminating the effects of ensemble averaging, single nanocrystal spectroscopy should allow a conclusive determination of the nature of the emitting state.

Figure 9-15a shows the zero phonon line of a single nanocrystal under a range of electric fields. The peak can be seen shifting continuously over more than 60 meV. The slight change in zero-field energy over the series is due to spectral diffusion. Although the first absorbing state has not yet been characterized for single CdSe nanocrystals, even conservative estimates based on ensemble measurements suggest that these shifts are more than an order of magnitude greater than the width of this state [58]. Shifts of this magnitude support the potential of these nanocrystals for use in electro-optic modulation devices. Shifts observed in these experiments are even comparable to room-temperature single nanocrystal linewidths [23, 59], indicating the potential for use in non-cryogenic devices.

Figure 9-15b plots Stark shift as a function of field for the nanocrystal in Fig. 9-15a. The Stark data can be fit with the sum of a linear and quadratic function of the applied field, indicating the presence of both polar and polarizable character in the emitting state***. While a wide range of values for the excited state dipole along the applied field axis are measured between individual nanocrystals (Fig. 9-16a), all single nanocrystals demonstrate this dual character.

Previous ensemble Stark measurements of CdSe nanocrystals revealed a purely quadratic shift of the peak emission energy as a function of applied field [53]. In light of the current single nanocrystal experiments, this result can be explained as the product of ensemble averaging. While individual dots reveal a strong linear component in the Stark shift, the average dipole over the ensemble is zero (Fig. 9-16b). The result is that ensemble samples should have a quadratic shift of the peak emission as a function of field with an additional broadening due to the linear contribution of individual

*** What is measured in these experiments is actually the difference between the excited and ground state dipole (polarizability). For simplicity in discussion, however, we refer to this value as the excited state dipole (polarizability). This is a reasonable assertion since, to a first approximation, any structural (ground state) dipole (polarizability) should be relatively unaffected by the presence of a delocalized excitation.

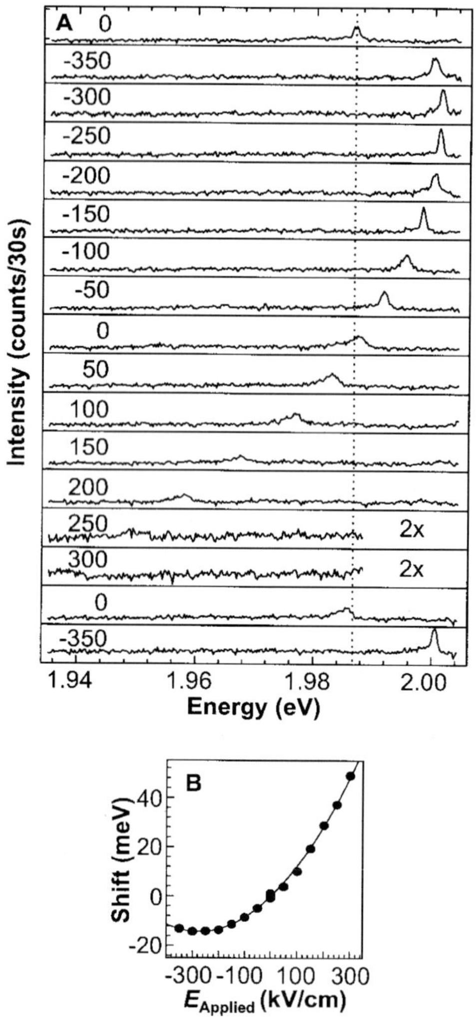

Figure 9-15. Single nanocrystal Stark spectroscopy. (a) Seventeen spectra of the same single 75 Å overcoated nanocrystal under a range of electric fields. Spectra were taken with an excitation intensity of 25 W/cm^2 and an integration time of 30 seconds. Inset numbers indicate the applied field in kV/cm and magnification of the y-axis. (b) Plot of Stark shift versus electric field for the single nanocrystal in (a) with fit to the sum of a linear and quadratic function of the applied field. Data in Fig. 9-15 are from ref. [25].

nanocrystals. Consistent with this, ensemble experiments reveal a broadening of the emission with increasing electric field. Ensemble studies are further complicated, however, by additional broadening due to changes in phonon coupling. In the past, ensemble emission broadening has been attributed entirely to changes in phonon coupling [53]. These single nanocrystal results, however, indicate that a polar contribution must also be considered.

It is important to note that the excited state dipole measured in these Stark experiments is completely unrelated to the transition dipole moment measured in the polarization experiments described above. For instance, the application of a large electric field can induce an excited state dipole in single nanocrystals which is oriented along the applied field axis. However, when a large electric field ($2.5 \cdot 10^5$ V/cm) is applied to a single nanocrystal sample, no change in the orientation of the transition dipole is observed (Fig. 9-17). This is not surprising, since the static excited state dipole is primarily a function of the electron and hole envelope wavefunctions [57], while the

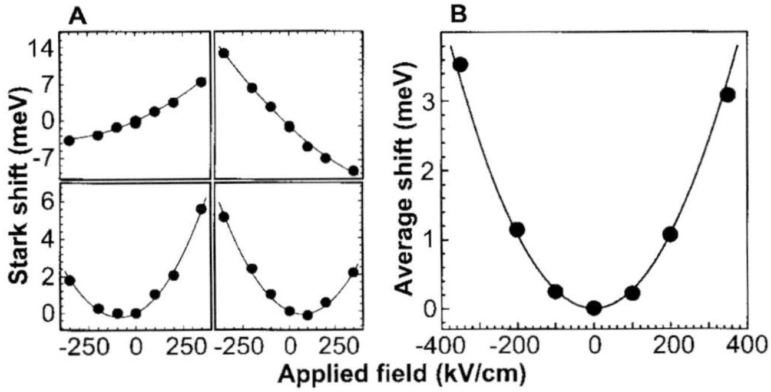

Figure 9-16. Single nanocrystal Stark shifts. (a) Stark shift of emission versus applied electric field for four different single 58 Å nanocrystals. Lines indicate fits to the sum of a linear and quadratic shift as a function of field. (b) Average shift versus field for 54 single nanocrystals from the same sample with a fit to a pure quadratic function of field. Data in Fig. 9-16 is from ref. [25].

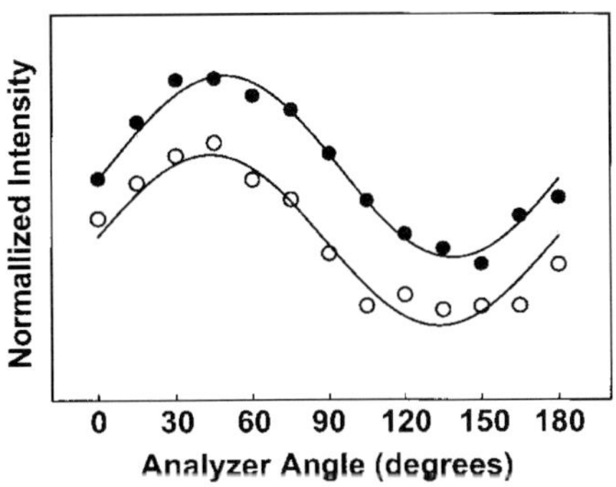

Figure 9-17. Emission polarization dependence on electric field. Emission intensity of a single nanocrystal as a function of analyzer angle in the presence and absence of a 2.5 × 0^5 V/cm electric field (open and closed circles respectively). The field is applied within the sample plane along the direction that is parallel to the zero degree line of the polarizer. The data has been normalized and offset for clarity, so that a direct comparison between the phase and degree of polarization can be made.

dynamic transition dipole is primarily a function of the unit cell wavefunction [60], and to a good approximation, these two components of the total exciton wavefunction are separable.

The observed excited state dipole is consistent with a highly polarizable excited state in the presence of a strong local electric field. A comparison between the changes observed in spectral diffusion and those induced by the presence of an applied electric field are found to be very similar (Fig. 9-18a). In order to quantify this, it is possible to compare phonon coupling vs shift for the two effects (Fig. 9-18b). An identical relationship suggests that spectral diffusion results from changes in the local electric field.

The presence of large local electric fields has several implications. First, the magnitude of the internal electric field extracted in all samples ($\sim 10^5$ V/cm) implies extensive state mixing near the band edge. This conclusion has also been reached on the basis of ensemble non-linear optical experiments, where state mixing has been observed directly [61]. At the same time, this state mixing may also help explain the anomalously large LO phonon couplings measured in these nanocrystals [58]. Accord-

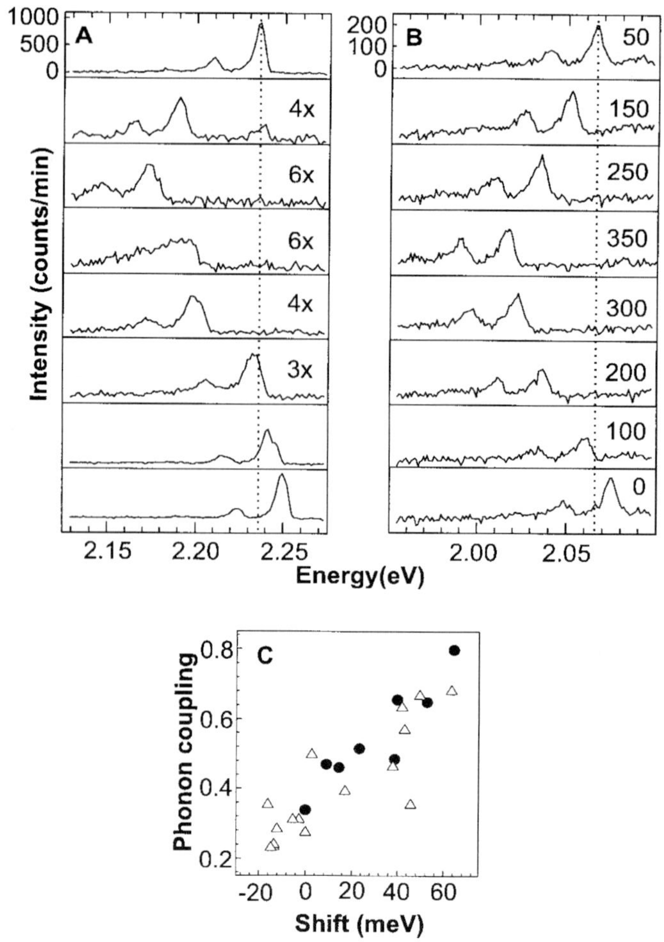

Figure 9-18. Stark shift versus spectral diffusion. (A) Eight sequential 60 second emission spectra of a single nanocrystal shifting due to spectral diffusion in the absence of an applied electric field. Insets indicate magnification of the y-axis. (B) Stark series for a single 24 Å ZnS overcoated nanocrystal in the absence of any significant spectral diffusion. Insets indicate applied field in kilovolts per centimeter. (C) Phonon coupling versus shift for Stark data (closed circles) and spectral diffusion (open triangles). Phonon coupling is measured as the ratio of the integrated intensity of the one longitudinal optical phonon line to the zero phonon line. Stark and spectral diffusion data in (c) were taken from different nanocrystals. Excitation intensity for data in (A) and (B) was 2500 and 285 W/cm^2, respectively. Data in Fig. 9-18 are from ref. [25].

ing to current theory, emission occurs from a delocalized "dark-exciton" state [5] which should only weakly couple to optical phonons [62, 63]. A strong local field breaks the inversion symmetry of the exciton wavefunction, creating a separation of charge within the ionic crystal. This charge asymmetry should then increase exciton-LO phonon coupling through a Frolich interaction, as described above (Fig. 9-14). Previous theoretical calculations of LO phonon couplings have concluded that the experimentally observed ensemble values could be accounted for by the presence of a local electric field equivalent to an electron on the surface of the nanocrystal [62]. This field magnitude is consistent with what is observed in these single nanocrystal experiments. In addition, different local electric fields around individual nanocrystals may also explain the wide range of phonon couplings observed in single nanocrystal experiments as seen in figure 8.

Another implication of changing local electric fields is that there should be a direct correlation between spectral diffusion shifts and changes in the magnitude of the induced excited state dipole. Consistent with this prediction, distinct changes in the single nanocrystal Stark shift can be seen accompanying spectral diffusion shifts. Figure 9-19a shows 11 consecutive spectra of the same single nanocrystal under different field conditions. Following a spectral diffusion shift in the sixth frame, there is an increase in the magnitude of subsequent Stark shifts. Over 50 minutes, distinct changes in the zero-field position are clearly observed (Fig. 9-19b). Accompanying each of these shifts is a corresponding change in the response to the applied field. During minutes 26–32, the Stark shifts become almost purely quadratic, indicating that the excited state dipole along the applied field has become very small. By measuring the complete Stark series for a single nanocrystal before and after a spectral diffusion shift, it is possible to determine which of the Stark parameters is changing. Figure 9-20a shows a single nanocrystal Stark series for a 52 Å overcoated nanocrystal. Figure 9-20b show a Stark series for the same nanocrystal after a spectral diffusion shift. A fit to the two curves reveals that while the polarizability remains the same, the magnitude of the excited state dipole changes by a factor of ~2.5. Analysis of the Stark

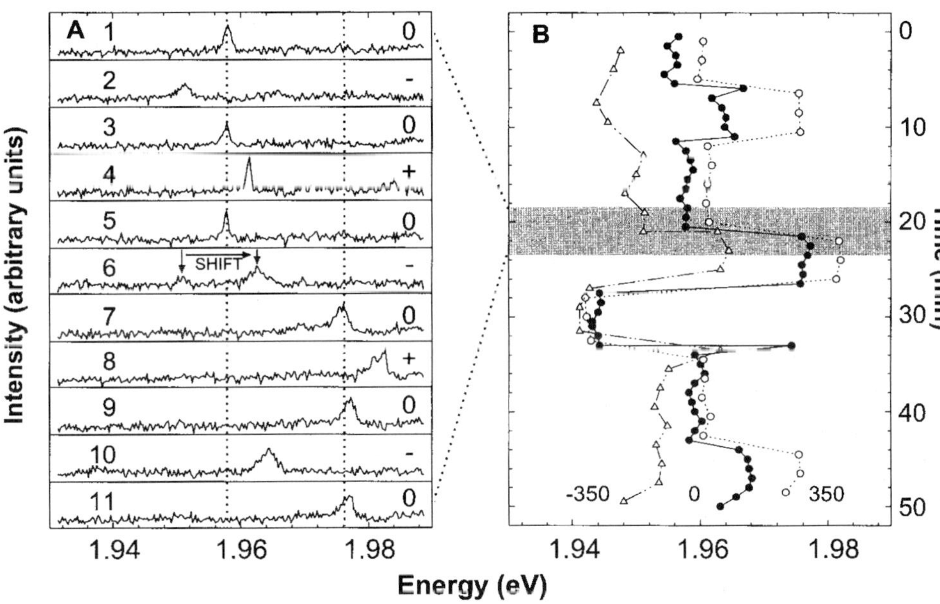

Figure 9-19. Effect of spectral diffusion on Stark shifts. (a) Eleven consecutive 30 second emission spectra from a single 75 Å overcoated nanocrystal in the presence of a repeating sequence of electric fields: negative field, zero field, positive field, zero field. Inset numbers and symbols represent the frame number and applied field respectively. (–), (0) and (+) represent –350, 0 and +350 kV/cm applied fields respectively. A spectral diffusion shift is indicated in frame 6. (b) Summary of 100 consecutive 30 second spectra of the same nanocrystal under the field conditions described above. Data are plotted as a function of time, peak energy, and electric field. Open triangles, closed circles and open circles indicate fields of –350, 0 and 350 kV/cm respectively. Excitation intensity for all spectra in Fig. 9-19 was 25 W/cm^2. Data for Fig. 9-19 are from ref. [25].

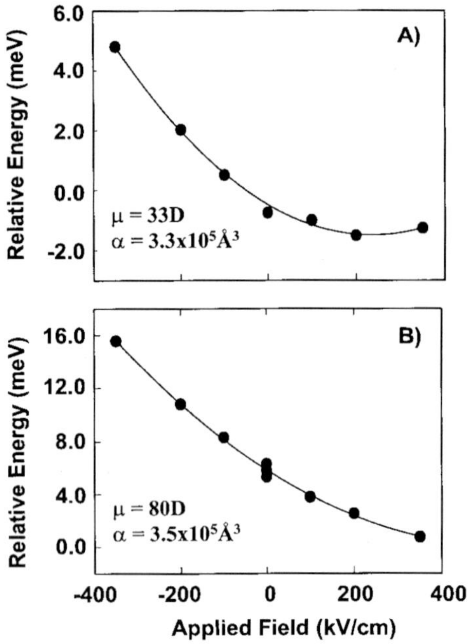

Figure 9-20. Effect of spectral diffusion on single nanocrystal Stark spectrum. (a) and (b) Stark series from a single 52 Å overcoated nanocrystal immediately before and after a spectral diffusion shift. Also included are the measured polarizability and excited state dipole for each series. Note that the offset of ~4 meV between the zero-field energy of (a) and (b) represents the magnitude of the spectral diffusion shift.

parameters between the spectral diffusion shifts in Fig. 9-19b also reveal no change in the measured polarizability over the 50minutes, however, the dipole contribution changes by almost a factor of 50. The correlation of spectral diffusion with changes in the magnitude of the induced excited state dipole supports the Stark model of spectral diffusion.

An alternative interpretation of the data in Fig. 9-19 is that it is not the local field that is changing around the individual nanocrystals, but rather that the nanocrystal is reorienting (rotating) relative to a static local field. While it is unlikely that reorientation of the nanocrystals at 10K would occur on such a fast timescale, it is very important to know for sure that this is not the case in order to correctly interpret the data in Fig. 9-19. Of course, information about the orientation of single nanocrystals is available through the use of polarization spectroscopy as described above.

Figure 9-21 a plots the total emission intensity as a function of analyzer angle for a single nanocrystal. Figure 9-21b show the simultaneous emission energy of the zero phonon line, plotted as a spectral trajectory over time. Over 40 minutes, many large spectral diffusion shifts can be seen, however, no change in the degree of polarization or phase is observed. Since the transition dipole is not affected by electric fields, the data in Fig. 9-21 suggests that spectral diffusion is the result of a dynamic local electric field and not reorientation of the nanocrystal within a static field.

Ensemble dielectric dispersion measurements have suggested the presence of a ground state dipole in CdSe nanocrystals [38]. However, while the data in Fig. 9-19 does not preclude a contribution to the excited state dipole from intrinsic structural or charge asymmetries in the ground state [38, 61], it does indicate that a large portion of the dipole arises from an extrinsic source.

If we assume that the excited state dipole is the result of local electric fields (either intrinsic or extrinsic), the magnitude of this average field is found to be quite large, comparable to that produced by a point charge (electron or hole) trapped on or near

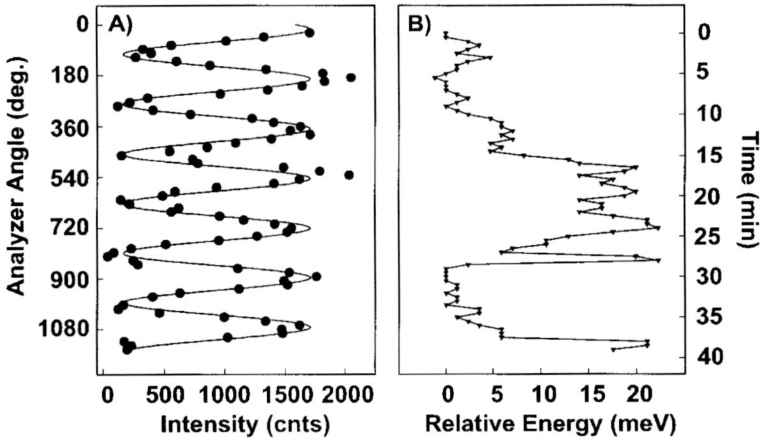

Figure 9-21. Polarization versus spectral diffusion. (a) Total emission intensity of a single 52 Å overcoated nanocrystal as a function of analyzer angle with fit. (b) Relative emission energy of the zero phonon line of the nanocrystal in (a) as a function of time. The data in (a) and (b) were taken simultaneously so that changes in emission energy and polarization could be monitored concurrently. Data in Fig. 9-21 was taken with an integration time of 30 seconds, an excitation intensity of 60 W/cm^2 and excitation polarization at 0 degrees. Data in Fig. 9-21 are from ref. [27].

the surface of the nanocrystal. As mentioned previously, photoionization has been proposed as the source of fluorescence intermittency in single CdSe nanocrystals at room temperature [23]. Photoionizatiuon leaves a charged nanocrystal core that may not relax radiatively upon further excitation. This is due to the fact that the presence of a single charge in the CdSe core creates a very rapid, non-radiative Auger relaxation pathway [64]. Emission resumes when the core is neutralized by the return of the ejected charge. As discussed above, a similar on-off behavior is observed at liquid helium temperatures [22, 29]. At 10 K, however, there is little thermal energy available to promote the return of an external charge. Instead, neutralization may occur through an additional ionization event, resulting in an emitting nanocrystal in the presence of a potentially large and randomly oriented local electric field. Additional ionization or recombination events, as well as relocalization of external charges, could results in changes in both the zero-field energy as well as the excited state dipole of a single nanocrystal. This is consistent with what is observed in Fig. 9-19.

Stark measurements of different size samples reveal some general trends in the average polarizability, excited state dipole and calculated internal electric field as a function of size (Fig. 9-22). The increase in polarizability with size is consistent with ensemble measurements [53] and with the increase in volume of the nanocrystals. The observed increase in excited state dipole with size is a result of the corresponding increase in polarizability, since the average internal electric field actually decreases with size. A decrease in the internal field with size is consistent with the proposed ionization model. As the size of the nanocrystal is increased, the average distance of an external charge from the center of the exciton wavefunction increases. The result is a decrease in the average internal electric field.

To reinforce this result, a ZnS overcoated sample with the same size CdSe core as one of the non-overcoated samples was also studied. The measured values of the polarizability for the overcoated and non-overcoated samples were found to be statistically identical, consistent with minimal delocalization of the exciton into the ZnS

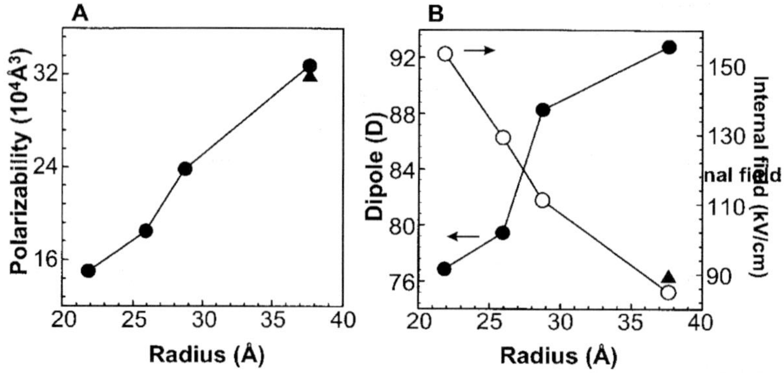

Figure 9-22. Size dependence of Stark characteristics. (a) Measured average polarizability as a function of size for non-overcoated and overcoated nanocrystals (closed circles and triangle respectively). (b) Measured average excited state dipole and calculated internal electric field (closed and open circles respectively) as a function of size for non-overcoated nanocrystals and measured average excited state dipole for overcoated nanocrystals (closed triangle). Data in Fig. 9-22 was taken with a 30 second integration time and 250 W/cm^2 excitation intensity from (47) 37.5 Å overcoated nanocrystals, (57) 37.5 Å, (83) 29 Å, (74) 26 Å and (16) 22 Å non-overcoated nanocrystals. The calculations include screening by the CdSe core and ZnS shell as necessary. Data in Fig. 9-22 are from ref. [25].

shell [35]. At the same time, however, a significant decrease in excited state dipole is observed. This is consistent with the ZnS shell forcing external charges to reside farther from the exciton wave function, while simultaneously screening external fields. The ZnS shell also acts as a barrier to ionization [23], potentially decreasing the number of charges around the nanocrystal.

Within the framework described above, the large spectral shifts seen in figure 9-13 may result from ionization and recombination of carriers, while the small shifts in figure 10 may result from slight field modulations due to movement of charges between local trap sites. For electric fields on the order of the local fields measured in these experiments, Fig. 9-16b indicates that the dependence of the Stark shift (ΔE) on electric field (F) is in the quadratic regime ($\Delta E \propto F^2$). This implies that the magnitude of the change in energy [$\Delta(\Delta E)$], and therefore the observed linewidth, due to a fluctuating field (ΔF) will also depend on the average field present [$\Delta(\Delta E) \propto F{\cdot}(\Delta F)$]. Consistent with this prediction, Stark shifted emission spectra like those in Fig. 9-15a are seen to broaden as they shift to lower energies (higher total internal fields). Figure 9-23a plots the linewidth of a single nanocrystal as a function of relative shift in the presence of an applied field. This data can be roughly fit with a function that varies as the square root of the observed shift, consistent with the Stark model. In addition, Fig. 9-22b indicates that overcoated nanocrystals have a smaller average internal electric field than non-overcoated nanocrystals (as evidenced by a smaller measured excited state dipole). As a result, overcoated nanocrystals should have narrower average linewidths resulting from the same local field fluctuations than non-overcoated nanocrystals. This is also found to be true (Fig. 9-23b). A comparison between the average linewidth of single overcoated and non-overcoated nanocrystals with the same size CdSe core as a function of excitation intensity reveals that the non-overcoated sample has consistently broader linewidths under the same experimental conditions.

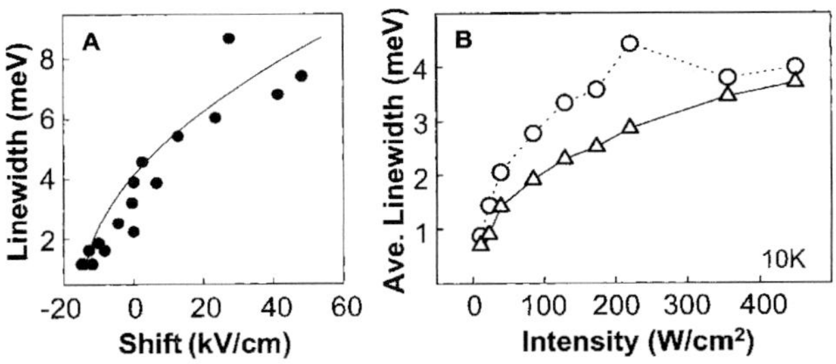

Figure 9-23. Effects of a fluctuating local electric field on single nanocrystal linewidths. (a) Linewidths vs shift from the zero-field energy for a single 75 Å overcoated nanocrystal in response to an applied electric field. The solid line represents the fit to a square root function of the shift magnitude. (b) Average single nanocrystal linewidth as a function of excitation intensity for 34 overcoated nanocrystals and 25 non-overcoated nanocrystals (triangles and circles respectively) with the same CdSe core size (56,5 Å). Data for Fig. 9-23 was taken with a 30 second integration time. The excitation intensity for Fig. 9-23a was 25 W/cm^2. Data for (a) and (b) are from ref. [25] and [26] respectively.

As a final comment on spectral diffusion, it has been proposed here that the changes in the local electric field responsible for spectral diffusion are the result of charge carriers near individual nanocrystals. One thing that has not been addressed, however, is the nature of the trapped charges (i.e. where are they?). To our knowledge, no evidence currently exists which can conclusively locate these charges, however, some speculation can be made. Both spectral diffusion and the Stark results shown in Fig. 9-19, have been observed in nanocrystals embedded in a polymer matrix as well as ones deposited onto a quartz substrate with no surrounding matrix. Since these effects are present in the absence of a matrix, this suggests that charges may reside in trap sites on the surface of the NC. While this is just speculation, it is consistent with the experimental evidence currently available.

9.9 Conclusion

In conclusion, current work being performed in single quantum dot spectroscopy has greatly increased our understanding of CdSe nanocrystals. Many new and previously unobserved or unexpected phenomena have been revealed including fluorescence blinking, ultra-narrow transition linewidths, a range of phonon couplings between individual nanocrystals, and spectral diffusion of the emission spectrum over a wide range of energies. Eletric field studies have conclusively revealed both polar and polarizable character in the emitting state. The polar component has been attributed to an induced excited state dipole resulting from the presence of local electric fields. These fields, which are thought to result, in part, from trapped charge carriers on or near the surface of the nanocrystal, are seen to change over time, resulting in spectral diffusion. Careful observation of spectral diffusion has shown that on the timescale of a typical single nanocrystal spectrum, lineshapes are primarily the result of spectral shifting and not the intrinsic physics of the nanocrystal. The effects of integration time and excitation intensity on observed single nanocrystal linewidths also strongly demonstrate the sensitivity of single nanocrystal spectroscopy to experimental conditions. Finally, polarization spectroscopy suggests that band edge emission occurs through a 2 dimensional "dark axis" transition dipole, which may allow the determination of the 3 dimensional orientation of individual nanocrystals.

Experiments such as these and other single nanocrystal studies have introduced a new level of understanding of both the theoretical aspects as well as the potential for device application of this unique physical system. The use of single chromophore spectroscopy along with corresponding ensemble experiments is a powerful tool for elucidating the microscopic details of ensemble systems. It is the hope of the authors that the general interest in these and other single chromophore experiments will continue and that the techniques of single chromophore detection will someday become a common analytical tool used in laboratories throughout the physical sciences.

S.A.E thanks the Eastman company and the Lester Wolfe Foundation for graduate fellowships. This work was funded in part by the NSF-Materials Research Science and Engineering Center program (DMR-98-08941) and by the AT&T foundation. We thank the Massachusetts Institute of Technology Harrison Spectroscopy laboratory (NSF-CHE-97-08265) for support and for use of its facilities.

References

[1] L. Brus, *Appl. Phys.* A, **1991**, *53*, 456.
[2] A.P. Alivisatos, *Science* **1996** *271*, 933.
[3] D.J. Norris and M.G. Bawendi, *Phys. Rev. B* **1996**, *53*, 16338.
[4] D.J. Norris et al., *Phys. Rev. B* **1996**, *53*, 16347.
[5] Al.L. Efros et al., *Phys. Rev. B* **1996**, *54*, 4843.
[6] C.B. Murray, D. J. Norris, M. G. Bawendi, *J. Am. Chem. Soc.* **1993**, *115*, 8706.
[7] Al.L. Efros, A.L. Efros, *Sov. Phys. Semicond.* **1982**, *16*(7), 772.
[8] L.E. Brus, *J. Chem. Phys.* **1984**, *80*, 4403.
[9] M. Danek et al., *J. Cryst. Growth* **1994**, *145*, 714.
[10] M. Danek et al., *Chem. Mater.* **1996**, *8*(1), 173.
[11] C.R. Kagan et al., *Phys. Rev. Lett.* **1996**, *76*(9), 1517.
[12] C.B. Murray, C. R. Kagan, M. G. Bawendi, *Science* **1995**, *270*, 1335.
[13] X. Peng, T.E. Wilson, A.P. Alivisatos, P.G. Shultz, *Angew. Chem. Int. Ed. Engl.* **1997**, *36*(1–2), 145.
[14] D.J. Norris, M.G. Bawendi, *J. Chem. Phys.* **1995**, *103*(13), 5260.
[15] A.P. Alivisatos et al., *J. Chem. Phys.* **1989**, *90*(7), 3463.
[16] U. Woggon et al., *Phys. Rev. B* **1993**, *47*, 3684.
[17] D.M. Mittleman et al., *Phys. Rev. B* **1994**, *49*(20), 14435.
[18] H. Giessen et al., *Appl. Phys. Lett.* **1996**, *68*(3), 304.
[19] V. Jungnickel, F. Henneberger, *J. Lumin.* **1996**, *70*, 238.
[20] W. E. Moerner, *Science* **1994**, *265*, 46; and references therein.
[21] T. Basche, W.E. Moerner, M. Orrit and U.P. Wild, *Single-molecule Optical Detection, Imaging and spectroscopy* **1997**, VCH Weinheim.
[22] S.A. Empedocles, D.J. Norris, M.G. Bawendi, *Phys. Rev. Lett.* **1996**, *77*(18), 3873.
[23] M. Nirmal et al., *Nature* **1996**, *383*, 802.
[24] S.A. Blanton, M.A. Hines, P. Guyot-Sionnest, *Appl. Phys. Lett.* **1996**, *69*(25), 3905.
[25] S.A. Empedocles, M.G. Bawendi, *Science* **1997**, *278*, 2114.
[26] S.A. Empedocles, M.G. Bawendi, *Influence of Spectral Diffusion on the Lineshapes of Single CdSe Nanocrystal Quantum Dots*, (submitted).
[27] S. Empedocles, R. Neuhauser, M. Bawendi, *Polarization spectroscopy of single CdSe nanocrystal quantum dots* (in preparation).
[28] S.A. Blanton et al., *Chem. Phys. Lett.* **1994**, *229*, 317.
[29] J. Tittel, W. Göhde, F. Koberling, T. Basché, A. Kornowski, H. Weller, A. Eychmuller, *J. Phys. Chem. B* **1997**, *101*(16), 3013.
[30] R. Leon et al., *Science* **1995**, *267*, 1966.
[31] M. Grundmann et al., *Phys. Rev. Lett.* **1995**, *74*(20), 4043.
[32] H. Hess et al., *Science* **1994**, *264*, 1740.
[33] D. Gammon, E. Snow and D. Katzer, *Appl. Phys. Lett.* **1995**, *67*, 2391.
[34] M.A. Hines, P. Guyot Sionnest, *J. Phys. Chem.* **1996**, *100*, 468.
[35] B.O. Dabbousi et al., *J. Phys. Chem. B* **1997**, *101*, 9463.
[36] J.J. Macklin et al., *Science* **1993**, *272*, 255.

[37] A.G.T. Ruiter et al., *J. Phys. Chem. A* **1997**, *101*, 7318.
[38] S.A. Blanton, R.L. Leheny, M.A. Hines, P. Guyot-Sionnest, *Phys. Rev. Lett.* **1997**, *79*(5), 865.
[39] M. Kuno et al., *J. Chem. Phys.* **1998**, *108*(10), 4242.
[40] F. Guttler, S. Sepiol, T. Plakhotnik, A. Mitterdorfer, A. Renn, U.P. Wild, *J. Lumin.* **1993**, *56*, 29.
[41] F. Guttler, M. Croci, A. Renn, U.P. Wild, *Chem. Phys.* **1996**, *211*, 421.
[42] Al.L. Efros, *Phys. Rev. B* **1992**, *46*(12), 7448.
[43] W. Lukosz, *J. Opt. Soc. Am.* **1981**, *71*(6), 744.
[44] W. Lukosz, *J. Opt. Soc. Am.* **1979**, *69*(11), 1495.
[45] W. Lukosz, *Phys. Rev. B* **1980**, *22*(6), 3030.
[46] Ch. Fattinger and W. Lukosz, *J. Lumin.* **1984**, *31&32*, 933.
[47] M. Berg et al., *Chem. Phys. Lett.* **1987**, *139*(1), 66.
[48] J.K. Trautman, J.J. Macklin, L.E. Brus, E. Betzig, *Nature* **1994**, *369*, 40.
[49] J. Koedijk, R. Wannemacher, R. Silbey, S. Volker, *J. phys. Chem.* **1996**, *100*(51), 19945.
[50] E. Geva, J.L. Skinner, *J. Phys. Chem. B* **1997**, *101*(44), 8920.
[51] W. Pfluegl, F.L.H. Brown, R.J. Silbey, *J. Chem. Phys.* (in press).
[52] M. Kuno, J.K. Lee, B.O. Dabbousi, F.V. Mikulec, M.G. Bawendi, *J. Chem. Phys.* **1997**, *106*(23), 9869.
[53] A. Sacra, Ph. D. thesis, Massachusetts Institute of Technology **1996**.
[54] V.L. Colvin, A.P. Alivisatos, *J. Chem. Phys.* **1992**, *97*(1), 730.
[55] V.L. Colvin, K.L. Cunningham, A.P. Alivisatos, *J. Chem. Phys.* **1994**, *101*, 7122.
[56] F. Hache, D. Richard, C. Flytzanis, *Appl. Phys. Lett.* **1989**, *55*, 1504.
[57] A. Sacra, D.I. Norris, C.B. Murray, M.G. Bawendi, *J. Chem. Phys.* **1995**, *103*(13), 5236.
[58] D.J. Norris, Al.L. Efros, M. Rosen, M.G. Bawendi, *Phys. Rev. B* **1996**, *53*, 16347.
[59] S.A. Empedocles, D.J. Norris, M.G. Bawendi, *Materials Research Society Symposium Proceedings – Fall* **1996**, *452*, 335.
[60] M.G. Bawendi, *Confined Electrons and Photons*, New York, Plenum Press, **1995**, p. 339.
[61] M.E. Schmidt, S.A. Blanton, M.A. Hines, P. Guyot-Sionnest, *J. Chem. Phys.* **1997**, *106*, 5254.
[62] S. Nomura, T. Kobayashi, *Phys. Rev. B* **1992**, *45*(3), 1305.
[63] Al.L. Efros, *Phonons in Semiconductor Nanostructures*, Boston, Kluwer Academic Publishers, **1993**, p. 299.
[64] Al.L. Efros, M. Rosen, *Phys. Rev. Lett.* **1997**, *78*(6), 1110.

10 Nanomagnetism

Walt A. de Heer

10.1 Introduction

When an object becomes so small that the number of surface atoms is a sizable fraction of the total number of atoms then obviously surface effects will be important. In general, a property will depend on the size of an object if its size is comparable to a dimension which is relevant to that property. In the simplest case above, the parameter is the ratio of the atomic radius to the size of the object (which approximately equal to the surface to bulk atom ratio). However there are many other length scales that are relevant depending on the physical property under investigation. In magnetism for example, a typical size of a magnetic domain is of the order of 1 μm and particles much smaller than that will be mono-domain. Another scale involves exchange coupling effects, which affect the magnetic polarization of neighboring ferromagnetic particles in non-magnetic hosts, have a range of the order of several nanometers. Here we will concentrate on several ferromagnetic systems which are so small that their properties are size dependent [1, 2, 3].

It is important to realize that the "nano" in nanomagnetism is not uniquely defined. For many workers active in nanomagnetism, nano does not signify nanometer scale, but rather sub-micron or merely small (and all these interpretations are justifiable). Bulk magnetic materials arrange in magnetic domains which are of this small size scale and consequently virtually all magnetic systems can be considered to be "nanostructured".

Properties of very small metal particles have been extensively studied in the past 50 years [4], and much is known about their electronic and geometrical structure. Nevertheless, the field is still rapidly growing due to the vast number of different systems and properties which are magnetic and the subtle complexities of material magnetism.

While interest in many systems on the nanoscopic scale is academic to a great extent, small magnetic particles have been of industrial importance since the nineteen fifties primarily stimulated by the recording industry's desire for ever denser and more reliable recording media. The quest for smaller particles which can be used for recording purposes [5, 6] continues since smaller particles implies higher data storage densities [7, 8]. There are of course material limits to how small a magnetic particle can be and still be useful. For example, not only should the particle be ferromagnetic, it should retain its magnetic orientation in ordinary conditions for many years.

New magnetic materials are also using for reading and recording magnetically encoded information. Recently, it has been found that nanocomposites composed of ferromagnetic particles embedded in noble metal matrices exhibit giant magnetoresistance effects [9] related to those found in magnetic/non-magnetic superlatices [10]. These new materials are important for magnetic sensors: giant magnetoresistance reading heads are used to read magnetically stored data.

Magnetic nanostructures are also important objects of investigations for basic materials research [1, 2]. Magnetic systems have complex electronic structures [11, 12, 13] and consequently it is a computational challenge to determine even the ground state of the smallest of ferromagnetic particles (i.e. clusters of only a few atoms) from first principles [14]. The complications mainly arise because the electronic spin (which is the carrier of the magnetic moment), clearly cannot be ignored as it can be in more simple metallic systems [4]. To bypass these complications, numerous approximations have been devised which highlight specific aspect of magnetic response, however, approximations usually compromise predictive power [11].

Many interesting magnetic properties are related to the dynamics of the spin system [11, 12, 13]. At low temperatures, the magnetic moment (or total electronic spin) of the system may spontaneously change direction in a quantum tunneling process [15, 16]. At higher temperatures, the spin direction will fluctuate due to thermal agitation [17]. At the same time the magnitude of the magnetic moment will reduce, and eventually vanish at high enough temperatures (at the Curie temperature). These processes are well known and reasonably well understood, however spin dynamics in magnetic systems is still a topic of great theoretical interest and controversy [11, 12, 13].

Some basic knowledge of the properties of ferromagnetic systems is useful to appreciate issues involved in nanomagnetism. The following Section 10.2 is intended to provide some of this background, much of which can be found in various standard textbooks on ferromagnetism. Section 10.3 discusses magnetism in systems with reduced dimensions; Section 10.4 illustrates some characterization methods and Section 10.5 discusses several nanomagnetic systems in some detail.

10.2 Basic concepts in magnetism

10.2.1 Atomic magnetism

Most atoms have net electronic angular momentum **J** which is a combination of the total spin **S** and the total orbital angular momentum **L**. In fact, these quantities can be determined for a specific atom using Hund's rules (see i.e. [18]). The electronic structure of the partially occupied atomic shell is found by maximizing the spin **S** for the valance shell of the atom, while respecting the Pauli principle. Next the orbital angular momentum **L** is found by maximizing it (without violating the Pauli principle). Finally, **S** and **L** are combined to give the total angular momentum **J** in such a way as to maximize the total $\mathbf{J} = |\mathbf{L} \pm \mathbf{S}|$ in units of $\hbar$, see Table 1.

For example, for the 3d shell fills in the 3rd row: (Sc:$3d^14s^2$; Ti:$3d4s^2$; V:$3d^34s^2$; Cr:$3d^54s^1$; Mn:$3d^54s^2$; Fe:$3d^64s^2$; Co:$3d^74s^2$; Ni:$3d^84s^2$; Cu:$3d^{10}4s^1$). Hence, for iron S = 4/2 since there are 5 electrons in the up state and one in the down state according to Hunds rules. Likewise $L = 2+2+1+0-1-2 = 2$, and $J = L+S = 4$. The magnetic moment if the state (SLJ) is given by [18]

$$\boldsymbol{\mu} = -g(LSJ)\mu_B \mathbf{J} \tag{10-1}$$

where g is the Lande g factor:

$$g = 3/2 + 1/2\ (S(S+1) - L(L+1)/J(J+1). \tag{10-2}$$

The energy of a paramagnetic atom ($\mu \neq 0$) in a magnetic field is

$$E(B) = \boldsymbol{\mu} \cdot \mathbf{B} \tag{10-3}$$

which means that its energy depends on its orientation. Consequently, if it is subjected to a field gradient, there will be a net force on it:

$$\mathbf{F} = -\nabla(\boldsymbol{\mu} \cdot \mathbf{B}). \tag{10-4}$$

for example, if a paramagnetic atom is subjected to a field gradient $\delta B_z/\delta z$, the force in the z direction is

$$F_z = g\mu_B m_J \delta B_z/\delta z \tag{10-5}$$

where m_J is the z component of **J** (quantized in units of $\hbar$) and μ_B is the Bohr magneton. A consequence of this effect is that the force on an atom due to a field gradient is quantized and is either attracrtive or repulsive depending on the orientation of J with respect to the field. This is the reason for the symmetric and quantized deflection patterns that are observed in Stern-Gerlach atomic beam deflection experiments.

On the other hand, if the atom is in contact with a heat bath then the angular momentum can relax in the direction of lowest energy which is aligned with the magnetic field, thereby causing a net attractive force towards the direction of increasing field strength. In other words, paramagnetic atoms in a matrix are attracted to magnets. The magnetization M is the time average of the projection of the magnetic moment in the direction of the magnetic field B [19]

$$M - \sum m_J g_J \mu_B \exp(m_J\, g_J \mu_B B/kT)/\sum \exp(m_J g\ _J \mu_B B/kT). \tag{10-6}$$

Defining $\mathbf{x} = J g_J \mu_B B/kT = \mu\ B/kT$ we obtain the Brillouin function [19]

$$M = g_J \mu_B (2J+1)/2J \coth\{(2J+1)\mathrm{x}/2J\} - (1/2\mathrm{J})\coth(x/2J) \tag{10-7}$$

For large J, this reduces to the well known Langevin function [19]

$$M = g_J \mu_B \coth(x) - 1/x \tag{10-8}$$

10.2.2 Ferromagnetism

When metal atoms are combined one at a time to form a particle (and eventually the bulk), they arrange in such a way to minimize the total energy [18]. The atomic bonding is mediated by the valence electrons which diffuse into the solid and are shared by several or, in the case of conduction electrons, by all the atoms in the solid [11, 12, 13]. To illustrate the bonding and its consequences for magnetism, we consider the simplest case, which is the bond between two hydrogen atoms.

The ground state spatial wave functions of a diatomic molecule can be approximated as linear combinations of the atomic ground states: following Heitler and London, for the H_2 molecule [18]:

$$\Psi_{s/t}(r_1,r_2) = \phi_1(r_1)\phi_2(r_2) \pm \phi_1(r_2)\phi_2(r_1); \tag{10-9}$$

where $\phi(\mathrm{r})$ is the (one electron) electronic ground state wave function of the hydrogen atom and Ψ is the two electron wave function Ψ_s is symmetric under interchange of the electronic coordinates, r_1 and r_2, and Ψ_t is anti-symmetric. The total electronic wave function Ψ must be anti-symmetric under particle exchange and hence the spin wave function of Ψ_s must be anti-symmetric. This means that the two spins are oppositely aligned and hence in a singlet state, while the spin wave function of Ψ_t must be symmetric (consequently, the spins are aligned to give a triplet state). The energy difference between these two states is the exchange energy J_{st} (not to be confused with the total angular momentum J)

$$J_{st} = \langle \Psi s|H|\Psi_s\rangle - \langle \Psi_t\,|H|\Psi_t\rangle, \tag{10-10}$$

where H is the total Hamiltonian ($\Psi_{s,t}$ are assumed to be normalized). The energy difference between the $S = 0$ and $S = 1$ state is essentially due to the electronic Coulomb interaction.

For most diatomic molecules, as for H_2, the singlet state is lower in energy than the triplet state (Ψ_s is then the bonding state and Ψ_t is the antibonding state of the molecule).

Hunds rules dictate that atoms with partially filled electronic shells will have relatively large spins. Those rules can be traced to the exchange interaction which in atoms favors aligned spins, since the anti-symmetry of the associated spatial wave functions produces vanishing probabilities for both electrons to occupy the same place at the same time (in accord with the Pauli principle). Consequently, the Coulomb energy is reduced compared with the symmetric case.

10.2.3 The Heisenberg Hamiltonian and localized magnetic moments

If the electronic wave function of a system is known then the term in the Hamiltonian which operates on the spins and gives the energy related to the spin configuration is given by [12]:

$$H_S = -\sum J_{ij}\mathbf{S}_i \cdot \mathbf{S}_j \tag{10-11}$$

where S_i is the spin operator of the ith electron and the J_{ij} are the exchange energy constants. To illustrate this, we apply this operator to the singlet and triplet hydrogenic states above. Realizing that $2\mathbf{S}_1 \cdot \mathbf{S}_2 = (\mathbf{S}_1 + \mathbf{S}_2)^2 - (\mathbf{S}_1)^2 - (\mathbf{S}_2)^2$, so that $S_1 \cdot S_2 = 1/2(1*2 - 1/2*3/2 - 1/2*3/2) = 1/4$ in the triplet state and $S_1 \cdot S_2 = 1/2(0 - 1/2*3/2 - 1/2*3/2) = -3/4$ in the singlet state then the singlet and triplet energies are

$$\begin{aligned} E_s &= \langle \Psi_s|H_S|\Psi_s\rangle = -3/4J_{12}, \\ E_t &= \langle \Psi_t|H_S|\Psi_t\rangle = +1/4J_{12} \\ E_t - E_s &= J_{12} \end{aligned} \tag{10-12}$$

hence the triplet-singlet energy difference equals the exchange energy. Hence the Heisenberg Hamiltonian (Eq. 10-11) allows one to extract the energies of the various spin configurations. However, the many body wave function must be found determined, but for many purposes, approximations are made.

In certain cases, for example for the ferromagnetic rare earth metals [18, 20] (i.e. Gd, Dy, Sm), the magnetic moments (i.e. the S_i) are strongly localized at the atomic positions and the Heisenberg operator can be directly applied simply by replacing the spin operators by the local spin values. Hence the total energy of the spin configuration becomes

$$E_S = -\sum J_{ij}\mathbf{S}_i \cdot \mathbf{S}_j \tag{10-13}$$

In the simplest approximation, the energies can be evaluated assuming only nearest neighbor interactions. If all the J_{ij} are positive, then clearly the ground state is the one where all the spins are parallel to each other, hence the total spin of the system with n spins is $S = nS_1$. This Hamiltonian is useful to describe spin excitations (magnons) and thermal properties of spin systems. For example, at high temperatures the spin directions randomize so that $S = 0$ as it should above the Curie temperature.

The case where the J_{ij} are negative is also interesting since its ground state consists of a lattice where the spins orientation reverses between nearest neighbors. This represents an anti-ferromagnetic ground state (not be confused with a non-magnetic state like for Na). In anti-ferromagnetic materials, the local spin densities are non-zero, yet the overall spin (and magnetic moment) vanishes. Examples of anti-ferromagnets are MnO, FeO, CoO, NiO. Chromium is a peculiar ferromagnet with non-vanishing local spin densities but with a periodicity which is incommensurate with the Cr lattice.

10.2.4 Molecular field approximation

The interaction of localized moments can also be described using the molecular field approach [12, 19], where it is assumed that the interaction of spin $\mathbf{S}$ with all the other spins can be approximated with an effective magnetic field $\mathbf{B}_{mf}$ due to those neighboring spins. For example, assuming that this spin has z identical nearest neighbors then $\mathbf{B}_W$ is determined from

$$gM_B z\langle\mathbf{S}\rangle \cdot \mathbf{B}_W = J_{ns} \tag{10-14}$$

where $\langle S\rangle$ is the average spin of the nearest neighbors and J_{ns} is the exchange energy (i.e. the energy required to reverse the direction of the spin).

Applying the Langevin equation (Eq. 10-8) with B_W as the applied field allows the magnetization to be determined a function of temperature. For example the Curie temperature is predicted to be

$$T_C = zJS^2/(3k) \tag{10-15}$$

where $S^2 = S(S+1)$, while the magnetic susceptibility ($\chi(T) = M/B$, c.f. Eq. 10-8) for $T > T_C$ is

$$\chi(T) = n(g\mu_B S)^2/3k(T-T_C) \tag{10-16}$$

where n is the density. Despite its simplicity, the mean field model gives reasonable predictions for the temperature dependence of the susceptibility, saturation magnetization and heat capacity of several ferromagnetic metals (Fig. 10-1).

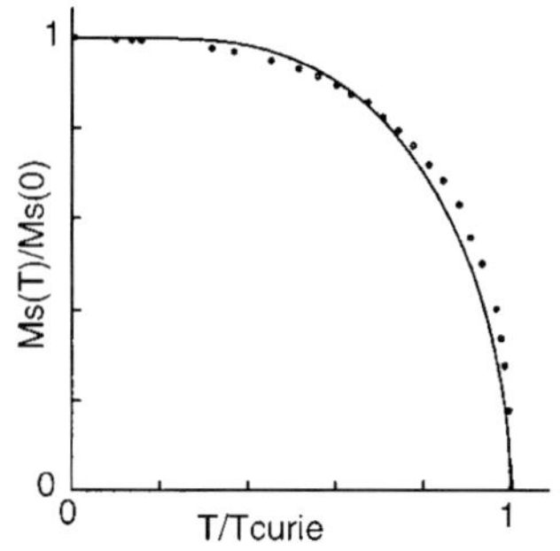

Figure 10-1. Magnetic moment (normalized with respect to the maximum moment) versus temperature (normalized with respect to the Curie temperature), as calculated in the molecular field approximation and as measured for nickel.

10.2.5 Itinerant ferromagnetism: the iron group metals

In a metal which is constructed with paramagnetic atoms, the bonding is mediated by the valence electrons which are often also responsible for the atomic paramagnetism. In Na for example, the single 2s valence electron ($S = 1/2$) gives rise to the conduction band, which is know to be almost free electron like [18]. This means that these electrons hardly feel the atomic potentials but instead an almost featureless average potential. In this state, the local spin density vanishes everywhere (spin up states and spin down states are equally populated everywhere) [4]. In other words the bulk is non-magnetic. For $Al(3s^23p^1)$, the atom is $^2P_{3/2}$. In the bulk the three electrons contribute to the conduction band which again is almost free electron like. The atomic orbital angular momentum is quenched and there is no imbalance in the spin up and spin down spin densities (not even locally).

This is not true for itinerant ferromagnetic metals such as the 3d transition metals (Fe, Co, Ni) [19]. For example for $Ni([Ar]3d^84s^2)$, the atom ground state ($^3F_4$) is paramagnetic with spin $S = 1$. The bulk is ferromagnetic where each atom contributes 0.54 μ_B. This is due to an imbalance in the up-down spin densities (orbital angular momentum contributions are small). If the bulk is imagined to be constructed by gradually reducing the distance between separated atoms, then each of the atomic orbitals broadens into bands. For nickel, the five electrons per atom which occupy the 3d majority spin band (spin up) are below the Fermi level as are about 4.54 of the electrons which occupy the minority (spin down) band. The 3d minority band and the unpolarized 4s bands intercept the Fermi level and hence are partially occupied: the total occupation of the 4s and 3d bands is 10. The spin polarization hence comes from a spin up/spin down imbalance, which in the bulk amounts to about 0.54 electrons per atom corresponding to a magnetic moment of about 0.54 μ_B per atom (Fig. 10-2). Consequently the local spin densities at the Fermi level of ferromagnetic metals do not vanish. This is important in the characterization of magnetic materials.

The above example suggests that some of the atomic character of the 3d states is retained even in the bulk (for example the large exchange splitting between the up and down spin states). The non-integer occupation of the bands is caused by the itinerant nature of the electrons and can be regarded as the fraction of time that those electrons spend in corresponding atomic states. This picture is reasonable since the 3d atomic orbitals are rather tightly bound, hence, even though the 3d electrons are itinerant, when they are near a core they momentarily fall into atomic orbits and acquire atom-like properties.

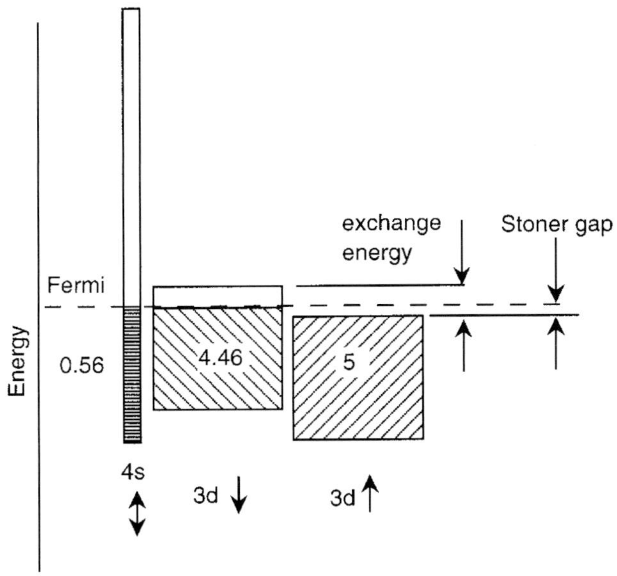

Figure 10-2. Schematic band structure of an itinerant ferromagnet (Ni). The majority spin band in entirely below the fermi level and hence fully occupied. The minority band intercepts the fermi level and is partially occupied. The spin imbalance gives rise to the magnetic moment (the s band is not spin polarized there is no exchange splitting of this band). The non-integer spin imbalance is characteristic for itinerant ferromagnets.

In the band picture ferromagnetism is explained in terms of the by exchange splitting of the d-bands that causes an imbalance in up- and down-spin occupations [12]. Hence the solid has a net spin which directly implies that it has a magnetic moment (Eq. 10-1). The band picture emphasizes the fact that the spin carrying electrons are itinerant and not localized at specific sites. In this picture it is natural to assume that the reduction, and the eventual disappearance of ferromagnetism at high temperatures, is caused by the thermal excitations of electrons from the filled down bands to the partially filled up bands causing a reduction in the spin imbalance. This Stoner mechanism however vastly overestimates the Curie temperature. In fact the ferromagnetic to paramagnetic transition at the Curie temperature in itinerant ferromagnets involves collective spin density effects (spin waves) which are not represented in the band picture [12]. The itinerant nature of the 3d electrons in ferromagnetic metals has been demonstrated in several experiments (see i.e. [21]).

10.2.6 Localized magnetism: the lanthanide metals

Strictly speaking, electrons in solids are always delocalized, however electrons in high angular momentum states like the f electrons in the rare earth metals [18, 20], may for all practical purposes be considered to be localized on specific atoms (Fig. 10-3) since the overlap of their electronic wave functions with those of neighboring atoms is very small. According to Hunds rules, the f shells of rare earth atoms (i.e. Sm, Gd) have large spins (3, 4 respectively) and large magnetic moments (8 μ_B; 10.6 μ_B respectively), which includes the orbital angular momentum contribution [18]. In the

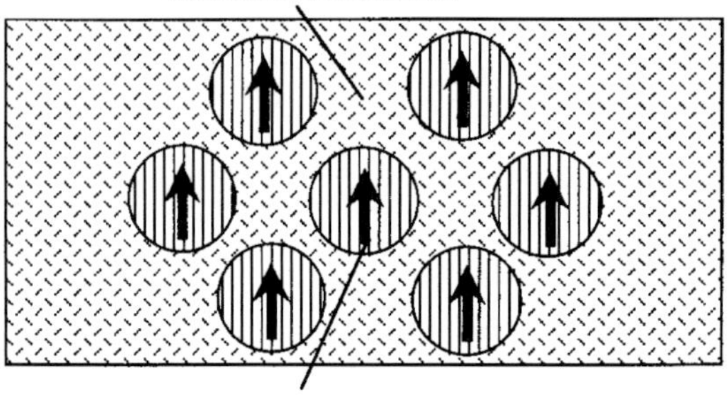

Figure 10-3. Schematic diagram of a localized ferromagnet. The magnetic moments are due to incomplete of high angular momentum atomic shells (i.e. f-shells), which in the bulk have a very small overlap with corresponding electrons in neighboring atoms. The spins are indirectly coupled via the conduction electrons, in contrast to the direct coupling in the itinerant magnets.

solid, the magnetic moment per atom (7.63 μ_B; 10.2 μ_B) are almost as large as the isolated atom, due to weak inter-atomic interactions of the f electrons. These weak interatomic interactions further cause low Curie temperatures (292 K; 88 K) compared with itinerant ferromagnets.

10.2.7 Exchange

Several exchange type interactions are responsible for magnetism in metals [11, 12, 13, 18]. One is the intra-atomic exchange which causes atoms to attain a net spin (see Section 10.2.1) that underlies Hunds rules. Exchange is also responsible for spin order (ferromagnetic or antiferromagnetic, Section 10.2.2). That interaction is the interatomic exchange coupling which can be either indirect or direct as described in the H_2 example. Indirect exchange occurs for example in the rare earth metals (Section 10.2.6) where the localized spins (from the partially filled f-shells) interact essentially entirely via the mutually shared conduction electrons. Due to the small overlap of the f-wave functions, this interaction is more important than the direct exchange in these cases.

The indirect exchange is described in a theory by Rudrman, Kittel, Kasuya and Yusida (the RKKY) [11, 12, 13, 18] model which finds that the coupling between S_i and S_j is given by

$$V_{ij} = -J_{sd}^2\,(2mk_F^4/\pi^3)F_0(2k_FR_{ij})S_i \cdot S_j \tag{10-17}$$

where

$$F_0(x) = -\cos(x)/x^3 + \sin(x)/x^4 \tag{10-18}$$

and J_{sd} is the indirect (s – d) exchange coupling constant, m is the electron mass, k_F is the Fermi wavevector, R_{ij} is the distance between the localized spins S_i and S_j.

The function $F_0(x)$ shows that for short distances the coupling is ferromagnetic, however it reverses sign and oscillates as the distance increases. These RKKY oscillations in the exchange coupling are important for the magnetic properties of surfaces and small particles.

10.3 Magnetism in reduced dimensional systems

10.3.1 Surface magnetism

The magnetic properties at the surfaces of ferromagnets (in particular of itinerant magnets) are significantly altered from the bulk for several reasons. For example, since the number of nearest neighbors is reduced a valence electron tends to spend more time at each ionic site, due to reduced coordination compared with the bulk. Weaker bonding to neighboring sites causes the ions to have more isolated atom character, i.e. the electrons (and hence the moments) are more localized than in the bulk [2, 22]. Figure 10-4 shows the evolution of the electronic structure as the dimensionality is increased from the atom to the surface to the bulk. In general, increasing the dimensionality causes a reduction of the spin imbalance as indicated in Fig. 10-4. Related considerations are important for ferromagnetic clusters [23] and thin films [24].

10.3.2 Magnetic anisotropy and domains in small particles

In extended ferromagnetic systems, long range magnetic effects are important [11, 12, 13, 18, 19]: if the magnetic moment of the system is maximized (i.e. all the spins are aligned), an extended magnetic fields will result (as in permanent magnets). There is a substantial amount of magnetic field energy in this configuration. In the bulk, energy is significantly reduced through the formation of domains [19, 25, 26]. Within a domain, the magnetization is uniform in a one direction, and the direction varies from one domain to the next. The net effect is a great reduction in the magnetic field strength and hence of the magnetic field energy. Domain walls (Bloch walls) are however energetically costly, since in traversing awall, the spin orientation changes at the expense of the exchange coupling which prefers aligned spins. Consequently, domains tend to be relatively large (on the order of 1 μm). Small magnetic particles therefore, are typically monodomain, since the energy cost to form a domain wall outweighs the reduction in magnetic energy (surface effects may complicate matters). Typically critical sizes for monodomain particles are in the range of the range 20–2000 nm and depends on the ferromagnetic material under consideration [1].

The direction of the magnetic moment in a small particle is determined by both the shape of the particle and by its crystal structure through the anisotropy effects [11, 12, 13]. Shape anisotropy accounts for preferential orientation of the magnetic polarization (or magnetization) along the long axis for elongated particles. This is caused by the fact that the total magnetic field energy of the system is minimized for this polarization.

The magnetization also prefers alignment along specific crystallographic axes. This is the crystalline anisotropy effect, which in fact dominates over the shape anisotropy in very small particles [11–13]. Physically the spin system couples to the crystal lattice via

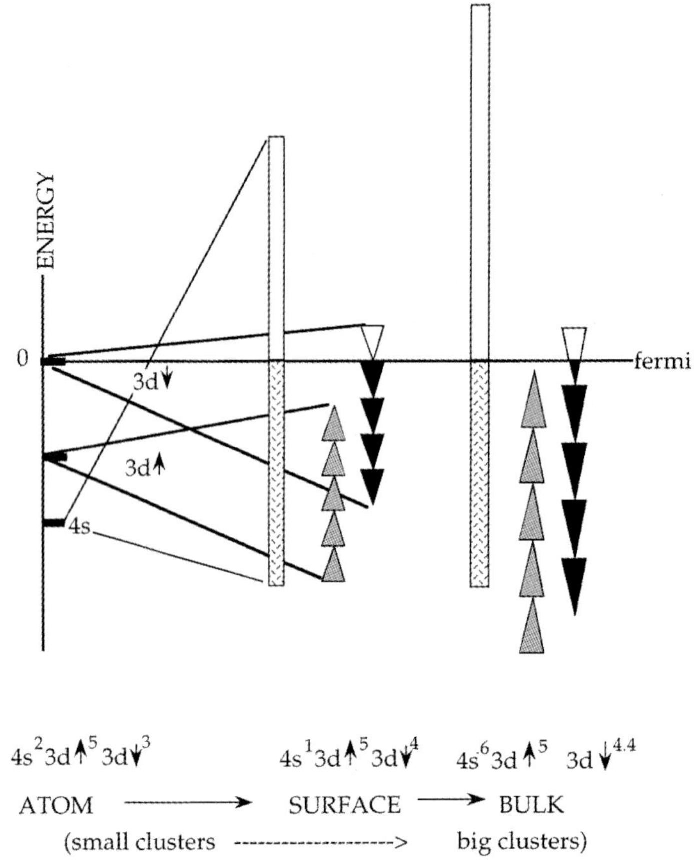

Figure 10-4. Schematic diagram of the evolution of electronic bands of an itinerant ferromagnet with increasing dimensionality (atom to bulk) for nickel. In the atom, the electronic system is organized in electronic shells with discrete energies. The 3d states are split by the exchange interaction so that the minority (spin down) states are higher in energy than the spin up states, as dictated by Hund's rules. The ground state configuration is [Ar] $4s^2 3d^5\uparrow 3d^3\downarrow$ with a spin imbalance of 2. Further increasing the coordination causes the discrete levels to broaden into continuous bands whose width is a measure of the frequency with which the electrons hop from one atom to the next. The s band broadens most rapidly causing one electron to transfer to the d bands. Consequently the surface which has a spin imbalance of 1. Increasing the coordination continues this trend so that ultimately the spin balance is reduced to 0.6 in the bulk electronic configuration. The electronic structure of small particles follows similar trends due to the large surface to volume ratio.

the spin orbit interaction. Axes of preferential magnetization are called the easy axes of magnetization. For example, if the z axis is the preferred axis then (to lowest order) the energy per unit volume will depend on direction of magnetization and vary as

$$E = K\cos^2\theta \tag{10-19}$$

where K is the anisotropy energy per unit volume and θ measures the angle between the z axis and the axis of magnetization.

10.3.3 Superparamagnetism

Consider a single domain ferromagnetic particle [27] without crystalline anisotropy. The dynamics of the orientation of the spin system at temperature T (far below the Curie temperature) in a magnetic field B is just that of a single large spin system with spin S and magnetic moment μ in a magnetic field. The average magnetic moment in the direction of the applied field is given by the Langevin equation for atomic paramagnetism (Eq. 10-8). The size of the spin in this case can be many orders of the magnitude greater than for a single atom. This analogy inspired the name superparamagnetism [17, 28] to describe these systems. In particular:

$$M/\mu = \coth(\mu \cdot \mathrm{B}/kT) - kT/\mu \cdot B \qquad (10\text{-}20)$$

where μ is the magnetic moment and M is the magnetization.

Consequently:

$$M/\mu = \mu \cdot B/3kT \text{ for } \mu\mathrm{B} << kT \qquad (10\text{-}21)$$
$$M/\mu = 1 \text{ for } \mu\mathrm{B} >> kT$$

If an assembly of identical uniaxial superparamagnetic particles is initially polarized along the easy axis, then the magnetization will reduce with increasing time:

$$M(t) = M_0\exp(-t/\tau) \qquad (10\text{-}22)$$

The relaxation time τ is given by [17, 29]

$$\tau = \tau_0\exp(KV/kT) \qquad (10\text{-}23)$$

where τ_0 is of the order of 10^{-10}–10^{-13}sec.

The approach to thermal equilibrium is due to random thermal effective torques or fields as explained by Neel [28] and Brown [29]. The relaxation rate has an Ahrenius like behavior, where τ_0^{-1} is the attempt frequency and the KV represents the barrier height, where V is the particle volume, and K is a material constant (the anisotropy energy density). Estimates of the relaxation times for 11.5 nm diameter spherical iron particles at 300 K is 10^{-1} sec, while for a 15.0 nm particle it is 10^9 sec, which demonstrates the extreme sensitivity with size: the smaller particle instantly relaxes while the larger one is stable for 30 years. The blocking temperature measures the temperature for which $\tau = 10^2$ sec which corresponds to an energy barrier $KV = 25\ kT$. Note that the experimental relaxation rates for small particles may be significantly lower than the values extrapolated from bulk parameters (see i.e. [30–32]).

An important property of superparamagnetic particle systems is that they are non-hysteretic and their magnetization curves scale with H/T. These effects were first demonstrated by Bean and Jacobs for 2.2 nm iron particles suspended in Hg. at 77 K and 200 K (see i.e. [13]).

The model above assumes that the spin reversing process is by uniform rotation of the magnetization over the anisotropy barrier: in the process the mutual alignment of the contributing spins remains intact. Other modes occur where the spin reversal proceeds with a distortion of the spin alignment (i.e. by buckling or curling) [33]. These modes are important and reduce the critical field for magnetization reversal (i.e. the coercive field H_c) [12].

10.4 Microscopic characterization of nanoscopic magnetic particles

Measurements of magnetic properties of nanoscopic systems requires tools which measure magnetic properties and simultaneously provide topographic information on a very small scale. Various probes have been developed to this end and several of those are discussed here. We will not discuss the plethora of methods that apply primarily to bulk magnetic materials but we will concentrate only on several of those that measure magnetic properties of small systems.

10.4.1 SEMPA

Scanning electron microscopy with polarization analysis (SEMPA) (see i.e. [34]) is one of the most powerful microscopy methods in nanomagnetics. In this method, the spin polarization of secondary electrons emitted from a target in a scanning electron microscope is determined using a spin polarization analyzer (Fig. 10-5). Hence, in the topographic and structural information contained in the conventional SEM image is superimposed on spin polarization information. The spin of the emitted electrons is closely related to the state of magnetization as explained in Section 10.2.5. Moreover, the method is surface sensitive (due to the escape depth of the secondary electrons) and contains information of the top 1 nm of the sample. The spatial resolution of this method is about 10 nm.

While this method is quite powerful it has mainly been applied to magnetic surfaces and thin films [35]. An example of a SEMPA image is shown in Fig. 10-6 which shows an image of a magnetically polarized bits on a storage medium for two perpendicular polarization directions [36].

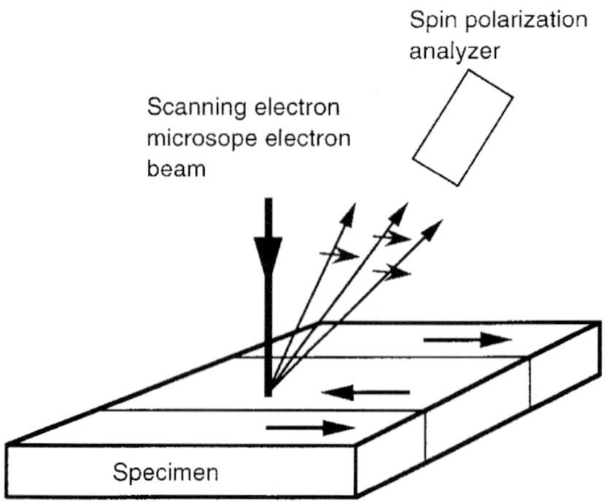

Figure 10-5. Schematic diagram of the SEMPA method. The collimated electron beam from a scanning electron microscope is scanned across the surface of the magnetized specimen. Secondary electrons from the surface are ejected in the process. In magnetic materials, the electrons which are ejected from states near the Fermi surface are spin polarized. The electrons are detected with an electron detector which is sensitive to the spin polarization of the electrons. Contrast is obtained by measuring the degree of spin polarization along a specific direction (i.e. by measuring $(I_+-I)/(I_+ +I)$ where $I_\pm$ are the signal intensities in the aligned and anti aligned directions).

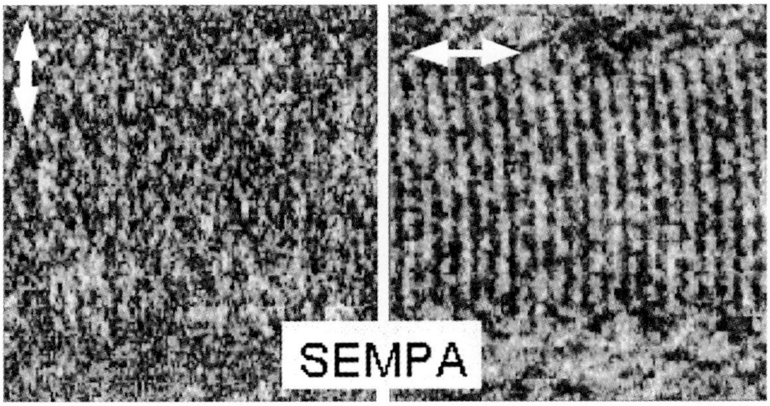

Figure 10-6. An example of a SEMPA image of magnetized bits of a magnetic medium [36]. The images are recorded for the two spin polarizations as indicated in the image. The magnetic information is recorded with the magnetization in the horizontal direction so that only this direction produces contrast. The individual bits (vertical stapes) are approximately 125 nm wide.

10.4.2 Lorentz microscopy

Lorentz microscopy is a form of transmission electron microscopy that depends on the presence of highly localized magnetic fields in the sample (see i.e. [37]). Observed contrast features are related to the deflections of the electron beam due to the Lorentz force: $\mathbf{F}_L = e\, \mathbf{v} \times \mathbf{B}$, where $\mathbf{v}$ is the electron velocity and B is the local magnetic field. Even though the transverse momentum imparted to the passing electrons by the local fields of magnetic particles is very small they can nevertheless be observed in certain TEM imaging modes (i.e. Fresnel and Foucault). In the Fresnel mode the defocusing effect due to the Lorentz force produces contrast so that domain walls are imaged as bright and dark lines. In the Foucault mode (Fig. 10-7), imaging involves placing an aperture in the same plane as the diffraction pattern and using the edge of the aperture to obscure part of the diffraction spot (Fig. 10-11). Thus, differently magnetized domains appear shaded dark or light (Fig. 10-8). A recent application of this

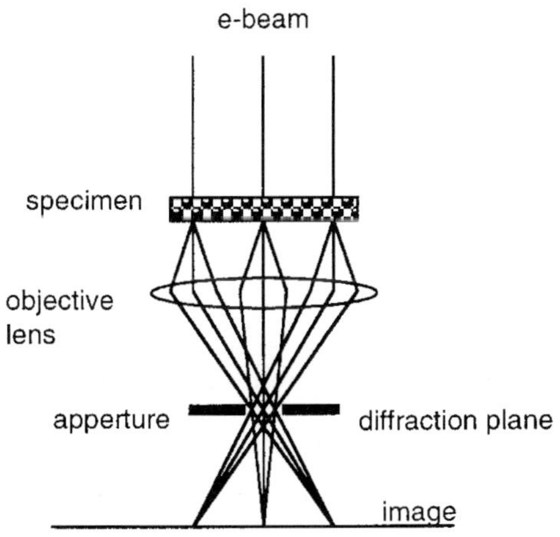

Figure 10-7. Schematic diagram of transmission electron microscope in the Foucault configuration for Lorentz microscopy. The electron beam passes through the specimen. The objective lens projects an image of the specimen. The aperture in the diffraction plane produces contrast for those electrons which are deflected in the sample due to the Lorentz force on the electrons as they pass through the magnetic fields at the sample.

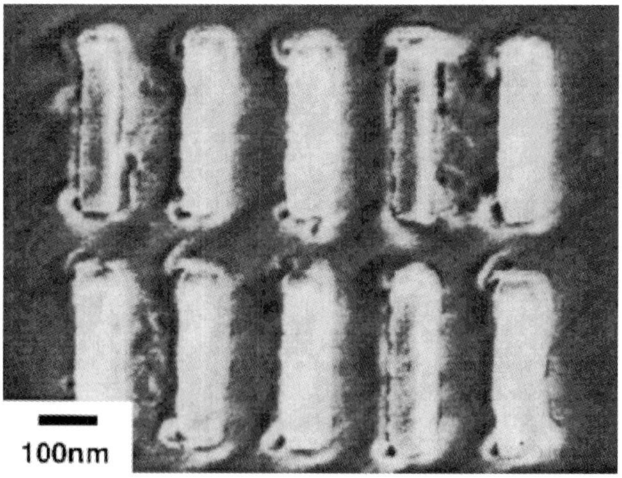

Figure 10-8. Foucault image of magnetized 300 × 80 nmCo elements. Dark and light shaded contrast reflect opposite magnetization directions, indicating that three of the ten elements (i.e. 1, 4 and 9, counting from top left) have switched relative to the others (from Ref. [37]).

method produced images of 5 nm Fe_3O_4 and $SmCo_5$ superparamagnetic monodomain nanoparticles [31]. Analysis of the observed spontaneous magnetization reversals indicate unexpected long relaxation times.

10.4.3 Holography

Transmission electron microscopy can be extended to measure magnetic fields by obtaining contrast caused by the interference of two coherent electron beam which pass by the sample. This electron optical feat is accomplished using an electronic prism to split the electron beam [38]. The interference pattern contains information regarding the relative phase shift of the two beams with respect to each other (Fig. 10-9).

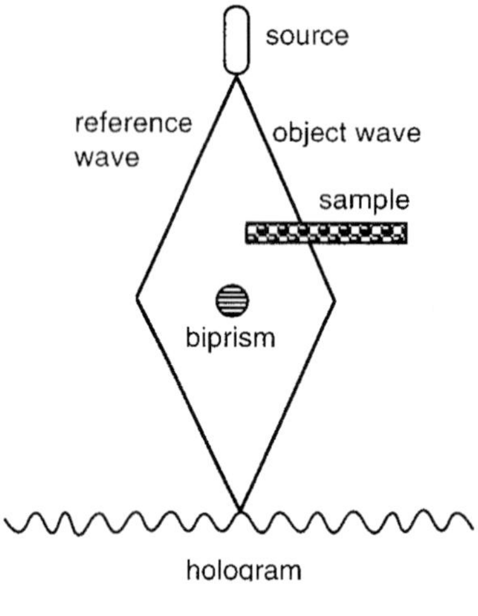

Figure 10-9. Schematic diagram of electron microscope configuration for holography. Half of the e-beam passes through the specimen and the other half passes by it. The two beams are combined and produce an interference pattern which is recorded. This pattern contains phase information that can be imaged by after appropriate processing. Since the phase shifts are caused by the magnetic fields the hologram provides magnetic field information.

The electrons in either beam accumulate phase when they pass through regions of space with non-zero magnetic vector potentials (due to magnetized objects). The phase shift between the two beams (as reflected in the contrast) can then be related into the amount of magnetic flux which passes through the area between the two beams, (due to the Aharonov Bohm effect). In particular the phase shift is given by

$$\Delta\Phi = e/h \sum B.\mathrm{d}s \tag{10-24}$$

where the integral is over the surface enclosed by the two interfering beams (see Fig. 10-9). The hologram in Fig. 10-10 shows the interference fringes caused by magnetic Ni nanostructures. The fringes can then be translated into a phase-shift pattern

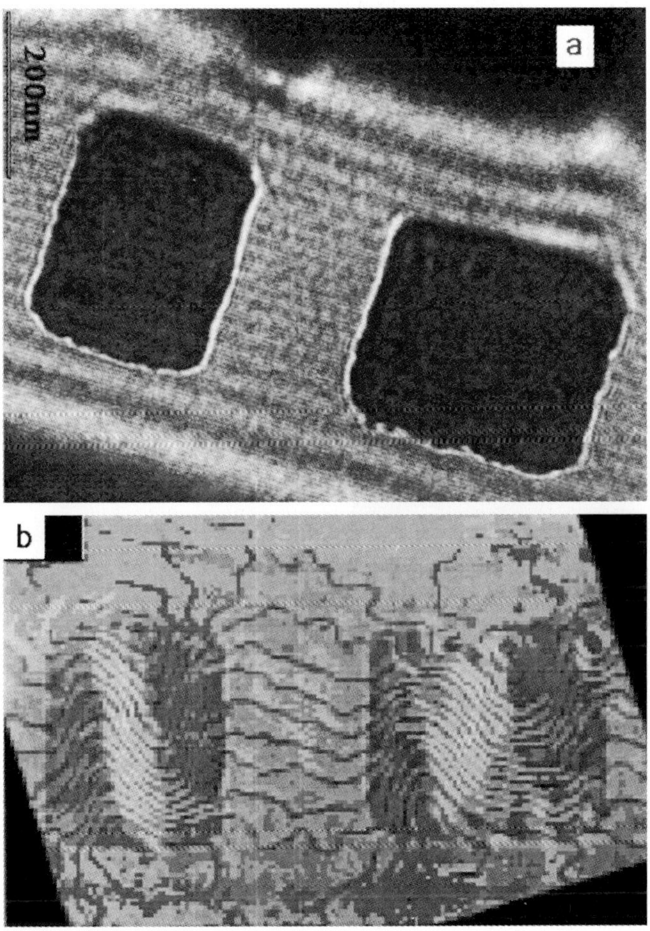

Figure 10-10. Example of electron holography. a) Electron hologram showing two rectangular patterned Co nanostructures from a linear chain of elements. Note the interference fringes. b) The magnetic contribution to the phases of the hologram. Here, the contrast reflects the magnetization direction (the original hologram is colored), whereas the phase contours (lines) are parallel to the lines of constant magnetization. The contour spacing is 0.21 π and is proportional to the magnetic field strength integrated along the e-beam path (from Ref. [58]).

that is related to the local magnetic fields in the specimen (Fig. 10-10b). This method has also been applied to demonstrate quantization of magnetic flux in magnetic nanowires [39].

10.4.4 SP-STM

Spin polarized scanning tunneling microscopy is an extension of scanning tunneling microscopy where the scanning tip emits spin polarized electrons [40]. Magnetized ferromagnetic tips (for example, Fe coated W tips) are used for this purpose. In STM, the tunneling current is related to the density of states of occupied and unoccupied states (depending on the electric polarity of the tip with respect to the sample). Since the electronic density of states in ferromagnetic materials is spin dependent (i.e. near the Fermi level one spin direction dominates over the opposite one), a spin polarized STM image will provide magnetic polarization information. Furthermore scanning tunneling spectroscopy (STS) can also be performed by scanning the bias voltage and recording the tunneling current. In the same way that STS provides information on the local density of states in non-magnetic systems, so SP-STS provides information on the local spin densities [41] in magnetic systems. Accurate experimental spin density information provides deep insight into the electronic structure of the spin polarized bands that ultimately are responsible for the ferromagnetic state. Figure 10-11 shows a SP-STM image of magnetic domains on a Ga film providing a striking example of the potential of this relatively new method [41].

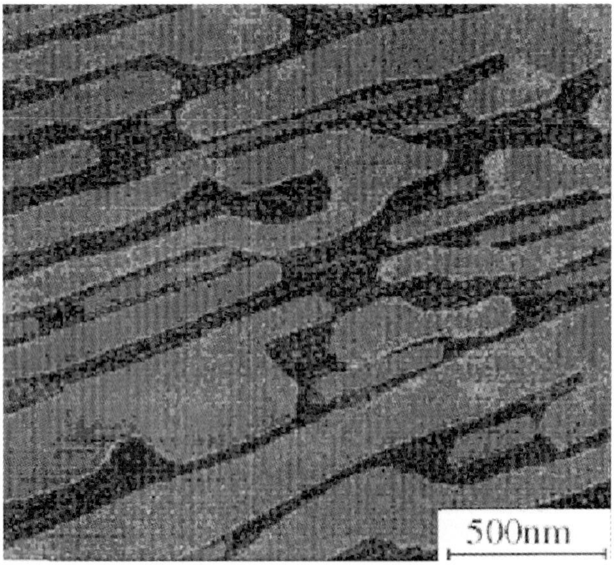

Figure 10-11. Spin polarized scanning tunneling microscope image of a Gd thin film showing a domain structure. The contrast in this image is caused by the fact that the electronic density of states of the minority spin is different from that of the majority spin state. Since the magnetized iron tip injects polarized electrons into the sample, the tunneling current will depend on the state of magnetization of the sample relative to the tip. Hence the domain structure (which represents regions of opposite spin polarization) produces contrast (after Bode et al. [41]).

10.4.5 MFM

Magnetic force microscopy is a scanning probe method which measures the force of a magnetized scanning force microscope tip on a magnetized sample (Fig. 10-12) [42, 43]. This method is very simple to apply and provides a direct probe for the magnetic fields which are near the surfaces of magnetic nanostructures (10 nm resolution has been achieved). The force between two magnetized objects (tip and sample for example) is related to the product of their magnetic field gradients. Hence the MFM is sensitive to sample field gradients (i.e. fringe fields) rather than the magnetic fields themselves. While the method provides immediate evidence for the presence of magnetic fields (and hence for ferromagnetism), it is in fact difficult to extract quantitative data concerning the strength and distribution of magnetic fields. The MFM is very useful to locate magnetic domain edges (Bloch walls) and other features involving magnetic gradients, and for this reason is very useful to characterize magnetic media. However serious objections have been raised concerning the interpretation of MFM derived information [44]. Fig. 10-13 shows MFM images of an array of iron nanoparticles [45].

10.4.6 Micro SQUID and Micro-Hall

Superconducting quantum interference devices rely on the effect that the critical current is a periodic function of the magnetic flux which passes through the SQUID loop (the cycle is determined by flux quantization in the loop). SQUID loops with sub micron dimensions, facilitate measurements of magnetic properties of individual particles within the loop [46, 47]. The method has been demonstrated [47] on individual Ni particles in order to examine the dynamics of magnetization reversal phenomena. The micro Hall method is related to the micro-squid and consists of a Hall probe cross of microscopic dimensions that is lithographically patterned on an insulating substrate [45]. The magnetic field of a magnetic particle on the cross causes a voltage difference to appear on two arms of the cross in response to a current through the other two arms as for a conventional Hall probe [19].

10.4.7 Near field methods

Near field scanning optical microscopy methods on magneto-optical materials allow magnetic features to be imaged with a resolution of at least 30 nm [25, 48]. The method relies on the Kerr effect (i.e. the rotation of the polarization axis of reflected light due to the magnetic polarization of the sample). Related methods include magneto-optical near-field scanning tunneling microscopy, which uses a semiconducting tip as a photodiode for light intensity measurements with NSOM [49]. These promising magnetic microscopy methods are still in early stages of their development.

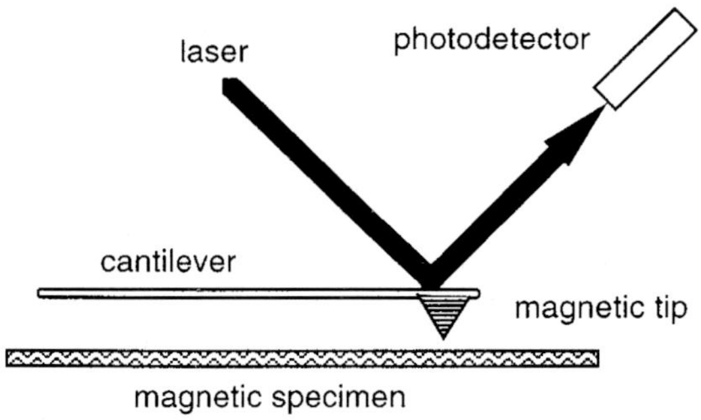

Figure 10-12. Schematic diagram of a magnetic force microscope. A small magnetized tip is attached to a small cantilever. The attractive forces between the tip and a magnetic sample causes the cantilever to deflect. The deflection is monitored by measuring the angular displacement of a laser beam which is reflect by the mirroring top surface of the cantilever. The MFM is sensitive to field gradients above the sample which are caused by spatial variations in the magnetization of the sample (i.e. due to domain structure).

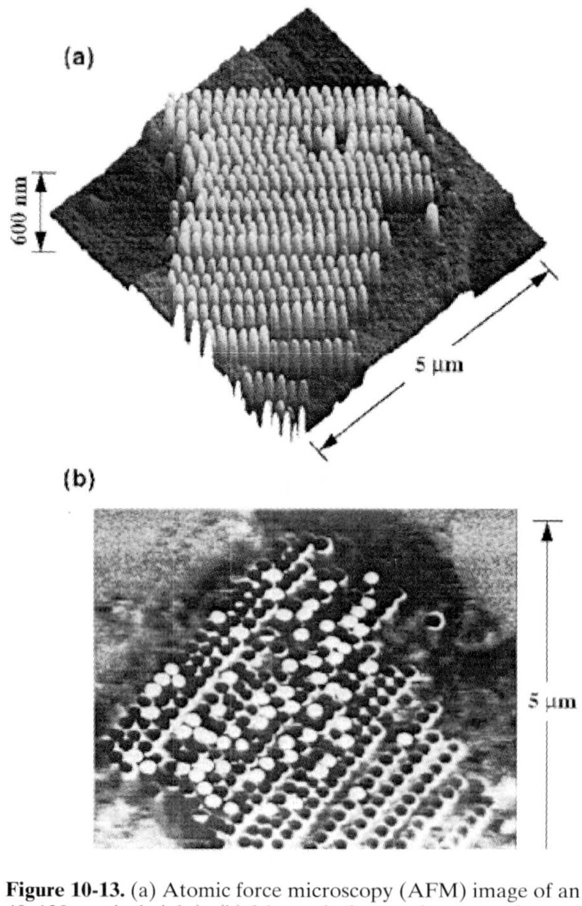

Figure 10-13. (a) Atomic force microscopy (AFM) image of an array of iron particles (< 40 nm diameter 40–120 nm in height). (b) Magnetic force microscopy image of the same array which shows "bits" in several states of magnetic polarization as indicated by the contrast. The bit density of this array corresponds to about 45 Gbit/in^2 (after Gider et al. [45]).

10.5 Magnetic properties of selected nanomagnetic systems

10.5.1 Ferromagnetic clusters in magnetic beams

Molecular beams offer a unique opportunity to study small magnetic clusters X_N where N ranges from the atom to clusters with several thousand atoms. The experimental method [4] involves (Fig. 10-14): (1) production of small ferromagnetic cluster by laser ablation of a metal target in a helium gas; (2) forming a cluster beam by expanding the cluster / gas mixture out of a nozzle into high vacuum; (3) collimating the beam using narrow apertures into a well defined beam; (4) passing the beam through the inhomogeneous magnetic field of a Stern-Gerlach magnet causing the magnetic particles to deflect; (5) photoionizing the particles with a pulsed ultraviolet laser; (6) measuring the mass of the deflected clusters using time-of-flight mass spectrometry. The method allows the mass, the deflection and the velocity of the particles to be determined. With further information on the magnetic field strength in the Stern-Gerlach magnet (at the position of the beam), the cluster temperature, allows determination of the magnitude of the magnetic moments of the clusters [50].

10.5.1.1 Magnetic moments of 3d transition metal clusters

Stern-Gerlach measurements of the 3d transition metal clusters (Fe, Co, Ni) have been made at several temperatures for a large range of cluster sizes [51]. An overall decreasing trend of the magnetic moment with increasing cluster size is observed in all three cases, with some oscillating fine structure features (Fig. 10-15).

This decreasing trend can be understood from general arguments, relating to the cluster surface. The magnetic moment of an atom at the surface is generally larger than in the bulk. This is due to the lower coordination of these atoms. Electrons tend to dwell on near the atomic cores for relatively longer times before hopping to a neighboring site, compared with electrons at the higher coordinated interior atoms (Fig. 10-4). Hence the lower coordination causes the surface atoms to have more of a free atom like character than the interior atoms. The electrons are more confined near these sites than at the interior sites and the arguments which lead to Hund's rules (i.e. large spins) also produce relatively large surface spins.

Note that for small sizes the magnetic moments per atom for Fe, Co and Ni are approximately 3 μ_B, 2 μ_B and 1 μ_B respectively. (Co is slightly higher, probably due to orbital effects.). These values correspond to the maximum spin that can be obtained with 7, 8 and 9 electrons in a d orbital. For example a nearly free iron atom has 8 valence electrons of which 1 is in the 4s orbital and 7 in the 3d orbital. Of those 7, 5

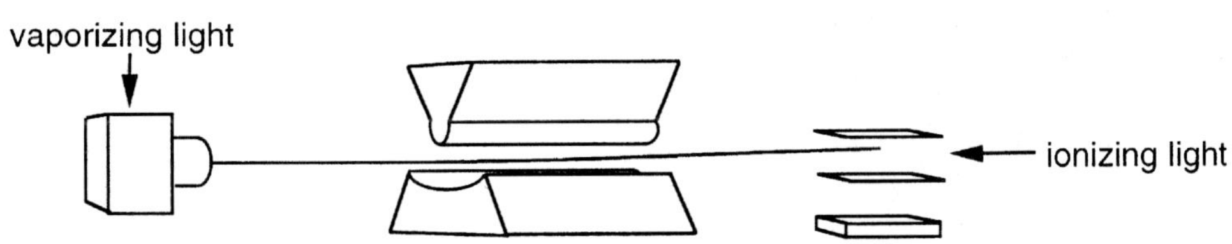

Figure 10-14. Schematic diagram of a molecular beam apparatus to measure magnetic properties of free ferromagnetic clusters.

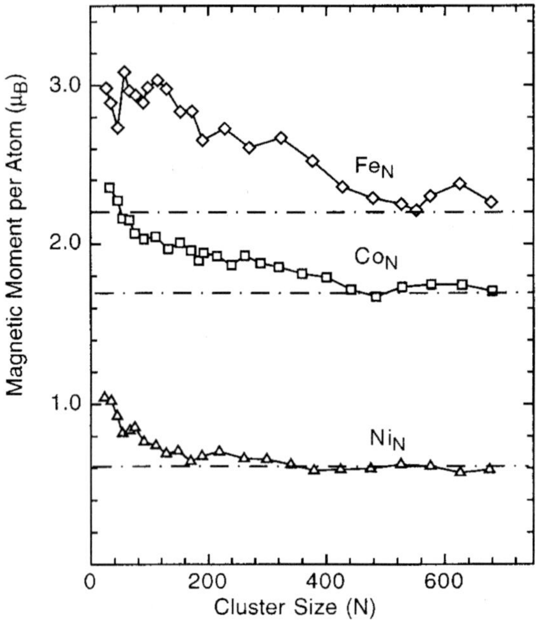

Figure 10-15. Measured magnetic moments per atom for iron, cobalt and nickel clusters X_N. The magnetic moments of the small clusters are close to the maximum values for the spin according to Hund's rules (i.e. 3 μ_B, 2 μ_B and 1 μ_B for iron, cobalt and nickel respectively), while the magnetic moments converge towards their bulk values for large clusters (i.e. 2.2 μ_B, 1.65 μ_B and 0.6 μ_B for iron, cobalt and nickel respectively). The deviations from the smoothly decreasing trend are probably due to cluster specific effects. Note that bulk like magnetic moments are observed for rather small clusters which for which the surface represents a relatively large fraction of the total number of atoms in the cluster. This effect is explained using a magnetic shell model (see Billas et al. [51]).

are in spin up states (forming the majority spin band) and 2 in spin down states (the minority spin band). Hence the net magnetic moment per atom due to the 3d orbitals is 3 μ_B (the s band is not spin polarized). This is roughly the situation for a surface atom, where the majority spin band is entirely below the Fermi surface. Analogous arguments predict 2 μ_B for Co surface atoms and 1 μ_B for Ni surface atoms.

Increased coordination of the interior atoms causes broadening of the bands (as a consequence of more frequent hopping) so that the minority band partially rises above the Fermi level (from which some electrons leak into the half empty 4s band), causing a reduction in the spin imbalance. Consequently, the magnetic moments of the interior atoms are reduced compared with surface atoms. In fact for Fe, $\mu_{bulk} = 2.2$ μ_B; for Co, $\mu_{bulk} = 1.6$ μ_B; for Ni, $\mu_{bulk} = 0.6$ μ_B.

This simple picture suggest a model: for a hypothetical spherical particle, the ratio of the surface atoms to total atoms is about $3N^{-1/3}$, so that for $N = 1000$ about 30% of the atoms are on the surface. Hence, one can estimate that the size dependence of the magnetic moment is

$$\mu \approx \mu_{bulk} + 3(\mu_{surf} - \mu_{bulk})N^{-1/3} \tag{10-25}$$

Measurements show that the convergence to the bulk value occurs much more rapidly than this prediction and that clusters with as few as 600 atoms already appear to have bulk-like magnetic moments. Moreover, the evolution to the bulk is not smooth. This indicates that the picutre is too simplistic, however its merit is that the extremes are reasonably well described. An attempt to explain the rapid convergence as a consequence of RKKY oscillations using a magnetic shell model is given in [51].

It is interesting to note that measurements of Cr clusters ($N > 10$) have demonstrated that these clusters do not deflect. This indicates that they are non-magnetic or more likely that they are anti-ferromagnetic (as in the bulk).

10.5.1.2 The temperature dependence of the magnetic moments

Loss of ferromagnetic order occurs at the Curie temperature where thermal motion overcomes the order imposed by the inter-atomic exchange interaction [11, 12, 13]. In itinerant magnetism there are two distinct pictures: in the band picture, the magnetic moment reduction is caused by thermally induced electronic excitations from the top of the majority spin band to the Fermi level of the minority spin band (Stoner excitations, c.f. the Stoner gap in Fig. 10-2) which reduce the total moment. In the localized moment picture, the global moment is reduced through local misalignments, however the local moments remain intact. The Curie temperature predicted by the Stoner model is too large, while the local moment picture misrepresents the itinerant nature of the 3d electrons. (It rests on fairly solid ground for rare earth metals where the moment carrying electrons are well localized).

The molecular beam method favors measurements of magnetic moments as a function of temperature over a wide range of temperature [50], ranging from at 80 K and to 1000 K (Fig. 10-16). In this way the ferromagnetic to paramagnetic phase transition can be probed. The magnetic moments have been measured as a function of temperature for several sizes and it is clear that they decrease with increasing temperature.

Several interesting general observations can be made. The magnetic moments reduce with increasing temperature, however they to not vanish at a well defined temperature. This property may be expected since finite systems cannot have sharp phase

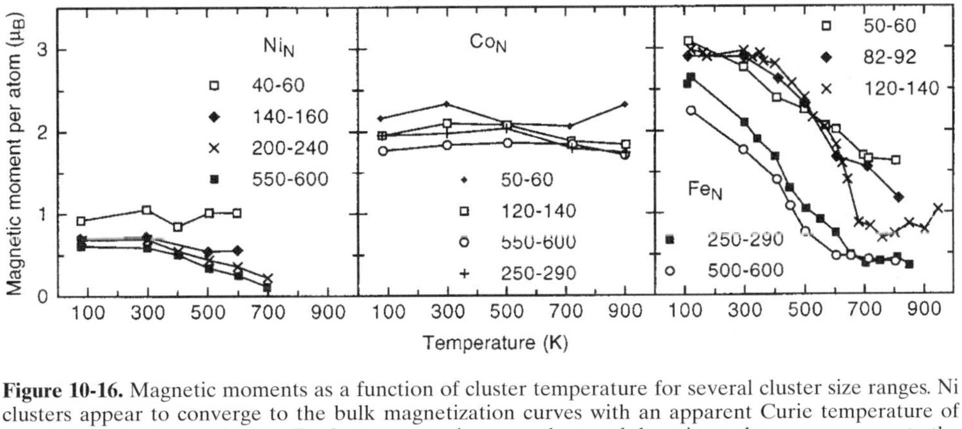

Figure 10-16. Magnetic moments as a function of cluster temperature for several cluster size ranges. Ni clusters appear to converge to the bulk magnetization curves with an apparent Curie temperature of about 700 K for larger clusters. Fe clusters are quite anomalous and there is no clear convergence to the bulk magnetization curve, in fact the apparent Curie temperature of larger clusters is far below the bulk value of 1041 K.

transitions. It is also observed that the for Ni and Co clusters, the magnetization curves appear to converge to their respective bulk behaviors. However Fe is anomalous and no obvious trend can be discerned. We suspect that this may be related to the various crystallographic phases of Fe which differ in their magnetic properties. In particular, bulk Fe is bcc. In the bulk this structure is stabilized compared with the fcc structure due to magnetic interactions. Small clusters usually prefer icosehedral structures (to minimize the surface energy). These steric considerations cause further complications in the structure and consequently in the magnetic properties as appears to be revealed by the measurements.

At sufficiently low temperatures it is expected that the magnetic moment becomes pinned to a preferred axis in the cluster giving rise to blocking effect [17]. In supported cluster systems this is manifested as an abrupt vanishing of the magnetization at and below the blocking temperature, when the magnetic moment is no longer able to align to the applied field because the anisotropy barrier is too high (or, equivalently, when the relaxation time is too long).

It is not clear how the superparamagnetic blocking effect manifests itself in free clusters, since the cluster as a whole can still orient with respect to the applied field. However it is unlikely that the response is still given by the Langevin equation. In fact it has been predicted that below the blocking temperature, the magnetization is still given by a Langevin-like equation however they are uniformly reduced by a factor of 2/3 [52].

In contrast, low temperature measurements (down to 80 K) do not show obvious anomalies although a slight reduction is observed in the Co data. Perhaps the blocking temperatures are lower than 80 K for these cluster systems in the measured size range. This information places upper bounds to the strength of the crystalline anisotropy coupling. (Shape anisotropies are not expected to be important for these sizes.)

10.5.2 Monodisperse cobalt nanocrystal assemblies

A remarkable recent development describes the synthesis of monodisperse cobalt clusters using solution phase reduction of cobalt ions in the presence of surfactants [53]. This chemical path produces particles with a rather large distribution of sizes (i.e. 2–6 nm or 7–11 nm) which are subsequently fractionated to produce essentially monodisperse colloidal solutions.

These nanocrystal solutions are then dried on surfaces to make rather uniform rafts of cobalt particles which are protected from coalescing by virtue of the surfactant layers (Fig. 10-17). Three dimensional super lattices can be built up by appropriate evaporation procedures which allow condensation in the hcp structure.

Magnetization curves for zero field cooled samples of arrays of these particles clearly show the various behaviors expected for such systems (Fig. 10-18). The low temperature magnetization is strongly reduced due to the anisotropy energy which inhibits the magnetic moment to orient along the direction of the applied field. Above the blocking temperature the system becomes paramagnetic with a characteristic $1/T$ temperature dependence of the magnetization, which reflects the paramagnetic susceptibility. In the ideal superparamagnetic system, the magnetization should vanish below the blocking temperature and the transition to paramagnetic behavior at the blocking temperature should be sharp. Deviations from this ideal indicate that there are residual inter particle interactions (due to dense packing) or dispersion in the individual particle properties.

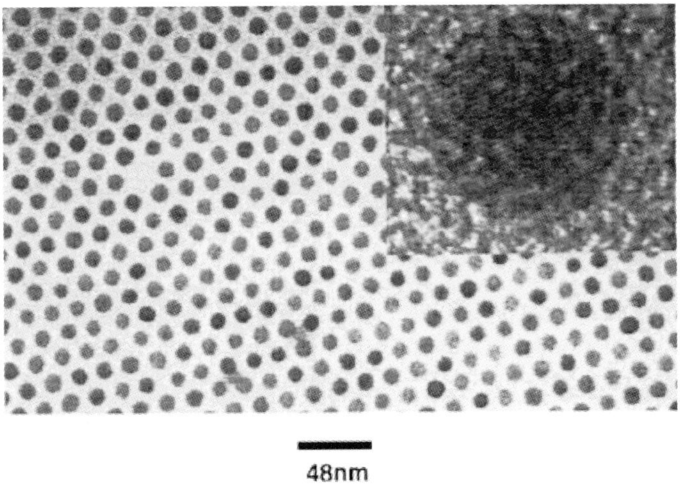

Figure 10-17. Transmission electron microscopy image of a two dimensional assembly of 9 nm cobalt nanocrystals. Inset: High resolution image of a single particle (from Sun and Murray [53]).

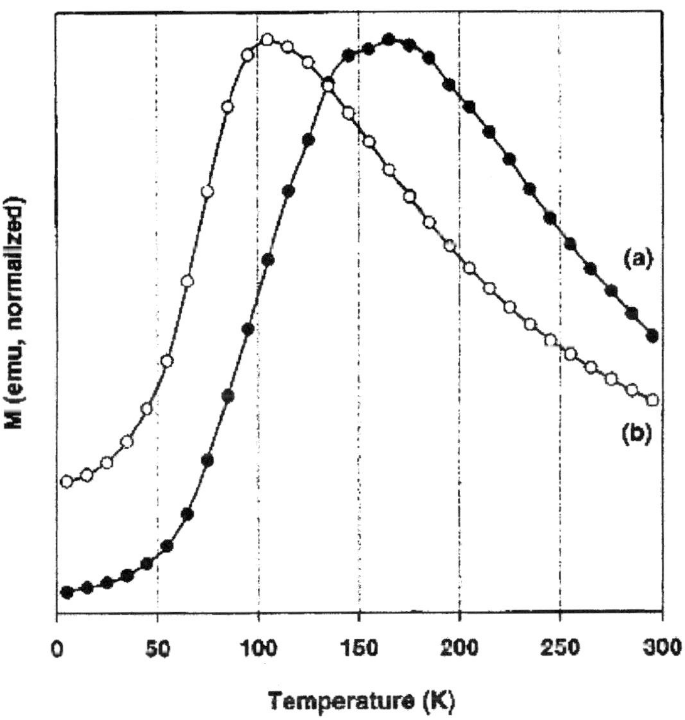

Figure 10-18. Magnetization versus temperature of single domain 9 nm diameter Co particles. (a) magnetization curve of an assembly of as-synthesized particles; (b) magnetization curve of diluted nanoparticles. Note the approximate $1/T$ behavior (consistent with Langevin behavior) for temperatures above the blocking temperature. For the diluted particles the blocking temperature is increased due to the reduced inter-particle interactions. For completely isolated particles (i.e. highly diluted) the transition the transition to the superparamagnetic state (at the blocking temperature) should be sharp (from Sun and Murray [53]).

Note that since the blocking temperature is below 300 K, these samples are too small to be considered for magnetic storage applications, however the processes developed here are certainly very important steps in the direction of nanoscopic magnetic storage media.

10.5.3 Quantum tunneling of the magnetization

Often the magnetization is treated as a classical vector and this is in many cases adequate, even to describe for dynamic effects. However, bona fide quantum mechanical effects may be important in particular for very small particles at low temperatures, when quantum mechanical reorientation of the magnetization may occur. This process proceeds through a tunneling mechanism which is perhaps best understood in analogy with the ammonia molecule, where the N atom may be above or below the plane defined by the three H atoms. The wavefunction for the up state is $\Phi\uparrow$ and for the down state is $\Phi\downarrow$. These states are not stationary states since the ground state and first excited states are linear combinations of them [18]:

$$P_{0,1} = 1/\sqrt{2}(\Phi\uparrow \pm \Phi\downarrow), \tag{10-25}$$

Hence if $\hbar\omega_{0,1}$ is the energy difference between the P_0 and P_1 states, then if a molecule is initially prepared for example in the up state, it will oscillate between $\Phi\uparrow$ and $\Phi\downarrow$ with frequency $\omega_{0,1}$ (the tunneling splitting).

A related effect may occur for the magnetization vector. The magnetocrystalline anisotropy in the particle will efine a preferred direction along which the magnetization will be oriented at very low temperatures. Equal and opposite directions are energetically degenerate. If there is only one anisotropy axis, then the ground state will be degenerate, and the magnetization will be either parallel or anti parallel to the magnetization axis. However, in the more general case, if there two or more anisotropy axes, then the ground state will be a superposition of states where the magnetization is oppositely aligned (compare this with the NH_3 molecule). Hence, if the magnetization is initially aligned along the easy axis (for example by cooling it in a strong magnetic field), then it will oscillate at the tunneling frequency between the spin up and spin down state.

This effect has been predicted and has been studied extensively theoretically [15]. Experimental evidence for quantum tunneling of the magnetization in several systems has been found in high spin molecules at very low temperatures [16, 54].

10.5.4 Giant magnetoresistive nanostructures

Very large magnetoresistances have been observed in multilayered structures consisting of ferromagnetic layers which are separated by non-magnetic layers [1, 55]. The prototype system is layers of Co separated by spacer layers of Cu. The in plane resistance of this system depends sensitively on the magnetic field applied to the layers. The origin of the effect is a strong exchange coupling between the ferromagnetic layers. This coupling is mediated by the Cu conduction electrons, which is surprising since Cu is non-magnetic. In fact the interlayer coupling oscillates between ferromagnetic and antiferromagnetic with increasing spacer thickness due to RKKY oscillations in the spin polarization [56]. Electrons which pass through the structure (parallel to the film) are scattered by the magnetically polarized layers and depends on the relative magnetization between the layers which can be changed with external magnetic fields. The effect not only occurs in magnetic films but also in nanoscopic granular materials, for example with nanoscopic Co particles and Ag particles [57]. Films are produced by pulsed laser deposition in which the two metals are co-sputtered to form metallic granular films. The giant magnetoresistance effect is utilized in reading heads that read magnetically encoded information.

References

[1] FG Himpsel, JE Ortega, GJ Mankey, RF Willis, *Adv. in Phys.* **1998**, *47*, 511–587.
[2] A Hernando, Eds., *Nanomagnetism*, vol. 247 Series, Dordrecht, Boston, E Kluwer Academic, **1993.**
[3] B Barbara, *J. Mag. Mag. Mat.* **1996**, *156*, 123–127.
[4] WA de Heer, *Rev. Mod. Phys.* **1993**, *65*, 611.
[5] D Speliotis, *J. Mag. Magn. Mat.* **1999**, *193*, 29–35.
[6] C Mee, *The Physics of Magnetic Recording*, Amsterdam, North-Holland, **1994**.
[7] H Bertram, J-G Zhu, *Solid State Phys.* **1992**, *46*, 271.
[8] D Lambeth, E Velu, G Bellesis, L Lee, D Laughlin, *J. Appl. Phys.* **1996**, *79*, 4497–4501.
[9] Berkowitz, et al., *Phys. Rev. Lett* **1992**, *68*, 3745.
[10] R Meservey, P Tedtrow, *Phys. Rep.* **1994**, *238*, 1133.
[11] J Crangle, *Solid state magnetism*, New York, Van Nostrand Reinhold, **1991**.
[12] D Mattis, *Theory of magnetism*, Springer, **1985**.
[13] G Rado, H Suhl, Eds., *Magnetism*, New York, Academic Press, **1963**.
[14] T Oda, A Pasquarello, R Car, *Phys. Rev. Lett.* **1998**, *80*, 3622.
[15] L Gunther, B Barbara, Eds., *Quantum tunneling of the magnetization*, vol. 301, Dordrecht, Boston, Kluwer Academic, **1995**.
[16] C Sangregorio, T Ohm, C Paulsen, T Sessoli, D Gatteschi, *Phys. Rev. Lett.* **1997**, *78*, 4645–4648.
[17] C Bean, J Livingston, *J. Appl. Phys.* **1958**, *30*, 120.
[18] N Ashcroft, N Mermin, *Solid state physics*, Forth Worth, Saunders publishing company, **1976**.
[19] C Kittel, *Introduction to solid state physics*, New York, Singapore, John Wiley & Sons, **1996**.
[20] J Jensen, A Mackintosh, *Rare earth magnetism*, Oxford, New York, Oxford University Press, **1991**.
[21] R Feder, S Alvarado, S Tamura, E Kisker, *Surf. Science* **1983**, *127*, 83.
[22] E Kisker, *J. Phys. Chem* **1983**, *87*, 3597.
[23] I Billas, A Chatelain, Wd Heer, *Science* **1994**, *265*, 1682.
[24] Y Li, K Baberschke, *Phys. Rev. Lett* **1992**, *68*, 1208.
[25] P Bertrand, L Conin, C Hermann, G Lample, I Peretti, *J. Appl. Phys.* **1998**, *83*, 6834.
[26] J-U Thiele, L Folks, M Toney, D Weller, *J. Appl. Phys.* **1998**, *84*, 5686–5692.
[27] J Dormann, D Fiorani, Eds., *Magnetic properties of fine particles*, Amsterdam, North-Holland, **1992**.
[28] L Neel, *Ann. Geophys.* **1949**, *5*, 99.
[29] W Brown, *J. Appl. Phys.* **1959**, *30*, 130.
[30] G Xiao, S Liou, A Levy, J Taylor, C Chien, *Phys. Rev. B* **1986**, *34*, 3113.
[31] S Majetich, Y Jin, *Science* **1999**, *284*, 470.
[32] Q Chen, A Rondinone, B Chakoumakos, Z Zhang, *J. Mag. Magn. Mat.* **1999**, *194*, 1–7.
[33] M Lederman, S Schultz, M Ozaki, *Phys. Rev. Lett.* **1994**, *73*, 1986–1994.
[34] R Allenspach, *J. Mag. Mag. Mat.* **1994**, *129*, 160–85.
[35] A Berger, H Oeper, *J. Mag. Mag. Mat.* **1993**, *121*, 102–104.
[36] From: http://physics.nist.gov/Divisions/Div841/Gp3/epg_files/mfm_proj.html, **1995**.
[37] K Kirk, J Chapman, C Wilkinson, *J. Appl. Phys.* **1999**, *85*, 5237.
[38] Wd Ruijter, J Weiss, *Ultramicroscopy* **1993**, *50*, 269–283.
[39] C Beeli, B Doudin, P Stadelmann, *Phys. Rev. Lett.* **1995**, *75*, 4630.
[40] R Weisendanger, H-J Gunterodt, G Guntherodt, R Gambino, R Ruf, *Phys. Rev. Lett.* **1990**, *65*, 287.
[41] M Bode, M Getzlaff, R Wiesendanger, *Phys. Rev. Lett.* **1998**, *81*, 4256.
[42] H Hug, et al., *J. Appl. Phys. 1998*, *83*, 5609–6520.

[43] M Hehn, K Ounadjela, J Bucher, F Rousseaux, D Decanini, B Bartenlian, C Chappert, *Science* **1996**, *272*, 1782–1785.
[44] P Grutter, R Allenspach, *Geophys. J. Int.* **1994**, *116*, 502–504.
[45] S Gider, J Shi, D Awschalom, P Hopkins, K Chapman, A Gossard, A Kent, Sv Molnar, *J. Appl. Phys.* **1996**, *69*, 3269–3271.
[46] L Narasimhan, C Patel, M Ketchen, *Appl. Phys. Lett.* **1998**, *73*, 993–995.
[47] W Wernsdorfer, K Hasselbach, D Mailly, B Barbara, A Benoit, L Thomas, B Suran, *J. Mag. Magn. Mat.* **1995**, *145*, 33–39.
[48] E Betzig, J Trautman, R Wolfe, E Gyorgy, P Finn, M Kryder, S Chang, *Appl. Phys. Lett.* **1992**, *61*, 142–144.
[49] R Schad, S Jordan, M Stoelinga, M Prins, R Groeneveld, Hv Kempen, *Appl. Phys. Lett.* **1998**, *73*, 2669.
[50] I Billas, J Becker, A Chatelain, WA de Heer, *Phys. Rev. Lett.* **1993**, *71*, 4067.
[51] I Billas, A Chatelain, WA de Heer, *Science* **1994**, *265*, 1682–1684.
[52] G Bertsch, K Yabana, *Phys. Rev. A* 49, 1930–1932.
[53] S Sun, S Murray, *J. Appl. Phys.* **1999**, *85*, 4325.
[54] J Friedman, M Sarachik, J Tejeda, R Ziolo, *Phys. Rev. Lett.* **1996**, *76*, 3830.
[55] J Moodera, L Kinder, *J. Appl. Phys.* **1996**, *79*, 4724.
[56] S Parkin, *Phys. Rev. Lett.* **1991**, *67*, 3598.
[57] W Zhang, I Boyed, M Elliott, W Herrenden-Harkerand, *J. Mag. Magn. Mat.* **1997**, *165*, 330–333.
[58] R Dunin-Borkowski, M McCartney, B Cardinal, C Smith, *J. Appl. Phys.* **1998**, *84*, 374.

11 Metal-oxide and -sulfide Nanocrystals and Nanostructures

A. Chemseddine

11.1 Introduction

The synthesis and characterization of nanocrystals is a very exciting emerging field of research which has received important scientific and technological attention [1–5]. At this time however, this research should be initially classified as being fundamental in nature, since there is an increasing awareness of scientists in chemistry, physics and engineering to investigate and understand the fundamental details which relate chemical reactivity to the nucleation and growth of solids. This effort will enable current research to establish a scientific basis for the chemical synthesis and processing of high quality, or perfect nanocrystals and nanostructured materials.

For decades, solid materials were made with techniques that have remained fundamentally the same. Now, processes leading to solids need to be controlled at the earliest stages of production and arrested at desired stages to isolate pieces of solid with desired size, shape, core internal structure and a well defined surface structure.

The fundamental physical properties of nanoscale materials are also of interest, since their small size and large surface area can lead to unexpected or dramatically differing properties [6, 7]. The need for high quality systems in order to achieve a quantitative understanding of size effects on the physical properties of sulfides and oxide nanocrystallites is another motivating factore in this research [1, 8–10].

Chemical and physical research on nanomaterials could provide breakthroughs in building a new class of highly structured materials at the nanoscale, where the physical properties can be fine-tuned for specific applications to create high-performance devices in electrochromic, nonlinear optics, electronics, magnetism, electrocatalysis, catalysis and for solar energy conversion where the quest for novel photovoltaic and carrier confining structures is stressed [1, 11].

At this time, research on nanocrystals can only be portrayed as a growing body of knowledge, since we are at the beginning of understanding and controlling the synthesis and properties of a very reduced number of traditional materials at the nanoscale, and also because of the ongoing spectacular progress taking place in this interdisciplinary field. Important advances in this field were made possible by the ability and ease to do analysis and characterization, with a variety of techniques, at an unprecedented level of resolution [12].

This chapter will successively introduce wet chemical techniques for the synthesis and processing of materials in general and point out some important aspects, which should be considered when nanocrystals are to be made and characterized. The effects of different synthetic factors will be reported and discussed, by taking CdS as a model system for sulfides and titania for the oxides. Even then, it is important to point out that each material will require specific physico-chemical conditions for the control of its synthesis and specific techniques for its characterization.

11.2 Nanocrystals processing by wet chemical methods – general remarks on synthesis and characterization

Soft chemistry, sol-gel process, wet chemical technique, solution technique, colloidal method are all different names that describe the process of starting from molecular precursors in a solution, to making a condensed phase [13–15]. As shown in Fig. 11-1, the nature of the condensed phase in solution depends on physico-chemical conditions, such as the precursor type, concentration, stoichiometry, reaction time, solvent system, the presence of additives, temperature, pressure, the way of mixing and in some cases light [13, 14]. In general, if a reaction takes place, it may lead to the precipitation of amorphous or crystalline powders, gels, turbid sols, transparent colloidal solution, or a solution containing small clusters or molecules. In many cases, reaction leads to a mixture of precipitates in the form of large particles, polymers or aggregates

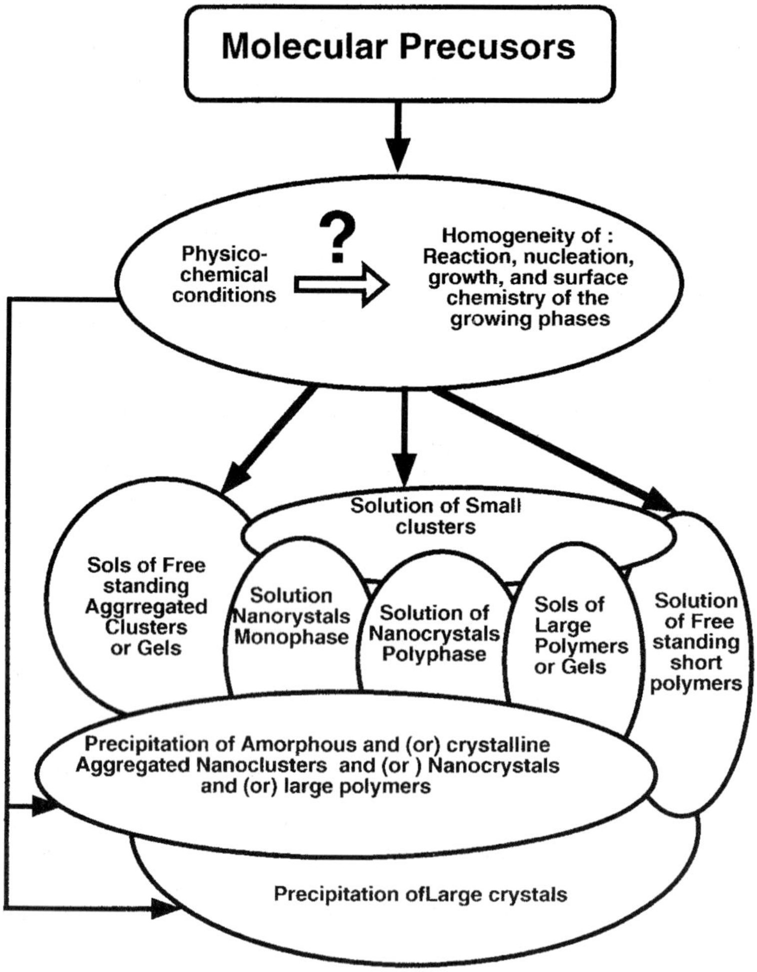

Figure 11-1. Wet chemical methods and a variety of possible reaction products which my form, and particularly in the nanosize regime.

and in the same time to clusters or particles in the form of free standing species in solution. Depending on the technique used to probe these samples, it is helpful to keep in mind these scenarios and simple details to achieve a reliable characterization of the whole samples.

These wet techniques are very promising for a controlled synthesis of nanoparticles and nanostructured materials. In techniques such as CVD, MOCVD, and simple thermal pyrolysis, molecular precursors are allowed to react at rather high temperatures in the gas phase, or as molecular solids. Control over particle size, size distribution, shape and crystalline orientation is inherently difficult to achieve. In contrast, in wet chemical methods, a solvent is used as the reaction medium or vehicle, and the reaction is performed at low temperature. There are many advantages in using wet chemistry for the synthesis and characterization of nanomaterials: (a) Controlling the mixing at the molecular level of the different precursors. (b) Controlling the reaction kinetics by choosing the right solvent or by other additives. (c) Reaction, nucleation and growth can be easily monitored and influenced (d) The growth can be arrested by controlling the stoichiometry or by using appropriate ligands (electronic and structural matching with condensed phase core) [11, 17]. Furthermore, even small clusters are fully relaxed, while bare clusters require a diameter of about 250 Å because of surface contributions (interfacial tension and density of clusters are radius dependent) [18]. (e) In contrast to growing clusters on solid substrates or in solid matrices, in the liquid phase the mobility of the growing phase is assured, allowing structural arrange-

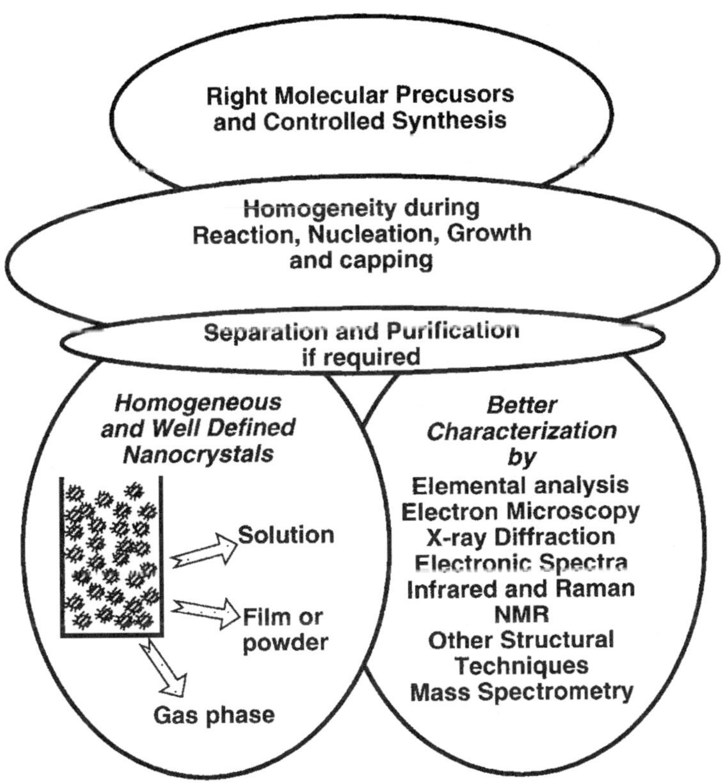

Figure 11-2. Importance of homogeneity in characterizing nanocrystals

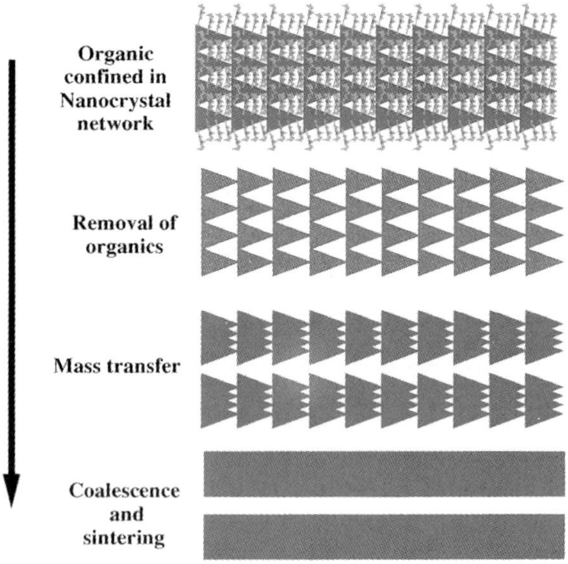

Figure 11-3. Nanocrystal route to materials tailoring: Control of Reactivity through organization of nanocrystals.

ment of the core, and growth termination by surface restructuring and capping. (f) By variation of the synthesis conditions or reactants and characterization of the materials obtained, the production of nanocrystals can be rationalized. (g) Nanocrystals can be isolated according to size in the form of a soluble, processable manageable powders, or used as precursors in the processing of films. A variety of characterization techniques to probe nanocrystals in solution, in transparent films and in the gas phases can be used (Fig. 11-2). (h) Furthermore, a major line of development of nanocrystal wet chemistry concerns the spontaneous generation of a well defined architecture by self-assembly of suitably designed and functionalized nanocrystals [19]. The molecular interaction of capping ligand tails plays a basic role in the generation of assemblies with different structures in solution and the solid state. (i) This issue of ordered nano-structured arrays of functional nanocrystals, is of central importance for the evaluation of the competitiveness of wet chemistry, i.e. sol-gel or solution techniques, since through size, shape, and organization of particles, organic species can be confined within well-defined inorganic structures, allowing control over their thermal removal, coalescence and sintering of particles. Therefore, organized assemblies can be used as precursors [20] for morphological engineering of films, membranes or powders of inorganic solids Fig. 11-3 [21].

11.3 Sulfides nanocrystals

11.3.1 CdS as a model system

Although cadmium sulfide has been known and investigated for a long time, a full understanding of its growth and formation has not yet been fully achieved [22]. This traditional material is a great model system for practicing synthetic chemistry of nanocrystals and for illustrating the importance of understanding the chemistry and growth history of nanocrystals. Generally, heterogeneity in many samples is a serious obstacle to a reliable characterization. Therefore, changing synthesis conditions and collecting data by investigating the different products, leads to better characterization and to rationalization of the synthesis in terms of chemical reactivity, nucleation and growth processes. In fact, crystals of CdS form spontaneously after mixing a sulfur source (like H_2S, Na_2S) and a cadmium source (cadmium salt) in a solvent like water or alcohol. This great advantage provided flexibility to traditional colloidal chemistry to achieve the production of particles in the nanometer range, by simply adjusting stoichiometry, working at very low concentrations to regulate the reaction rate, adding stabilizers like SiO_2 particles [5, 9] or sodium polyphosphate [5, 9], or by lowering the temperature [23]. These techniques provided samples for a relative quantitative study of the already known size effects on the electronic, optical and dynamical properties of nanoparticles [8, 10, 23]. In earlier studies on the size and shape dependent optical and electronic properties of AgI and CdS nanocrystals, Berry was limited by the lack of a preparative technique for controlling CdS growth, to enable a systematic study of size effects on the physical properties. Progress was achieved using colloidal techniques in providing nanoparticles of CdS, PbS, HgS, In_2S_3 and ZnS. Details on the physicochemical properties of these colloidal solutions are described in a series of reviews [5, 9, 10, 24].

Major problems in these nanometric suspensions are the size distributions obtained during the preparation of samples, which makes a detailed analysis of the various phenomena very difficult. In addition, ageing of the sols is always accompanied by changes in the absorption and fluorescence properties of the nanoparticles. The instability of the sols is in many studies one of the causes for the poor reproducibility of the optical properties [5]. Furthermore, the effect of shape, already mentioned in Berry's work [6] and contributions of the surface structure and chemistry, have not been addressed. Nanocrystals with generally irregular shapes were assumed spherical, ignoring geometrical and crystallographic contributions to electronic structure and chemical and photochemical reactivities. Although the physics and the theoretical aspects related to these investigations [7, 23, 25–27] have been one of the stimulating factors for the ongoing development of new synthetic techniques to furnish stable nanocrystals with the desired crystal structure and a well-defined size, shape and surface structure.

11.3.2 Importance of the precursors

Due to the quantum size effect, the size of cadmium sulfide particles is directly related to the absorption wavelength. Therefore the growth of this material is usually investigated in solution by many groups using optical spectroscopy. In these studies, the spectra recorded immediately after injection of H_2S, or the addition of Na_2S, always exhibit long tails, indicating a heterogeneous growth which is a direct consequence of the mechanical mixing [28–30]. The sulfurising agents are very reactive towards cadmium ions, and it is difficult to fully control the size distribution of the species present in the solution during mixing and diffusion of the reactants. Species with a wide size distribution are formed due to concentration gradients. If the conditions are adjusted to work in the small clusters domain [22, 31], an aging or heat-treatment of the solution, leads to the formation of the thermodynamically stable cluster as suggested by the appearance of maxima [22, 31]. In this size regime, the thermodynamically stable clusters form at the expense of others which dissolve or coalesce [22, 31]. In the presence of thiolate ligands, these clusters are stabilized at basic pH and can be isolated as a solid by crystallization [32]. Therefore, in this traditional small cluster regime, the ultimate method for structural characterization of clusters is the determination of crystal structures by X-ray diffraction.

However to make relatively larger nanocrystals, a less reactive sulfur source like thiourea is required. The first advantage in using it is in achieving homogenous mixing of the reactants at the molecular level. Then, the sulfur is generated by a thermally activated chemical reaction, enabling uniform nucleation and growth of CdS clusters or nanocrystals [33]. Second, thiourea molecules are themselves coordinating ligands for cadmium [34] and therefore in intimate contact allowing one to investigate effectively and elegantly, the effect of Cd/S ratio on nucleation and growth of cadmium sulfide. Furthermore, it allows one to adjust the stoichiometry to generate nanocrystals which once the growth has reached the final size of the nanocrystal, share a bidentate capping ligand with its neighbor, leading to a self-connecting process and the formation of networks or organized assemblies [19, 34].

11.3.3 Solvent and capping ligands

The final size, shape, crystal structure and surface composition of nanocrystals depend on stoichiometry, solvent and particularly on the type of capping ligand or combination of ligands used to arrest the growth and stabilize them. The issue of understanding the role of these ligands or coordinating solvent molecules during the reaction, nucleation and their interaction with the growing crystal is of central importance. Advances in controlling the shape and crystal structure of nanocrystals depend on the design of specific ligands targeted for binding to the core without disrupting its structure at the interface. In some extent, this is achieved in the case of cadmium sulfide by using an organothiolate anion, RS^-, which can replace S^{2-} to stabilize the structure on one hand [35], whilst on the other hand, the substituent R can be chosen or manipulated to confer to the nanocrystal a different solubility, or to isolate it as powder with the desired rheological properties [19, 33, 34]. Furthermore, the building of nanocluster assemblies with different dimensionalities can be achieved by having chemical groups in R, which can be engaged in specific interactions [19, 34].

In order to visualize the importance of the structural matching RS^- with CdS, Fig. 11-4 shows the naked CdS unit cell and then after substitution of S^{2-} by RS^-. The product of this substitution is $Cd_4(SR)_{10}{}^{2-}$ which has been known for many years to organometallic chemists [35–37]. If the growth is taking place in the 111 direction, a whole family of tetrahedral clusters can be made where thiolate ligands function as termini at the edges and vertices [35]. The next cluster formed is $S_4Cd_{10}(SR)_{16}{}^{4-}$. which has also been made by Dance et al. [37, 38]. The same author pointed out the rich and complicated chemistry involving cadmium, thiolate ligands and sulfur. Other clusters such as, $[SCd_8\,(SR)_{16}]^{2-}$ and $[S_4Cd_{17}(SPh)_{28}]^{2-}$ were also isolated and characterized crystallographically [39, 40]. This pioneering work did show and discuss the variety of structure-types formed in the thiophenolate chemistry of cadmium [35]. Lately, stopped-flow [31] and NMR spectroscopy were used to show the effect of thiol concentration on the formation processes of polynuclear cadmium-thiolate complexes and CdS clusters. FT-IR and Raman spectroscopies [41] and electro-spray mass spectrometry [42] have provided interesting information on the capping and functionalization of of small CdS clusters.

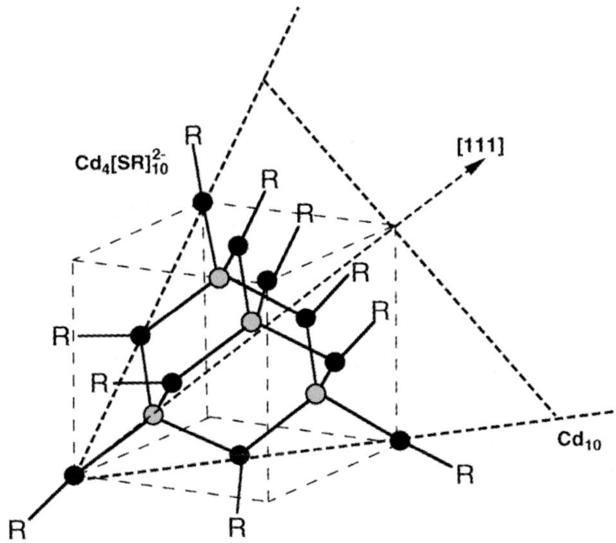

Figure 11-4. Illustration: of the importance of capping ligands by substitution of sulfide (S^{2-}) with thiolates (RS^-) in CdS unit cell.

Herron and Wang used $[(CH_3)_4N^+]_4[SCd_{10}(SPH)_{16}]^{4-}$as a precursor in DMF to prepare the relatively larger cluster $Cd_{32}S_{14}(SPh)_{36}$ by heat-treatment (43). This cluster was elegantly characterized crystallographically and the core exhibits an intermediate structure in which the atomic packing of Cd and S is neither the 4-coordinate cubic phase nor the hexagonal phase [43]. Lately a cluster with the same structure was made by Katsikas in aqueous solution using 1-Mercapto-2-propanol instead of thiophenol [44]. The later synthetic technique was also used to prepare $Cd_{17}S_4(RS)_{26}$, comparable to the above mentioned Dance cluster [32].

In this series of small clusters, thiolate ligands strongly influence or determine the structure of the small inorganic core as well as their whole electronic structure. This is clearly illustrated in the difference between the absorption spectra of $Cd_{32}S_{14}(SPh)_{36}$ and $Cd_{32}S_{14}(SCH_2CH(OH)CH_3)_{36}$, where the first electronic transition energy appears respectively at 360 nm and 325 nm [43, 44]. According to the so-called quantum size effect, these energies are relatively low as expected, compared to larger, biogenic, short-peptide capped CdS particles of 20 Å diameter [45] (ca. 80 Cd atoms) [46], with the lowest excited state occurring between 300 and 315 nm [45]. According to these examples, particle size determination from UV/vis absorbance specra can be erroneous.

The dramatic structural effect of thiolate on the core structure was shown by EXAFS in a comparative study with ZnS clusters [47]. In this small clusters regime, the coordination environment is CdS_4 like in the bulk cadmium sulfide, however the long range order, the effect of thiolate ligand on the core and electronic structures, as well as the chemical composition should be taken into account when their physicochemical properties are to be compared with the bulk CdS.

An understanding of the solution chemistry of cadmium thiolate is, however, of great relevance to the development of synthetic techniques for larger nanocrystals. The nature of the species formed after mixing thiols and cadmium salts in a solution of precursors will determine the type of nucleation and growth of cadmium sulfide and particularly the homogeneity of the nucleation and growth processes. The nature of the intermediate species in these solutions depends on many factors such as the substituent R of the thiols, on the Cd/RS ratio, the solvent and the presence of other additives [48, 49].

Analysis of the literature shows that depending on the physico-chemical conditions, the mixing of cadmium and thiols may lead to the formation of small polynuclear complexes such as $[Cd_{10}(SCH_2CH_2OH)_{16}]^{4+}$ [50, 51], or to a mixture of larger polymeric complexes between cadmium and thiols [48, 49, 52, 55]. The difference in reactivity of these species toward sulfide ions, is generally one of the factors responsible for the heterogeneity in size, shape and crystal structure, generally obtained in solutions of thiophenolate capped CdS nanocrystals [29, 58]. Using high resolution electron microscopy and X-ray diffraction, Vogel et al investigated CdS samples and confirmed this heterogeneity [58]. Generally, optical spectroscopy is the best technique to have a quick information on nanocrystal size distribution in the whole sample.

The learning from this chemistry is the coordination ability of thiolate sulfur and the role of the substituent R. There are 3 principal modes of ligation of thiolates, terminal, μ^2 – bridging and μ^3 – bridging [35]. The substituent will influence the strength of thiolate bonding as a capping ligand to the inorganic core and its ability to bind to crystal vertices, edges and surfaces, and therefore to stabilize nanocrystal shapes.

11.3.4 Optical spectroscopy and growth of CdS nanocrystals

The effect of experimental conditions on nanocrystal growth and on their size distribution, can be elegantly investigated by optical spectroscopy in the case of CdS. Fig. 11-5 shows optical spectra of cadmium sulfide nanocrystals obtained by changing the

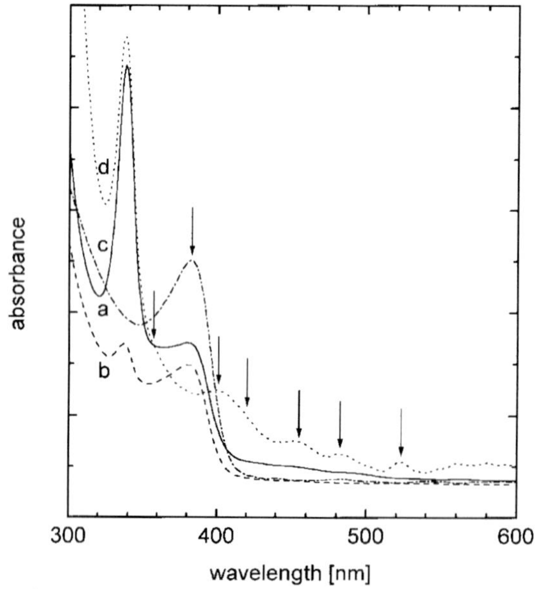

Figure 11-5. Change in the absorption spectrum (a→b→c→d) with decreasing the amount of thiols 0.92 < Cd/RS < 1,42.

composition of the starting solution. In this case, different solutions of thiolate capped CdS nanocrystals were obtained at different values of the Cd/S ratio by changing only the thiol concentration (thioglycerol), and keeping the concentration of cadmium acetate and thiourea constant in pure N,N-dimethyl-formamide [34, 59].

The spectral development with the thiol concentrations indicate the formation of a mixture of well defined species corresponding to the different absorption maxima in Fig. 11-5. The concentration in each species depends on the thiol concentration in the starting solution and no changes in the absorption was observed with ageing, and by keeping the solutions under argon in the dark. A decrease in thiol concentration lead to the appearance of maxima, corresponding to the formation of well defined and large nanocrystals as indicated by arrows on spectrum d. These spectral changes are a consequence of a homogeneous mixing at the molecular level where the sulfur in the form of thiourea and cadmium acetate as well as thiolates, are in intimate contact. The thermally induced decomposition of thiourea leads to nucleation and growth of CdS clusters, where thiolate and acetate play the role of substrate.

The size of the growing core will depend on the thiolate concentration as a capping ligand, and of course the stoichiometry will determine the number of species formed in solution. The stoichiometry can be adjusted to favor the formation of one type of species (see later) [19, 34]. The growth mechanism and cluster formation still need to be investigated in real time. However, it is clear that in the presence of thiols and by using DMF as an aprotic solvent, the growth do not occur ion by ion via a mechanism such as Ostwald ripening. Such growth will lead to a continuous shift of the absorption maxima during the growth, or by gradually changing stoichiometries. In contrast, Fig. 11-5 shows a discontinuous growth, reflected by the occurrence of well-defined absorption maxima appearing reproducibly, always at the same wavelengths and even for larger nanocrystals. This is a direct evidence of cluster-cluster growth with certain agglomeration numbers. Through a thiolate- inorganic core interaction, only clusters with well defined size and structure are thermodynamically favored, and these can fuse leading to larger and also well defined nanocrystals.

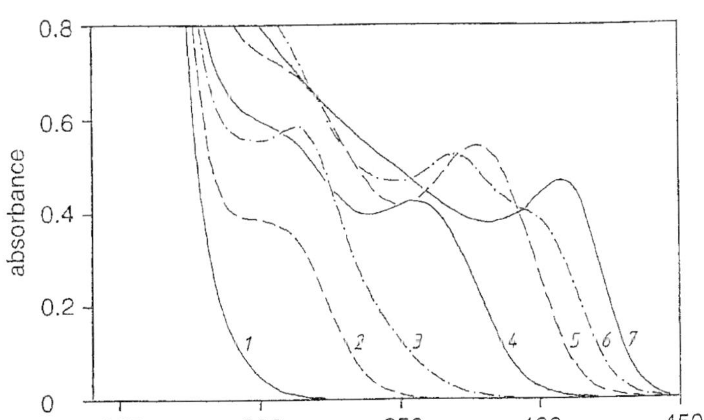

Figure 11-6. Optical spectrum 1 shows the reaction mixture in DMF before heating. Spectra 2-5 were taken after different times of heating. Spectrum 6 was taken after addition of water and further heating. This mixture leads finally to sample e. Spectrum 7 was recorded after prolonged heating without adding water. This route leads directly to a solution of sample f.

Most studies were performed in protic solvents such as water or alcohols to overcome solubility problems [29, 58], as it is difficult to obtain information on growth due to heterogeneity, generally reflected by the absence of structural features in the optical spectra.To show the solvent influence on thiolate capped CdS nanocrystals, Fig. 11-6 shows the growth of clusters in pure DMF, but by using cadmium acetate dihydrate and by adding an excess of water [22, 33].

The presence of water, even in small quantities, in the precursor solutions, leads to optical spectra where the absorption bands are relatively broad compared to the spectra obtained in an absolute absence of water (Fig. 11-5). The maxima indicate the formation of clusters with a narrow size distribution and a high concentration of a well defined cluster, which can be purified by size-selective precipitation [22, 33]. However, the growth becomes continuous with increasing size. Through a protonation process, in the presence of water, thiolate ligands become good living groups, destabilizing small clusters, which dissolve providing cadmium and sulfur ions. On the other hand, surface etching processes of larger particles also take place, allowing further growth via the mechanism of Ostwald ripening. The consequence of water or other protic solvents on the bonding between thiolate ligands and the particle core, is the absence of a well defined shape through the above mentioned etching process and the presence of chemical and structural defects on the surface. This and the presence of a large size distribution make the interpretation of optical properties and particularly the photoluminescence spectra, very difficult.

11.3.5 Size distribution, size, shape and crystal structure

The need for highly monodispersed particles is usually stressed in the ceramics literature [60]. Different chemical techniques have been developed by many groups, particularly by Matijevic and his colleagues, to grow monodisperse spheres and polyhedra, of a variety of materials in the μm size regime [60–62]. The size distribution was generally evaluated using electron microscopy, a technique which gives generally only a partial information on the sample. The variation in particle size which is in the nanometer scale, can be negleted relative to the large size of the particles. However, the size dependent properties of nanoparticles, like optical absorption in the case of CdS, is a great advantage to probe precisely the size distribution in the whole sample. Furthermore, in this size regime, the relatively strong size dependence of the solubility of nanoparticles, is a great advantage for size separation and therefore for achieving a better characterization [22, 33, 63, 64]. Finally, the capping via well chosen ligands confer to nanocrystals molecular properties and great stability which can be separated according to size, purified, and in some cases crystallized into superlattices [19, 64]: thus allowing the use of small angle X-ray or neutron diffraction.

Size-selective precipitation was used for the first time to separate thioglycerate capped CdS nanocrystals, using DMF as a solvent and acetone as the non solvent [33].

Samples with diameter of approximately 13, 14, 16, 19, 23 and 39 Å have been prepared as fully redispersable powders [22, 33]. Fig. 11-7 clearly illustrate the shift of the excitonic transition toward higher energies with decreasing cluster size. In some extent the major contribution to this shift can be related to size variation, since all nanocrystals are capped with the same ligand, however, other contributions due to changes in stoichiometry, and particularly in coverage and composition of nanocrystal surfaces, still need to be evaluated, since a decrease of the relative amount of 1-thio-

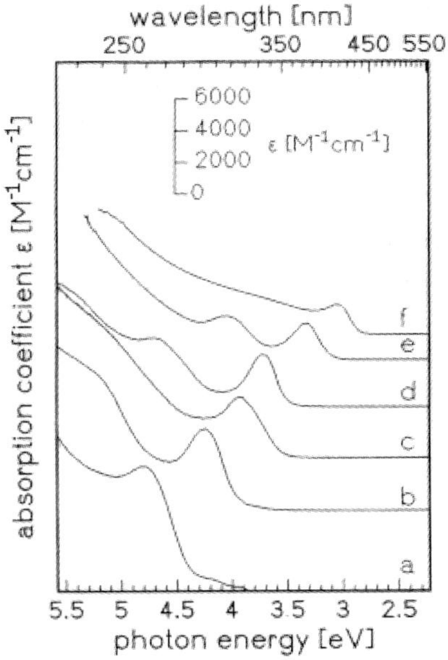

Figure 11-7. Absorption spectra of size-separated thiolate capped CdS clusters. The red shift of the excitonic transition can be related mainly to an increase in cluster size. However, contributions due to changes in surface composition need to be evaluated.

glycerol decreases with increasing cluster size, found by elemental analysis and thermogravimetric analysis [22]. This decrease in coverage was also found by 1H NMR studies of thiophenolate capped cadmium sulfide nanocrystals [65, 66].

In addition to optical spectroscopy, the samples were characterized by small and large angle powder X-ray diffraction, electron microscopy [22] and scanning tunnel microscopy for size determination [34]. Figures 11-8 and 11-9 shows an overview of

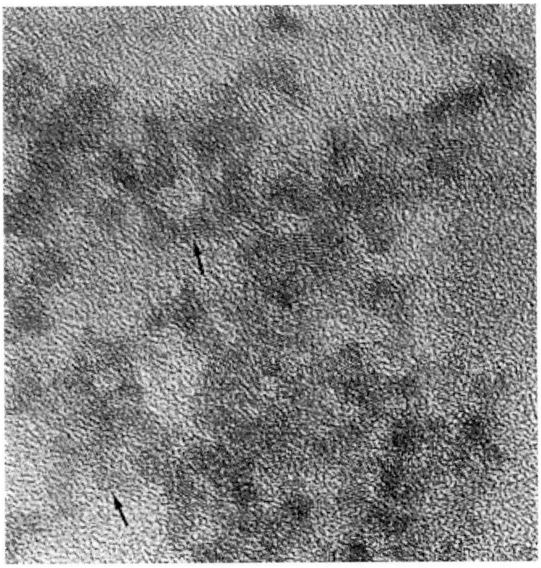

Figure 11-8. High-resolution transmission electron microscopy of CdS nanocrystals (sample f). the arrows point to CdS particles containing 14 ± 3 lattice planes. Edges and vertices are clearly observed.

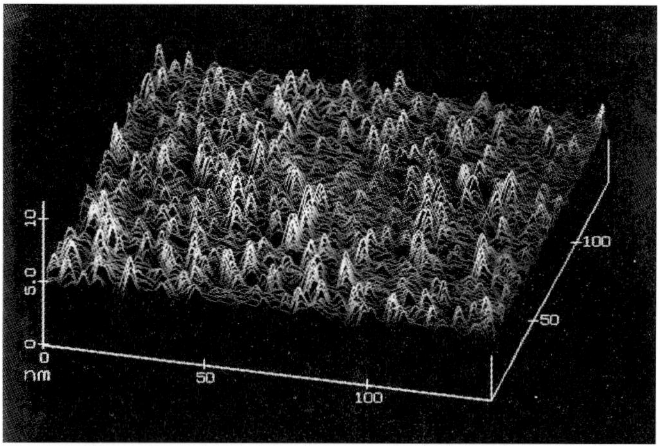

Figure 11-9. STM image of CdS nanocrystals capped with thiolate and deposited on a $MoTe_2$ substrate. Particle height is 42 ± 3 Å.

1-thioglycerate capped nanocrystals (spectra **f**, Fig. 11-7) using respectively HRTEM and STM. In contrast to most observations, the nanocrystals are not spherical but clearly exhibit edges and vertices, illustrating the role of thiolates as good stabilizers of nanocrystal shape and surface structure. The lateral dimensions measured on the TEM pictures are in agreement with the average height, around 41 Å, as measured on the STM pictures [34]. However, it is clear to observe a distribution in shape on the TEM micrograph, while in the case of smaller particles like in sample **e,** the pyramidal shape seems to be predominant. In this size regime between sample **e** (absorption at 370 nm) and **f** (400 nm) of 1-thioglycerol stabilized CdS nanocrystals, some important observations must be made. a) a continuous shift of the absorption corresponding to a continuous growth is taking place. b) the dependence of solubility on size becomes weak, making the size selective precipitation more difficult and time consuming. c) a distribution in shape is observed. d) a dramatic decrease in thiolate coverage is observed. e) the EXAFS peak corresponding to the Cd-Cd shell starts to develop only for particles with diameter of about 40 Å.

Thiolate seems to stabilize tetrahedral shapes as generally observed in the case of small clusters (diameter < 23 Å). In this size regime, the growth is in the 111 direction and the nanocrystal core adopts a structure close to a super-tetrahedral fragment of the cubic (sphalerite) lattice. According to EXAFS experiments, the structure is however disrupted probably through a strong interaction with thiolates [47]. Further growth is accompanied with less thiolate coverage, probably for sterical reasons and the weakening of core-thiolate bonds. The development of other growth directions also takes place, leading to different shapes, Fig. 11-8a. Distribution of shape was also observed in the case of thiophenolate capped CdS nanocrystals by Vogel. et al. [58] According to this careful study, combining HRTEM, electron diffraction on single nanocrystals and XRD, it becomes probable that a mixture of cubic and hexagonal phases of thiolate capped CdS can be mistaken for a pure hexagonal phase, by considering only the broad XRD peaks.

Control over nanocrystal structure depends on our understanding of the effect of many parameters on nucleation and growth. i) In his study of the effect of crystal structure on the absorption of AgI nanocrocrystals, Berry was able to direct the growth toward cubic or hexagonal phase by simply changing the Ag+/I ratio [6]. In the same time, cadmium sulfide films made in the presence of an excess of thiourea lead mainly to a hexagonal phase [67] ii) The effect of additives on CdS growth, which is found in many investigations, still has to be investigated in a systematic way. The four coordinate cubic phase forms for instance in the presence of sodium polyphosphate, or in some cases with thiolates [22, 58]. The hexagonal phase forms generally in the absence of additives and in some cases, in the presence of thiols. Even the rock-salt structure, generally made at high pressures, forms in solution in the presence of certain additives, such as surfactants, using biomimetic in-situ synthesis [68], or a mixture of thiolate and acetate [47]. The coordination of Cd^{2+} and the type of interaction with the functional groups of a substrate (ligands, polymers or solid surface), will determine the type of nucleation and growth. In many synthetic techniques, capping ligands were added after, or during nanocrystal formation to avoid solubility problems, or to have favorable conditions for nucleation and growth [29, 58]. In these cases, the effects of capping ligands on the crystal structure, did not take place since it was not involved in the nucleation process as a substrate, to direct atomic packing and growth.

Generally, nucleation and growth processes are accompanied by changes in the composition of solutions, leading to new chemical environments. New coordination types may generate different nucleation processes and the formation of new populations of nanocrystals. Therefore mixture of nanocrystal phases may be produced, like in the case of thiophenolate capped CdS nanocrystals [58]. In addition to stoichiometry, kinetic and thermodynamic factors need also to be taken into account in nucleation and growth processes.

11.3.6 Other sulfide nanocrystals

ZnS nanoparticles were also made by colloidal techniques, using polyphosphate or colloidal silicon dioxide as stabilizers [9, 69], or at low temperature without using surfactants [23]. These samples were characterized mainly by optical spectroscopy. The particles tend to agglomerate into loose structures making the TEM investigations very difficult. In addition, in the absence of adequately chosen capping ligands, the particles seem to aggregate on a carbon grid or even in solution. Generally, the particles are composites of much smaller subunits [23, 70]. There is no evidence of the crystal structure of the nanoparticles made at room or lower temperatures, since no powder XRD experiments are reported. Lattice planes, generally observed by HRTEM, can also be product of coalescence and crystallization of smaller subunits under the heat of an electron beam. However, according to X-ray diffraction, larger particles precipitate at relatively higher temperatures which exhibit the sphalerite crystal structure [70].

Little is known about the growth of ZnS, even though its similarity in the chemistry and crystal structure with CdS is often mentioned in the literature. In addition, the chemistry of small metal thiolate clusters is also in some extent similar [37, 71]. Analogous clusters such $Zn_4(SPh)_{10}^{2-}$, $S_4Zn_{10}(SPh)_{16}^{4-}$ were isolated and characterized [37]. The formation of a variety of Zinc thiolate $[Zn(SPh)_2]$ based species and polymers were also reported, as well as their solvent-dependent polymerization and depolymerization behavior [72].

Preliminary experiments show a difference in behavior between zinc and cadmium, with regard to chemical reactivity, growth and the structure of thiolate capped nanocrystals [47, 73]. In fact, zinc acetate was used instead of cadmium acetate in the above mentioned experiments to grow 1-thioglycerate capped ZnS nanocrystals in DMF, using thiourea as sulfur source. The reaction and growth were found to be very slow, 39 hours reflux (140 °C) was required to grow particles of about 30 Å in diameter. A continuous shift in the absorption maximum was observed, indicating an Ostwald-ripening type growth.

Although this synthetic technique still needs to be modified and optimized, some relatively monodisperse 1-thioglycerate capped ZnS nanocrystals can be isolated by controlling the heating time and by using size dependent solubility to purify the particles and isolate them as a soluble powder. Optical spectroscopy, small and large angle powder XRD, HRTEM and EXAFS were combined to characterize and to evaluate the size of the nanocrystals. The red shift with size in the optical absorption is shown in Fig. 11-10 for nanocrystals with respectively 22, 25, 27 and 30 Å in diameter. Size was determined by Small angle X-ray scattering and TEM. The TEM determinations were difficult due to a poor contrast, and in some cases, to agglomeration and coalescence of the particles. This behavior has been observed for other thiolate capped

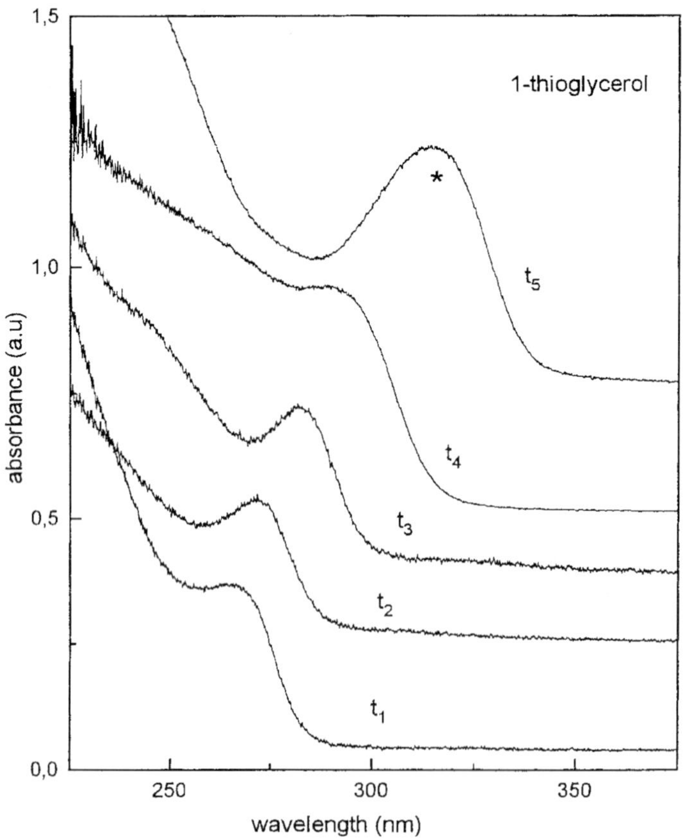

Figure 11-10. Absorption spectra of size-separated thiolate capped ZnS nanocrystals. Spectrum t5 was taken from aggregated ZnS nanocrystals.

nanocrystals [22, 74]. For the same reason, it was not possible to elucidate the shape of the nanocrystals. Thiolates seem to be weakly bonded to the ZnS and this weak interaction explains on the one hand the instability of the shape under the electron beam. On the other hand, it also explains the less important structural effect in ZnS compared to CdS nanocrystals found by EXAFS [47]. The large angle X-ray data displayed relatively broad peaks indicating the sphalerite cubic phase [73].

The high reactivity of lead salts toward H_2S or Na_2S made the control of the growth of PbS nanocrystals very difficult. In the presence of an excess of 4-hydroxythiophenol or 4-aminothiophenol in methanol, and by lowering the reaction temperature to –88 °C, it was possible to partially regulate the growth of PbS and to make solutions of nanocrystals [75, 76] However, the samples exhibit a large size distribution and little is known about growth mechanism, shape and crystal structure of the particles.

Heterogeneity in size, shape and the presence of other reaction products such as very small clusters, not generally detected by optical absorption, X-ray diffraction or HRTEM, generally make the investigations of these colloidal solutions difficult.

In the study of photoelectrochemical reactions on semiconductor particles using such colloidal solutions, the presence of these different species may also be involved in reactions, leading to changes in absorption and photoluminescence spectra. The presence of such species is also responsible for aging effects generally observed in many reported works [5]. Finally, these species can be involved in reactions performed to grow more complicated sulfide based composite nanoparticles such as, CdS/AgI, CdS/TiO2, CdS/PbS [77], CdS/PbS [78], ZnS/CdSe [79], ZnS/CdS [80] and HgS/CdS [81], CdSe/CdS [82]. These systems are mainly synthesized by an ion by ion displacing method, which consists of adding ions to grow the semiconductor coating. During these investigations, addition reactions are generally followed mainly by optical absorption. It is not clear whether spectral changes are the result of an homogeneous coating reaction of all particles, or via other possible scenarios. Since the reactivity or kinetics depends strongly on size, probably a total displacement reaction can take place with small clusters. Alternatively, only certain types of crystallographic surfaces of the nanocrystals are involved in the coating reaction. These considerations should be taken into account if a quantitative analysis of the reaction and a better identification of the expected free standing nanocomposites are to be done. In addition, size, size distribution, shape and surface structure of the precursor particles needs to be controlled in order to understand and control the coating reaction and to achieve a better characterization of the coated particles, therefore to providing a better interpretation of the unusual and interesting optical and electronic phenomena induced by the coating. Some progress in demonstrating the complexity of making and characterizing such nanocomposite particles is reported by Dabbousi et al, investigating (CdSe)ZnS core-shell quantum dots [83]. According to this investigation performed on relatively monodisperse CdSe nanocrystals, it is not possible to differentiate between an homogeneous coherent epitaxial growth, leading to a continuous film on one hand, and islands of growth, followed by coalescence and formation of grain boundaries on the other. Finally, the controlled synthesis and characterization of these composite nanoparticles is a challenging and interesting research area, which will open the door for designing well defined quantum confining heterostructures.

11.4 Connecting and assembling sulfide nanocrystals

11.4.1 Why to assemble and connect nanocrystals?

Much ongoing effort is devoted to the synthesis, characterization and the properties of individual nanocrystallites as free standing dots in solution, deposited on substrates [84] or embedded in polymers [85, 86] or glass matrices [87, 88]. Another research direction is taking advantage of strong interactions which may exist in the nanosize regime (15–100 Å), between nanocrystal inorganic core and organic capping ligands, since many properties depend on the chemistry of the ligand tails. In particular the interactions nanocrystal-nanocrystal [34, 64], nanocrystal-solvent [33] or nanocrystal-solid surface [89], depend strongly on functional groups in the ligand tails. This chemistry can be influenced to confer to nanocrystal base materials new rehological and adhesion properties for films, fine powders and membrane processing. In some cases, fiber drawing can be achieved as well. Furthermore, a major line of development of this chemistry, lies in the spontaneous generation of a well defined architecture, by self-assembly from suitably designed and functionalized nanocrystals. The molecular interactions of capping ligand tails, play a basic role in the generation of assemblies with different structures in solution and in the solid state [19, 34]. This issue of ordered nanostructured arrays of functional nanocrystals, is of central importance for the evaluation of the competitiveness of wet chemistry, since through size, shape and organization of particles, organics are confined into well defined inorganic structures, allowing control over their thermal removal. Through the organization of inorganic cores, mass transfer, coalescence and the sintering of particles can be controlled, leading to morphological engineering of films, membranes or powder of inorganic solids. This approach is illustrated in Fig. 11-3.

In addition to the processing of structured materials, an exciting aspect of this capping chemistry is the ability to connect nanocrystallites on a molecular basis, forming nanocrystallite networks. These assemblies of connected nanoclusters can be processed into transparent films for optical studies. They provide well defined systems in which to observe inter-dot interactions and their effects on electron-hole recombination processes, using optical absorption and photoluminescence spectroscopy.

Finally, nanocrystals placed along a periodic array like a crystal can be better characterized by diffraction techniques [19, 64].

11.4.2 Sythesis and structures of CdS nanocrystal networks

The synthesis consists of using hexamethylsilyl sulfide at low temperature or thiourea as a sulfur source, to achieve a homogeneous mixing of the reactants at the molecular level [34]. Then, sulfur is generated by a thermally activated chemical reaction, enabling uniform nucleation and growth of CdS clusters. In the presence of bridging ligands and by adjusting the stoichiometry, the growth can be followed by a self-connecting and self-organizing process of identically made clusters, leading to stable networks. The network formation requires, of course, a solvent in which the stability of the bridges between the clusters is given [34].

Cluster growth and network formation, can be followed by optical spectroscopy in the case of cadmium sulfide. One example is shown in Fig. 11-11. The development of spectra with time, shows a first maximum at 300 nm indicating the generation of a

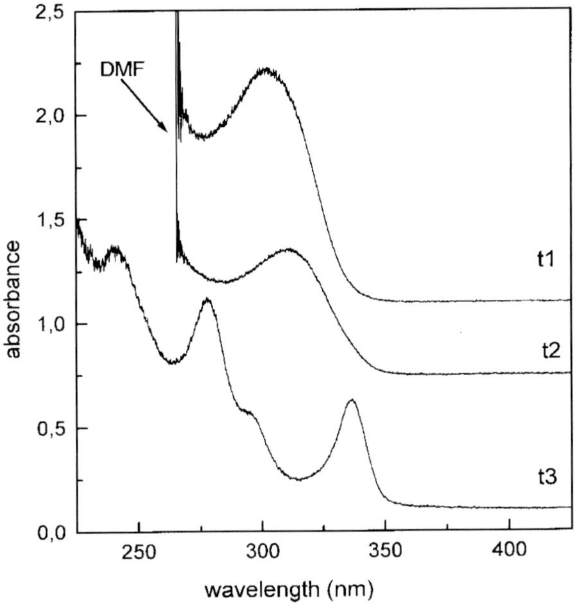

Figure 11-11. Optical spectra taken during cluster growth and assembly formation.

well-defined intermediate species (spectrum t1). As the reaction continues, this maximum shifts to a value of about 321 nm, and an asymmetric broadening of the band takes place (spectrum t2). These changes in optical absorption are immediately followed by the appearance of turbidity and the formation of assemblies of connected nanocrystals, spectrum t3. The four bands at 340 nm, 293, 277 and 240 nm are features that are not usually observed for isolated CdS crystallites and should be analyzed by taking into account the electronic structure of the whole nanocrystallite network [34].

The structure of the assemblies is influenced by ligands tails. Fig. 11-12 shows optical spectra of 3 types of assemblies obtained by using thiourea as a sulfur source and respectively 1-thioglycerol, 1-mercapto-2-propanol and 3-mercapto-2-butanol as capping ligands. Substitution of thiourea by hexamethylsilyl sulfide, leads also to the same optical spectra [90]. The interesting observation is the difference in structures of these assemblies, even though they exhibit the same optical spectra. In the 1-mercapto -2propanol case the nanocrystals are connected into single chains, forming a one dimensional assembly, Fig. 11-13a, b while in the two other cases, two dimensional assemblies are formed, Fig. 11-14a, b and 11-15a, b. In all cases, lattice planes are observe indicating the presence of small CdS crystallites. To be sure that these are not the product of coalescence processes usually observed under electron beams for these organically capped inorganic cores, X-ray diffraction analysis was also done. Experiments were performed on transparent films made by orienting mechanically the assemblies using a simple doctor-blade technique on mica. X-ray diffraction patterns in reflection and transmission geometry are shown in Fig. 11-16 in the case of 1-thioglycerol. Analogous spectra are obtained for 3-mercapto-2-buthanol. However, no anisotropy was observed in the case of 1-mercapto-2propanol. The broad peak is more resolved in the case of powders and corresponds to the (111) and (200) reflections of the six coordinate cubic phase [91]. Change in X-ray diffraction patterns and EXAFS

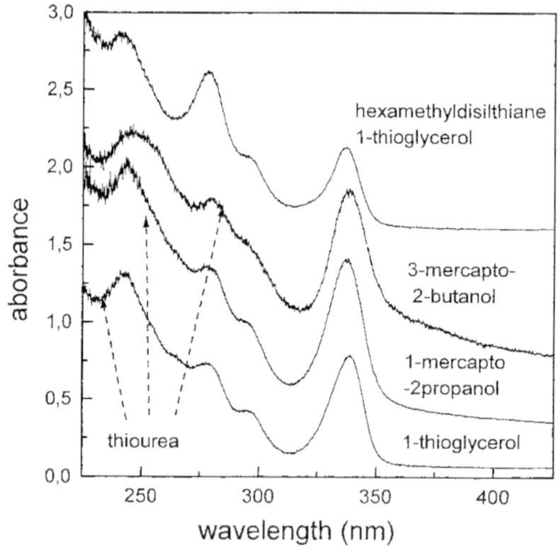

Figure 11-12. Optical spectra of connected CdS nanoclusters obtained using different sulfurizing agent and different thiols.

spectra during film heat-treatment, revealed clearly the presence of the rock salt structure Fig. 11-17 [47]. A phase transition leading to the four coordinate phase is observed at about 200 °C, a temperature corresponding to the departure of organic ligands [47]. In fact, the clusters are capped with a combination of thiolates and acetate. Cadmium is known to be six coordinated in solution in the presence of a combination of two ligands [92]. In this case, addition of thiols to cadmium acetate in DMF

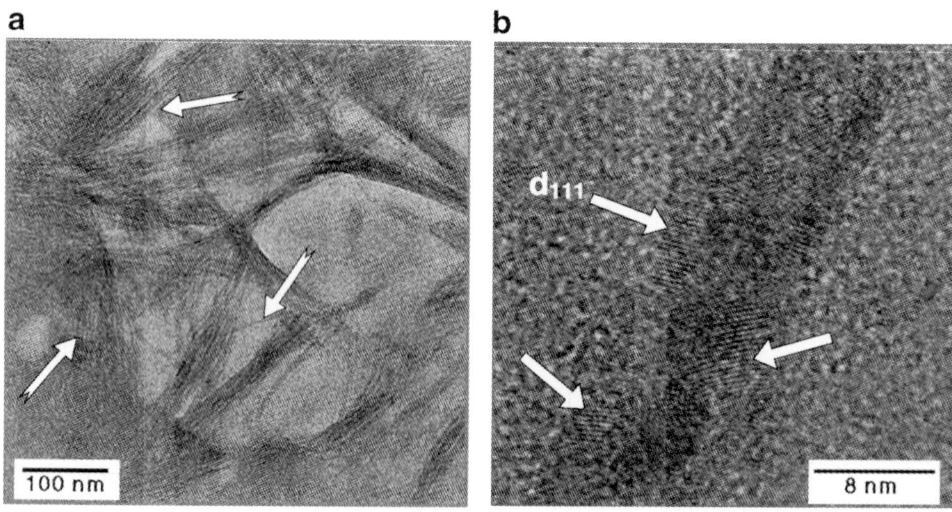

Figure 11-13. a) Transmission electron micograph showing CdS nanocrystals obtained with 1-mercapto-2-propanol. The clusters are connected on a molecular basis forming chains (1-dimensional assemblies); b) At high resolution (higher electron beam), some nanocrystals coalesce along the chains into larger particles. The interplanar distance measured on different particles (3.2 <d<3.48 Å) may correspond to the 6 coordinate cubic phase (d_{111}=3.16 Å) and (or) to the 4 coordinate one (d_{111}=3.36 Å) (see Fig. 11-17).

Figure 11-14. Transmission electron micograph of 2D assemblies made with 1-thioglycerol (see Ref. 19, 20 and 34).

and the presence of thiourea, leads probably to the same coordination. This octahedral environment is conserved during thiourea decomposition, and the presence of acetate and thiolate stabilizes this coordination during nucleation and directs the growth

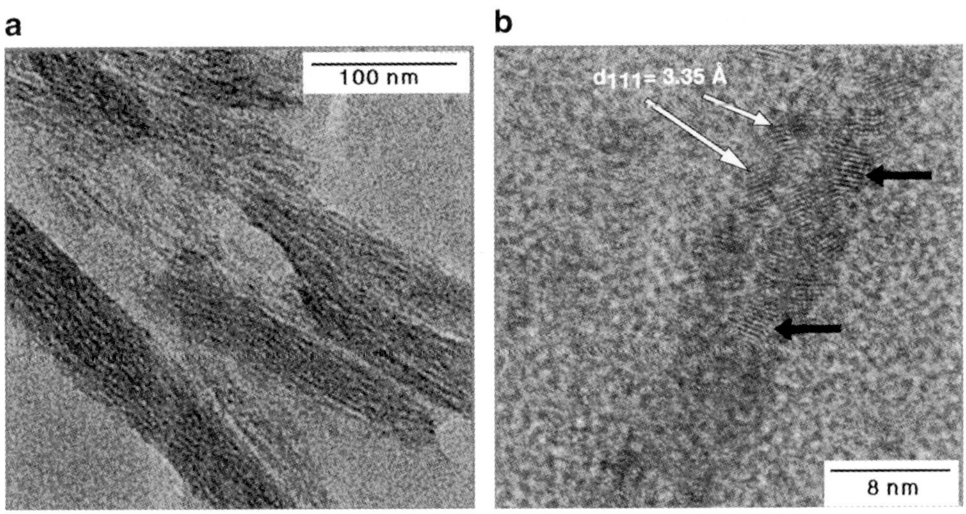

Figure 11-15. a) Transmission electron micograph of 2D assemblies made with 3-mercapto-2-butanol; b) HRTEM micrograph show single CdS nanocrystals with a preferred orientation explaining the anisotropy observed by X-ray diffraction.

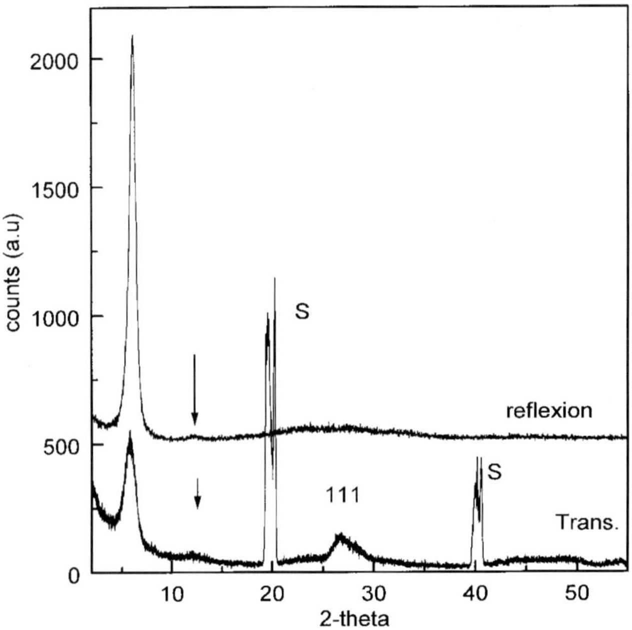

Figure 11-16. X-ray diffractograms of connected nanocrystals forming a two dimensional assemblies. These assemblies form a multilayer film with the nanocrystals having a preferred orientation explaining the difference between reflection and transmission geometry.

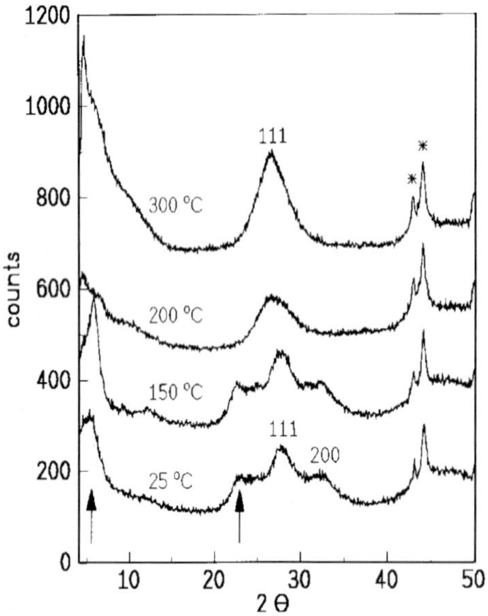

Figure 11-17. Change in X-ray diffraction patterns during film heat-treatment. At 200 °C, organic are removed followed by a phase transition from six to four coordinate cubic phase.

to the formation of the rock-salt structure. It is also possible that the growth of clusters followed by condensation into a network, is a consequence of this type of growth. In other words, the rock salt structure is probably stabilized by the formation of networks. This crystal phase, which is generally obtained at high pressures [93, 94], was found to form at low temperature in the presence of some functional groups [68]

The presence of only the (111) reflection in the XRD transmission spectrum of the two 2D assemblies, is a consequence of both the mechanical orientation and the preferred orientation of the nanocrystals in the assemblies. The nanocrystal chains are disposed parallel to each other and interact strongly, forming a 2D crystal. These 2D crystals lie parallel to the substrate, forming a layered structure. In these structured films, peaks are suppressed in reflection geometry and relatively strong in transmission, due to the preferred orientation of the nanocrystals. The peaks at low angular range with the spacing 29.35, 14.76, 7.26 and 3.634, observed for the 2D assemblies, reflect the layered structure [19]. The reflection intensities, as a function of diffracted angle, can be explained by the differences in the electron density of atomic layers [95].

11.4.3 Self-assembling process and optical properties of nanocrystal networks

The spectral evolution of Fig. 11-11 reflects the growth of clusters which, spontaneously self-connect at a given concentration into networks. The red shift and the broadening of the 300 nm peak observed just before precipitation (spectrum b, Fig. 11-11), can be related to the cluster condensation and the growth of the assemblies. This process is reversible in the case of the 1D assembly where the nanocrystals are capped with 1-mercapto-2propanol. In contrast to 2D assemblies, they exhibit a better solubility, allowing the investigation exclusively of the self-assembling process and its reversibility by optical spectroscopy and scanning tunnel microscopy. Figure 11-18 shows change in optical absorption with cluster concentration. At very low concentration, only a large band around 300 nm is present. An increase in concentration leads to the appearance of a second band around 340 nm and a decrease of intensity of the 300 nm band. At a given concentration, the 300 nm cluster is in equilibrium with the 340 nm cluster assembly.

The formation process of these chains was followed by STM. At very low concentration, only single particles are observed. An increase in concentration leads to an alignment and the formation of chains, Fig. 11-19. Acetate and thiolates were both found to be involved in the formation of the network by using infrared spectroscopy, Fig. 11-20, and by investigating the chemical reactivity of both bridging ligands [34]. The nanocrystals of the network are generated through a condensation process of clusters absorbing around 300 nm. The larger cores produced through this condensation process are stabilized by acetate and bridged by thiolates [34].

Connecting CdS nanocrystals on a molecular basis has very interesting electronic consequences. Fig. 11-21 shows the blue shift (340 nm $\rightarrow$ 324 nm) observed immediately after breaking thiolate bridges using 1-methylimidazole [34]. In addition, the band becomes sharper. The dramatic effect is, however, observed in the luminescence spectra. Figure 11-22 shows the evolution of the photoluminescence spectra with temperature. At room temperature, the direct-recombination band at the absorption edge (emission 1), which has never been observed for thiolate capped single nanocrystals, is clearly observed when such nanocrystals are connected on a molecular basis. With decreasing temperature, this band shifts toward higher energies and increases in intensity. At the same time, the red shifted one from the deep-trapped states (emission 2),

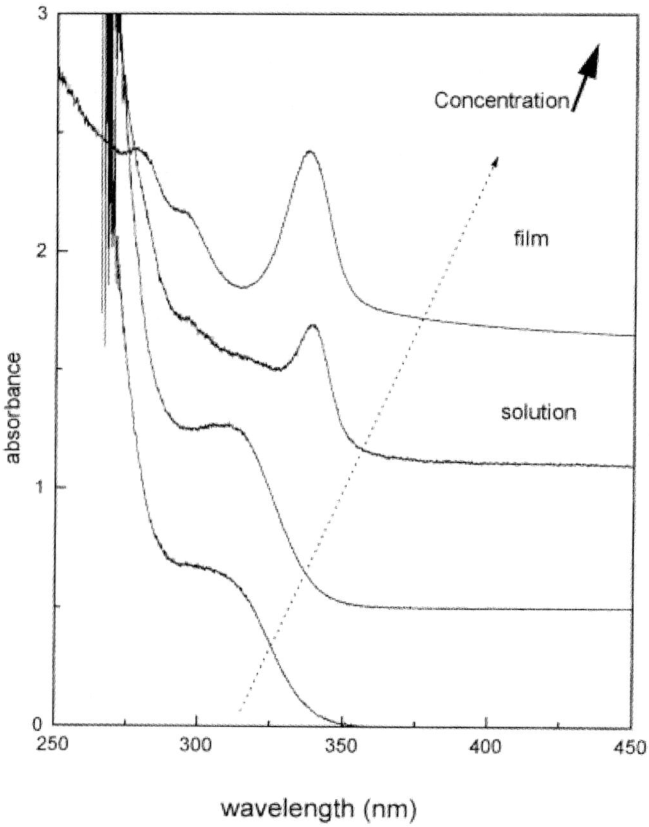

Figure 11-18. Change in optical spectra showing effect of concentration on growth and aggregation of Nanocrystals capped with 1-mercapto-2-propanol.

which is very weak at room temperature, strongly increases in intensity with decreasing temperature. This kind of behavior is observed for the first time in this new class of compounds. In fact, single isolated nanocrystallites in this regime do not show direct-recombination emission, since holes are more easily trapped by surface thiolate groups [29, 32, 43, 96]. Connecting nanocrystals however, leads to direct-recombination emission. At this stage and in the absence of studies of the luminescence dynamics as a function of tempertaure it is difficult to have a single and clear explanation for this behavior and a good quality theoretical description. However the shift in energy before and after forming the network (340 $\leftrightarrow$ 325 nm) is very substantial indicating a binding energy between clusters. This binding is probably responsible for opening energy transfer channel between clusters. Therefore there are two competing channels for exciton quenching in networks, energy transfer and trapping on surface sites. The temperature dependent shift is due to the lattice period adjustment and can be seen in both absorption and fluorescence spectrum. The bandgap temperature dependence can be explained by eliminating of the nonradiative phonon-assistant channel in single nanocrytal (critical temperature ~100 K = 200 cm^{-1} (phonon frequency)). The temperature dependence of the trap emission involves an activation

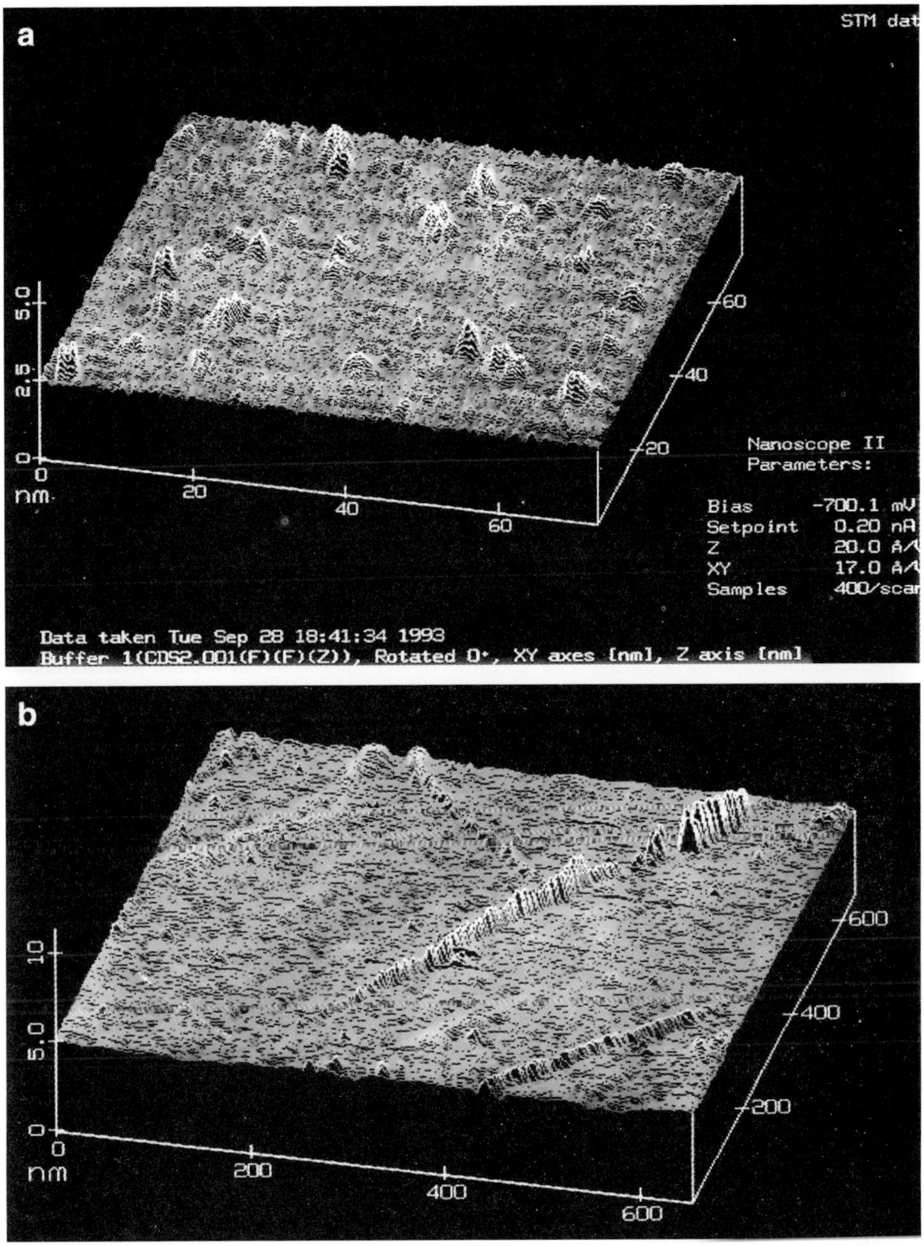

Figure 11-19. STM images showing the self-assembling process a) single nanocrystals at low concentration; b) Alignment at higher concentration.

process stronly dependant on temperatue [97a]. The presence of coherently coupled nanocrystals leading to excitonic superradiance similar to the one observed in molecular aggregates needs also to be considered [97b].

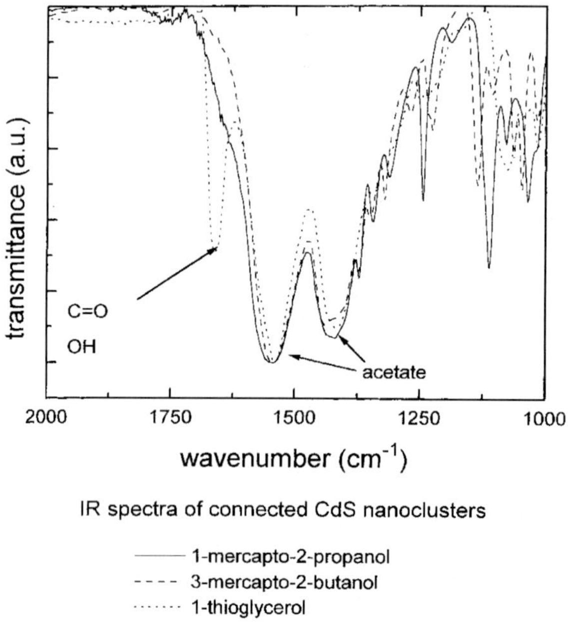

Figure 11-20. Infrared spectra of the 3 assemblies shows clearly that the acetate are involved in stabilizing the nanocrystals. The sharpness of the bands is a consequence of the homogeneity of the samples.

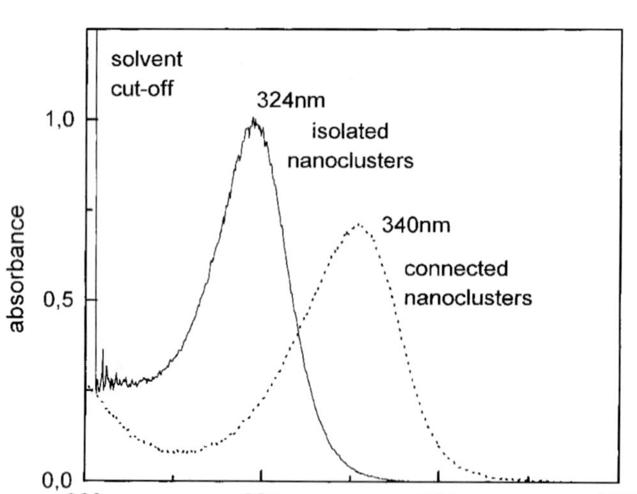

Figure 11-21. Optical absorption of isolated and connected nanocrystals. The important red shift is a consequence of connecting the nanocrystals on a molecular basis (Ref. 34)

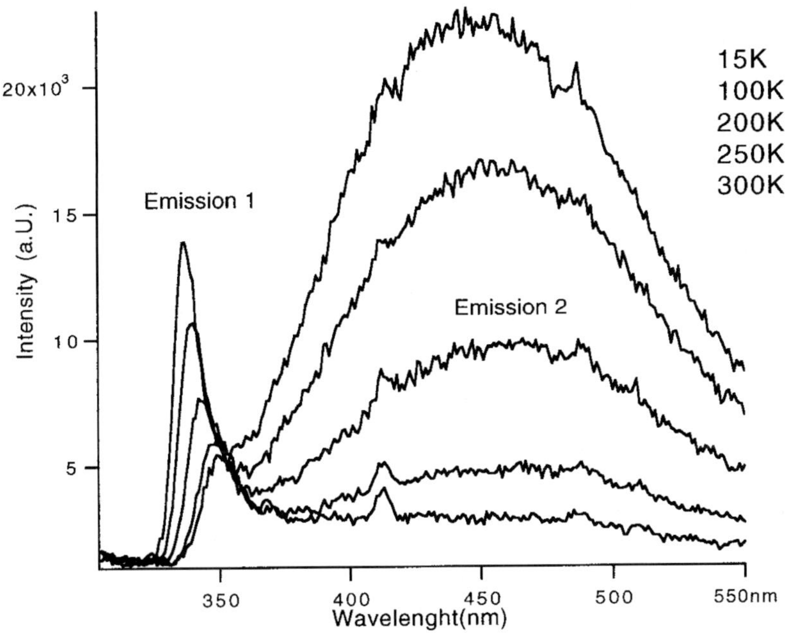

Figure 11-22. Luminescence spectra of connected nanocrystals were measured by lowering the temperature. For clarity, only 5 spectra are presented.

11.5 Oxide nanocrystals: synthesis and characterization

11.5.1 General remarks

The use of organic ligands to assist and direct the growth of some II–VI chalcogenides, or some metals, has been successful in making high quality nanometer-scale crystallites [22, 63, 64]. As already mentioned, the growth of the desired crystallographic phase in these cases is thermodynamically favored at the early stages of growth, allowing control over size, size distribution and shape. However, in the case of oxide, the chemical reactivity of available precursors such alkoxides or chloride is complicated. Generally, the reactivity is to high to control hydrolysis and condensation processes in order to grow nanoclusters with the same structure as the bulk oxide. Despite an extensive literature, there is evident uncertainty as to the identity, structure and chemical nature of polycondensed colloidal products. For instance nanocrystals of TiO_2, WO_3 and SnO_2 are ambiguous terms applied, to various particles or aggregated polymers simply precipitated through hydrolysis of alkoxides, chlorides or by changing the pH of the metal cation solution. The physical properties of these particles were compared with crystalline bulk material, only in terms of size effects and without having a reliable characterization of samples. Hydrolysis reaction of titanium chloride or alcoxide was assumed to lead to TiO_2 nanocrystals in solution at room temperature or even at 0 °C. Similarly, hydrolysis of tungsten chloride or acidification of tungstate were assumed to give WO_3 nanoparticles or nanocrystals. In both cases an extensive literature is available where such reaction products are identified as structurally different from bulk TiO_2 [98–104] or WO_3 [105, 109]. Small titania clusters which are partially amorphous and aggregated, do form in these solutions, but it is difficult to extract information from TEM observations since coalescence and crystallization processes of these aggregated, clusters take place under the electron beam. This may lead to confusion in size determination and the mistaking of aggregated particles in solution for single nanocrystals. Similarly, in the WO_3 case, only isopolytungstates or lamellar hydrates [105–109] such $WO_3 \cdot nH_2O$ (n = 2, 1, 1/3) [110] are produced in solution or under hydrothermal treatment. There is no evidence of WO_3 formation, a material which was shown to form only above 300 °C through solid state reactions [106–108]. The physical properties of these undefined particles such as band-gaps, absorption and fluorescence spectra and charge carrier dynamics were compared to bulk TiO_2 or WO_3 without taking into account the influence of fundamental parameters such as stoichiometry, aggregation or geometrical and crystallographic structure. Furthermore these parameters and others such as electronic structure, chemistry and the photochemistry of solids, are particularly strongly interrelated in the nanosize regime.

The study of oxide nanocrystals, and the factors which determine their structures and properties depends on our ability to control the preparation and to characterize them adequately in term of stoichiometry, geometric structure and surface chemistry. Some progress in our understanding of fundamental details which relate chemical reactivity to nucleation and growth of titania nanocrytals will be discussed in relatively more detail. Other very interesting results on Fe_2O_3, CoO, ZnO and MgO will be briefly reviewed.

11.5.2 Titania nanocrystals

Molecular precursors (alkoxides, chlorides etc.,) used in different techniques for titania, preparation are similar and generally the same. In techniques such as CVD, MOCVD or simple thermal pyrolysis, these molecular precursors are brought to react at generally higher temperatures in the gas phase or as molecular solids. During the reaction, nucleation, growth and coalescence processes take place simultaneously generally leading to crystallites with different sizes and without defined shapes, making the extraction of structural information on nanocrystal surfaces very difficult. On the other hand, one of the problems encountred in making titania using wet chemical techniques is the high reactivity of the available precursors, such alcoxides or chlorides, toward water [13, 14]. The reaction invariably leads to a mixture of different polymeric species which assume a variety of structures and sizes, and finally to an amorphous solid [98–104]. Although certain chemical modifications have been successful in controlling the reactivity of titanium alcoxides, and have permitted the isolation of intermediate polyoxoalcoxides, [117, 118] none of these species resemble the structure of bulk dioxide and any further hydrolysis again leads also to gels and amorphous precipitates. In both cases, a peptization process is required to convert these polymers into crystalline particles. Control over size, size distribution and shape, is difficult during these dissolution-growth processes [99, 103, 109].

Two major factors made the control over polycondensation and growth toward Titania bulk structure difficult. i) hydroxylation and polycondensation processes are governed by kinetics rather than thermodynamics. ii) Under the experimental conditions usually used, hydrolysis of the four alcoxide groups is not achieved simultaneously during the primary steps of the reaction, making the remaining alcoxide groups play an extremely important role by directing the packing of titanium and oxygen into structures different from the bulk TiO_2. Probably a lack of matching (structural, steric and coulomb interactions) between an alcoxide capping ligand and, for instance, an anatase inorganic core is a realistic explanation. To overcome these difficulties, a promising approach consists of performing reactions in purely aqueous medium in the presence of a base, so as to achieve a complete hydrolysis and to provide conditions similar to those used in the aforementioned peptization process. Thus, hydrolysis and polycondensation of titanium alcoxide ($Ti(OR)_4$) was performed in the presence of tetraalkyl-ammonium hydroxide (NR_4OH). This base both catalyzes the reaction [13, 14] and provides an organic cation that is known to stabilize polyanionic cores in basic media, e.g. polyanions of tungsten, molybdenum and vanadium. These processes can be described by the following reactions:

$$Ti(OR)_4 + H_2O \longrightarrow (RO)_3Ti(OH) + ROH \quad (a)$$

$$Ti{-}OH + XO{-}Ti \longrightarrow Ti{-}O{-}Ti + XOH \;(X = H, R) \quad (b)$$

$$Ti{-}OH + NMe_4OH \longrightarrow Ti{-}O^- + NMe_4^+ + H_2O \quad (c)$$

$$Ti_xO_yH_z + NMe_4OH \longrightarrow (Ti_xO_yH_{z-1})^- + N(CH_3)_4 + H_2O \quad (d)$$

At the high titanium concentrations used in this approach, it is difficult to separate the hydrolysis (Eq. a) and condensation (Eq. b) processes. These two primary processes are accelerated by the presence of NMe_4OH as a result of the formation of the highly nucleophilic TiO^- according to Eq. c. The cation also plays a role in preventing further condensation and stabilizing condensed species according to Eq. d.

It is generally accepted that early in the condensation process, the coordination number of titanium changes from 4 to 6, and that the basic building unit in the polymerization process is an octahedral arrangement of six oxygen atoms around a titanium centre [112, 118]. These octahedra link together by sharing edges, so as to form dimers. According to the partial charge model, these dimers condense to form skewed chains as a result of olation process as illustrated in Fig. 11-23 [112, 118]. In the presence of NMe_4^+, the condensed species can be stabilized allowing their isolation, or

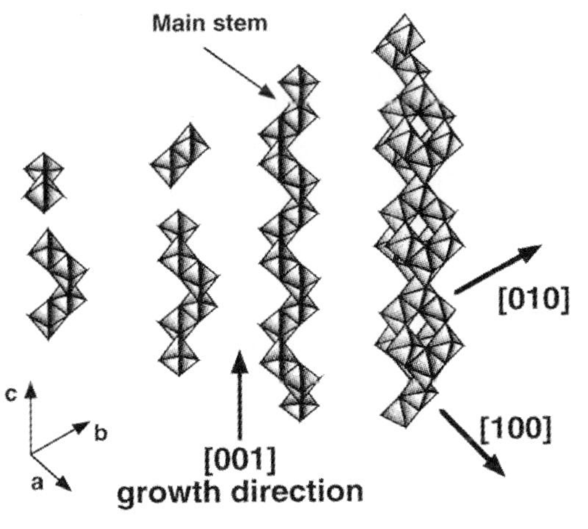

Figure 11-23. Model to explain condensation and growth of titania anatase; The mode of linking dimers imposes restriction on the growth direction leading to cis-skewed chains (main stem) and this is favoured by the physicochemical conditions of the present reaction; [100] and [010] are the first branching directions then, depending on the organic cation present and ist concentration, the secondary branching can be parallel to [101] or can remain parallel to [100] and [010].

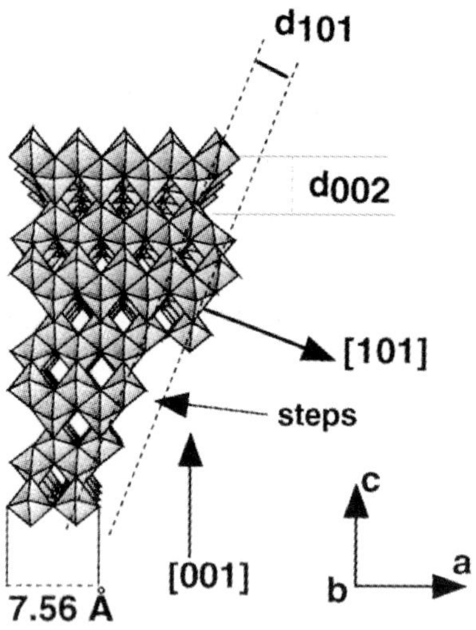

Figure 11-24. Polyhedral and packing drawing of anatase, TiO_2 in the [010] projection. [101] and [001] growth directions and a stepped [101] surface are shown.

may be forced to condense into larger particles by further heat-treatment. In the early stages of condensation, the linear growth (polymerization) leads to skewed chains [112, 118] which corresponds to growth in the [001] direction of the anatase phase (main stem). Then, through an oxolation process, the nuclei can grow in one of the two equivalent directions [112, 118]. The shape of the particle will be determined by the difference in growth velocities in the [101] and [001] directions, as indicated in Fig. 11-24. At low concentrations of $N(CH_3)_4^+$, the growth is favored in the [101] direction Fig. 11-25a, b, while at high concentrations the growth velocity is favored in the [001] direction, Fig. 11-26a, b. Therefore, the shape of nanocrystals can be fine-tuned by changing the ratios $Ti/N(CH_3)_4OH$, denoted as R [112].

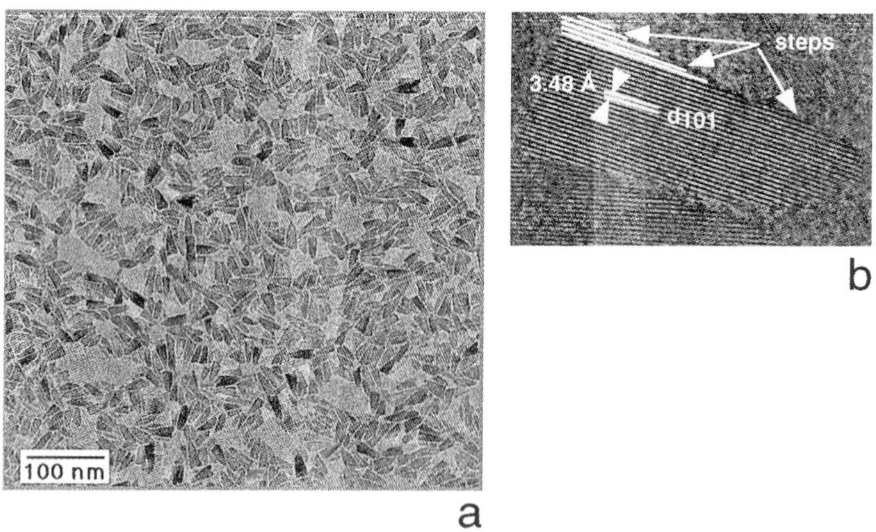

Figure 11-25. a) TEM micrograph of TMA -capped TiO2 particles obtained for R = 3, the particles are homogenous in size and triangular in shape; b) HRTEM image shows [101] planes of the anatase crystal structure, arrows indicate stepped [101] surfaces.

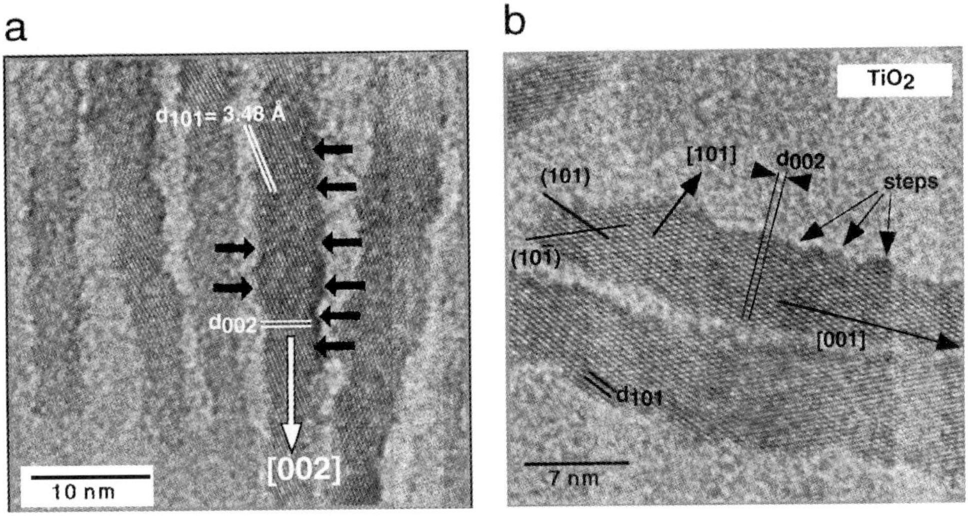

Figure 11-26. a) Elongated nanocrystals made at R = 1.25; b) HRTEM of nanocrystals lying in the [010] plane and elongated in the [001] direction. The corrugated sides shown by arrows are due to oscillatory repetition of [101] and [10-1] planes.

The advantages in investigating crystal growth in the case of nanocrystals is that the specimens are already thin to the point of electron transparency, and are thus readily amenable to TEM examination. High resolution is quite easily achieved nowdays and image processing is straightforward, allowing a stimulating interplay between synthesis and TEM observation [119]. A careful analysis of the micrographs can provide unique periodical and morphological information to aid in the understanding of the effects of additives on crystal growth [112].

Figure 11-25a shows for instance, an overview image of titania nanocrystals formed at low R values. The particles are uniform in shape and are not aggregated. The particles are single nanocrystals and exhibit a triangular prismatic shape with a relatively narrow size distribution. The high resolution microscopy in Fig. 11-25b clearly shows lattice planes. All images of the sample show particles with an interplanar distance of 3.52 Å in the tetragonal [101] planes, which corresponds to the anatase phase. The lateral dimensions of the particles can be estimated by counting the number of lattice planes (about 30 LP in this case, which amounts to 10.5 nm). The average length measured directly from the micrograph is about 20 ± 1 nm. A careful analysis of the particles reveals structural features which can be utilized for the study of surface morphologies of small particles and the changes that occur during growth. Thin edges of a TEM sample represent surfaces that can be viewed in profile by HRTEM [120]. The steps on the particle edges are in the (101) crystallographic directions and the height of each step can be taken as an indication of a cluster-cluster growth. The triangular shape is in fact due to the presence of stepped [101] surfaces Fig. 11-25b.

Figure 11-26a shows a typical TEM image at low magnification of titania particles with an homogeneous fibrous or rod like texture. Figure 11-26b shows a selected high-resolution image of elongated nanocrystals. The nanocrystal is viewed along the [010] direction. The surface profile image shows stepped terraces on the [101] surface, which are marked with arrows. Its height is measured as a multiple of 3.52 Å in accordance with the $d_{[101]}$. The corrugated sides are structural features that have not previously been observed in the case of titania and are a consequence of cluster-cluster growth and a condensation of skewed octahedra in the case of the anatase. The two sets of lattice fringes gives two interplanar distances, which correspond to the [101] and the [002] of the anatase crystal structure. The elongated particles lie in the [010] plane.

The HRTEM micrograph shown in Fig. 11-26b and the illustration in Fig. 11-24 relate to titania nanocrystals where the [101] and [010] are the predominate surfaces. The corrugated sides are due to oscillatory repetition of [101] and [10-1]. In terms of the mechanism discussed above, the main stem of the nanocrystal is seemingly in the [001] direction. The [100] and [010] directions are equivalent in the case of the anatase crystal phase, and both can be primary branching directions. Small slabs and very thin elongated particles in the [001] crystallographic direction are observed at the beginning of the hydrothermal treatment and when an excess of $N(CH_3)_4^+$ is used. Figure 11-25b also shows stepped [101] surfaces. The surface steps are of a height equivalent to the cluster size. Growth will take place by the building-up of these steps so as to form new cluster layers, and the rate of this process will depend on the concentration of $N(CH_3)_4^+$. The crystal surface in contact with the building units can be influenced by $N(CH_3)_4^+$. This cation seems to interact preferentially with the [101] surface, leading to the favoured development of these surfaces, as shown in Fig. 11-25 and 11-26.

X-ray diffraction patterns shown in Fig. 11-27 show clearly a decrease of nanocrystal size with R. At a value of about 1, slabs disposed parallel to one another along the [001] direction are observed at low beam intensities. At high magnification (higher electron fluxes), a coalescence (condensation) is induced by the loss of the organic cations. The nanocrystals are viewed along the [010] axis oriented parallel to the electron beam. The coalescence of these slabs (building units) leads to particles with corrugated sides or stepped [101] surfaces. In addition to these thin slabs, spherical particles are formed, Fig. 11-28. Particles of this type form at R=1 and they show a propensity to self-assemble into organized structures. Their concentration increases, at the expense of their elongation with the TMA concentration [112].

The particles are seen to be organized in an ordered structure (2D superlattice). TEM provides a real-space imaging of the superlattice. In addition, small-angle electron diffraction can be used to probe the superlattice structure on the micrometer scale, while small-angle XRD allows insight into the structures of individual titania/$N(CH_3)_4^+$ nanocrystals on the Ångström scale. Figure 11-29a shows an array of nanocrystals with a mean inter-particle (center-to-center) distance of 15.7 nm, determined by direct imaging and confirmed from the power spectrum. The high magnification Fig. 11-29b reveals well-faceted hexagonal crystals that form an hexagonal close-packed sheet. The crystals measure 13.5 nm in diameter. The high regularity in terms of size and shape, and long range translational and orientational order, is confirmed by the relatively sharp spots seen in the power spectrum (see inset in Fig. 11-29a) [112].

These clusters and nanocrystals are negatively charged and their electrostatic repulsion accounts for the stability of the solutions inasmuch as it prevents flocculation. On the other hand, these interactions and the monodispersity are responsable for the long-range ordering observed after evaporation of the solvent (Fig. 11-29). The charged titania nanocrystals are aligned in a periodic array, rather like in a crystal [121].

Additional heat-treatment of these clusters induce further condensation leading to slabs, which seem to influence and orient the self-assembling process, leading to arrays exhibiting a less regular geometries on a large scale. Changes in particle shape and in

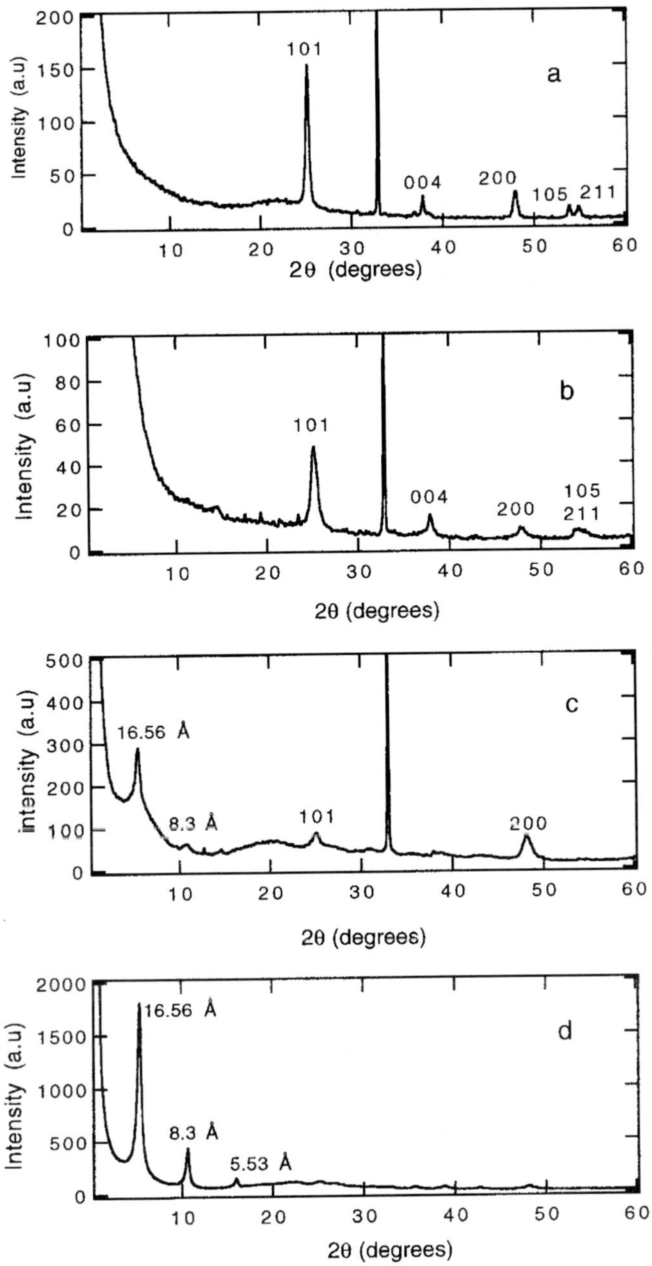

Figure 11-27. X-ray powder diffractograms of TiO2 nanocrystals made with various R=Ti/NMe_4^+ ratios R; Figures b, c and d correspond to solutions A (R = 4), B (R = 1.25) and E (R = 0.84). Figure a was obtained after autoclaving B under a saturated vapor pressure of water at 200 °C and 2500 kPa; large-angle peak assignments are based on anatase titania; peak broadening indicates a decrease of nanocrystal size with increasing R; in addition to the perfect organization in the lateral dimensions shown in Fig. 11-29, the series of harmonics in d indicates ordering of the growing anatase cluster in the nanocrystals forming a layered nanostructure; the sharpness of the peaks indicates a high degree of structural communication between the nanocrystals

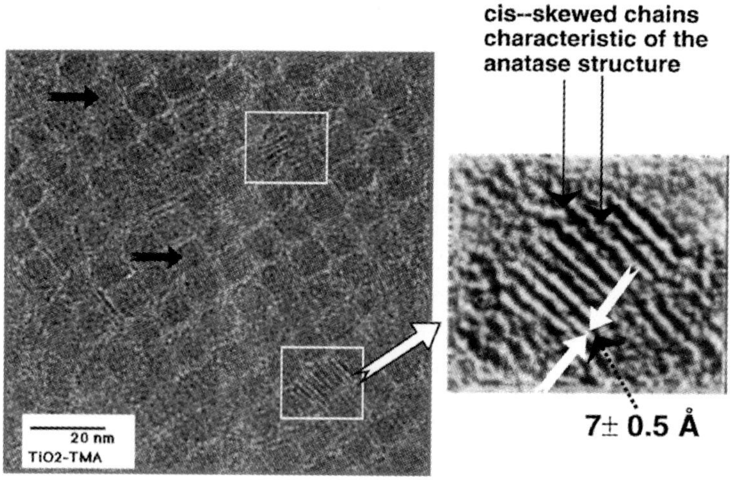

Figure 11-28. Nanocrystals obtained by additional heat-treatment of the particles shown in Fig. 11-29. Arrow shows nanocrystal composed of slabs disposed parallel to one another; high magnification indicate structural features that can be related to the cis-skewed chains in the [001] direction of the anatase crystal structure. The neighboring nanocrystals do not exhibit the same internal structure; arrows point to small clusters that probably did not condense to give slabs; a competing structure, where the slabs are differently oriented, is present.

their organization, as well as in the internal structures of the nanocrystals are observed. Instead of the hexagonal shape observed above, the drying leads to a rectangular shape and the nanocrystals are differently packed (Fig. 11-28). Furthermore, two types of internal structures are observed. This can be explained, in terms of the presence of two competing structures, since one type of nanocrystal is seen to be composed of slabs disposed parallel to one another, while the other type does not have any discerning features. Another explanation is that in a few nanocrystals a condensation process takes place, leading to slabs. These local transformations will affect the self-assembling process, thereby leading to new structures. A very intresting observation is that the high magnification images (Fig. 11-28) reveal that the newly formed slabs adopt a zig-zag structure. This is an indication of the presence of cis-skewed chains, characteristic of the anatase crystal structure. In fact, the anatase slabs are the condensation products of smaller ones (Fig. 11-29), the anatase structure of which was shown by EXAFS and XRD experiments.

Further structural information could be gleaned from the XRD pattern of cluster films prepared by simple deposition of solutions onto a silicon wafer folloed by drying in air (Fig. 11-27 d). A similar procedure was used to obtain TEM measurements, where one drop of the solution was applied to a carbon. A series of peaks at low angular range with the spacing 16.56 Å, 8.3 Å and 5.53 Å, was observed, attributable to [001], [002] and [003] reflections, respectively, of a spacing of 16.56 Å. These XRD data suggest that the clusters are arrangec inside the nanocrystal such as to form a layered structure. Packing of this type has been observed in the case of colloidal nanosheets made from exfoliated titanate. The exfoliation/reassembling process in the presence of tetrabutylammonium hydroxide NBu_4^+ leads to a comparable XRD pattern [122]. In both cases drying of the aggregate leads to textured films in which reflections are suppressed due to the preferred orientation of the nanosheets or, in

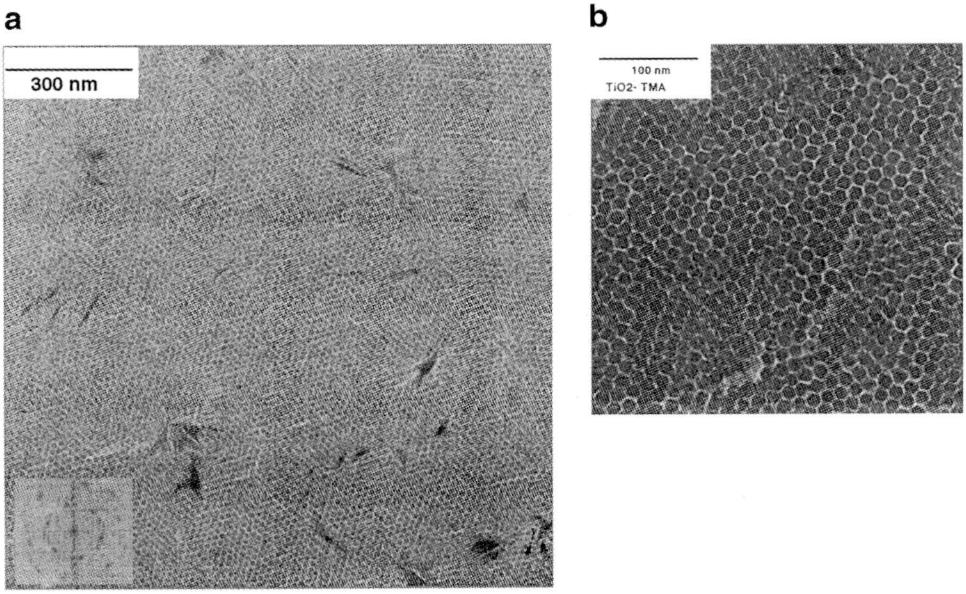

Figure 11-29. a) Transmission electron micrograph of a titania/NMe_4^+ nanocrystals self-assembled into a superlattice with dimensions in the micrometer length scales. The power spectrum (Inset) confirms the long-range translational and orientational order characteristics of the superlattice; b) High magnification shows well faceted hexagonal nanocrystals.

this case, clusters or slabs. The thickness of the nanosheets is 7.5Å which is comparable to the thickness of the slabs, while their lateral dimension is in the range of 0.3 – 1 μm [122]. Figure 11-28 shows nanocrystals composed of slabs disposed parallel to one another with a repeating distance of about 16 Å ± 1Å. This distance falls within the range of the spacing estimated from the XRD pattern. The exact internal structure of the superlattice still has to be determined, and this will be adressed once single crystals are obtained. However, EXAFS results, XRD, HRTEM and elemental analysis indicate the presence of anatase clusters and $N(CH_3)_4^+$ cations.

This method allows one to control the synthesis and stabilization of small clusters and to build larger and larger nanocrystallites all the way to the bulk titania with exclusively the anatase crystal structure. Small titania clusters or polytitanate anions have been prepared and stabilized in solution for the first time. Cluster-cluster growth has been shown to take place, leadint to larger nanocrystals. The growth can be influenced by controlling the hydrolysis and condensation with $N(CH_3)_4OH$. There is an apparent preference for interaction between [101] surfaces and $N(CH_3)_4^+$, wich would account fot the differential growth velocities in the [101] and [001] directions. The monodispersity of the negatively charged clusters renders them suitable for the progressive formation of titania/$N(CH_3)_4^+$ superlatices. Thin film growth allows control over orientations of superlattices, individual nanocrystals and individual clusters. The polycondensation and growth processes observed and discussed in this study are in accordance with theoretical predictions [118].

11.5.3 Other oxide nanocrystals

The thermal decomposition of metal carbonyls $Fe(CO)_5$ in a high boiling point solvent and in the presence of a surfactant as capping ligands, were successful in making $Fe_{1-x}C_x$ precursors particles [113, 114]. These amorphous particles were converted by slow oxidation in air or by heat treatment into Fe_2O_3 nanocrystals uniform in size. HRTEM shows for the first time, particles ordered into three-dimensional close-packed structures called super crystals. In these studies, a mixture of antiferromagnetic and ferrimagnetic Fe_2O_3 was revealed by X-ray, electron diffraction and Mössbauer spectroscopy. The later technique was shown to be very sensitive for studying a variety of small magnetic particles of Fe_2O_3 made in aqueous solutions [115].

Following the same principle, Yin and Wang were successful in growing CoO nanocrystals through a simultaneous decomposition and oxidation of $Co_2(CO)_8$ in toluene under an oxygen atmosphere [116]. Nanocrystals were size and shape selected using a novel magnetic field separation technique. These purified nanocrystals which are capped with bis(2-ethylhexyl) sulfosuccinate also have the propensity to self-assemble into superlattices.

ZnO particles with the wurtzite crystal structure have the advantage of forming spontaneously at even low temperature (0 °C) [123, 124].These nanocrystals can be prepared in two steps, firstly by heating an ethanolic solution of zinc acetate to prepare an intermediate species through hydrolysis and condensation [124]. In this reaction, acetic acid generated during the heat treatment reacts with ethanol leading to additional water through an esterification reaction. One of the intermediate species formed by hydrolysis and condensation was suggested to be $Zn_{10}O_4(Ac)_{12}$ [125]. Secondly, LiOH is added to induce or catalyze further growth. In these solutions a concentration dependent aggregation or cluster-cluster growth takes place and was found to be reversible by optical spectroscopy [124]. This type of behavior which has been known for many years in classical colloidal chemistry is strongly manifested in these studies because of the high concentrations used. However, the nature of the species involved and the type of bonding involved in this aggregation or growth mechanism still needs to be elucidated. Furthermore, the issue of shape also needs to be addressed since until now particles were found or assumed to be spherical [124].

Some progress in investigating the effects of ageing and other parameters on the growth or etching kinetics of ZnO nanoparticles such as pH, concentration, additives (acetate or acedic acid, LiOH) and composition and temperature, show very interesting and helpful results in understanding the chemistries of growth or etching. This work provides information for the next logical step, namely the development of an approach to stabilize and isolate the intermediate species by creating the right physico-chemical conditions, and by using adequately chosen capping ligands. Structural and morphological investigations of growth will enable one to adjust the chemistry to have control over the shape and surface chemistry of nanocrystals.

Some progress in understanding surface chemistry and the role of complexing ligands in the stability of nanometric suspensions of cerium oxide and zirconia has been achieved by combining a surface force apparatus and atomic force microscopy techniques [125]. These studies demonstrate the important effect of surface complexing agent such as acetylacetonate or acetate on adhesion between oxide nanocrystals. These results are of great significance for the processing of stable nanometric suspensions on one hand, but also for the processing of highly adherent nanoparticulate films [125]. These complexing agents and others such as paratoluene sulfonic acid are also used to regulate the growth of zirconia [126] and recently of titania nanoparticles

[127]. The bonding between metals and these additives were investigated by infrared, 13C, 17O, 1H NMR spectroscopies. In these studies, protection of particles towards aggregation is achieved, however, the issue of controlling size distribution, shape and surface structure of particles still need to be addressed [126, 127].

11.6 Applications, prospects and concluding remarks

One of the objectives of material chemists has been the control of atomic arrangements during the formation of condensed phases in order to prepare materials with particular topologies or to design low-dimensional solids. Now, this issue is primordial for the developments of new methods to create perfect manometer-scale crystallites and nanostructured materials. Progress in this area depends on our ability to use working knowledge's in molecular, solid state and surface chemistries to understand fundamental details which relates chemical reactivity to nucleation, growth and surface chemistry of condensed Phases. The time is right because of the ability and ease to probe reactions and do analysis at an unprecedented level of resolution.

The performances of materials in general and nanostructured materials in particular can be optimized by specific microstructural and macrostructural control of materials morphologies. For instance, one of the important goal in improving catalytic activities and selectivity of oxide, is reducing size and size distribution of highly quality oxide and sulfide nanocrystals [128] Other morphological requirement can be fulfilled in paints, cosmetics and electroceramic industries. Nanostructuring of porous materials is of great interest in building membranes for gas phase separations, nanoporous electrode to develop new types of solar cells [129], active layer in building electrochromic devices [130]. Recently TiO_2-coating on glasses have exhibited interesting antifogging and self-cleaning properties [131]. Cadmium sulfide nanocrystals can be used as precursors for the processing of buffer/windows layers for thin film solar cells [20]. The particle size effect of CdS nanocrystals was used for the sensitization of TiO_2 electrodes to build a solar cell [132]. This principle is motivating many groups to build a nanostructured and all solid state injection cell. The development of high performance fuel cells will depend on our ability to structure highly porous nanocomposite materials.

Acknowledgments

The author thank Dr. John A. Stride for reading the manuscript and all of the coworkers who have contributed to the work referenced to and or presented in this chapter. The author is grateful for generous support of the research mentioned in this chapter by the Hahn-Meitner-Institute.

References

[1] R. P. Andres et al. *J. Mater. Res.* **1989**, *4*, 704.
[2] G. A. Ozin *Adv. Mater.* **1992**, *4*, 612.
[3] H. Gleiter, *Adv. Mater.* **1992**, *4*, 474.
[4] See e.g.: Nanomaterials: Synthesis, Properties and Applications; Edelstein, A. S. Cammarata, R. C. Eds., Bristol, Institute of Physics Publishing **1996**.
[5] A. Henglein, *Topics Cuur. Chem.* **1988**, *143*, 80
[6] C. R. Berry, *Phys. Rev.* **1967**, *161*, 848.
[7] R. Rosseti, S. Nakahara, L. E. Brus, *J. Chem. Phys.* **1983**, *79* (2), 1086. L. E. Brus, *J. Chem. Phys.* **1984**, *80* (9), 4403.
[8] Y. Wang, H. Herron, *J. Phys. Chem.* **1991**, *95*, 525.
[9] H. Weller. *Angew. Chem.* **1993**, *195*, 43.
[10] H. Weller, *Adv. Mater.* **1993**, *5*, 89.
[11] A. Zunger, S. Wagner, P. M: Petroff, *J. Electronc. Mater.* **1993**, *22* (1), 3.
[12] this book.
[13] M. Henry, J. P. Jolivet, J. Livage, *Aqueous Chemistry of Metals Cations, Hydrolysis, Condensation and Complexation in Structure and Bonding*, vol. 77, Berlin, Springer-Verlag, **1992,** pp. 155–206.
[14] C. J. Brinker, G. W. Scherer, *Sol-Gel Science*, New York, Academic Press, Inc. **1990**.
[15] M. Figlarz, *Mater. Science Forum*, **1994**, *152–153*, 55–65.
[16] I. G. Dance, personal communication.
[17] A. Chemseddine, *J. Sol-gel science Technol.* **1994**, *2*, 685.
[18] K. L. Chopra, *Thin film Phenomena*, McGraw-Hill, Inc. **1969.**
[19] A. Chemseddine, *Chem. Phys. Lett.* **1993**, *216*, 265.
[20] A. Chemseddine, M. L. Fearheily, *Thin Sold. Films* **1994**, *247*, 3–4.
[21] A. Chemseddine, *Mat. Res. Soc. Sympo. Proc.* **1994**, *346*, 881.
[22] T. Vossmeyer et al., *J. Phys. Chem.* **1994**, *98* (31), 7665.
[23] R. Rossetti, R. Hull, J. M. Gibson, L. E. Brus, *J. Chem. Phys.* **1985**, *82* (1), 552.
[24] A. Henglein, *Chem. Rev.* **1989**, *89*, 1861–1873.
[25] P. E. Lippens, M. Lannoo, *Phys. Rev. B,* **1989**, *39* (15), 10935.
[26] Y. Kayanuma, *Solid Stat. Comm.* **1986**, *59*, 405.
[27] L.E. Brus, *J. Chem. Phys.* **1983**, *79*, 5566.
[28] L. Spanhel, M. Haase, H. Weller, A. Henglein, *J. Am. Chem. Soc.* **1987**, *109*, 5649.
[29] N. Herron, Y. Wang, H. Eckert, J. *Am. Chem. Soc.* **1990**, *112*, 1322.
[30] N. Chestnoy, T. D. Harris, R. Hull, E.L. Brus. *J. Phys. Chem.* **1986**, *90*, 3393.
[31] Y. Nosaka, H. Shigeno, T. Ikeushi, *J. Phys. Chem.* **1995**, *99*, 8317.
[32] T. Vossmeyer, G. Reck, L. Katsikas, E. T. K. Haupt, B. Schulz, H. Weller, *Science* **1995**, *267*, 1476.
[33] A. Chemseddine. *New Approaches for Controlling Size Distribution of Quantum Dots in Solution and in Glasses.* Presented at Eurogel, Colmar. France 1. 4. 1992.; A. Chemseddine, H. Weller. *Ber. Bunsenges. Phys. Chem.* **1993**, *97*, 636.
[34] A. Chemseddine, H. Jungblutt, S. Boolmaaz, *J. Phys. Chem.* **1996**, *100*, 12546.
[35] I. G. Dance, *Polyhedron*, **1986**, *5*, 1037–1104.
[36] A. Choy, D. Craig, I. G. Dance, M. Scudder, *J. Chem. Soc. Chem. Comm.* **1982**, 1246.
[37] I. G. Dance, A. Choy, M. L. Scudder, *J. Am. Chem. Soc.* **1984**, *106*, 6285.
[38] G. S. H. Lee, K. J. Fisher, A. M. Vassallo, J. V. Hanna, I. G. Dance, *Inor. Chem.* **1993**, *32*, 66–72.
[39] G. S. H. Lee, K. J. Fisher, D. C. Craig, M. L. Scudder, I. G. Dance, *J. Am. Chem. Soc.* **1990**, *112*, 6435.
[40] G. S. H. Lee, D. C. Craig, I. Ma, M. L. Scudder, T. D. Bailey, I. G. Dance, *J. Am. Chem. Soc.* **1988**, *110*, 4863.
[41] T. L. Graham, A. Bowmaker, J. M. Seakins, R. P. Cooney, *Chem. Mater.* **1997**, *9*, 967.
[42] T. Lover, W. Henderson, G. A. Bowmaker, J. M. Seakins, R. P. Cooney, *Chem. Mater.* **1997**, *9*, 1878.
[43] N. Herron, J. C. Calabrese, W. E. Farneth, Y. Wang. *Science* **1993**, *259*, 1426.
[44] T. Vossmeyer, G. Reck, G. Schulz, L. Katsikas, H. Weller, *J. Am. Chem. Soc.* **1995**, *117*, 12881.
[45] C. T. Dameron et al. *Nature*, **1989**, *338*, 596.
[46] C. T. Dameron, I. G. Dance, *Biomimetic Materials Chemistry*, Ed. S. Mann, New York, VCH Publishers, Inc. **1996**, p 69.
[47] A. Chemseddine, M. Fieber-Erdmann, E. Holub-Krappe, S. Boulmaaz, *Z. Phys. D.* **1997**, *40*, 566.
[48] G. Schwarzenbach, K. Gautshi, J. Peter, K. Tunaboylu, *Trans. Royal. Inst. Technology Sweden* **1972**, 295.
[49] H. F. De Branbander, L. C. Van Poucke, *J. Coord. Chem,* **1974**, *13*, 301–306.
[50] P. Strickler, *Chem. Comm.* **1969**, 655.

[51] R. A. Haberkorn, L. Que, W. O. Gillum, R. H. Holm. C. S. Liu, R. C. Lord, *Inorg. Chem.* **1976**, *15*, 2408.
[52] V. D. Craig, I. G. Dance, R. Garbutt, *Angew. Chem.* **1986**, *98*, 178.
[53] I. G. Dance, R. G. Carbutt, D. C. Craig, M. L. Scudder, T. D. Bailey. *J. Chem. Soc. Chem. Comm.* **1987**, 1164.
[54] I. G. Dance, R. G. Garbutt, D. C. Craig, M. L. Scudder *Inorg. Chem.* **1987**, *2*, 4057.
[55] I. G. Dance, M. L. Scudder, R. Secomb, *Inorg. Chem.* **1983**, *22*, 1794.
[56] H. F. De Brabander, G.G. Herman, L. C. Van Poucke, *Thermochimica Acta,* **1974**, *10*, 385.
[57] T. Vossmeyer, G. Reck, L. Katsikas, E. T. K. Haupt, B. Schulz, H. Weller, *Inorg. Chem.* **1995**, *34* (19), 4929.
[58] W. Vogel, J. Urban, M, Kundu, S. K. Kulkarni, *Langmuir*, **1997**, *13*, 827–832.
[59] A. Chemseddine, Unpublished Results.
[60] E. Matijevic, *Acc. Chem. res.* **1981**, *14*, 22–29.
[61] T. Ishikawa, E. Matijevic, *Langmuir*, **1988**, *4*, 26–29.
[62] I. Ul Haq, E. Matijevic, K. Akhtar, Chem. Mater. **1997**, *9*, 2659.
[63] C. B. Murray, D. J. Norris, M. G. Bawendi, *J. Am. Chem. Soc.* **1993**, *115*, 8706.
[64] C. B. Murray, C. R. Kagan, M. G. Bawendi, *Science*, **1995**, *270*, 1335.
[65] J. R. Sachleben, E. W. Wooten, L. Emsley, A. Pines, V. L. Colvin, P. Alivisatos, *Chem. Phys. Lett.* **1992**, *198*, 431–436.
[66] C.R. Bowers, T. Pietrass, E. Barash, A. Pines, R. K. Grubbs, A. P. Alivisatos, *J. Phys. Chem.* **1994**, *98*, 9400–9404.
[67] M. L. Breen, J. T. Woodward, D. K. Schwartz, A. W. Apblett, *Chem. Mater.* **1998**, *10*, 710, and refrences herein.
[68] J. Lin, E. Cates, P. A. Bianconi, *J. Am. Chem. Soc.* **1994**, *116*, 4738.
[69] A. Henglein, M. Guterrez, Ch. H. Fisher, *Ber. Bunsenges. Phys. Chem*, **1984**, *88*, 170.
[70] D, M. Wilhelmy, E. Matijevic, *J. Chem. Soc, Faraday Trans. 1*, **1984**, *80*, 563–570.
[71] R. Bertoncello, M. Bettinelli, M. Casarin, C. Maccato, L. Pandolfo, A. Vittadini, *Inorg. Chem.* **1997**, *36*, 4707.
[72] I. G. Dance, *J. Am. Chem. Soc.* **1980**, *102*, 3445.
[73] A. Chemseddine, unpublished results.
[74] H. Inoue, N. Ichiroku, T. Torimoto, T. Sakata, H. Mori, H. Yoneyama, *Langmuir*, **1994**, *10*, 4517.
[75] T. Torimoto, H. Uchida, T. Sakata, H. Mori, H. Yoneyama, *J. Am. Chem. Soc.* **1993**, *115*, 1874.
[76] T. Torimoto, T. Sakata, H. Mori, H. Yoneyama, *J. Phys. Chem.* **1994**, *98*, 3036.
[77] K. R. Gopidas, M. Bohorquez, P. V. Kamat, *J. Phys. Chem.* **1990**, *94*, 6435.
[78] H. S. Zhou, I. Honma, H. Komiyama, J. W. Haus, *Mat. Res. Soc. Symp. Proc.* **1993**, *283*, 858.
[79] A. R. Kortan, R. Hull, R. L. Opila, M. G. Bawendi, M. L. Steigerwald, P. J. Carroll, L. E. Brus, *J. Am. Chem. Soc.* **1990**, *112*, 1327.
[80] H. Youn, S. Baral, J.H. Fendler, *J. Phys. Chem.* **1994**, *98*, 934.
[81] A. Mews, A. Eychmüller, M. Giersig, D. Schoos, H. Weller, *J. Phys. Chem.* **1994**, *98*, 934.
[82] X. Peng, M. C. Schlamp, A. V. Kadavanish, P. Alivisatos, *J. Am. Chem. Soc.* **1997**, *119*, 7019.
[83] B. O. Dabbousi, J. Rodriguez-Viejo, F. V. Mikulec, J. R. Heine, H. Mattoussi, R. Ober, K. F. Jensen, M. G. Bawendi. *J. Phys. Chem. B* **1997**, *101*, 9463–9475.
[84] I. Daruka, A. L. Barabasi, *Appl. Phys. Lett.* **1998**, *72* (17), 2102.
[85] Y. Wang, S. Suna, J. Mcttugh, E. F. Hilinski, P. A. Lucas, R. D. Johnson, *J. Chem. Phys.* **1990**, *92* (11), 6927.
[86] Y. Wang, N, Herron, *Chem. Phys. Lett.* **1992**, *200*, 71.
[87] N. F. Borelli, D. W. Hall, H. J. Holland, D. W. Smith, *J. Appl. Phys.* **1987**, *61*, 5399.
[88] A. Chemseddine, H. Weller, *Eurogel 91*, ed. S. Vilminot, R. Nass. H. Schmidt, Amsterdam, Elsevier. Sci. Publ. **1992**, p. 371.
[89] V. L. Colvin, A. N. Goldstein, A. P. Alivisatos, *J. Am. Chem. Soc.* **1991**, *114*, 5221.
[90] A. Chemseddine, unpublished results.
[91] A. Chemseddine, *Mat. Res. Res. Soc. Symp. Proc.* **1994**, *346*, 881.
[92] P. J. Barrie, A. Gyani, M. Motevalli, P. O'Brien, *Inorg. Chem.* **1993**, *32*, 3862.
[93] J. Osugi, K. Shimizu, T. Nakamura, A. Onodeva. *Rev. Phys. Chem. Japan* **1986**, *36*, 59.
[94] S. H. Tolbert, A.P. Alivisatos, *Ann. Rev. Phys. Chem.* **1995**, *46*, 595.
[95] A. Matsude, M. Sugi, T. Fukui, S. Lizima, M. Miyahara, Y. Otsubo, *J. App. Phys.* **1977**, *48*, 771.
[96] N. Herron, A. Suna, Y. Wang, *J. Chem. Soc. Dalton Trans.* **1992**, 2329.
[97] a) A. Chemseddine, S. Luganov, unpublished results. b) F. C. Spano, S. Mukamel, *J. Chem. Phys. 1989*, *9*, 683.
[98] M. Visca, E. Matijevic, *J. Colloid. Interf. Science*, **1979**, *68*, 308.
[99] B. L. Bischoff, M. A. Anderson, *Chem. Mater.* **1995**, *7*, 1772.
[100] N. Serpone, D. Lauless, R. Khairutdinov, *J. Phys., Chem.* **1995**, *99*, 16646.

[101] J. H. Jean, T. A. Ring, *Langmuir*, **1986**, *2*, 251.
[102] Y. Oguri, R. E. Riman, H. K. Bowen, *J. Mat. Sci.* **1988**, *23*, 2897.
[103] J. R. Bartlett, J. L. Woolfrey, *Chemical Processing of Advanced Materials*, Ed. by L.L. Hench, J. K. West, New York, John Wiley and Sons Inc. **1992**.
[104] B. A. Morales, O. Novaro, T. Lopez, E. Sanchez, R. Gomez, *J. Mater. Res.* **1995**, *10*, 2788
[105] A. M. Morley, *J. Chem. Soc.* **1930**, 1987.
[106] G. Biehler, *Ann. Chim.* **1947**, *12*, 489.
[107] A. Chemseddine, M. Henry, J. Livage, *Rev. Chim. Miner.* **1984**, *21*, 487.
[108] A. Chemseddine, F. Babonneau, J. Livage. *J. Non-cryst. Solid* **1987**, *91*, 271.
[109] T. Nanba, Y. Nishiyama, I. Yasui, *J. Mater. Res.* **1991**, *6*, 1324.
[110] M. S. Whittingham, J. Guo, R. Chen, T. Chirayil, G. Janauer, P. Zavalij, *Solid. State Ionics*, **1995**, *75*, 257.
[111] T. Moritz, J. Reiss, K. Diesner, D. Su, A. Chemseddine, *J. Phys. Chem, B.* **1997**, *101*, 41, 8052.
[112] A. Chemseddine, T. Moritz, *Eur. J. Inorg. Chem.* **1999** 235.
[113] J. Van Wonterghem, S. W. Morup, S. Charles, S. Wells, J. Villadsen, *Phys. Rev. Lett.* **1985**, *55*, 410.
[114] M. D. Bentzon, J. Van Wonterghem, S. W. Morup, A. Thölen, C. J. Koch, *Phil. Mag. B*, **1989**, *60*, 169.
[115] J. P. Jolivet, C. Chaneac, P. Prené, L. Vayssières, E. Tronc *J. Phys. IV France* **1997**, C1–573. (and references herein); E. Tronc, *Il Nuovo Cimento*, **1996**, *18D* (2, 3) 163
[116] J. S. Ying, Z. L. Wang, *J. Phys. Chem. B* **1997**, *101*, 8979.
J. S. Ying, Z. L. Wang, *Phys. Rev. Lett.* **1997**, *79*, 2570.
[117] V. W. Day, T. A. Eberspacher, Y. Chen, J. Hao, W. G. Klemperer, *Inorg. Chim. Acta.* **1995**, *229*, 391.
[118] M. Henry, J. P. Jolivet, J. Livage, *Aqueous Chemistry of Metals Cations, Hydrolysis, Condensation and Complexation in Structure and Bonding*, vol. 77, Berlin, Springer-Verlag, **1992**, pp. 155–206.
[119] R. Sinclair, *MRS Bulletin/June* **1994**, 26–30.
[120] S. Ijima, *J. Electron. Microsc.* **1985**, *34*, 249–265.
[121] P. Pieranski, *Contemp. Phys.* **1983**, *24* (1), 25–73.
[122] a) T. Sasaki, M. Watanabe, H. Hashizume, H. Yamada, H. Nakazawa, *J. Chem. Soc. Chem. Commun.* **1996**, 229–230. b) T. Sasaki, M. Watanabe, H. Hashizume, H. Yamada, H. Nakazawa, *J. Am. Chem. Soc.* **1996**, *118*, 8325–8335. c) T. Sasaki, M. Watanabe, *J. Phys. Chem. B* 1**997**, *101*, 10159–10161.
[123] E. A. Meulenkamp, *J. Phys. Chem. B* **1998**, *102*, 5566.
[124] L. Spanhel, M. A. Anderson, *J. Am. Chem. Soc.* **1991**, *113*, 2826.
[125] O. Spalla, P. Kékicheff, *J. Colloid. Interface Sci.* **1997**, *192*, 43–65
M. Nabavi, O. Spalla, B. Cabane, *J. Colloid. Interface Sci.* **1993**, *160*, 495–471.
V. Peyre, O. Spalla, L. Belloni, M. Nabavi, *J. Colloid. Interface Sci.* **1997**, *187*, 184–200.
[126] M. Chatry, M. Henry, M. In, C. Sanchez, J. Livage, *J. Sol-Gel Sci. Technol*, **1994**, *1*, 233.
[127] E. Scolan, C. Sanchez, accepted in *Chemistry of Materials, Special Issue on Frontiers in Solid State Chemistry*, **1998**.
[128] A. Corma, *Chem. Rev.* **1997**, *97*, 2373–2419. Kraeutler, A. J. Bard, *J. Am. Chem. Soc.* **1978**, *100*, 2239. A. Chemseddine, H.P. Boehm, *J. Mol. Catal.* **1990**, *60*, 295.
[129] B. O. Regan, M. Graetzel, *Nature* **1991**, *353*, 737.
C. J. Barbé, F. Arendse, P. Comte, M. Jirousek, F. Lenzmann, V. Shklover, M. Grätzel. J. *Am. Ceram. Soc.* **1997**, *80* (12), 3157–71
[130] C. G. Granqvist, *Handbook of inorganic Electrochromic Materials*, Amsterdam, Elsevier Science, **1995**.
[131] R. Wang, K. Hashimoto, A. Fujishima, *Nature*, **1997**, *388*, 431.
[132] H. Gericher, M. Lübke, *J. Electroanal. Chem.*, **1986**, *204*, 225.; R. Vogel, K. Pohl, H. Weller, *Chem. Phys. Lett.* **1990**, *174*, 241

12 Electron Microscopy of Fullerenes and Related Materials

G. Van Tendeloo and S. Amelinckx

12.1 Introduction

12.1.1 Why electron microscopy, electron diffraction?

Shortly after the discovery of the fullerenes and related molecular crystals electron microscopy has played a major role in elucidating their structures and microstructures. This was due on the one hand to the fact that initially only minor quantities of pure crystallized materials were available and on the other hand because of the nanometer sizes of the objects: graphite nanotubes and onions. The small size of available fullerene crystals restricted the possibility of single crystal diffraction experiments to electron diffraction. Moreover electron microscopy offers the unique possibility to study direct and reciprocal space of the same object and also to determine the relationship between both.

Another aspect which favoured the use of electron microscopy for a structural study of fullerene related materials is the fact that the carbon nanotubes as well as the bucky onions are not crystalline materials in the strict sense, but are more like textures. This implies that some of the reflections in reciprocal space are not sharp, but spread out along lines or planes. Because the atomic scattering factor of carbon is small and because tubes as well as onions are sub-micron size objects electron diffraction is about the only diffraction technique which is able to produce useful structural information; mainly because the electron-sample interaction is very strong.

12.1.2 Short survey of materials which will be considered

In this survey we will review the main results obtained by the combined use of microscopy and diffraction. We will in particular discuss the structure and microstructure of molecular crystals of C_{60} and C_{70} and of some of their derivatives obtained by the reaction with alkali metals (e.g. Rb) or by the intercalation with halogen molecules (in particular I_2). The study of various morphologies of graphite derived particles such as nanotubes and onions relies most heavily on electron microscopy. The characteristic diffraction effects produced by nanotubes of various kinds will be discussed in some detail and related to the morphology.

12.2 Molecular crystals of fullerenes

12.2.1 Single crystals of C_{60}

The first electron diffraction experiments on C_{60} crystals as well as theoretical considerations concerning the packing of the molecules (either HCP or FCC) had left some doubt as to the room temperature structure. It was soon realized that the room temperature structure depends on the method by which the crystals were prepared, as well as on the purity of the substance. Electron diffraction suggested that crystals grown from solution in an organic solvent were predominantly hexagonally packed, but under examination in the electron microscope vacuum they transformed into a faulted face centered cubic structure [1]. This transformation was attributed to the loss of the solvent molecules, which stabilized the HCP packing and it was concluded, in agreement with X-ray powder diffraction experiments [2], that the stable room temperature structure of pure C_{60} was FCC. Sublimation grown crystals of sufficiently pure C_{60} always exhibit a room temperature FCC structure.

Due to the shape factor of the quasi-spherical C_{60} molecules together with the particular value of the ratio r_m/a of the radius r_m of the molecule to the lattice parameter a in the FCC phase, extinction occurs for all h00, 0k0 and 00ℓ reflections. This was first found by X-ray powder diffractometry [2] but it was more strikingly illustrated by single crystal electron diffraction in a careful tilting experiment [3]. These experiments confirmed that in the room temperature phase the molecules can to a good approximation be considered as randomly oriented so that they can be modelled as thin spherical shells of diffracting material. Room temperature high resolution images confirm the FCC structure, but reveal that the crystals are frequently coherently twinned on (111) and contain numerous stacking faults, which lead to streaking of the diffraction spots along [111] directions. In some cases the concentration of faults is so large that through each node in reciprocal space continuous streaks along <111> result. In electron diffraction patterns this may give rise to weak extra spots violating the FCC diffraction conditions where Ewald's sphere intersects these streaks. Tilting the specimen causes a continuous shift of these weak spots revealing in this way their origin.

Additional spots at positions which are extinct in an FCC lattice (i.e. positions which correspond to FCC reflections with indices of mixed parity) result on cooling below 250K (see Fig. 12-1) [2, 4]. These additional spots correspond to a primitive cubic lattice. They reveal the low temperature structure in which the molecules are still situated on a FCC lattice but in which the molecules are orientation ordered. The diffraction conditions deduced from electron diffraction experiments show that, in agreement with the X-ray diffraction results, the spacegroup is Pa$\bar{3}$ which is a subgroup of the spacegroup Fm$\bar{3}$m of the high temperature FCC phase. In this primitive structure each molecule is oriented in such a way that electron rich regions of the surface of one molecule (corresponding to edges of the pentagons) face electron poor regions of the surfaces of the adjacent molecules (i.e. in the centres of the hexagon). Such an arrangement minimizes the electrostatic energy.

The phase transition FCC $\rightarrow$ P is accompanied by a decrease in rotation as well as in translation symmetry. The low temperature phase will thus break up into a domain structure consisting of orientation and of translation domains. The fourfold axis of the pointgroup of the FCC phase is reduced to a two fold axis in the pointgroup of the primitive phase leading to orientation domains related by 90° rotations about one of the $<100>_{FCC}$ axis. The FCC lattice further breaks up into four sublattices each sublattice being occupied by molecules of a given orientation related by translations of the

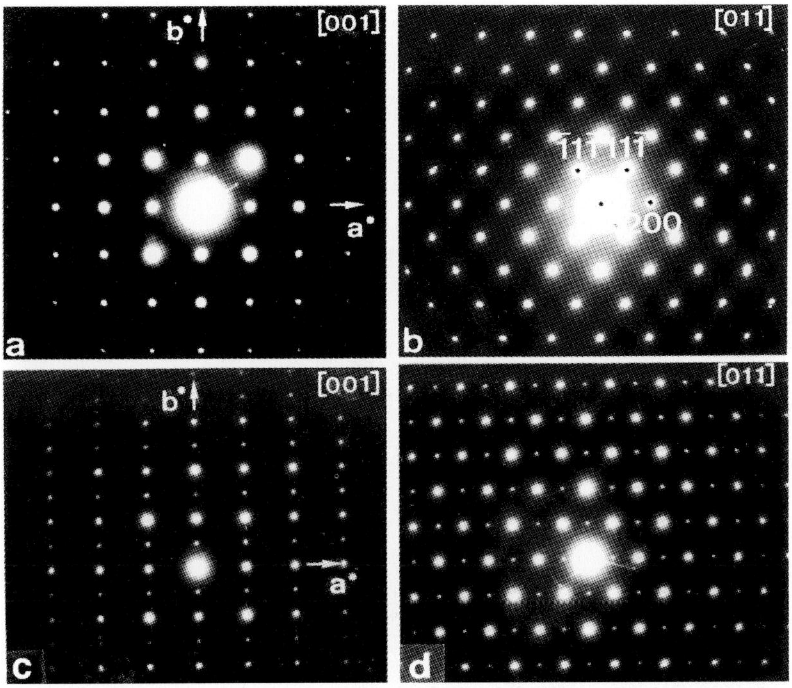

Figure 12-1. Diffraction patterns of C60 crystallite along [001] (a, c) and along [011] (b, d) at room temperature (a, b) and at 100K (c, d). Note the appearence of extra reflections at low temperature related to a lowering of symmetry due to orientation ordering.

type $\frac{1}{2}$<110>$_{FCC}$ which are symmetry translations in the FCC phase but not in the P-phase. The domain structure can thus contain four translation variants. Dark field electron microscopy has given direct evidence for this domain structure [4, 5].

12.2.2 Single crystals of C_{70}

Pure C_{70} is prepared by an arc discharge between graphite electrodes under a reduced pressure of He and purified by repeated liquid column chromatography, before crystallizing in a closed quartz tube. Single crystals are formed by sublimation at 550–600°C [6]. Within a single batch two different crystal structures may occur: either FCC or HCP, exhibiting also the corresponding differences in morphology [7]. This suggests that the energy difference between the two phases is quite small, which is consistent with the fact that the theories based on total energy for Van der Waals bonded crystals are inconclusive [8] in predicting the crystal structure.

However the stacking mode may also be determined by the growth mechanism rather than by energetic considerations [9]. It was shown that the FCC stacking is favoured over the HCP stacking if it is assumed that steps associated with faults emerging in the surface are responsible for the nucleation of growth layers. Such faults occur frequently in crystals of molecules or atoms interacting by Van der Waals forces. High resolution electron microscopy has confirmed the occurrence of two different room temperature stacking modes (see Fig. 12-2). It is not clear to what extent these alternative stackings are related to the degree of purity of the C_{70} material.

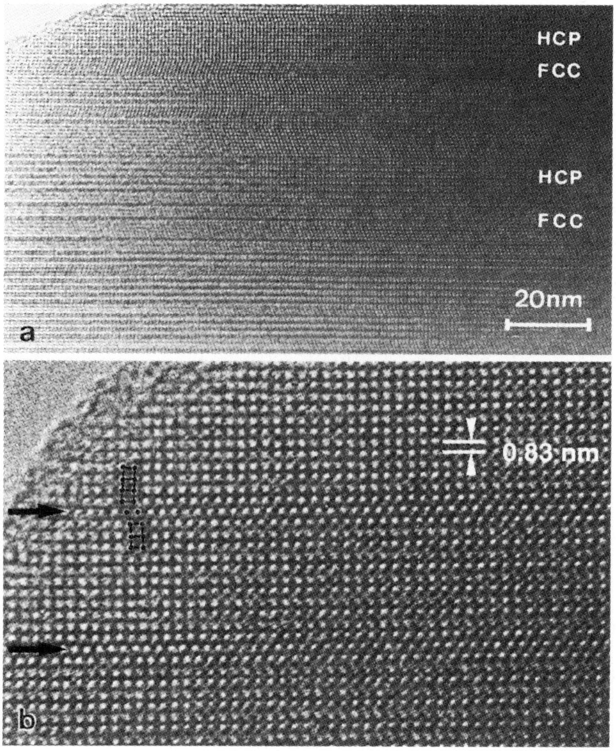

Figure 12-2. HREM of C70, showing a mixture of hexagonal and cubic stacking.

The HCP stacked crystals adopt orientation ordered superstructures on lowering the temperature as revealed by electron diffraction of single crystal fragments. On cooling ($T < 270$K) the *c*/*a* ratio is found to be equal to that for the ideal stacking of spherical objects (i.e. $c/a = 1.63$) corrected for the aspect ratio of the elongated C_{70} molecules which leads to $\frac{c}{a} \approx 1.82$. This suggests that the molecules occur in layers in which their long axis are parallel to the c-axis; the unit cell contains two such layers, but exceptionally a second superstructure containing four layers is found (Fig. 12-3). Moreover it is found that one edge of the lozenge shaped unit mesh in the basal plane of the hexagonal unit cell is doubled in length. This is a consequence of the orientation ordering within the layers. The two-layer structure can be referred to an orthorhombic unit cell containing four molecules. The azimuthal orientations of the molecules are governed by the same principle as in primitive C_{60}; electron rich parts of the cage (edges) face electron poor regions (face centers) of the adjacent molecules.

As the temperature increases first the superstructure spots in electron diffraction patterns disappear showing that the orientation correlation breaks down and suggesting that the molecules are now rotating or librating along their length axis, whilst remaining parallel to the c-axis. On further heating also the c/a ratio decreases and approaches that characteristic of the HCP stacking of spheres. This means that the molecules no longer have their long axis parallel to *c* but presumably their length axis precesses about the *c*-axis. At sufficiently high temperature the ED pattern is that typical of a HCP structure with $c/a = 1.63$, suggesting that the C_{70} molecules now behave as spheres i.e. they are rotating isotropically.

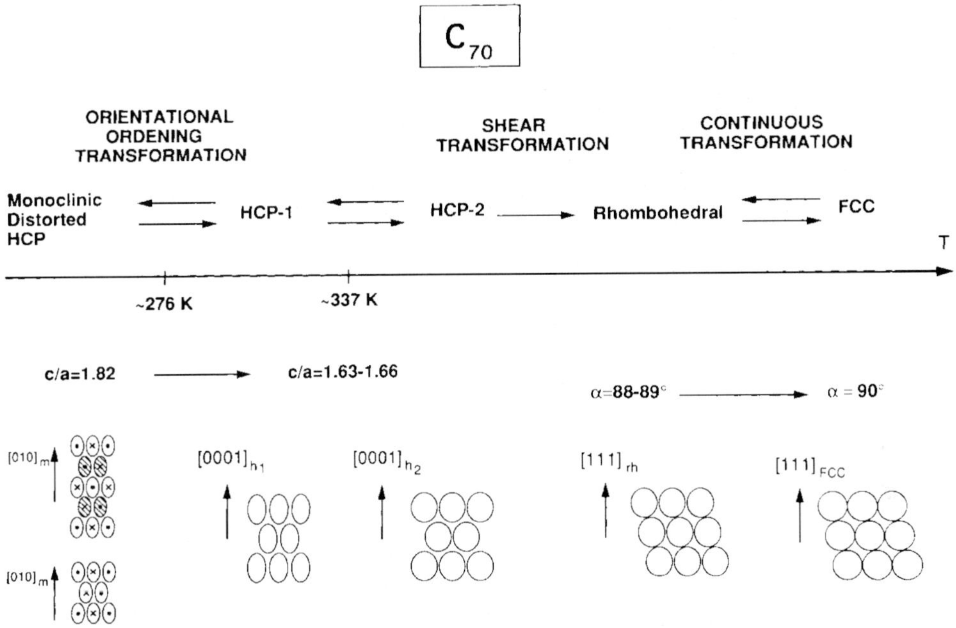

Figure 12-3. Schematic representation of the different phases and phase transitions occurring in C70.

Like in the case of C_{60} cooling leads to a decrease in symmetry and hence to the fragmentation of single crystals in orientation and translation domains. Also the FCC stacked crystals become orientationally ordered on lowering the temperature below ~250K. The long axis of the ellipsoidal molecules become gradually parallel to a <111>$_{FCC}$ direction, leading to a rhombohedral structure in which the length axis of the molecules are parallel to the threefold axis. At even lower temperature azimuthal ordering of the molecules leads to a monoclinic structure [10–12]. Since there are four different <111> directions in the FCC structure a domain structure consisting of four orientation variants is to be expected in the rhombohedral phase. Such a domain structure has in fact been observed [11] by optical microscopy. No electron microscope observations are available as yet.

12.2.3 Defects in single crystals

In high resolution images of C_{60} columns of molecules parallel to the viewing direction are imaged as single dots. Under certain critical imaging conditions these are ring shaped (Fig. 12-4). However individual carbon atoms cannot be resolved; they are too closely spaced, they do not form columns along the electron beam and moreover the molecules are quasi freely rotating at room temperature.

High resolution images have revealed that single crystal fragments of C_{60} very often contain coherent (111) twin bands as well as intrinsic (...ABCBCABCABC...) and extrinsic (...ABCABACABC...) stacking faults. All types of defect configurations which commonly occur in cubic face centered metallic alloys with low stacking fault energy also occur in C_{60} (Fig. 12-5). The C_{60} molecules play the same role as the atoms

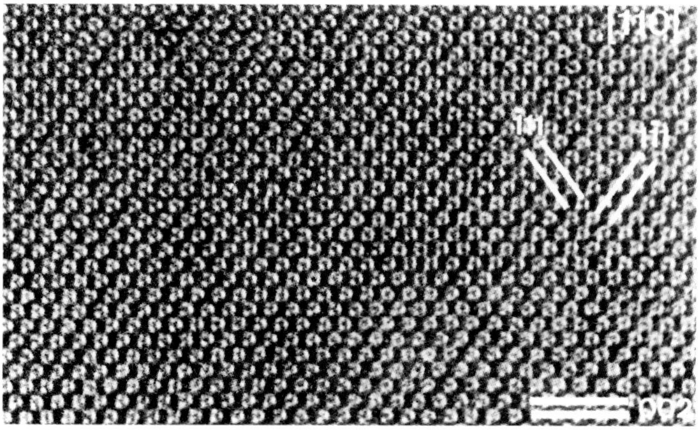

Figure 12-4. [110] zone HREM image of the perfect structure of C60. Note the ringshaped bright dots.

in the metallic alloys. In [13] evidence has been presented for the presence of Frank partials, Shockley partials and Lomer-Cottrell barriers. A remarkable configuration of intersecting intrinsic stacking faults is shown in Fig. 12-5a. Along the intersection line a row of "partial" vacancies is present; a model for these intersecting stacking faults is presented in Fig. 12-5b [14, 15].

"Single" crystal fragments of C_{70} often consist of bands of FCC phase and HCP phase; the (111) close-packed planes of FCC being parallel to the (0001) planes of the HCP phase (see Fig. 12-2). The FCC parts exhibit the same type of faults as the FCC phase of C_{60}. The HCP part often contains single layer faults of the type ...ABABCBC... as well as double layer faults of the type ...ABABCACA... At low temperature (< 2°C) the rhombohedral phase of C_{70} is finely twinned on $\{110\}_{FCC}$ type planes, so as to minimise the strain energy resulting from the orientational phase

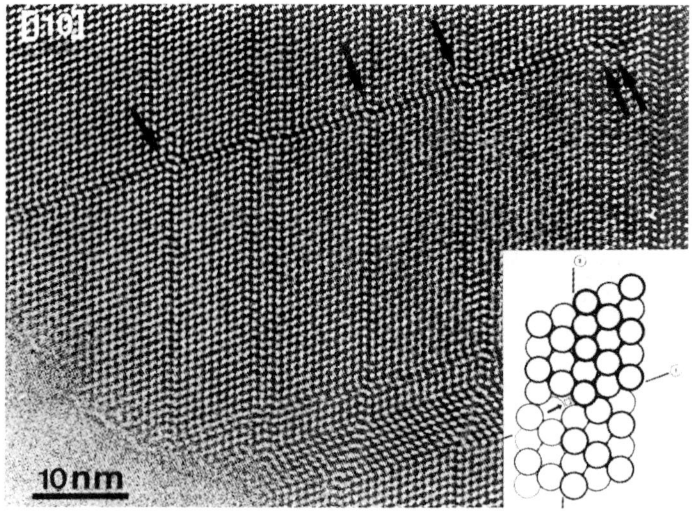

Figure 12-5. [110] HREM image of a set of intersecting intrinsic stacking faults in C60; (arrows) a model is shown as inset.

transition FCC → rhombohedral. The twins result from the cooperative alignment of the molecular length axis along different <111>$_{FCC}$ directions in different domains. The coherent twin planes are the $\{110\}_{FCC}$ and $\{100\}_{FCC}$ type mirror planes relating the different <111>$_{FCC}$ directions of alignment.

12.2.4 Nanoparticles of multiply twinned crystals

Small particles of pure C_{60} and pure C_{70}, as well as of the 50:50 mixture of C_{60} and C_{70} can be obtained by a variety of methods; we have studied nanosize particles prepared by the aerosol method [16]. Transmission electron microscopy and scanning electron microscopy were applied in the study of particles obtained by the atomization of a solution in toluene using nitrogen as a carrier gas. The solvent was eliminated by evaporation on passing the stream of droplets through a hot wall generator held at temperatures in the range from 400°C to 700°C. Particles were collected on microscope grids at the exit of the generator.

The morphology and size distribution of the particles are most easily obtained by SEM (Fig. 12-6), whereas their internal structure is studied by means of HREM (Fig. 12-7). The particle sizes vary from 10–100 nm, their distribution depending on wall temperature during preparation. At wall temperatures of 500°C or higher the particles have well defined polyhedral shapes. Most particles are platelike having trigonal or hexagonal outlines (Fig. 12-6a). Such particles are usually multiply twinned on a single (111) plane of the FCC phase. They preferentially grow laterally because of the presence of reentrant angles along the lateral facets, which are not parallel to the twin plane, and thus acquire a platelike shape. However a fair fraction of the particles have a more complicated shape; they are either icosahedral or decagonal, similar to vapour grown small particles of FCC metals such as gold and silver. High resolution images, such as Fig. 12-7, show that these complicated shapes result from the juxtaposition of tetrahedral regions of FCC structure. For instance the decahedral shape of Fig. 12-7 results from the juxtaposition of five tetrahedrally shaped regions along a common $[110]_{FCC}$ edge and separated by $(111)_{FCC}$ twin planes. As viewed along the common $[110]_{FCC}$ edge this crystal aggregate has fivefold symmetry.

Both types of composite particles (C_{60} as well as C_{70}) have to be internally strained as can most easily be illustrated by the decagonal particle. In the ideal FCC structure the angle between two {111} planes is 70.52°. Along the common $[110]_{FCC}$ edge in the

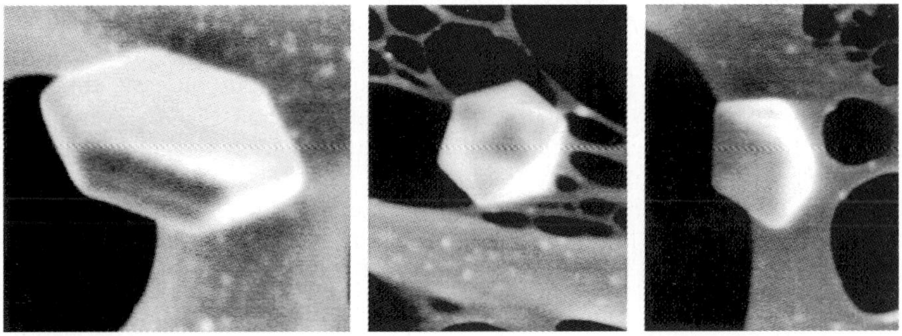

Figure 12-6. SEM image of polyhedral C70 particles a) hexagonal plate shaped b) icosahedral shaped c) decahedral shaped

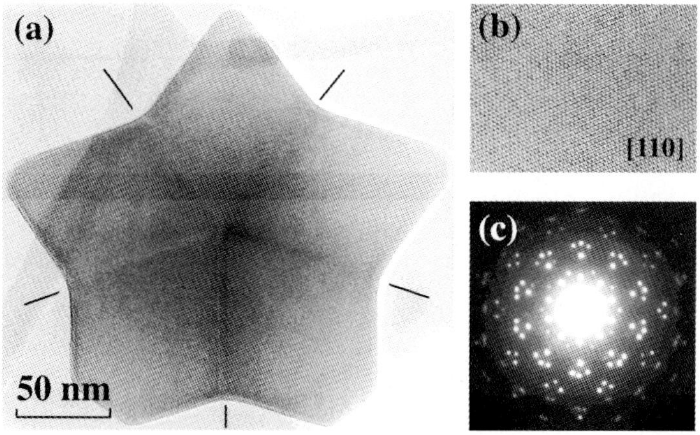

Figure 12-7. [110] HREM low magnification of a single decahedral multiply twinned C60 particle imaged along the common [110] direction; the corresponding pseudo fivefold diffraction pattern is shown as an inset, together with a HREM image of one of the variants.

decahedral particle the five wedge angles have to be fitted without leaving a gap. This requires the wedge angles to be 72° (360°/5). This is clearly not possible if the structures are assumed to be ideal FCC; an angular gap of about 6° would be left in that case. Closing this gap causes a stress pattern. Intuitively it is clear that the outer rim of the composite particle must be under tensile stress whereas the inner part must be under a compressive stress a "neutral" surface separating both regions. These stresses can be relieved either elastically, i.e. by local changes in the intermolecular separation or plastically i.e. by the introduction of lattice defects such as dislocations which can locally change the orientation of lattice planes. The insertion of supplementary half planes in the outer parts of the particle could efficiently relieve the stresses. However such configurations are not observed in HREM images. Often several stacking faults and microtwins in radial planes parallel to the {111} section boundaries are observed close to these boundaries. Such intersecting stacking faults produce rows of "partial" vacancies along the intersection lines. These intersection lines are concentrated in the central part of the particle because the faults are close to the sector boundaries (Fig. 12-8). These vacancy lines may contribute in relieving the compressive stresses in this region. Also the icosahedral particles consist in fact of 20 tetrahedrally shaped regions of FCC structure related by twinning operations. Interpretation of HREM images of these particles is less straightforward because of unavoidable overlap between different twin variants.

The ellipsoidal shape of the C_{70} molecules allows another mode of stress-relief. A layer of molecules of which the parallel length axis are normal to the layer plane becomes thinner when the molecules become inclined with respect to the layer plane. Simultaneously the lattice parameters of the unit mesh within the layer plane become larger. These considerations are consistent with the observed values of the *c*/*a* value in the HCP stacked phases of C_{70} [4, 6]. The angular gap in multiply twinned particles of C_{70} can thus be accommodated by fitting the degree of orientation to the local strain. On cooling the orientation pattern may actually be stress-induced. In HREM images occasionally parts of a sector exhibit a double period along $(111)_{FCC}$ planes parallel to the sector boundary. This can be attributed to the alternation of two different (111) layers of molecules the sense of inclination being different in successive layers [17].

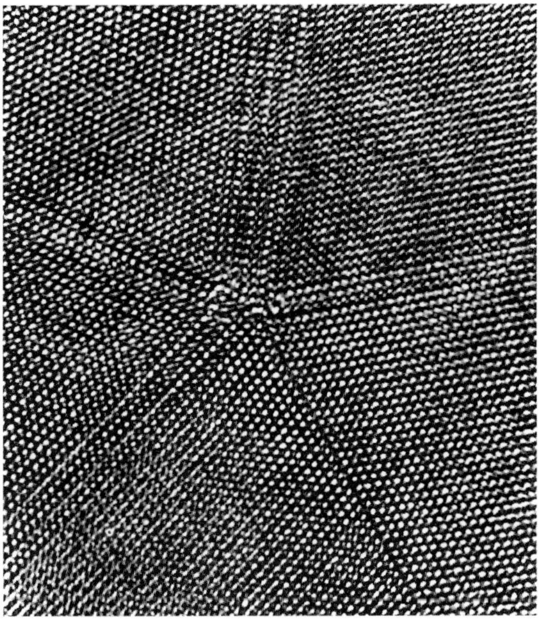

Figure 12-8. [110] HREM image of the node of a C70 particle, showing the presence of several stacking faults intersecting near the core.

12.2.5 Polymers of C_{60}

Pure C_{60} crystals can polymerize under laser irradiation [18]. Individual C_{60} molecules (monomers) can be covalently linked together to form a polymeric structure when submitted to strong visible light [18]. The optical spectra show extra lines, indicating a reduction of the molecular symmetry. Similar spectroscopic changes were noted in high pressure studies, which led in 1993–1994 to the discovery of several one and two dimensional high pressure polymerised phases. At pressures above 9.5 GPa even three dimensional polymerised C_{60} was produced; for a review see e.g. [19]. The polymers remain stable when subsequently the pressure is removed (at least at room temperature or below). Specimens of the latter type have been examined by electron diffraction. After long exposure times and reducing the illumination conditions it is possible to obtain patterns exhibiting superlattice diffraction spots, suggesting a fourfold period and a decrease of cubic symmetry.

12.2.6 Epitaxial layers on different substrates

Many materials can be obtained as single crystal thin films by epitaxial growth from the vapor phase on a single crystal substrate. The unit mesh of the exposed substrate surface has to be sufficiently similar to that of the desired film and the temperature of the substrate has to be adequate so as to allow sufficient mobility of the growth units along the surface. The fullerenes C_{60} and C_{70} have been grown successfully on various substrates such as mica, silicon, and various metal substrates. In particular GeS is well suited [20]. The quality of such layers and their relation to the substrate geometry can conveniently be studied by HREM and electron diffraction.

The relationship between film and substrate is particularly interesting in the case of a GeS substrate. This material has an orthorhombic deformed sodium chloride-like layer structure (a = 0,429 nm, b = 0,364 nm, c = 1,047 nm) and cleaves easily along the (001) wavy layer planes. Such cleavage surfaces exhibit grooves parallel to [100]. The separation of these grooves is such that C_{60} (and C_{70}) molecules can fit in every other groove and form a close-packed fullerene layer with the usual unit mesh in C_{60} ($d_{\bar{1}01}$ = 0,86 nm i.e. $2 \times d_{100}$ in GeS) (see Fig. 12-9a). This one-dimensional epitaxial orientation relationship is sufficient to promote the growth of large areas of monocrystalline C_{60} film. The diffraction pattern of specimens consisting of film + substrate is more complicated than the mere superposition of the two diffraction patterns (Fig. 12-9b). Each diffracted beam acts as an incident beam for the second layer and produces a diffraction pattern of this layer. The epitaxial relationship can nevertheless be deduced unambigeously from such composite patterns:

C_{60} [111] parallel to GeS [100]

C_{60} [$\bar{1}$01] parallel to GeS [010]

For C_{70} this orientation relationship is more complex and incommensurate patterns are formed because of the larger size of the C_{70} molecule

The fullerene layers exhibit defects which are a direct consequence of the epitaxial relationship. Since only one out of two substrate grooves is filled by a row of fullerene molecules it is possible that a different set of parallel grooves is filled in different parts

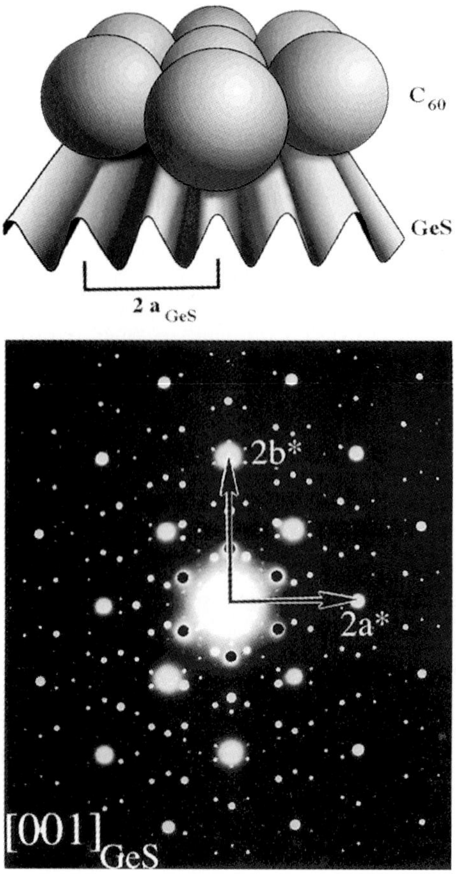

Figure 12-9. a) Schematic representation of a fullerene film on a GeS substrate
b) complex diffraction pattern allowing to deduce the orientation relationship between the GeS substrate and the C70 thin film.

of the film. This gives rise to antiphase defects in the initial fullerene layers, which may on further growth lead to out-of-phase boundaries [21]. In thick films (111) stacking faults such as those occurring in bulk material, are common features.

12.3 Crystals of C_{60} derived materials

12.3.1 Iodine intercalated $C_{60}(I_2)_{2-x}$

Iodine can be intercalated in crystalline C_{60} as neutral I_2-molecules. The intercalated structure was determined by means of X-ray diffraction [22]. The framework of the C_{60} molecules is primitive hexagonal the I_2 molecules occupying trigonal prismatic interstices in this framework. Not all interstices can be occupied; the composition is $C_{60}(I_2)_{2-x}$. The intercalation process is difficult to study "in-situ" in an electron microscope. However the de-intercalation is much less detrimental to the microscope and it occurs gradually as the crystal looses iodine in the microscope vacuum; it was studied in detail in [23]. It was shown that as the intercalate looses iodine the framework is sheared progressively over $c/2$ (c being the hexagonal lattice parameter) along prismatic planes of the hexagonal intercalate. Hereby the primitive hexagonal stacking is transformed into the FCC stacking of the pristine phase. High resolution imaging allows to observe the shear planes as well as the shear displacement, whereas the corresponding electron diffraction patterns reveal the changes in the geometry of the lattice in particular the relaxation of the prismatic planes on de-intercalation.

The transformation proceeds by the migration of partial dislocations which form the borderline of the intercalated areas of the shear plane; it takes place in the wake of the migrating dislocation. This mechanism is consistent with the orientation relationship between intercalate and pristine phases [23].

12.3.2 Rb_xC_{60} compounds

Alkali metal atoms can be intercalated in C_{60} as ions and form compounds A_xC_{60} with $x = 1, 2, 3, 4$ or 6. Different superstructures are formed depending on the value of x; a number of them exhibit superconductivity. The framework of the C_{60} molecules remains topologically face centered cubic the alkali-ions occupying the interstices. Depending on the composition this structure is slightly deformed and the actual symmetry can be either cubic or tetragonal. Intercalated phases of the type A_1C_{60} (A=Cs, Rb, K) exhibit several phase transitions as a function of temperature. At high temperature A_1C_{60} has a rocksalt type structure the A-atoms occupying the octahedral interstices in the close packed arrangement of the C_{60} spherical molecules. On slow cooling the room temperature form of A_1C_{60} remains topologically a rocksalt structure but bonds are formed between C_{60} molecules leading to the formation of linear chains of C_{60} molecules, cations occupying the remaining interstices. The polymer bonds have been proposed to be four membered carbon rings.

Electron microscopy of alkali metal fullerides such as Rb_xC_{60} is strongly hampered by the chemical reactivity of these compounds in air. In a normal ambient atmosphere the materials oxidise readily and e.g. the superconducting properties of A_3C_{60} are seriously affected. The preparation of samples usable for electron microscopy can therefore only take place under a carefully controlled atmosphere. Samples of Rb_xC_{60} can

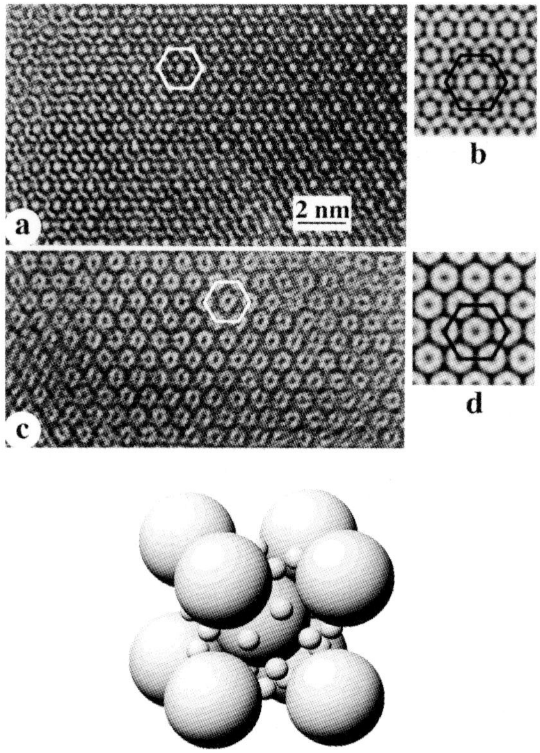

Figure 12-10. [111] HREM images under different imaging conditions in (a) and (c), respectively, together with the projected structure model. The calculated images based on this model are shown as insets in b and d. A schematic model of the intercalated structure is shown below.

however be successfully prepared and studied inside a transmission electron microscope using a double inflatable transparent plastic glove bag. The flexible bag can be linked directly to the microscope so as to protect the specimen under a N_2 atmosphere during crushing, mounting and its insertion into the microscope [24].

The electron diffraction patterns of Rb_6C_{60} are indexed on a body-centered cubic (*bcc*) lattice with lattice parameter $a = 11.56$ Å, in agreement with the structure determined by X-ray diffraction [25]. The filling of the octahedral interstices of C_{60} by rubidium atoms causes the face centered cubic cell to become body-centered cubic. Stresses are therefore generated causing the disintegration of the C_{60} crystals so as to give rise to fine particle sizes of only a few nm.

In the *bcc* structure the molecules are arranged in close packed columns along the <111> direction; the rubidium clusters, too, project along three columns parallel to the <111> direction (Fig. 12-10c). In the [111] HREM image of Fig. 12-10a, the centres of C_{60} columns show up as the largest bright dots while the rubidium columns are imaged as dark dots. Simulated images for different thicknesses and defocus values Δf are reproduced as Fig. 12-10b and Fig. 12-10d. The correspondence of the simulated and observed images is quite striking, confirming the structure as proposed in [26].

12.4 Graphite nanotubes

12.4.1 General considerations

Graphite nanotubes have been discovered by means of HREM and a model for their microstructure was suggested by the image and by the corresponding electron diffraction patterns [27].Various types of graphite fibres were already known for some time but their microstructures were incompletely characterized due to the limited resolution of the methods available at the time of their description. Moreover their microstructures and morphologies are variable apparently dependent on the method of preparation. A full characterization of the microstructures had to rely on the application of high resolution electron microscopy and electron diffraction of single fibres [28, 29]. Recently the electric-arc methods used for the preparation of fullerene molecules produced also carbon nanotubes as a by product. The microstructures and morphology of such fibres were more reproducible than those of previously known carbon fibers. The present survey is limited to this last type of fullerene related nanotubes.

In HREM the nanotubes are imaged as pairs of sets of parallel lattice fringes with a spacing roughly equal to the interlayer spacing of graphite. Since the numbers of fringes in the two sets are almost invariably equal it was concluded that they image concentric cylindrical sheets. Simulation shows that each dark fringe corresponds to a graphene sheet seen edge on (Fig. 12-11).

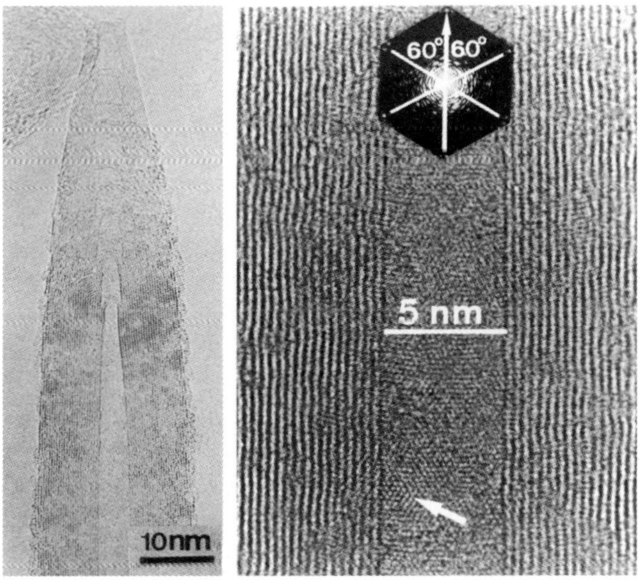

Figure 12-11. Direct space observation of a multishell nanotube; (a) at relatively low magnification, highlighting the interplanar (*00.l*) planes, (b) at higher resolution, also showing the (*hk.0*) interference between upper and lower parts in the centre.

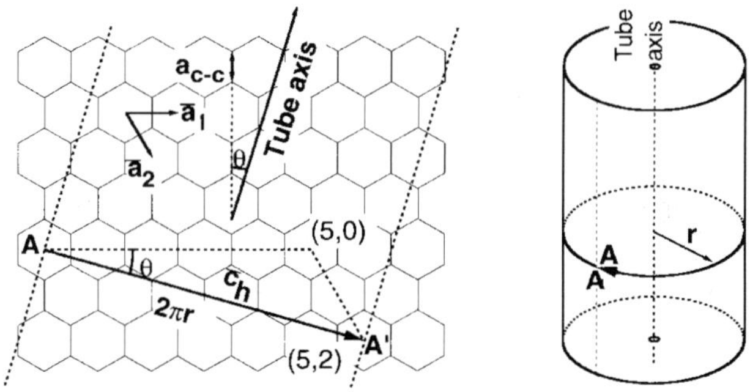

Figure 12-12. Schematic representation of the relation between r, θ and Hamada indices (L, M) for a cylindrical carbon nanotube (5,2). Wrapping up of the graphene sheet in such a way that A coincides with A′ leads to the formation of a chiral tube with wrapping vector c_h.

12.4.2 Geometry of nanotubes

As shown below the main geometrical characteristics of the diffraction pattern can be understood intuitively on the basis of the concentric cylinder model provided it is moreover assumed that the cylindrical graphene sheets may be helically wound.

A helically wound chiral cylinder is obtained from a planar sheet of graphene by applying the following operations. A line segment is chosen in the plane of the graphene sheet, it is characterized by two integers L and M (called Hamada-indices [30]) which are the numbers of hexagons spanned by the components of this line segment along two directions enclosing an angle of 60° (Fig. 12-12). The graphene sheet limited by two parallel lines normal to this line segment is subsequently rolled up in such a way that this line segment forms a circle i.e. making coincide A and A′. The radius r of the resulting cylinder is then:

$$r = \frac{\sqrt{3}}{2\pi} a \sqrt{L^2 + M^2 + LM} \tag{12-1}$$

(a = the c-c distance in a graphene hexagon) and the chiral angle θ as defined in Fig. 12-13 is then:

$$\theta = arctg \frac{M\sqrt{3}}{2L+M} \tag{12-2}$$

Non helical or achiral tubes are obtained for two special orientations. For $\theta = 0°$ the resulting tube is called a *zig-zag tube*; for $\theta = 30°$ the tube is called an *arm-chair tube*. This nomenclature is suggested by the configuration of carbon atoms along the terminating circular rims of the cylinders.

12.4.3 Diffraction pattern: observations

The electron diffraction pattern of a single multishell nanotube consists of sharp spots along a row perpendicular to the fibre axis and corresponding to the c-spacing of graphite; they are *000ℓ* spots. Moreover there are streaks parallel to the row of *000ℓ* spots through the positions of *hk.0* reflections of graphite. These streaks are sharply terminated inwards at the *hk.0* positions but they fade out gradually in the outward

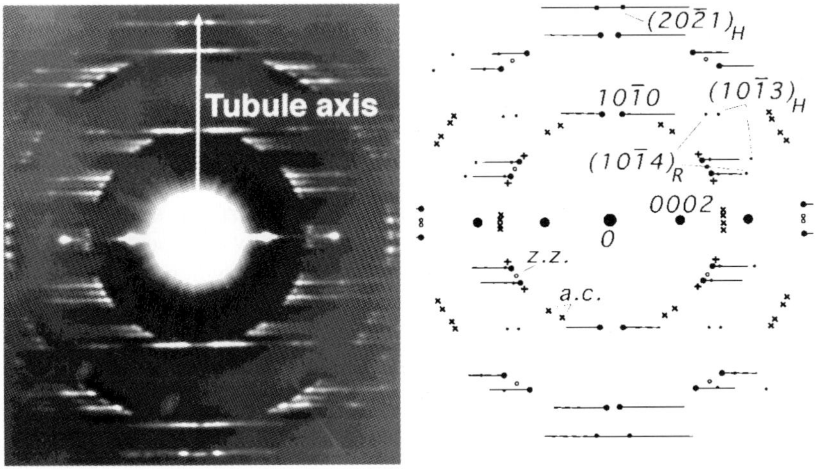

Figure 12-13. a) Experimental normal incidence diffraction pattern of a multichiral tubule. The pattern exhibits clearly *hk.l* ($l \neq 0$) reflections corresponding to the graphite AB (H) and the rhomboheral ABC stacking sequences.(R); b) Schematic representation of a). The *hk.l* spots are indicated.

direction i.e. away from the projection of the tube axis (Fig. 12-13). Very often the pattern contains several differently oriented hexagons of streaked *hk.0* spots. In such cases the spot positions form a pattern with overall planar symmetry 2mm. However taking into account also the intensities of the spots the inversion centre is in many cases the only symmetry element left in het planar pointgroup.

12.4.4 Geometric diffraction theories of nanotubes

12.4.4.1 Intuitive model: direct space

The diffraction pattern with the incident electron beam normal to the tube axis can intuitively be interpreted. The sets of graphene sheets produce the lattice fringes in the walls of the tube, where these are locally parallel to the electron beam; they also give rise to the set of *00.ℓ* diffraction spots. The sets of graphene sheets which are locally perpendicular to the incident beam produce the *hk.0* reflections. In helically wound tubes the sets of sheets along "top" and "bottom" of the tube give rise to the angular "doubling" 2θ of the hexagons of hko spots, which is a direct measure for the helical angle in the case of normal incidence only.

The streaking of *hk.0* reflections towards higher spatial frequences and the sharp cut-off at some well defined maximum interplanar distance of graphite can qualitatively be understood as related to the curvature of the graphene sheets which causes the "projected" spacings along the incident beam direction to decrease with increasing local inclination of the lattice planes belonging to a given spot. A more exact theory based on the diffraction space of cylindrically curved lattices is given below.

12.4.4.2 Disordered stacking model: reciprocal space

The concentric cylinder model implies some form of disorder of the local stacking of graphene sheets. If the intersheet distance is to be kept constant and equal to $c/2$ of graphite the circumferences of successive coaxial cylinders have to increase in length by $2\pi.c/2{=}\pi c$. Depending on the chiral angle θ the numbers of rows of hexagons in successive cylinders differ by a number of the order of $\pi c/a \simeq 8$–9. For circular cylinders this causes a gradual relative shift of the honeycomb patterns in adjacent cylinders. Moreover if successive sheets are nucleated independently the local stacking is likely to be randomly disordered. We will make this assumption when discussing the geometry of diffraction space.

The diffraction space of multishell tubes is obtained as the superposition of the "local" diffraction spaces of the disordered graphite structures assuming all conceivable orientations compatible with the model. This "local" diffraction space consists of a

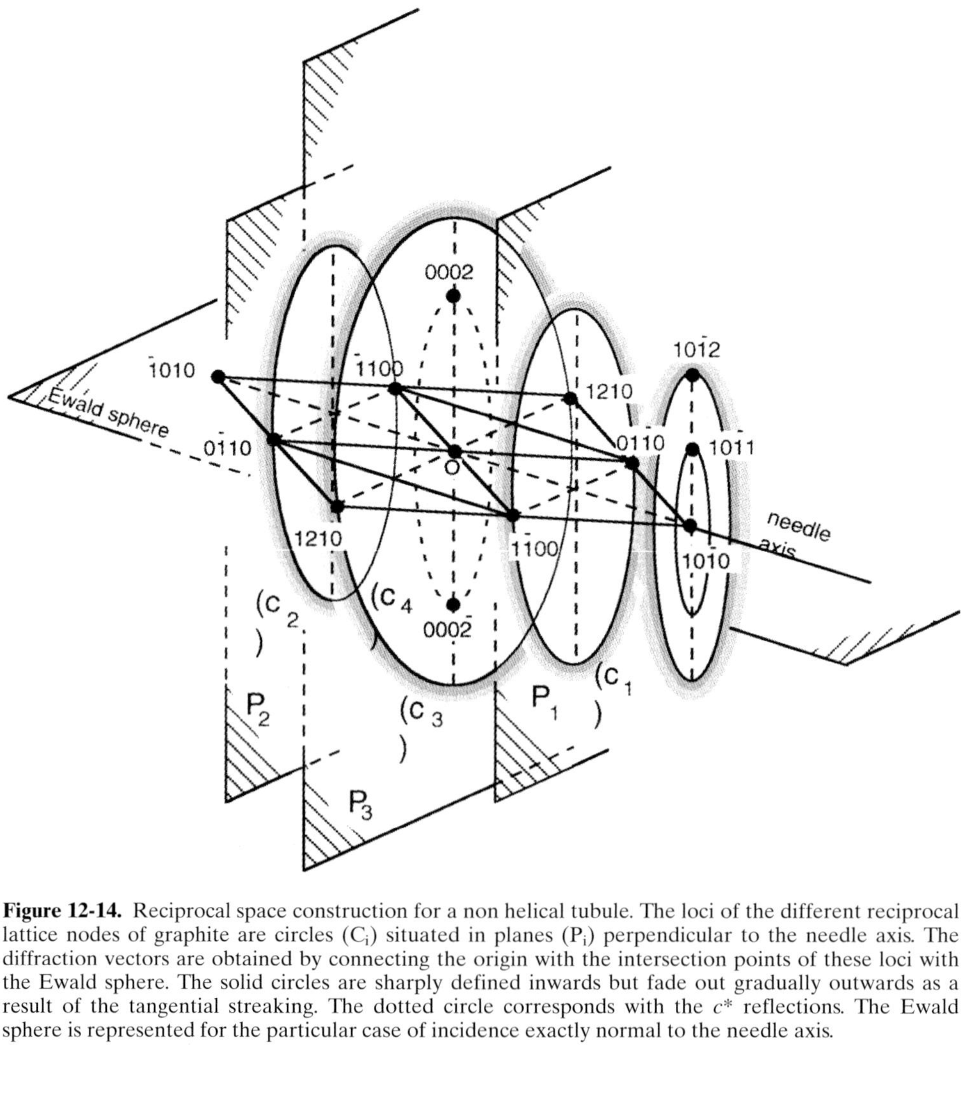

Figure 12-14. Reciprocal space construction for a non helical tubule. The loci of the different reciprocal lattice nodes of graphite are circles (C_i) situated in planes (P_i) perpendicular to the needle axis. The diffraction vectors are obtained by connecting the origin with the intersection points of these loci with the Ewald sphere. The solid circles are sharply defined inwards but fade out gradually outwards as a result of the tangential streaking. The dotted circle corresponds with the c^* reflections. The Ewald sphere is represented for the particular case of incidence exactly normal to the needle axis.

row of sharp nodes *00ℓ* through the origin (ℓ= 0, ±2, ±4, ...) separated by 2c* along a line normal to the local tangent planes to the cylinders and of nodes *hk.0* streaked along the local sheet normal (i.e. alongℓ). These streaks are due to the one-dimensional disorder which nevertheless leaves the *00.ℓ* nodes sharp, even if cylinders with different helicities are present.

Due to the cylindrical symmetry all conceivable orientations are obtained by rotating the reciprocal space of turbostratically and translationally disordered graphite about an axis through the origin and parallel to the fibre axis. Hereby the *00.ℓ* nodes describe sharp circles in planes through the origin and normal to the tube axis. Each streaked *hk.0* node generates a diffuse "corona" in a plane normal to the tube axis. The diffuse coronae are sharply limited inwards by a circle but fade gradually outwards. For chiral tubes each node generated a separate corona. For achiral tubes degeneration occurs; two symmetry related nodes generate a single corona (Fig. 12-14).

The diffraction pattern is now obtained as a quasi-planar section of diffraction space passing through the origin and normal to the incident beam direction. (the Ewald sphere is quasi-planar in electron diffraction!). The sharp circles give rise to well defined *00.ℓ* diffraction spots whereas the diffuse "coronae" lead to streaked reflections limited inwards by the *hk.0* spots (Fig. 12-14).

12.4.4.3 Homogeneous shear model (Fig. 12-15)

The interpretation of a number of observed features in diffraction patterns suggests the necessity of a more detailed theory, which no longer implies complete randomness of the relative shifts of successive cylinders. We now assume that along one generator of the cylinder a well defined stacking is present, for instance AAA ... or ABAB..., and we investigate how the homogeneous shear resulting from the increase in circumference of successive cylinders deforms the structure and hence affects the diffraction space.

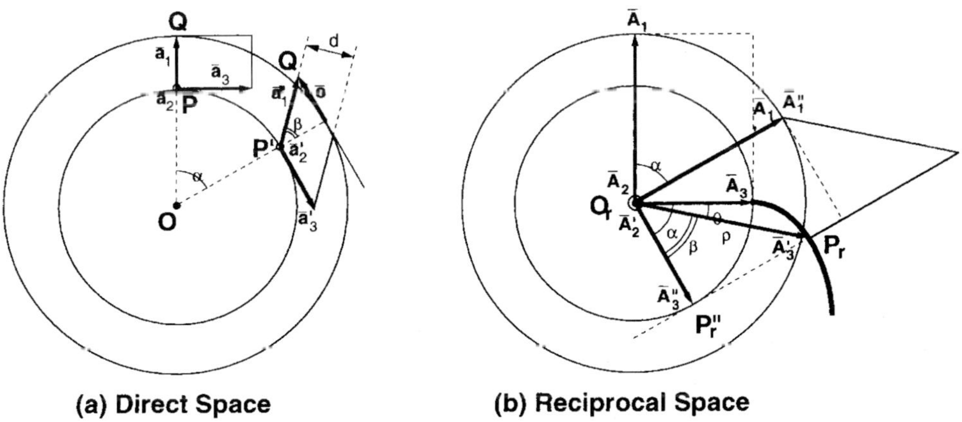

Figure 12-15. Two successive concentric tubes of a cylindrical seamless multishell nanotube as viewed along the tube axis. (a) Direct space: the homogeneous shear deforms the unit cell ($\mathbf{a}_1$, $\mathbf{a}_2$, $\mathbf{a}_3$) at $\alpha = 0$ into unit cell ($\mathbf{a}_1'$, $\mathbf{a}_2'$, $\mathbf{a}_3'$) at azimuth α. (b) Diffraction space: the same homogeous shear changes the reciprocal unit cell ($\mathbf{A}_1$, $\mathbf{A}_2$, $\mathbf{A}_3$) into ($\mathbf{A}_1'$, $\mathbf{A}_2'$, $\mathbf{A}_3'$). The shear vector s is tangential to the cylinder; the shear angle is β.

We assume the structure which the material would adopt in an unbent crystal to be based on the lattice with mutually perpendicular base vectors ($\bar{a}_1$, $\bar{a}_2$, $\bar{a}_3$). This structure is thus locally realized along the line with azimuth $\alpha = 0$. For simplicity we assume the tube to be achiral (non-helical !) and to have a circular cross section. In Fig. 12-15a we have represented in cross section two successive cylinders in a multishell tube. In P the unit cell is the same ($\bar{a}_1$, $\bar{a}_2$, $\bar{a}_3$) as that of the planar structure, which is tangent to the cylinders. In P′ at azimuth α this unit cell is sheared in such a way that the arc PP′ is equal in length to the arc QQ′, since the graphene sheet is assumed neither to be extended nor compressed but merely to be bent; it now becomes ($\bar{a}_1'$, $\bar{a}_2'$, $\bar{a}_3'$). The corresponding reciprocal unit cells are then respectively ($\bar{A}_1$, $\bar{A}_2$, $\bar{A}_3$) and ($\bar{A}_1'$, $\bar{A}_2'$, $\bar{A}_3'$) (Fig. 12-15b).

In the first place we consider sets of lattice planes parallel to the tube axis i.e. equatorial reflections. The situation is then two dimensional since all relevant features are then confined to the plane of the drawing. We focus attention on successive shapes and orientations of the reciprocal lattice unit cell as a function of the azimuth α. The reciprocal lattice node corresponding to the set of lattice planes parallel to ($\bar{a}_2'$, $\bar{a}_3'$) with spacing d is A_3' such that $O_rA_3' = 1/d$. The following relations follow from simple geometrical considerations. The shear equals $2\pi|\bar{a}_1|$ for a full turn ($\alpha = 2\pi$) and hence at the point P′ with azimuth α, s becomes $2\pi|\bar{a}_1| \mathrm{x} \alpha/2\pi = |\bar{a}_1|\alpha$. The angular shear β is given by $tg\beta = s/|\bar{a}_1|$ and one can thus conclude that $\alpha = tg\beta$. This implies in turn that the line segment $P_r'P_r''$ has the same length as the arc $P_r''A_3$. The geometrical locus of P_r' is thus described by a point of the straight line $P_r'P_r''$ when this straight line rolls without sliding on the circle with radius O_rA_3. Such a curve is called the evolvente of the circle. The turning point A_3 is such that O_rA_3 is perpendicular to $\bar{O}_r\bar{A}_1$.

The reference stacking at P will be repeated at the point where the shear becomes an integer multiple of the lattice vector in the sheet parallel to s i.e. for $s = na_3$ (n = integer). This will occur for the first time (for $n = 1$) at an azimuth α_c given by $\alpha_c = s/a_1 = a_3/a_1$. The number of turning points N will thus be $N = 2\pi/\alpha_c = 2\pi a_1/a_3$ or in terms of reciprocal space parameters $N = 2\pi A_3/A_1$. This will only be an integer for certain ratios of A_3/A_1. For the considered achiral tube we have for the circumference $2\pi r = La_3$ (r = radius, L = integer). The length increase of circumferences of two successive tubes is then $\Delta\ell = 2\pi(r + a_1) - 2\pi r = (L + n)a_3 - La_3$ (n = integer). Thus $\Delta\ell = na_3$. If the interlayer spacing a_1 is to be maintained one has to have $na_3 = 2\pi a_1$ or $2\pi(a_1/a_3)$ = n. This condition means that $2\pi(a_1/a_3)$ should be an integer. We thus find that this condition implies that the number of evolutes is an integer $n = N$!

Non-equatorial reflections in a monochiral multitube, are produced by sets of lattice planes which are inclined with respect to the fibre axis; let the angle be γ (Fig. 12-16) where the parallel lines represent the traces of the lattice planes. The length increase due to the rolling up of concentric sheets remains the same since it only depends on the interlayer spacing a_1; it is still $2\pi a_1$ for a complete turn ($\alpha = 2\pi$). At azimuth α it is thus $s = 2\pi a_1 \frac{\alpha}{2\pi} = a_1\alpha$. The repeat condition is now that the critical shear s_c must be such that the set of parallel planes with interplanar spacing is brought into coincidence with itself i.e. $s_c = d/\cos\gamma$ and hence $\alpha_c = s_c/a_1 = d/a_1\cos\gamma$ and hence the number of turning points $N = 2\pi/\alpha_c = 2\pi a_1\cos\gamma/d$ and in terms of reciprocal parameter $N = 2\pi A_3\cos\gamma/A_1$; $A_3 = 1/d$. The evolutes one now obtained by rolling a straight line with equidistant points separated by A_1 on a circle with radius $A_3\cos\gamma$.

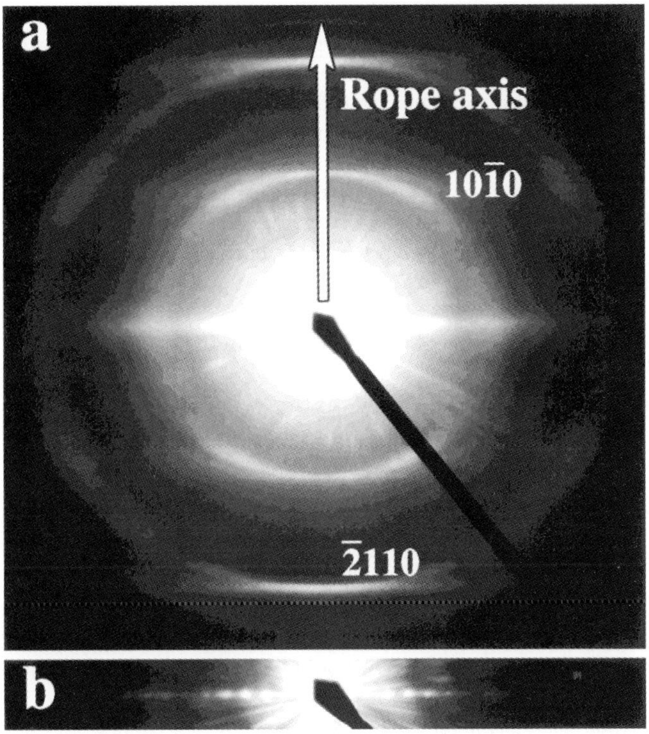

Figure 12-16. Electron diffraction pattern of an isolated rope of single shell tubes. (a) The complete pattern. (b) Shorter exposure of (a), showing the row of basal reflections.

12.4.5 Kinematical diffraction theory

12.4.5.1 General considerations

The kinematical theory of diffraction by a nanotube can be formulated analytically in a closed form [31, 32]. According to the kinematical approximation the scattered amplitude in a given direction is the sum of the complex amplitudes of the wavelets scattered once along that direction by all the scattering centers, taking properly into account the phase differences resulting from the different geometrical positions of the carbon atoms.

Quite generally the scattered amplitude $A(\bar{k})$ by an assembly of atoms at positions $\bar{r}_j$ is given by:

$$A(\bar{k}) = f_C(\bar{k}) \sum_j \exp i\bar{k}.\bar{r}_j \text{ with the scattering vector } \bar{k} = \bar{K} - \bar{K}_o$$

where $\bar{K}_o$ is the wavevector of the incident wave and $\bar{K}$ that of the scattered wave; $fc(\bar{k})$ is the atomic scattering amplitude of carbon. This expression gives the value of $\bar{A}(\bar{k})$ in all points of diffraction space (i.e. $\bar{k}$-space). The diffraction pattern is obtained as a quasi-planar section of $\bar{k}$-space with Ewald's sphere, which in electron diffraction can conveniently be approximated by a plane through the origin of $\bar{k}$-space, normal to the incident beam direction $\bar{K}_o$.

For scattering centers situated on a lattice $\bar{k}$-space consists of discrete nodes situated on the reciprocal lattice each node having a weight given by the structure amplitude $F(\bar{k})$. However if the scattering centers are *not* on a lattice the non-zero regions of the function $A(\bar{k})$ are no longer concentrated in nodes on a lattice but are distributed quasi continuously in $\bar{k}$-space. Also in this case the scattered beam direction $\bar{K}$ is obtained by the Ewald construction i.e. by joining the center of Ewald's sphere, which is at $-\bar{K}_o$, with its intersection point at $\bar{k}$ with the amplitude distribution $A(\bar{k})$ in $\bar{k}$-space. The corresponding amplitude of the diffracted beam is then proportional to the value of $A(\bar{k})$ at the intersection point. The knowledge of diffraction space thus allows to obtain the diffraction pattern for an arbitrary direction of the incident beam.

12.4.5.2 Diffraction space of a "primitive" helix

Using an orthogonal reference system with unit vectors $\bar{e}_x$, $\bar{e}_y$, $\bar{e}_z$, the scattering centers on a monoatomic "primitive" helix are situated at positions:

$$\bar{r}_j = R_o cos\phi_j \bar{e}_x + R_o sin\phi_j \bar{e}_y + z_j \bar{e}_z \text{ with } z_j = z_o + jp \text{ and}$$
$$\phi_j = \phi_o + 2\pi(p/P)j \; (j = \text{integer}) \tag{12-3}$$

where ϕ_o, z_o refer to the origin respectively of the azimuth and of the level z; p is the level difference between the z-positions of two successive scattering centers along the helix and P is the pitch of the helix; R_o is the radius of the cylinder on which the helix is wound.

The diffraction space of such a "primitive" helix was calculated, using Fourier methods, in the context of the X-ray diffraction study of DNA molecules in [33]. We shall briefly sketch the direct summation procedure following [31].

Introducing an orthogonal reference system in $\bar{k}$-space with unit vectors $(\bar{e}_{kx}, \bar{e}_{ky}, \bar{e}_{kz})$ such that $\bar{e}_{ki}.\bar{e}_j = \delta_{ij}$ one can write:

$$\bar{k} = k_x \bar{e}_{kx} + k_y \bar{e}_{ky} + k_z \bar{e}_{kz} \tag{12-4}$$

or introducing $k_\perp$ given in length by $|k_\perp|^2 = k_x^2 + k_y^2$ and in orientation by ϕ_k = arctan (k_y/k_x) one obtains:

$$\bar{k}.\bar{r}_j = k_\perp R_o \cos(\phi_k - \phi_j) + k_z z_j$$
$$\text{or } \bar{k}.\bar{r}_j = k_\perp R_o \cos\left[\phi_k - \phi_o - 2\pi(p/P)j\right] + k_z(z_o + pj) \tag{12-5}$$

Applying the Jacobi-Anger identity

$$\exp(ix \cos\varphi) = \sum_n i^n J_n(x) \exp(in\varphi) \tag{12-6}$$

one obtains finally after some lengthy algebra (see 29):

$$\begin{aligned} A(\bar{k}) &= (2\pi/p) f_C(\bar{k}) \exp ik_z z_o \\ &\times \sum_{m,n} J_n(k_\perp R_o) \exp\{in[\phi_k - \phi_o + (\pi/2]\} \\ &\times \delta\{k_z - [n(2\pi/P) + m(2\pi/p)]\} \end{aligned} \tag{12-7}$$

where the J_n are Bessel functions of order n.

The presence of the δ-function limits the non-vanishing values of $A(\bar{k})$ to discrete layer planes $k_z = n(2\pi/P) + m(2\pi/p)$ (m, n = integers). This can intuitively be understood by noting that the helical arrangement can be considered as consisting of centers with a spacing P situated on straight lines arranged on a circle with radius R_o and parallel to the z-axis, but differing in z-level by multiples of p. In this context it should be reminded that the Fourier transform (i.e. k-space) of a linear arrangement of equispaced (spacing d) scattering centers consists of diffuse planes (spacing $1/d$) perpendicular to the direction of the linear arrangements. This remains the case for the Fourier transform of a two-dimensional lattice or bundle of such parallel arrangements provided the longitudinal positions of the lines of scattering centers are randomly distributed. If the latter condition is *not* verified, which is the case for nanotubes, the diffuse planes acquire moreover a fine structure consisting of evolutes and circles as will be shown below.

12.4.5.3 Diffraction by a single shell tube

Any carbon nanotube can be constructed from one such primitive monoatomic helix by applying *screw displacements*, along the cylindrical surface with radius R_o, involving an azimuthal component $\Delta\varphi$ and a longitudinal componentΔz.

As a first step we consider a zig-zag chain of carbon atoms consisting of two "parallel" primitive helices related by a screw displacement $(\Delta\varphi_1, \Delta z_1)$ bordering a string of hexagons. The structure amplitude $A_2(\bar{k})$ of such a zig-zag chain can thus be written as the sum of two structure amplitudes $A(\bar{k})$ of the primitive helices, taking into account the phase shift due to the screw displacement. One obtains:

$$A_2(\bar{k}) = A(\bar{k})\left[1 + \exp i(k_z \Delta z_1 - n\Delta\varphi_1)\right] \tag{12-8}$$

The complete single shell tube consists of an integer number of equispaced parallel zig-zag helices. Their number is determined by the diameter of the tube. They are related by the screw displacements $(j\Delta\varphi_2,\ j\Delta z_2)$ (j = integer). These screw displacements introduce constant phase shifts between the beams diffracted by successive zig-zag chains. The $A_{ss}(\bar{k})$ for the complete single shell tube is thus obtained by the summation of a geometrical progression of which the terms are the structure factors A_2 of the single zig-zag helices, each multiplied by the corresponding phase factor.

The result of this three step summation for a single shell tube with Hamada indices (L, M) can be summarized by the following formula [31]:

$$A_{tube}(\bar{k}) = f_C(\bar{k}) \sum_{\ell} s_\ell(\bar{k}) \delta[k_z - (2\pi/T)\ell] \tag{12-9}$$

with $f_C(\bar{k})$ the atomic scattering factor of carbon and where $s_\ell(\bar{k})$ is given by:

$$\begin{aligned} s_\ell(\bar{k}) &= (4\pi C/a^2\sqrt{3})[\exp i(2\pi z_o \ell/T)] \sum_{s,m} J_{sL'-mM'}(k_\perp R_o) \\ &\times f_C(\bar{k})[1 + \exp i(2\pi/3)(s+2m)] \\ &\exp\left\{i\left[sL' - mM'\right]\left[\sigma(\phi_k - \phi_o) + (\pi/2)\right]\right\} \\ &\times \delta\{\left[s(L' + 2M') + m(2L' + M')\right]/N, \ell\} \end{aligned} \tag{12-10}$$

with $L' = \mathrm{L}$, $M' = M$, $\sigma = +1$ if $M \geq o$ (right handed helices)
$L' = L + M$, $\mathrm{M}' = -M$, $\sigma = -1$ if $M < o$ (left handed helices)
s, ℓ, m summation indices
φ_o, z_o are the cylindrical coordinates of the origin on the cylinder with radius R_o.
The meaning of the other symbols is as follows:
$T = C\sqrt{3}/N;\quad N = \ell.c.d.(2L' + M', 2M' + L')$
(ℓ.c.d. = largest common divisor of the two numbers between brackets); T is in fact the true period along z of the nanotube

$$C = a\sqrt{L^2 + M^2 + LM}\ (a = \text{lattice parameter of graphite}) \tag{12-11}$$

Equation (12-10) allows us to calculate the complex amplitude $A_{\text{tube}}(k)$ in each point $\bar{k}$ of diffraction space for a single shell tube, formed either by left or right handed primitive helices, the square modulus $I_{\text{tube}}(k) = A_{\text{tube}}(\bar{k})\ \mathrm{A}^*_{\text{tube}}(\bar{k})$ then gives the intensity. Normal incidence patterns are obtained for $\bar{K}_o$ vectors situated in the (k_x, k_y) plane.

Although each single shell nanotube can be considered as consisting exclusively of either right handed (or left handed) primitive helices, it is of interest to leave the option between these two descriptions open in view of the application to multishell tubes. If the chiral angle is limited to the interval $o < \eta \leq 30°$ one may have to describe certain multishell tubes as a mixture of clusters of righthanded an lefthanded monochiral tubes.

12.4.5.4 Diffraction by a multishell tube

The $A_{\text{MS}}(\bar{k})$ for a multishell monochiral tube is obtained by summing the contributions of the individual coaxial single shell tubes. An assumption must then be made concerning the stacking of these constituent single shell tubes. Requiring that the intertube separation of the coaxial seamless tubes remains (1/2)c restricts the values of the chiral angle θ in monochiral tubes.

The diffraction space for a polychiral multishell tube containing several monochiral clusters of coaxial tubes is the superposition of the diffraction spaces of the different clusters, since it is a reasonable approximation to ignore interference effects between different isochiral clusters.

12.4.5.5 Diffraction by a lattice of single shell tubes (ropes)

Single shell tubes most often occur as bundles of parallel tubes with the same radius, forming a close packed array. The axis of the tubes then form a two-dimensional hexagonal lattice with a lattice parameter equal to the tube diameter. The longitudinal and azimuthal positions of the tubes are presumably randomly distributed. If the periodicities along the length axis of the tubes are the same for all tubes they are also characteristic for the bundles. Non-vanishing values of the amplitude $A_{\text{B}}(\bar{k})$ are then limited to the same layer planes as for a single shell tube.

However the main feature of $A_{\text{B}}(\bar{k})$ is the presence in the zero level layer plane of a two-dimensional lattice of nodes, which is the reciprocal of the lattice of the bundle array. This lattice of nodes produces a row of equidistant sharp spots along the zero level line of the diffraction pattern; it allows to deduce the lattice parameter of the tube array (Fig. 12-16).

If tubes with different chiral angles are present in the bundle the streaked reflections of the individual tube become widened and form small arcs of the corresponding powder diffraction rings, exhibiting diffuse tails in the direction away from the tube axis (Fig. 12-16).

12.4.6 Computed diffraction patterns

We shall now comment on a few examples of computed sections of diffraction space. Ewald's construction implies that only cross sections containing the origin of diffraction space can be revealed directly as an electron diffraction pattern. Non-central sections can only be explored by tilting the specimen producing in this way line sections of the two-dimensional patterns.

12.4.6.1 Single shell tubes

Single shell nanotubes with a diameter of about 1 nanometer were first described and produced by Iijima and Ichihashi [34]. Such tubes are characterized by only two parameters (L, M). Their growth seems to be favoured by the introduction of metal in the graphite electrodes [34, 35] or by vapourising and laser heating a mixture carbon/cobalt/nickel [36]. Most tubes however tend to agglomerate in "ropes", i.e. bundles of parallel tubes which are arranged on a two dimensional triangular close packed lattice [36]. An image of a single shell nanotube is visible in the centre of Fig. 12-17, while ropes are present above and below. TEM images indicate that the lattice parameter of the triangular lattice in the ropes is 1.70 nm and the radius r of the tubes is about 0.7 nm (Fig. 12-18). Both the metallic behaviour of the material and the value of r suggest that the majority of the tubes are (10,10) armchair tubes [40]; diffraction experiments however can bring confirmation of this idea.

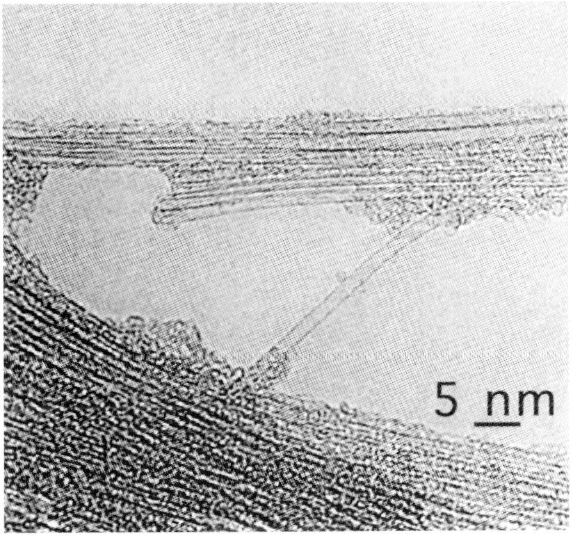

Figure 12-17. Single shell nanotubes produced by the arc discharge method; in the centre an isolated tube is visible.

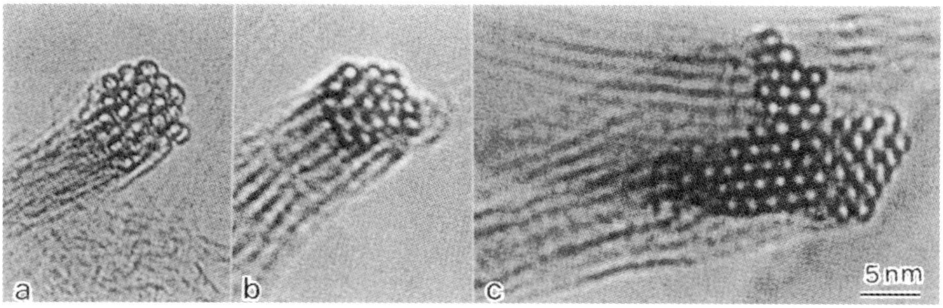

Figure 12-18. HREM view of different bundles of single wall nanotubes, bent in such a way that it is seen edge on. The bundles have a uniform width, packed according to a triangular lattice. (courtesy A. Loiseau).

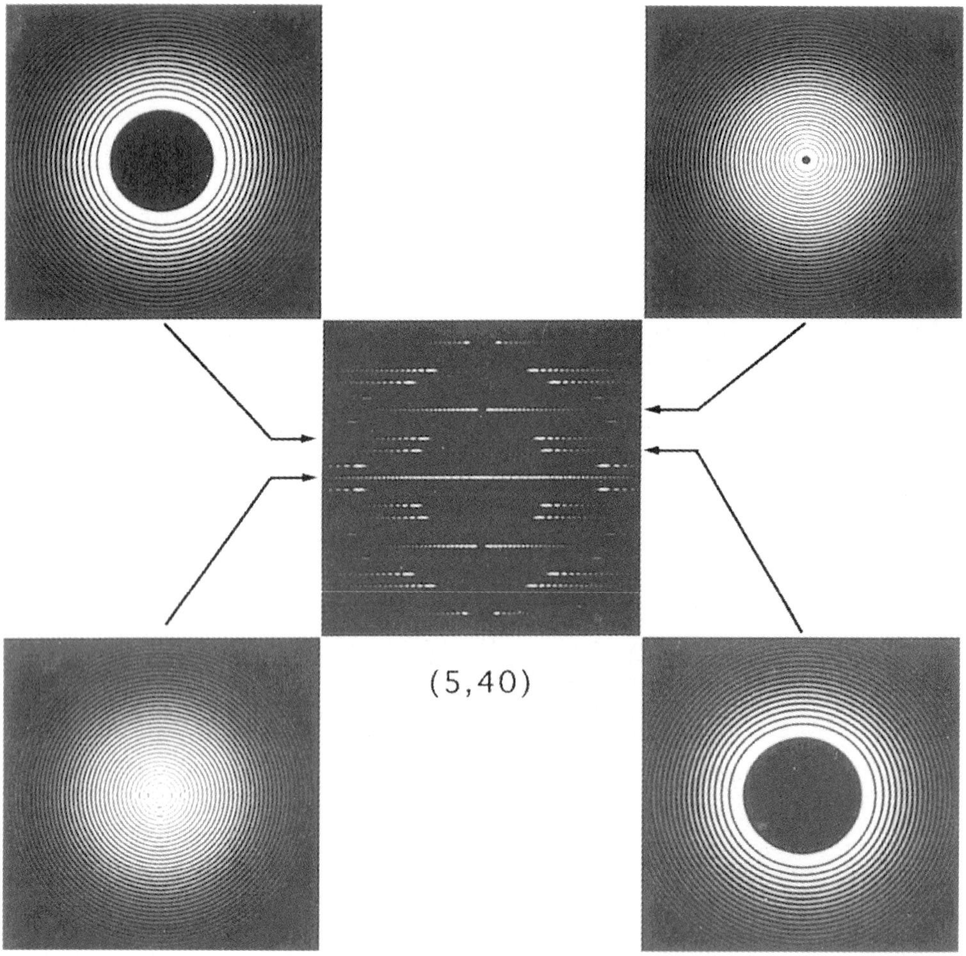

Figure 12-19. Simulated diffraction space of a (40, 5) single shell chiral tube. A normal incidence diffraction pattern with 2mm symmetry is shown in the centre, together with four sections in reciprocal space at the levels indicated by the arrows. Note the absence of azimuthal dependence of the intensity. The radii of the dark circles are given by the zeros of the sums of the Bessel functions.

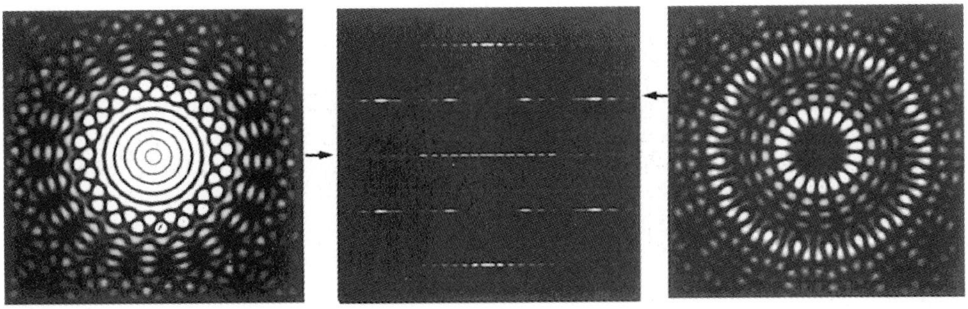

(10,10)

Figure 12-20. Simulated diffraction space of a (10,10) armchair tube. a) equatorial section; the pattern has 20 fold symmetry. b) Normal incidence pattern; note the absence of the *00.l* reflections. c) The section $k_z = g_{10.0}$ ($\sqrt{3/2}$). The pattern contains 20 radial "black" lines, i.e. extinction occurs for the corresponding azimuthal orientations of Ewald's plane.

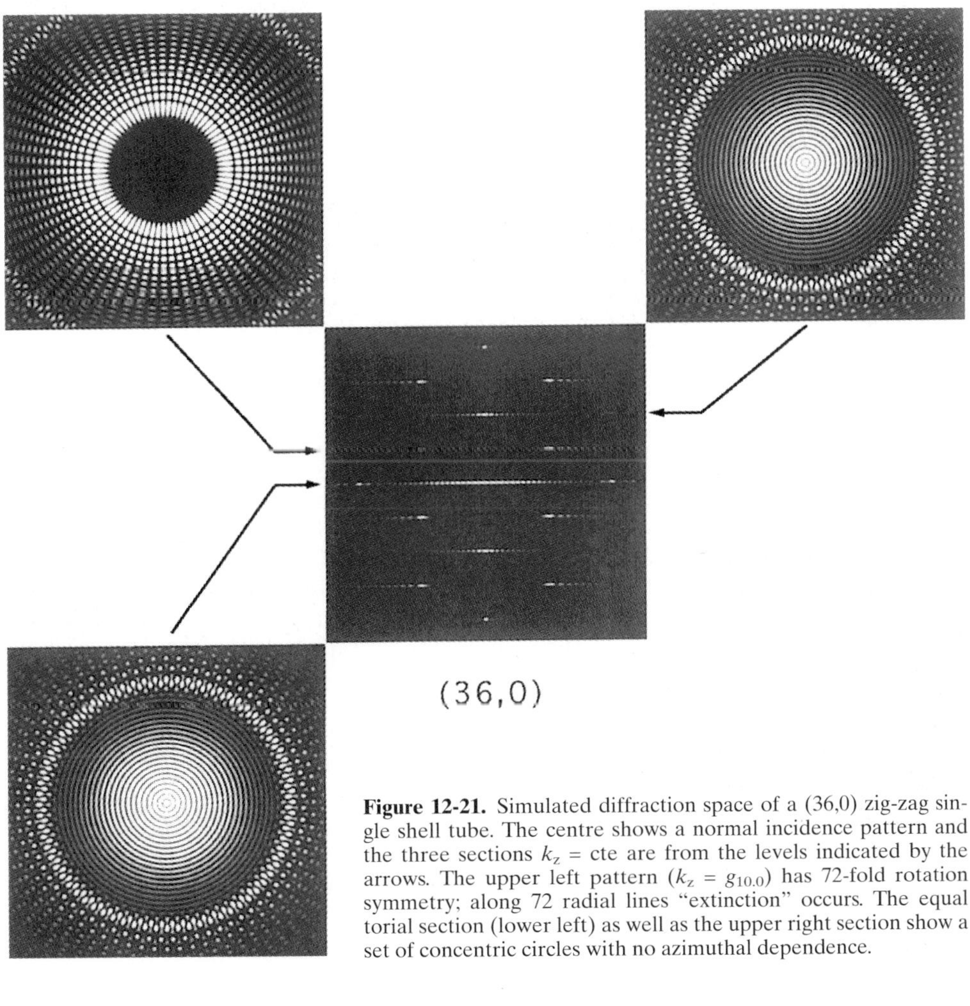

(36,0)

Figure 12-21. Simulated diffraction space of a (36,0) zig-zag single shell tube. The centre shows a normal incidence pattern and the three sections k_z = cte are from the levels indicated by the arrows. The upper left pattern ($k_z = g_{10.0}$) has 72-fold rotation symmetry; along 72 radial lines "extinction" occurs. The equal torial section (lower left) as well as the upper right section show a set of concentric circles with no azimuthal dependence.

Although it is difficult to obtain a well defined electron diffraction pattern of an isolated single shell tube it is of some interest to comment on the diffraction space to be expected for such objects. A simple characteristic feature is obviously the absence of $00.\ell$ reflections. Interference occurs between the waves diffracted by diametrically opposite parts of the cylinder. In chiral tubes this leads to a central row of spots with a separation which increases with decreasing R_o; it is independent of the atomic structure of the graphene sheet. Moreover sets of concentric circles are present limited inwards by the $10\bar{1}0$ (respectively $11\bar{2}0$) circles and with spacings given by the zero's of the Bessel functions. The pseudo period in $k_\perp$ increases with decreasing diameter of the tube since the argument of the Besselfunctions is the product $k_\perp R_O$.

Figure 12-19 shows the normal incidence pattern as well as four different sections k_z = constant of diffraction space of a (40–5) tube. The atomic structure of the cylindrical sheet does not give rise to a fine structure in the interference pattern because the chiral character of the cube does not allow for a well defined phase relation between waves scattered by diametrically opposite walls.

This is no longer true for achiral tubes. Now the honeycomb networks in diametrically opposite parts of the tube are in parallel orientations giving rise to an amplitude modulation as a function of the azimuth. For an armchair tube (n, n) this gives rise to $2n$ maxima along concentric circles of which the radii are as before determined by the radius R_o and by the considered reflections (either $10\bar{1}0$ or $11\bar{2}0$). For a zig-zag tube (n, o) the number of interference maxima along the concentric circles is also $2n$. Figure 12-20 is the normal incidence pattern as well as two sections for a 10.10) tube and Fig. 12-21 refers to a (36,0) tube.

In simulated sections of the diffraction space of double and triple tubules with a well defined stacking at φ_o such as AA, AB or ABC evolutes start appearing; they become quite clear in multishell tubules with more than three tubes.

12.4.6.2 Monochiral multishell tubes: geometrical constraints

In order to investigate the geometrical aspects of the diffraction space of monochiral multishell tubes, we first inquire under which conditions a monochiral multishell tube can be formed. As noted above the circumference increase of successive coaxial seamless cylindrical tubes is πc, assuming the separation between successive cylinders to remain $c/2$. If we require moreover that successive tubes should have the same chiral angle the Hamada indices of possible monochiral multishel tubes are severely restricted. The latter requirement implies that "wrapping" vectors of successive tubes should be parallel. Considering two successive tubes with Hamada indices (L, M) and $(L+l, M+m)$, such that this condition is satisfied we must have $(L+l)/M = (M+m)/M$ and hence also $l/L = m/M$. It is clear that the increase in length is given by $a(l^2 + m^2 + lm)^{1/2}$, where a is the lattice parameter of graphite.

The isochirality condition now becomes:

$S = l^2 + m^2 + lm \approx (\pi c/a)^2 = 72.76$

Since the left handed side is an integer and the righthand side is not, this relation cannot be be exactly satisfied. Reasonable approximations are (8,1), $S = 73$, $\gamma = 9.8°$; (6,4), $S = 76$, $\gamma = 26.3°$ and (7,2), $S = 67$, $\gamma = 17.8°$. The tube which best satisfies the isochirality condition is thus given by Hamada indices ($L = 8k$, $M = k$) (k = integer).

The knowledge of the chiral angle θ (from the diffraction pattern) and of the radius r (from the direct image) alllows to compute the Hamada indices from the relations:

$$L = \sqrt{\frac{x(y-1)}{y^2+3}} \qquad M = 2\sqrt{\frac{x}{y^3+3}}$$

with $x = 4\pi^2 r^2/a^2$; $y = \sqrt{3} tg\theta$

If no restriction of isochirality is imposed on the chiral angles two successive tubes with Hamada indices L_1, M_1 and L_2, M_2 can be coaxial and maintain the spacing $c/2$ provided:

$$\sqrt{L_2^2 + M_2^2 + L_2 M_2} = \frac{\pi c}{a} + \sqrt{L_1^2 + M_1^2 + L_1 M_1}$$

Given L_1 and M_1 this Diophantine equation has to be solved for integers L_2 and M_2. A graphical algorithm leading to acceptable values of L_2 and M_2 is given in [29].

12.4.6.3 Multishell tubes: simulations

The simulated patterns of Fig. 12-22 refer to a 10-layer monochiral tubule with Hamada indices $(8k, k)$ with $k = 5,6, \ldots 14$ ($\eta = 9°82$). This chiral angle was chosen in such a way that the normal intertube spacing remains $c/2$. The initial stacking along $\varphi_O = 0$ was the AA...A stacking. Figure 12-22a shows the normal incidence pattern along $\varphi_O = 0$; it can be compared with an experimental pattern such as Fig. 12-13.

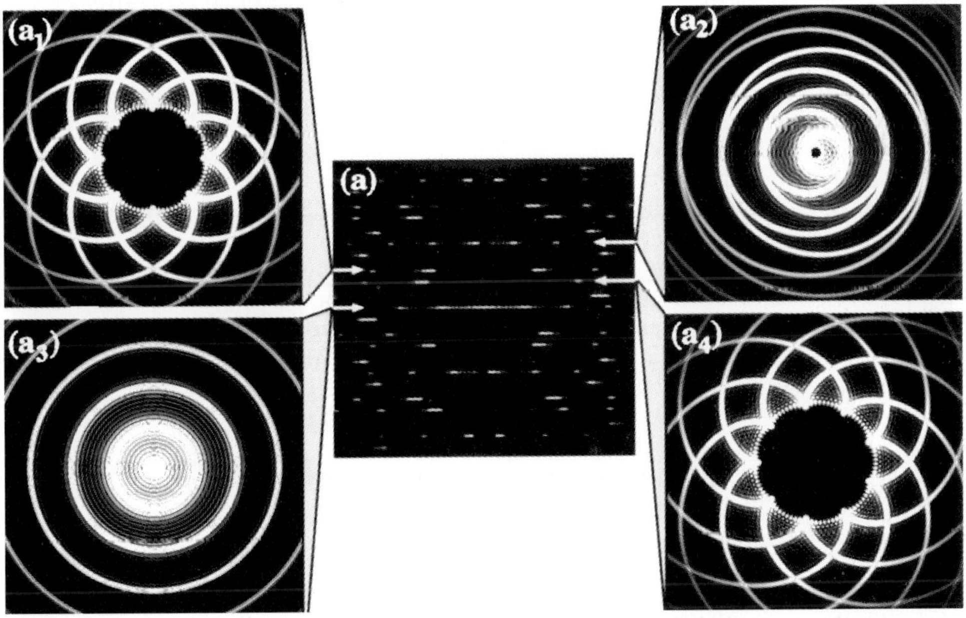

Figure 12-22. Diffraction space of a 10-layer monochiral tubule ($\eta = 9°82$). The initial stacking is AA. (a) Normal incidence pattern; only a centre of symmetry is present. (a_1) Corona giving rise to the second layer line; the basic circle results from a $10\bar{1}0$ type reflection; the number of cusps is 8, they are separated by arcs of length $2c^*$. (a_2) Corona giving rise to the fourth layerline; it is also based on a $10\bar{1}0$ type circle. The number of cusps is 1. (a_3) Equatorial cross section; the spacing between concentric circles is $2c^*$. (a_4) Corona giving rise to the first layerline; there are 9 cusps separated by arcs of length $2c^*$.

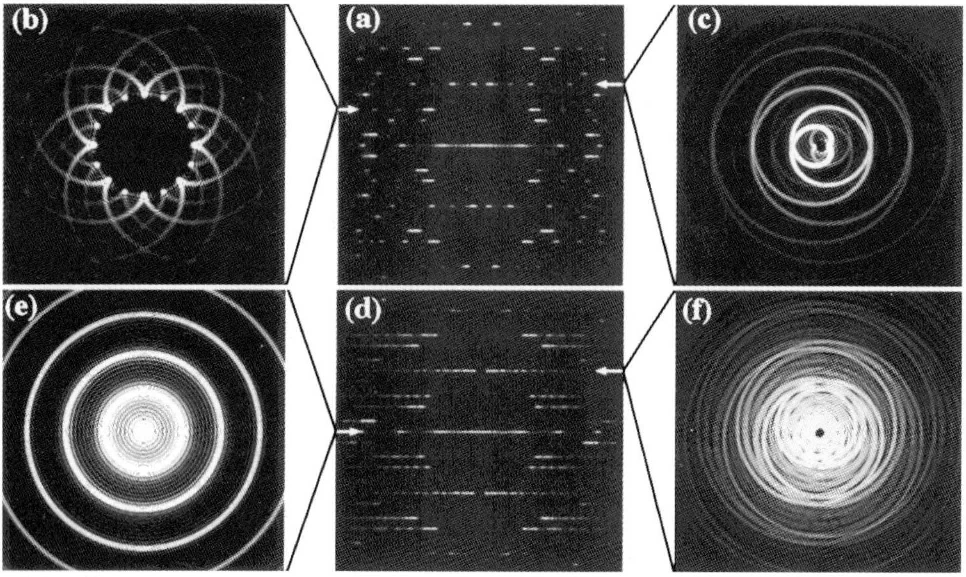

Figure 12-23. Diffraction patterns and sections of reciprocal space, calculated for a 10 layer monochiral tubule ($\eta = 9°82$).

Note the splitting of the spots on the $10\bar{1}0$ circle and the associated short streaks. The pattern has only a centre of symmetry. Four different sections k_z = constant are shown. The $k_z = 0$ section consists of concentric circles spaced $2c^*$ which produce the rows of sharp spots $00.\ell$ ($\ell = 2, 4...$). The sections a_1, a_3 and a_4 consist respectively of 9, 8 and 1 double evolute(s); the cusps being in each case spaced by $2c^*$ along the generating circles. It is clear from these sections that a cut along $k_y = 0$ will be symmetrical (2mm) for (a_1) but asymmetric for (a_4). (I)

Figure 12-23 shows different sections k_z = constant of the diffraction space of the same 10-layer tubule ($\eta = 9°82$) but with different initial stackings. In (a), (a_1) and (a_2) the initial stacking is ABAB... whereas in (b)(b_1) and (b_2) the initial stacking is assumed to be random in z_o and φ_o. In the latter case the cusps of the evolutes occur in random positions along the generating circles giving rise to diffuse coronae of the type described above. Note the 2mm symmetry of the pattern in (b), as well as the long streaks.

12.4.7 Alternative models for cylindrical tubes

12.4.7.1 Polygonized tubes

There has been some speculation concerning the exact nature of the microstructure of nanotubes. The coaxial cylinder model is undoubtedly a valuable approximation and most observations are consistent within this model. However the available evidence, mainly based on the observation of equal numbers of c-lattice fringes in both wall projections, is not unambigeous. It has been suggested that the cross section might be polygonal rather than circular; also a number of observations suggest that certain tubules are in fact in part scrolls. We will briefly discuss these different possibilities.

The observation of singular extra wide fringes in one of the walls was taken as evidence by Liu et al. [38] for the polygonal character of the tubes. Liu et al. [37] found a change in the apparent tubule diameter on tilting, which they interpreted as meaning that the tubules were not circular. However it should be pointed out that the pentagonal cross section proposed by these authors implies that for certain viewing directions *all* fringe spacings in one wall would be larger than those in the other wall which is not observed.

Electron microscopic cross-section views, which would allow to settle this question, are almost impossible to obtain and moreover it is difficult to distinguish between the image of a polyhedral particle and a tubule cross section. Attempts were nevertheless made without a clear cut conclusion however.

A different type of polygonization was proposed in [29], [39]. It was noted that the circumference increase between successive graphene cylinders implies the insertion in the outer cylinders of extra material schematically in the form of wedges with their cutting edges parallel to the tubule axis. The extra elastic energy associated with these wedges will be smallest if this extra material is distributed as uniformly as possible along the periphery. However the arrangement must also be compatible with the atomic structure of matter. The smallest wedge which can be considered is thus the edge of a single half plane of atoms as occurs in a pure edge dislocation. The optimum configuration will thus be realized if parallel edge dislocations with the smallest Burgersvector compatible with the graphite structure and all of the same sign are arranged equidistantly along the periphery. Such a configuration is induced by the mutual repulsion of such dislocations in the same "glideplane". Moreover the interaction between parallel edge type dislocations of the same sign in parallel "glide planes" tends to align the dislocation lines in vertical walls. Between coaxial graphene cylinders this would lead to the formation of tilt boundaries consisting of partial dislocations in radial planes. Multishell tubules would thus become polygonized as soon as several layers have been formed. The main driving force for this type of polygonization is the possibility to adopt low energy stackings such as hexagonal (AB ...) or rhombohedral (ABC ...) graphite within the flat facets between two walls. Electron diffraction has given evidence for the occurrence of such stackings in volume elements large enough to produce electron diffraction spots corresponding to such stackings.

Since the number of such radial dislocation walls is of the order of 18, depending slightly on the chiral angle the deviation from circular shape would be relatively small and difficult to deduce from morphological features.

12.4.7.2 Scroll model

In one of the walls of a number of straight tubules anomalously wider 00.2 fringe spacings are observed. This was attributed to the hypothetical pentagonal cross section of the tubule [40]. However the anomalous fringe spacings have also been explained by assuming the tubule to have in part scroll character. According to this model the anomalous spacing is associated with the edge on view of the supplementary "halfplane" of an edge dislocation parallel to the tube axis. Such anomalous spacings can be observed in Fig. 12-24, and a schematic representation is given in Fig. 12-25.

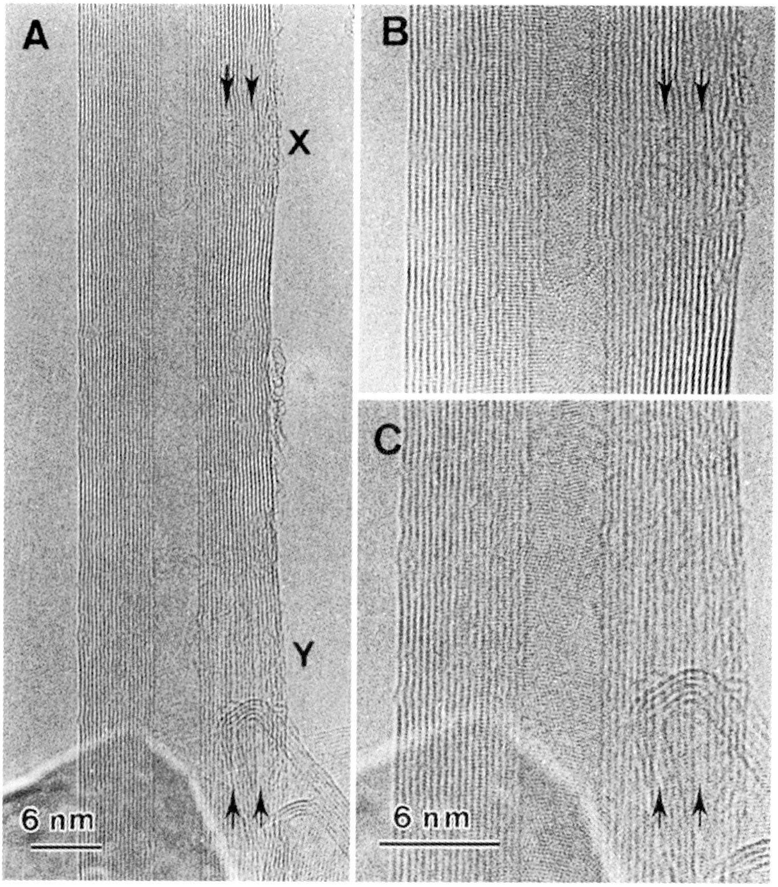

Figure 12-24. Carbon nanotubes showing singular spacings indicated by arrows between successive graphene planes.

12.4.8 Variously shaped carbon fibres

12.4.8.1 Helix shaped tubes

Nanotubes can be bent due to deformation (see further) or due to growth. The introduction of diametrically opposed pentagon-heptagon pairs causes bending of the tube over an angle of 30° [41, 42]. Such tubules in different forms have been observed experimentally in catalytically grown nanotubes and different examples are shown in Fig. 12-26. At low magnification the curvature seems to be continuous; at high resolution however the polygonised nature of the tubule becomes evident [43]. A model for the growth of such coiled nanotubes can be found in [44].

The equation of the axis of the helix shaped tubule can be represented in cartesian coordinates by $x = R\cos\psi$, $y = R\sin\psi$, $z = (p/2\pi)\psi$ where p is the pitch of the helix and R the radius of the cylinder on which the helix is wound. The inclination angle is $\mathrm{tg}\theta = p/2\pi R$. The local axis of the tubule describes a cone with semi apex angle $\frac{\pi}{2} - \theta$ about the axis of the helix.

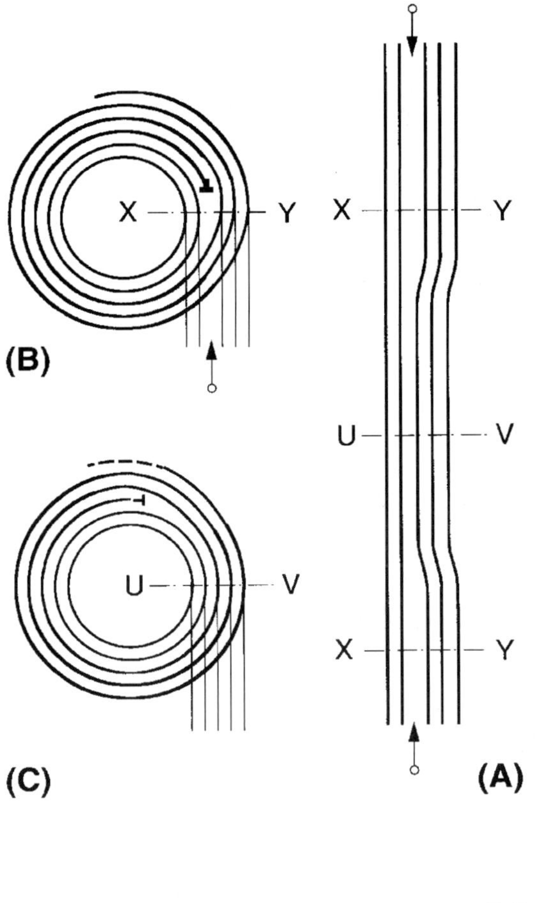

Figure 12-25. Schematic representation of a scroll type nanotube containing a dislocation, explaining the unusual spacing observed in Fig. 12-24.

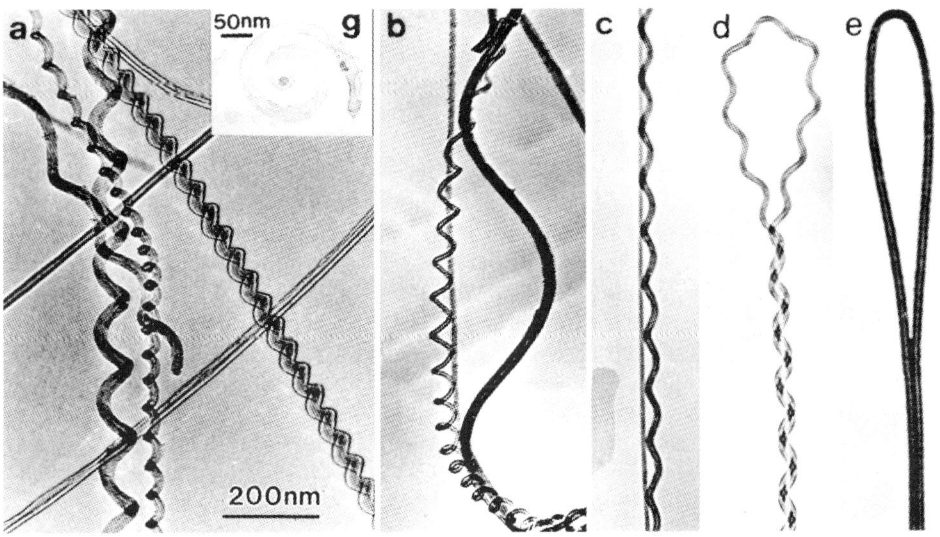

Figure 12-26. Different observations of coiled helix shaped nanotubes grown by the catalytical method.

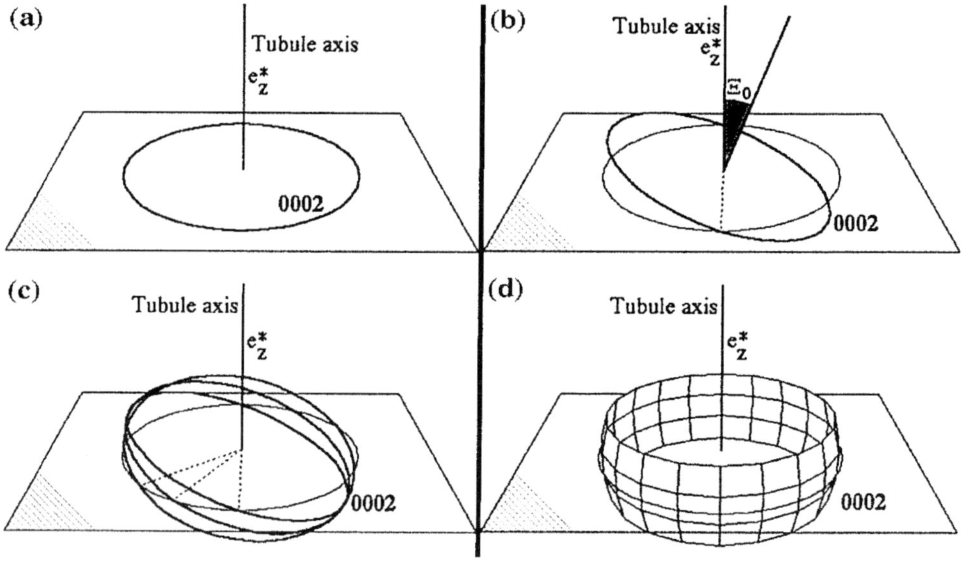

Figure 12-27. *(a) 0002* locus in reciprocal space of a straight tubule, perpendicular to the tubule axis. (b) Effect of rotation on the *0002* locus. (c) The 0002 locus for three different azimuths. (d) Final shape of the *0002* locus of a continously coiled helix, found from (c) by superposing the *0002* loci for all possible azimuths.

To discuss reciprocal space we locally replace a segment of the helix shaped tubule by a straight segment of which the axis has the same local orientation as that of the helix shaped tubule. We exclusively focus attention on the $00.\ell$ reflections which in the case of straight tubes produce sharp spots. The loci of the $00.\ell$ (ℓ= 2, 4 ...) nodes are circles with radii equal in length to $g_{00.l}$ and which are situated in planes normal to the local tube axis through the origin of reciprocal space. For a circular, continuously curved helix these circles completely fill a band with an angular width 2θ parallel to the equator of the sphere with radius $g_{00.\ell}$ centered on the origin (Fig. 12-27).

The diffraction pattern, with the electron beam normal to the helix axis, is now the intersection of this band with the Ewald plane. It thus consists of two continuous arcs with an angular width 2θ. The intensity distribution within the arcs is not uniform the extremities exhibiting a larger intensity than the central parts. Where only a single turn of the helix is selected it is found that the arcs are "spotty" the extremities exhibiting a larger density of spots than the central parts. The spotty nature of the arcs clearly suggests that the helix is not continuously curved but consists of small straight segments, i.e. the helix is polygonized. Also the direct space images support this view. The helix shaped tubule periodically exhibits sharp bends. Such a bend is formed each time a pentagon-heptagon pair occurs in the tube, the pentagon and the heptagon occupying diametrically opposite position. Moreover the line connecting the pentagon with the heptagon has to change systematically in azimuth. Mechanisms which generate such structures are discussed in [39].

12.4.8.2 Conically wound whiskers

Conically wound multishell fibres have been observed in synthetic serpentines [45]. Under rather special conditions also conically wound columnar fibres of carbon can be grown on a silicon carbide substrate by the cracking of carbon monoxide. The axis of the cone is then parallel to the fibre axis. Such fibres are scrolls i.e. they consist of a single wound up sheet of graphite. In the case of the serpentines the cones have an acute apex angle but in graphite the cones are rather flat the cone angle being roughly 140°.

The cones can be obtained by the following imaginary operation. In a sheet of graphite an ending slit is made and subsequently the two lids are superposed over a sector with opening angle θ. The result is a cone with a semi-apex angle φ given by the simple relationship $\theta = 2\pi \sin \varphi$. For certain values of θ corresponding to symmetry rotations of the sheet (θ = k 60°) the two lids of the sheet can be superposed in a normal graphite stacking. Other epitaxially favourable stackings, corresponding to a high density of coinciding lattice sites, can occur for certain θ values. Such θ-values lead to preferential semi-apex angles.

The sheet exhibits moreover two radial steps associated with the boundaries of the sector of overlap. Further lateral growth preferentially along the steps generates the conically wound columnar "crystal" which is in fact a single helicoïdal surface.

Each turn of this helicoïdal surface has a structure which differs in orientation with that of the two adjacent turns by an angle θ.

The most informative diffraction pattern is that obtained along the zone parallel to the cone axis i.e. roughly perpendicular to the conical cleavage plane. Such a pattern shows circles due tot $10\bar{1}0$ and $11\bar{2}0$ type reflections along which spots are regularly spaced. The pattern usually exhibits an N-fold rotation symmetry where N is a six multiple which may be as large as 126. It is in fact a composite consisting of the superposition of hexagonal patterns due to individual successive sheets which differ by a constant angle. This can be proved directly by showing that the orientation of the complete pattern of spots depends on the position of the selector aperture (i.e. on the selected area) with respect to the emergence point of the fibre axis.

12.4.9 Defects in tubes

12.4.9.1 Caps

Most tubes are terminated by a polyhedral cap or dome, which may ideally be one half of a fullerene cage. The closure of the cap requires the presence of six pentagonal meshes which introduce convex curvature. High resolution imaging has made it possible to visualize the geometry of these caps by revealing the angular bends in the graphene sheets caused by the pentagonal meshes. An example is shown in Fig. 12-28b. As is often the case the closure of the multishell tube takes place progressively i.e. concentric tubes close in successive stages. First the inner tubes close, afterwards the more outer ones. The tube presumably grows in length as long as the tip is open so that atoms can be added at the edges of the cylindrical sheets. As the growth conditions change the formation of pentagonal meshes instead of hexagonal ones may become probable and closure of the inner tube may take place hereby terminating the

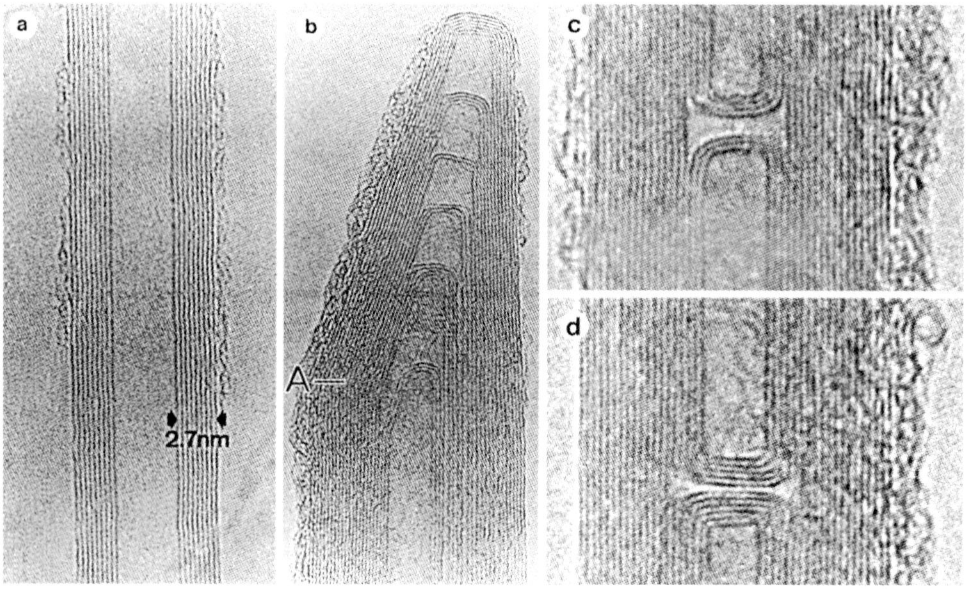

Figure 12-28. HRTEM images of different configurations of nanotubes a) a perfect, straight tube b) closure of a tube in different stages. c) "bamboo" like configuration of an inner closure of the tube.

length growth of these tubes. However the more outer tubes may continue to grow for a while hereby overshooting the already closed ones, but also close somewhat later. In this way complicated arrangements of successive caps may be formed.

12.4.9.2 "Bamboo" microstructure

A somewhat exceptional arrangement of closed inner tubes is shown in Fig. 12-28c. A number of inner tubes becomes closed but at the same level a similar number of inner tubes is "reformed, leaving a "bamboo-like" compartement dividing the multi-shell tube into two separated parts. The deviding wall is apparently common to the two sets of inner tubes. A plausible model for such a wall is represented in Fig. 12-29, assuming implicitely that along the periphery of this common wall sp_3-bonded carbon is present.

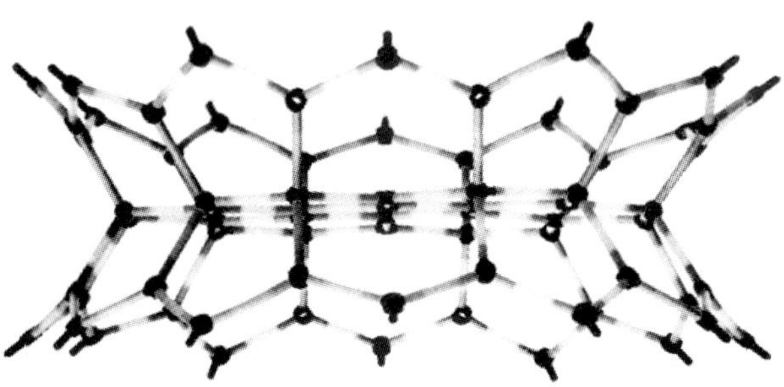

Figure 12-29. Schematic model for the "bamboo" like configuration observed in Fig. 12-28c.

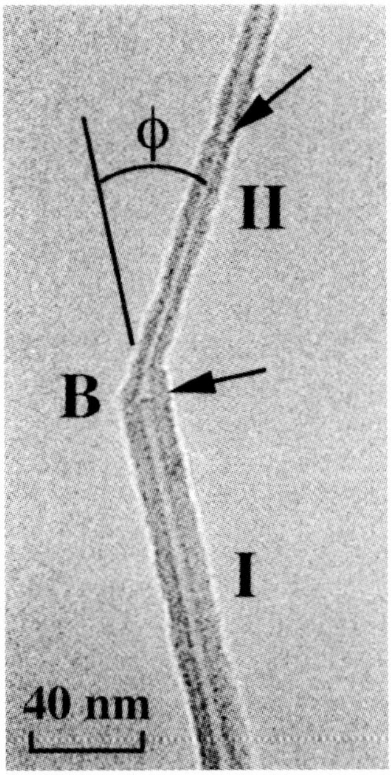

Figure 12-30. A 30° bend accompanied by a change in diameter.

12.4.9.3 Bends

Figure 12-30 represents a bend accompanied by a change in diameter in an otherwhise straight multishell tube. A model for such a bend consists of a pentagon-heptagon pair in diametrically opposite parts of the tube situated in the plane determined by the axes of the two tubes [41]. The theoretical angle, which ensures continuity of the graphene sheet where the latter is parallel to the plane of the bend, is 30°. The observation confirms this theoretical value [41, 42]. Molecular dynamics predicts a somewhat larger value for small size single shell tubes [46].

12.4.9.4 Deformed tubes

The HREM observation of bent nanotubes revealed the presence of numerous kinks along their inner rims, which is under compressive stress. However on removing the constraints the fibre was found to recover its original shape, showing that even a severe deformation is still elastic in nature.

Radial deformation of nanotubes can be revealed by measuring the variation of the 0002 fringe spacing in the images of juxtaposed parallel tube pairs interacting by Van der Waals focus. In the contact region the tubes are flattened, accompanied by a decrease in the c-spacing.

Complete flattening of isolated nanotubes seems to occur occasionally in multilayer tubes. The cross section then consists of a flat ribbon like part, in which the graphite layers in both walls are in parallel contact, termined by strongly curved borders. This shape is stabilized by the Van der Waals attraction between the parallel parts.

Low magnification CTEM was recently used to estimate Young's modulus of single multishell tubes [47]. Tubes clamped at one end only, as in a cantilever, vibrate constantly as a result of thermal agitation. The instantaneous positions of the fibre can of course not be observed however the extreme positions can nevertheless be recorded and hence the amplitude of vibration can be measured as a function of temperature. From such a measurement the magnitude of Young's modulus can be estimated. It is found that carbon nanotubes are much stronger than the carbon fibres used so far for the reinforcement in composite materials [47].

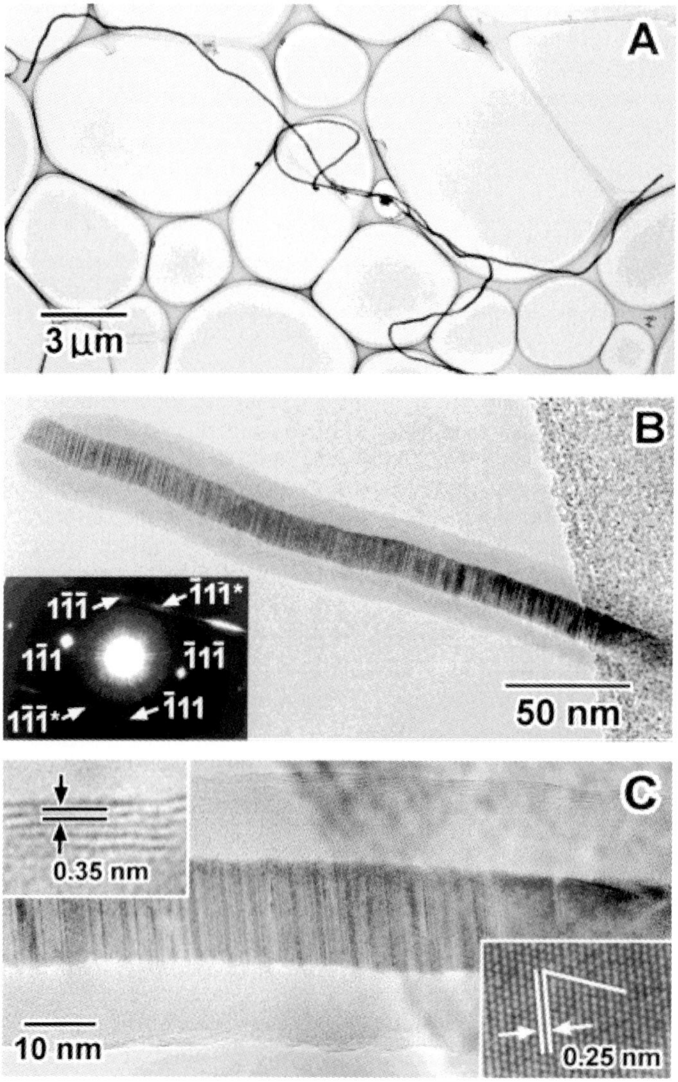

Figure 12-31. a) Low magnification TEM of nanocables b) A magnified image showing the crystalline core and an amorphous surface layer. The selected area diffraction pattern (inset) indicates the crystalline core is β-SiC with the <110> axis parallel to the electron beam. c) HREM image of the nanocable structure. The upper left insert shows the magnified lattice fringes of the outer carbon-BN layers. The lower right insert shows a magnified image of the β-SiC core. (courtesy S. Iijima [46])

12.4.9.5 Filled nanotubes

Tubes can be "opened" by the selective oxydation (i.e. burning) of the caps. A carbon nanotube then becomes a capillary tube with a nanometer size diameter. Where the end of such a tube is brought in contact with a liquid that wets the carbon surface very strong capillary forces cause the liquid to be sucked into the tube. The filling of nanotubes with liquid metals can be observed by means of CTEM. Ajayan and Iijima where the first to observe this effect [48]. A different method is to mix the material to be encapsulated in the rod used for arc discharge. At the present moment the latter technique gives the better results as to the formation of long nanowires. Not only metals can be incorporated but also semiconductors or compounds such as SiC. A nice example of modern nanotechnology is shown in Fig. 12-31 where a β-SiC is surrounded by a SiO_2 layer, within a graphitic nanotube (courtesy S. Iijima) [49].

12.4.10 Chiralty of straight tubes

Classical dark field diffraction contract images allow us to determine the sign of the chiral angle θ i.e. to determine whether the tube is right or left handed. The meaning of right and left handed is obvious for a single helix since it is an *intrinsic* property of the helix. However the same simplicity does *not* apply to carbon nanotubes. Any carbon nanotube can be considered as consisting of "parallel" helical arrangements of carbon atoms which can be chosen as either all right handed or all left-handed wound on the same cylinder. Through each point on the cylindrical surface of a chiral tube one can always find two carbon helices of one hand and one of the opposite hand. This is a consequence of the hexagonal symmetry of the cylindrical graphene sheet. The same ambiguity exists for all tubes formed from sheets with hexagonal symmetry. However *conventionally* one can attribute a hand to a tubule by considering a well defined parallel set of helices as representing the hand. One can for instance choose the set with the smallest pitch as representative of the hand.

For diffraction purposes it is more convenient to focus attention on the hand of a family of helicoidal surfaces, for instance one of those leading to the streaked reflections of the type $\{10\bar{1}0\}0$. By convention one can for instance consider as representative the family of helicoidal surfaces of which the corresponding diffraction vector encloses the smallest angle with the projected axis i.e. which produces the smallest corona.

On tilting in a known sense about an axis perpendicular to the tube axis, the streaked diffraction spots produced by the chosen corona can be made to coincide into a single streak situated on the projection of the tube axis. A dark field image made in this single streak then exhibits an asymmetry in the brightness of the wall images, one wall showing up brighter than the other one. The sense of this asymmetry, together with the known sense of tilting, is representative of the hand of the considered family of helicoidal surfaces. The observation technique, which is not simple to apply to polychiral multishell tubules is described and the contrast effects explained in detail in [50].

12.4.11 Onions

12.4.11.1 Formation

When heating carbonaceous material such as amorphous carbon, carbon nanotubes or even diamond at high temperature a sufficiently long time (tenths of minutes) by intense electron beam irradiation in the electron microscope so called “onions” are formed, as was first shown by Ugarte [51] using high resolution electron microscopy. The combined effect of heating and of increased mobility as a result of knock-on displacements seems to be responsible for the formation process. It suggests that onions are stable under the adequate conditions.

12.4.11.2 Morphology

The onions consist in fact of concentric closed cage fullerene molecules. With increasing size spheroidal fullerene cages tend to become polyhedral containing graphite like facets connected along twelve pentagonal meshes at the summits of twelve flat pentagonal pyramids. However in onions the concentric cages seems to be more spherial than predicted for the corresponding molecular cage of the same size. Electron diffraction patterns of single onions are Debye-Scherrer like with some small reinforcements along the graphite rings.

12.4.11.3 Onions as pressure cells

Onions consist of concentric all carbon cages the spacing between successive shells being roughly the $c/2$ spacing of graphite. In large multishell onions the interlayer spacing, as deduced from high resolution images, decreases towards the centre suggesting that the central part is under high pressure. This was shown directly by proving that in the centre of large onions tiny diamond crystals can be formed as deduced from their diffraction pattern and from the high resolution image. An image is reproduced in Fig. 12-32 [52, 53].

12.5 Conclusions

With the combined use of electron microscopy and electron diffraction it is possible to fully characterize the geometry of both single shell and multishell carbon nanotubes. The purely morphological features such as shape, diameter and length can be obtained by CTEM and HREM. The chiral angle, and hence the Hamada indices, can be derived from the diffraction pattern, which also contains the information required to correct this angle for inclined incidence of the electron beam.

Conventional dark field electron microscopy makes it possible to determine the sign of the chiral angle i.e. the hand of a representative family of helicoidal planes. It also allows in principle to determine the geometrical features of successive isochiral clusters of tubes in a multishell tube.

On the basis of the kinematical diffraction theory one can interpret a number of image features in terms of the local stacking of the graphene sheets. Also the geometry of ropes of single shell tubes can be studied using these techniques.

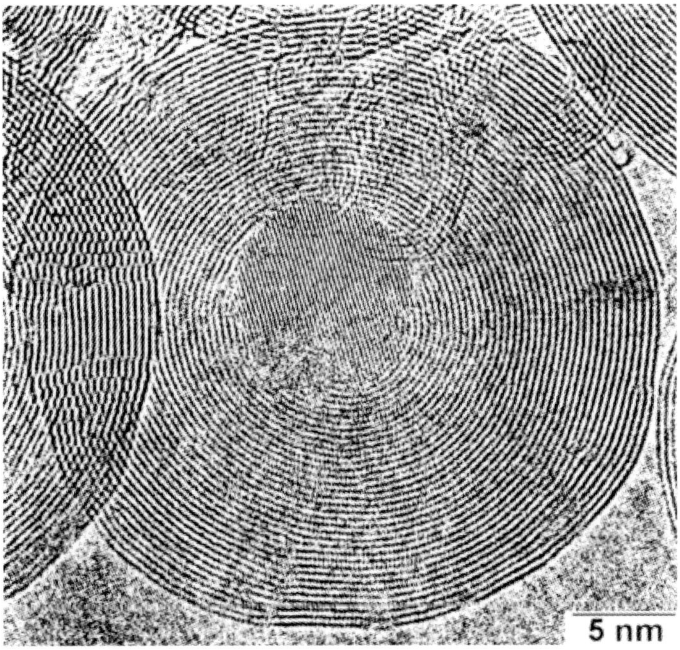

Figure 12-32. Spherical particle consisting of onion like shells with a monocrystalline diamond core, 10 nm in size. The core shows the (111) lattice fringes of diamond. The particle was generated under electron irradiation at 730°C. (courtesy F. Banhart [49])

High resolution electron microscopic images and the corresponding electron diffraction patterns have been of great use in studying the different phases, as well as the microstructures associated with the transitions between the different phases of C_{60} and C_{70} molecular crystals.

The successive stages in the de-intercalation process of iodine loaded C_{60} single crystal fragments have been followed "in situ".

Acknowledgements

The authors are grateful to D. Bernaerts, J. Van Landuyt, X.B. Zhang, X.F. Zhang M.A. Verheijen, G. Meijer, S. Muto, J. Bohr, W. Krätschmer, E. Kauppinen, A. Loiseau for the use of results from common publications. The authors with to thank F. Banhart, I. Iijima and A. Loiseau for the use of Fig. 12-18, 12-31, and 12-32 respectively. We wish to acknowledge IAUP 4/10 for making this research possible.

References

[1] G. Van Tendeloo, M. Op de Beeck, S. Amelinckx, J. Bohr and W. Krätschmer, *Europhys. Lett.* **1991**, *15* (3), 295–300.

[2] P.A. Heiney, J.E. Fisher, A.R. McGhie, W.J. Romanow, A.M. Denestein, J.P. McCauley and A.B. Smith III, *Phys. Rev. Lett.* **1991**, *66*, 2911.

[3] S. Amelinckx, C. Van Heurck, D. Van Dijck and G. Van Tendeloo, *Phys. stat. sol. (a)* **1992,** *131*, 589–604.

[4] G. Van Tendeloo, C. Van Heurck, J. Van Landuyt, S. Amelinckx, M.A. Verheijen, P.H.M. Van Loosdrecht and G. Meijer *Journal of Physical Chemistry*, **1992** *96*, 7424–7430.

[5] G. Van Tendeloo, S. Amelinckx, S. Muto, M.A. Verheijen, P.H.M. van Loosdrecht and G. Meijer, *Ultramicroscopy* **1993**, *51*,168–188.

[6] M.A. Verheijen, H. Meekes, G. Meijer, P. Bennema, J.L. De Boer, S. Van Smaalen, G. Van Tendeloo, S. Amelinckx, S. Muto and J. Van Landuyt, *Chemical Physics* **1992,** *166*, 287–297.

[7] G. Van Tendeloo, S. Amelinckx, J.L. De Boer, S. Van Smaalen, M.A. Verheijen, H. Meekes and G. Meijer *Europhys. Lett.,* **1993,** *21*, 329–334.

[8] Ph. Lambin, A.A. Lucas and J.P. Vigneron, *Phys. Rev. B* **1992,** *46*, 1794.

[9] B.W. van de Waal, "The FCC/HCP dilemma" Ph.D. thesis, University of Twente **1997**.

[10] G.M.B. Vaughan, P.A. Heiney, J.E. Fisher, D.E. Luzzi, D.A. Ricketts-Foot, A.R. McGhie, Y.-W. Hui, A.L. Smith, D.E. Cox, W.J. Romanow, B.H. Allen, N. Coustel, J.P. McCauley Jr., A.B. Smith III, *Science* **1991**, *254*, 1350.

[11] M.A. Verheijen, M.S. Couto, K.W.M. Koutstaal and W.J.P. van Enckevort, *Phil. Mag.* **1995**, *A72*, 1141.

[12] C. Christides, I.M. Thomas, T.J.S. Dennis, K. Prassides, *Europhys. Lett.* **1993**, *22*, 611.

[13] S. Muto, G. Van Tendeloo, S. Amelinckx, *Philosophical Magazine B*, **1993**, *67*, 443.

[14] S. Amelinckx in *Dislocations in Solids* vol. 2, ed. F.R.N. Nabarro, **1979**, p.71.

[15] A. Loiseau, G. Van Tendeloo and P. Bernier.in "Electronic properties of fullerenes" Eds. H. Kuzmany, J. Fink, M. Mehring and S. Roth, Heidelberg, Springer Verlag, **1993**, p. 223–227.

[16] J. Joutsensaari, E.I. Kauppinen, D. Bernaerts and G. Van Tendeloo, MRS Spring Meeting, San Fransisco **1998**.

[17] B. Pauwels, D. Bernaerts, S. Amelinckx, G. Van Tendeloo, J. Joutsensaari, E.I. Kauppinen, *Phys. Rev. B* (submitted).

[18] A.M. Rao, P. Zhou, K.-A. Wang, G.T. Hager, J.M. Holden, Y. Wang, W.-T. Lee, X.X. Bi, P.C. Eklund, D.S. Cornett, M.A. Duncan and I.J. Amster, *Science* **1993**, *259*, 955.

[19] V.D. Blank, S.G. Buga, G.A. Dubitsky, N.R. Serebryanaya, M.Yu. Popov, B. Sundqvist, *Carbon* **1998** in the press.

[20] G. Gensterblum, K. Hevesi, B.-Y. Han, L.-M. Yu, D. Bernaerts, G. Van Tendeloo, S. Amelinckx, G. Bendele, T. Buslaps, M. Foss, R. Feidenhans'l, R. L. Johnson, J.-J. Pireaux, P. A. Thiry, R. Caudano and G. Le Lay, *Phys. Rev. B* **1994,** *50*, 11981.

[21] D. Bernaerts, G. Van Tendeloo, S. Amelinckx, K. Hevesi, G. Gensterblum, L.M. Yu, J.-J. Pireaux, F. Grey and J. Bohr, *J. Appl. Phys.* **1996**, *80*, 3310.

[22] Q. Zhu, D.E. Cox, J.E. Fisher, K. Kniaz, A.R. McGhie, O. Zhou, *Nature* **1992**, *355*, 712.

[23] X.B. Zhang, X. F. Zhang, S. Amelinckx, H. Werner, *Appl. Phys. A* **1994,** *58*, 107.

[24] X.F. Zhang, X.B. Zhang, D. Bernaerts, G. Van Tendeloo, S. Amelinckx, J. Van Landuyt and H. Werner, *Ultramicroscopy* **1994**, *55*, 25–30.

[25] D.W. Murphy, M.J. Rosseinsky, R.M. Fleming, R. Tycko, A.P. Ramirez, R.C. Haddon, T.Siegrist, G. Dabbagh, J.C. Tully and R.E. Walstedt, *The Fullerenes*, Edited by H.W. Kroto, J.E. Fischer and D.E. Cox., Pergamon Press **1994**.

[26] R.M. Fleming, M.J. Rosseinsky, A.P. Ramirez, D.W. Murphy, J.C. Tully, R.C. Haddon, T. Siegrist, R. Tycko, S.H. Glarum, P. Marsh, G. Dabbagh, S.M. Zahurak, A.V. Makhija and C. Hampton, *Nature*, **1991**, *352*, 701.

[27] S. Iijima, *Nature* **1991**, *354*, 56.

[28] X.B. Zhang, X.F. Zhang, S. Amelinckx, G. Van Tendeloo and J. Van Landuyt, *Ultramicroscopy*, **1994,** *54*, 237–249.

[29] X.F. Zhang, X.B. Zhang, G. Van Tendeloo, S. Amelinckx, M. Op de Beeck, J. Van Landuyt, *Journal of Crystal Growth* **1993,** *130*, 368–382.

[30] N. Hamada, S.-I. Sawada, A.P. Oshiyama, *Phys. Rev. Lett.* **1992**, *68*, 1579.

[31] Ph. Lambin, A.A. Lucas, *Phys. Rev. B* **1997,** *56*, 3571.

[32] A.A. Lucas, V. Bruynickx, Ph. Lambin, *Europhys. Lett.* **1996**, *35*, 355.

[33] W. Cochran, F.H.C. Crick, V. Vand, *Acta Cryst.* **1952**, *5*, 581.

[34] S. Iijima and T. Ichihashi, *Nature*, **1993**, *363*, 603.

[35] D.S. Bethune, C.H. Kiang, M.S. de Vries, G. Gorman, R. Savoy, J. Vazquez, R. Beyers, *Nature* **1993**, *363*, 605.
[36] A. Thess, R. Lee, P. Nikolaev, H. Dai, P. Petit, J. Robert, C. Xu, Y.H. Lee, S.G. Kim, D.T. Colbert, G. Scuseria, D. Tomanek, J.E. Fisher and R.E. Smalley, *Science* **1996**, *273*, 483 .
[37] C. Journet, W.K. Maser, P. Bernier, A. Loiseau, M. Lamy de la Chapelle, S. Lefrant, P. Deniard, R. Lee, J.E. Fisher, *Nature* **1997**, *388*, 756.
[38] M. Liu and J.M. Cowley, *Ultramicroscopy* **1994**, *53*, 333–342.
[39] D. Bernaerts, S. Amelinckx, Ph. Lambin, A.A. Lucas, *Appl. Phys. A* **1998,** *67*, 53.
[40] M. Liu and J.M. Cowley, *Carbon*, **1994**, *32*, 393.
[41] B.I. Dunlap, *Phys. Rev. B* **1992,** *46*, 1933.
[42] B.I. Dunlap, *Phys. Rev. B* **1994,** *49*, 5643.
[43] D. Bernaerts, X.B. Zhang, S. Amelinckx, G. Van Tendeloo, J. Van Landuyt, V. Ivanov and J.B. Nagy, *Phil. Mag.A* **1995,** *71*, 605.
[44] S. Amelinckx, X.B. Zhang, D. Bernaerts, X.F. Zhang, V. Ivanov, J.B. Nagy, *Science* **1994**, *265*, 635.
[45] B. Devouard, A. Baronnet, G. Van Tendeloo and S. Amelinckx, *Eur. J. Mineral.*, **1997**, *9*, 539.
[46] Ph. Lambin, A. Fonseca, J.P. Vigneron, J.B. Nagy and A.A. Lucas, *Chem. Phys. Lett.* **1995,** *245*, 85.
[47] M.M.J. Treacy, T.W. Ebbesen and J.M. Gibson, *Nature* **1996**, *381*, 3676.
[48] P.M. Ajayan, S. Iijima, *Nature* **1993**, *361*, 333.
[49] Y. Zhang, K. Suenaga, S. Iijima, *Science* **1998** in the press.
[50] D. Bernaerts, M. Op de Beeck, S. Amelinckx, J. Van Landuyt and G. Van Tendeloo, *Phil. Mag. A*. **1996**, *74*, 723.
[51] D. Ugarte, *Nature*, **1992**, *359*, 707.
[52] F. Banhart, P.M. Ajayan, *Nature*, **1996**, *382*, 433.
[53] F. Banhart, T. Füller, Ph. Redlich, P.M. Ajayan, *Chem. Phys. Lett.* **1997**, *269***,** 349.

Index